D1520709

Reinhard Kirsch

Groundwater Geophysics

A Tool for Hydrogeology

Reinhard Kirsch

Groundwater Geophysics

A Tool for Hydrogeology

With 300 Figures

Editor

Dr. Reinhard Kirsch
Landesamt für Natur und Umwelt
des Landes Schleswig-Holstein
Hamburger Chaussee 25
24220 Flintbek
Germany

E-mail: rkirsch@lanu.landsh.de

ISBN 10 3-540-29383-3 **Springer Berlin Heidelberg New York**
ISBN 13 978-3-540-29383-5 **Springer Berlin Heidelberg New York**

Library of Congress Control Number: 2005938216

Springer is a part of Springer Science+Business Media
springeronline.com

Printed in Germany

Cover design: E. Kirchner, Heidelberg
Production: A. Oelschläger
Typesetting: Camera-ready by the Editor

Printed on acid-free paper 30/3111 54321 SPIN 11971627

Groundwater Geophysics – a Tool for Hydrogeology

Access to clean water is a human right and a basic requirement for economic development. The safest kind of water supply is the use of groundwater. Since groundwater normally has a natural protection against pollution by the covering layers, only minor water treatment is required. Detailed knowledge on the extent, hydraulic properties, and vulnerability of groundwater reservoirs is necessary to enable a sustainable use of the resources.

This book addresses students and professionals in Geophysics and Hydrogeology. The aim of the authors is to demonstrate the application of geophysical techniques to provide a database for hydrogeological decisions like drillhole positioning or action plans for groundwater protection.

Physical fundamentals and technical aspects of modern geophysical reconnaissance methods are discussed in the first part of the book. Beside "classical" techniques like seismic, resistivity methods, radar, magnetic, and gravity methods emphasis is on relatively new techniques like complex geoelectric or nuclear magnetic resonance. An overview of direct push techniques is given which can fill the gap between surface and borehole geophysics.

The applications of these techniques for hydrogeological purposes are illustrated in the second part of the book. The investigation of pore aquifers is demonstrated by case histories from Denmark, Germany, and Egypt. Examples for the mapping of fracture zone and karst aquifers as well as for saltwater intrusions leading to reduced groundwater quality are shown. The assessment of hydraulic conductivities of aquifers by geophysical techniques is discussed with respect to the use of porosity – hydraulic conductivity relations and to geophysical techniques like NMR or SIP which are sensitive to the effective porosity of the material. The classification of groundwater protective layers for vulnerability maps as required by the EU water framework directive is a relatively new field of application for geophysical techniques. Finally, the geophysical mapping of organic and inorganic contaminations of soil and groundwater is demonstrated.

I am indebted to Helga Wiederhold (GGA) for critically reading and finalising the manuscripts, to Anja Wolf and Christina Bruhn (both LANU) for skilful graphical work, and to Henriette von Netzer-Wieland (LANU) for corrections of the English texts.

Contents

Authors

Prof. Dr. Talaat A. Abdallatif
Geophysical Exploration Department
Desert Research Center, Cairo

Dr. Abdallah A. Abdel Rahman
Geophysical Exploration Department
Desert Research Center, Cairo

Prof. Dr. Nasser M. H. Abu Ashour
Geophysics Department, Faculty of Science
Ain Shams University, Cairo

Esben Auken, Associate Professor, PhD
Hydrogeophysics Group, University of Aarhus
Finlandsgade 8, DK-8200 Aarhus N; esben.auken@geo.au.dk

Dr. Norbert Blindow
Institut für Geophysik, Universität Münster
Corrensstr. 24, D-48149 Münster; blindow@nwz.uni-muenster.de

Dr. Frank Börner
Dresdner Grundwasserforschungszentrum e.V.
Meraner Str. 10, D-01217 Dresden; fboerner@dgfz.de

Anders V. Christiansen, PhD
Hydrogeophysics Group, University of Aarhus
Finlandsgade 8, DK-8200 Aarhus N; anders.vest@geo.au.dk

Dr. Peter Dietrich
Institut für Geowissenschaften, Universität Tübingen
Sigwartstr. 10, D-72076 Tübingen; peter.dietrich@uni-tuebingen.de

Prof. Dr. Kord Ernstson
Consulting Geophysicist and Geologist
Fakultät für Geowissenschaften, Universität Würzburg
Am Judengarten 23, D-97204 Höchberg; kord@ernstson.de

Dr. Gerald Gabriel
Institut für Geowissenschaftliche Gemeinschaftsaufgaben (GGA)
Stilleweg 2, D-30655 Hannover; gerald.gabriel@gga-hannover.de

Dr. Marian Hertrich
Fachgebiet Angewandte Geophysik, Technische Universität Berlin
Ackerstr. 71-76, D-13355 Berlin; hertrich@geophysik.tu-berlin.de

PD Dr. Andreas Hördt
Geologisches Institut, Fachrichtung Angewandte Geophysik, Uni Bonn
Nussallee 8, D-53115 Bonn; hoerdt@geo.uni-bonn.de

Dr. Markus Janik
geoFact GmbH
Reichsstr. 19b, D-53125 Bonn; geofact@t-online.de

PhD Flemming Jørgensen
Vejle County and Hydrogeophysics Group, University of Aarhus
Damhaven 12, DK-7100 Vejle; fj@vejleamt.dk

Dr. Reinhard Kirsch
Landesamt für Natur und Umwelt des Landes Schleswig-Holstein (LANU)
Hamburger Chaussee 25, D-24220 Flintbek; rkirsch@lanu.landsh.de

Dr. Heinrich Krummel
geoFact GmbH
Reichsstr. 19b, D-53125 Bonn; geofact@t-online.de

Dr. Carsten Leven
UFZ-Umweltforschungszentrum Leipzig-Halle GmbH
Department Grundwassersanierung
Permoserstrasse 15, D-04318 Leipzig; carsten.leven@ufz.de

Holger Lykke-Andersen, Associate Professor, PhD
Department of Earth Sciences, University of Aarhus
Finlandsgade 8, DK-8200 Aarhus N; hla@geo.au.dk

Prof. Dr. Mohamed Abbas Mabrouk
Geophysical Exploration Department
Desert Research Center, Cairo

Prof. Dr. Wolfgang Rabbel
Institut für Geowissenschaften, Universität Kiel
Olshausenstr. 40-60, D-24118 Kiel; rabbel@geophysik.uni-kiel.de

Dr. Bernhard Siemon
Bundesanstalt für Geowissenschaften und Rohstoffe (BGR)
Stilleweg 2, D-30655 Hannover; b.siemon@bgr.de

Peter B.E. Sandersen
Senior Consultant
Watertech a/s
Søndergade 53, DK-8000 Århus C; psa@watertech.dk

Kurt I. Sørensen, Associate Professor, PhD
Hydrogeophysics Group, University of Aarhus
Finlandsgade 8, DK-8200 Aarhus N; kurt.sorensen@geo.au.dk

Dr. Helga Wiederhold
Institut für Geowissenschaftliche Gemeinschaftsaufgaben (GGA)
Stilleweg 2, D-30655 Hannover; h.wiederhold@gga-hannover.de

Prof. Dr. Ugur Yaramanci
Fachgebiet Angewandte Geophysik, Technische Universität Berlin
Ackerstr. 71-76, D-13355 Berlin; yaramanci@tu-berlin.de

1 Petrophysical properties of permeable and low-permeable rocks

Reinhard Kirsch

Groundwater conditions at a location are mainly described through the distribution of permeable layers (like sand, gravel, fractured rock) and impermeable or low-permeable layers (like clay, till, solid rock) in the subsurface. To achieve a geophysical image of these underground structures, sufficient contrast of petrophysical properties is required. Seismic velocities (related to elastic properties and density), electrical conductivity, and dielectric constant are the most relevant petrophysical properties for geophysical groundwater exploration.

In this chapter, the influence of porosity, water saturation, and clay content on these petrophysical properties shall be explained.

1.1 Seismic velocities

Seismic velocities for compressional (V_p) and shear waves (V_s) are related to elastic constants like bulk modulus (k), Young´s modulus (E), and shear modulus (μ) by

$$V_p = \sqrt{\frac{3k + 4\mu}{3\rho}} = \sqrt{\frac{E \cdot (1 - \nu)}{\rho \cdot (1 + \nu) \cdot (1 - 2\nu)}} \tag{1.1}$$

and

$$V_s = \sqrt{\frac{\mu}{\rho}}$$

with ρ = density and ν = Poisson´s ratio.

Since elastic properties of rocks are highly influenced by porosity, e.g. highly porous material is more compressible than material of lower porosity, seismic velocities are also influenced by porosity.

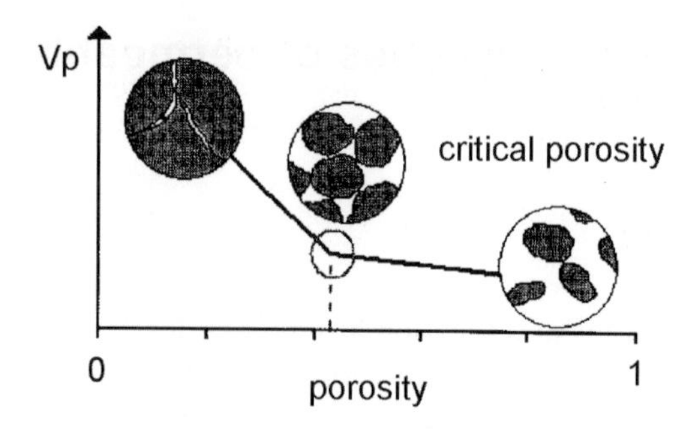

Fig. 1.1. Concept of critical porosity (after Nur et al. 1998)

The following seismic velocity – porosity relations are valid for porosities below the critical porosity threshold (Nur et al. 1998). For porosities above this threshold no grain contacts exist (Fig. 1.1). In that case, mineral grains or rock fragments and pore fluid form a suspension, in which the elastic properties are similar to a fluid. Soil liquefaction associated with earthquakes or landslides are such examples. The critical porosity for most sedimentary rocks is about 40%. As a consequence, seismic velocity – porosity relations are not always valid for structural aquifers formed by tectonic stress.

1.1.1 Consolidated rock

In a simple form, the seismic velocity – porosity relation for consolidated rocks is described by Wyllie et al. (1956) as "time average equation"

$$\frac{1}{V} = \frac{1-\phi}{V_{MATRIX}} + \frac{\phi}{V_{PORE}} \tag{1.2}$$

with V_{MATRIX} = seismic velocity of rock matrix or grains
V_{PORE} = seismic velocity of pore fluid
ϕ = porosity

This equation has been modified by Raymer et al. (1980) to:

$$V = (1-\phi)^2 \cdot V_{MATRIX} + \phi \cdot V_{PORE} \tag{1.3}$$

A very comprehensive compilation of elastic properties and seismic velocities of porous material is given by Mavko et al. (1998).

A large number of laboratory results on seismic velocities of porous material have been published. Mostly porosity changes were obtained by changes of confining pressure, whereas seismic velocities were measured in the kHz frequency range. Examples of seismic velocity - porosity relations for saturated sandstones found by different authors are (C = volumetric clay content):

Han et al. (1986) $V_p = 5.59 - 6.93 \cdot \phi - 2.18 \cdot C$

$V_s = 3.57 - 4.91 \cdot \phi - 1.89 \cdot C$

Klimentos (1991) $V_p = 5.87 - 6.33 \cdot \phi - 3.33 \cdot C$

and for unsaturated sandstone:

Kowallis et al. (1984) $V_p = 5.60 - 9.24 \cdot \phi - 5.70 \cdot C$ [km/s]

Some velocity-porosity relations found by field or laboratory experiments are shown in Fig. 1.2.

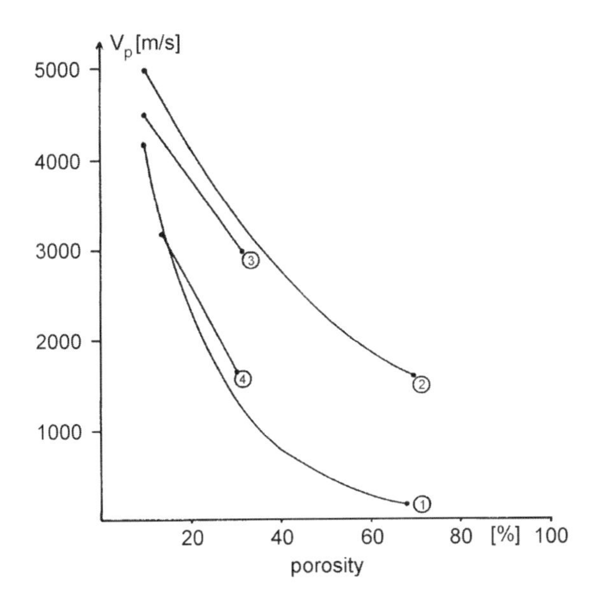

Fig. 1.2. Influence of porosity ϕ on p-wave velocities of sandstone, 1: Watkins et al. (1972), unsaturated rock, refraction seismic measurements, 2: Raymer et al. (1980), saturated rock, model calculations, 3: Klimentos (1991), saturated rock, laboratory measurements, 4: Kowallis et al. (1984), unsaturated rock, laboratory measurements; 1 and 2: clay free material, 3 and 4: clay content C = 20%

1.1.2 Unconsolidated rock

Seismic velocities of unconsolidated rocks (e.g. sand, gravel) are strongly influenced by porosity and water saturation. Fig. 1.3 shows the influence of the water saturation degree on p- and s-wave velocities. No influence of water saturation degree on seismic velocities is observed below a critical value of about 90% water saturation. A further saturation increase leads to a strong increase of p-wave velocity and a slight decrease of s-wave velocity.

Because the shear moduli of air and water are zero, increasing the saturation degree shall have no influence on s-wave velocity. The observed decrease of s-wave velocity can be explained by the increase of density when air is replaced by water as pore filling.

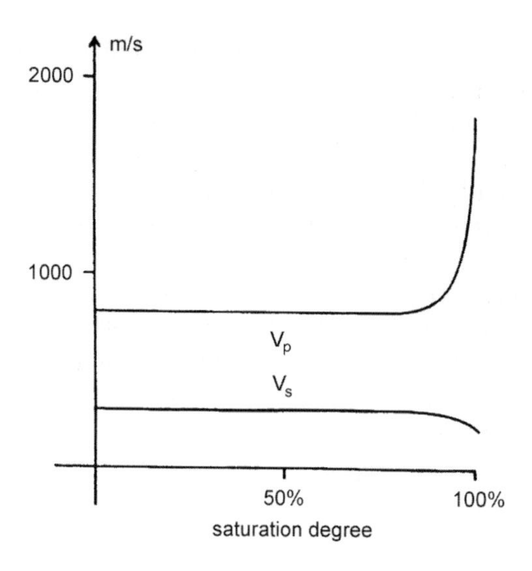

Fig. 1.3. Schematic view on the influence of water saturation on seismic velocities

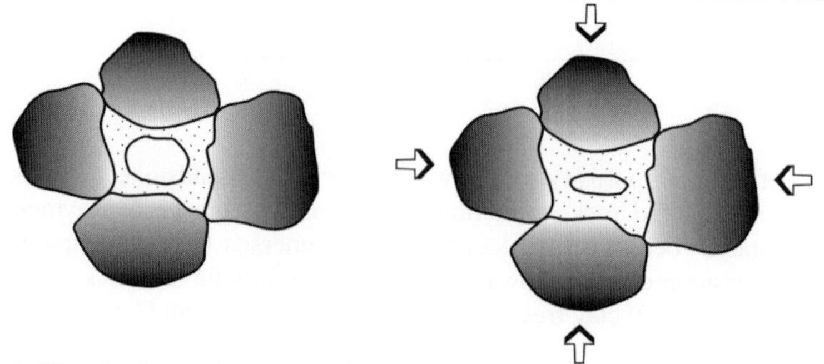

Fig. 1.4. Sketch of a partly saturated pore under compression

The crucial parameter for the p-wave velocity is the bulk modulus related to the compressibility of the material. In Fig. 1.4 a partly saturated pore has been sketched. Pore water is bound by adhesion on the grain surface. If the pore is compressed, the air in the pore space is easily compressible and the pore water cannot increase the bulk modulus of the material. Saturation variations for the partly saturated case below the critical saturation degree have no influence on the bulk modulus and, with the exception of slight density changes, on the p-wave velocity.

Only few field experiments on the influence of porosity on seismic velocities of dry unconsolidated material have been recorded. Watkins et al. (1972) made refraction seismic measurements on outcropping unsaturated hard rock as well as on unsaturated sands and found the following velocity-porosity relation:

$$\phi = -0.175 \cdot \ln(V_p) + 1.56 \tag{1.4}$$

As a consequence, p-wave velocities below sonic velocity (330 m/s) are possible and have been often observed. Bachran et al. (2000) found p-wave velocities as low as 150 m/s for dry beach sands with a velocity-depth increase as shown in Fig. 1.5. This increase can be described by a power law (depth to the power of 1/6). As a consequence, seismic ray paths in the shallow sub-surface are strongly curved.

P-wave velocities for water saturated sands are in the range of 1500 – 2000 m/s (seismic velocity of water: 1500 m/s). Hamilton (1971) measured p-wave velocities of marine sediments which are shown in Fig. 1.6. Morgan (1969) found the following seismic velocity – porosity relation for marine sediments (in km/s):

$$V_p = 1.917 - 0.566 \cdot \phi \tag{1.5}$$

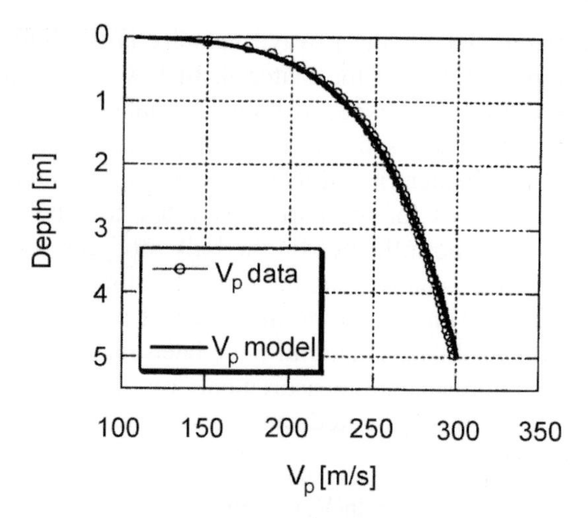

Fig. 1.5. Increase of p-wave velocity with depth (observed and calculated) in the shallow sub-surface (Bachran et al. 2000, with permission from SEG)

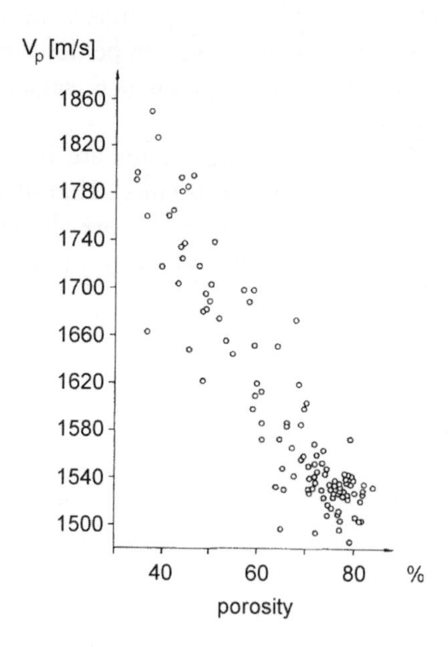

Fig. 1.6. P-wave velocities and porosities for marine sediments (after Hamilton 1971)

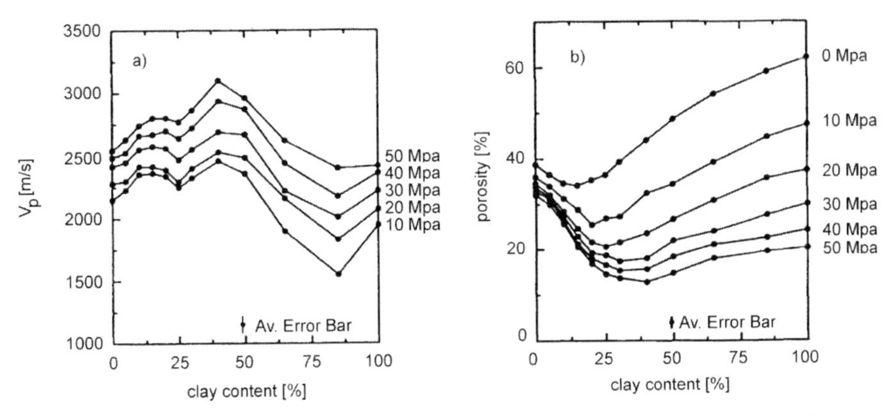

Fig. 1.7. P-wave velocity and porosity for sand-clay mixtures (Marion et al. 1992, with permission from SEG)

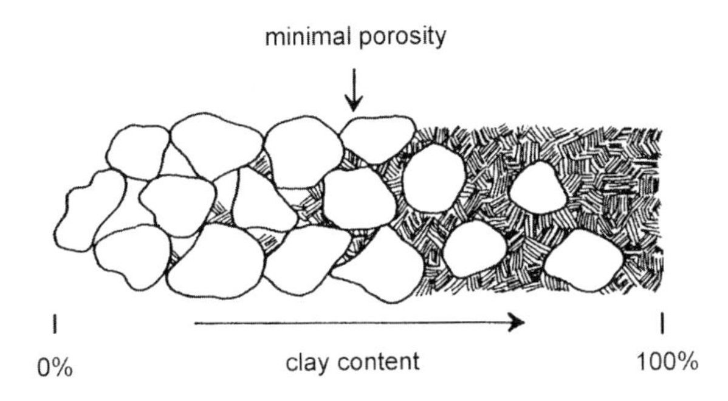

Fig. 1.8. Sketch of sand-clay distribution with increasing clay-content (after Marion et al. 1992)

1.1.3 Clay and till

Clay and till have low hydraulic conductivities. Their hydrogeological importance is that clay or till layers form hydraulic boundaries dividing aquifers.

Till is a mixture of sand, clay, and partly chalk with a wide variety of grain size distributions. The clay content influences the hydraulic conductivity significantly. To investigate the influence of porosity and clay content on seismic velocities, Marion et al. (1992) used artificial sand-clay mixtures for laboratory experiments. A maximum of p-wave velocities was found for clay contents of about 40% (Fig. 1.7).

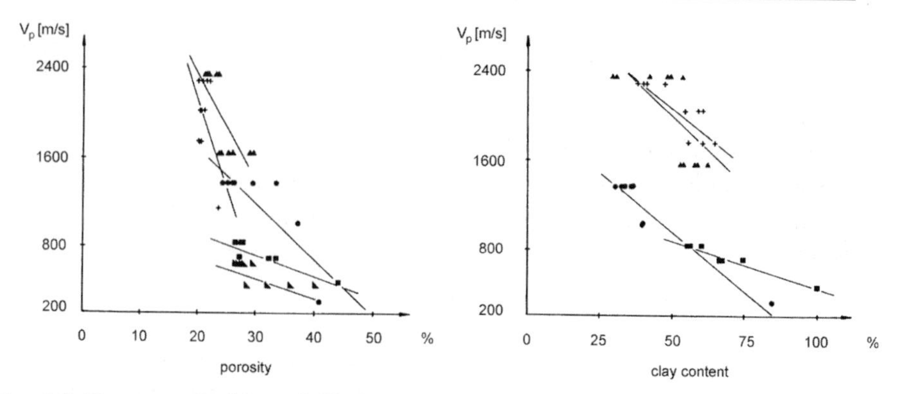

Fig. 1.9. P-wave velocities of tills in relation to porosity and clay content (Baermann and Hübner 1984, with permission from BGR)

An explanation is given in Fig. 1.8. Porosity of clay is about 60%, porosity of sand is about 40%. Small clay content in sands reduce porosity because clay particles fill the pore space. Increasing clay content reduces porosity, until the entire pore space is finally filled with clay. If the clay content is increased further, sand grains loose contact and are isolated in the clay matrix. From that point on, increasing the clay content leads to an increased porosity of the mixture due to the high porosity of clay. It must be taken into account that these results were obtained by using sand and clay of uniform grain size.

Under real field conditions, where tills show a wide variety of grain size distributions, results may not have been so clear. Field measurements on till soils (borehole measurements as well as refraction seismic measurements at steep coasts) by Baermann and Hübner (1984) show decreasing p-wave velocities with increasing porosity and clay content (Fig. 1.9). However, the obtained velocity/porosity or velocity/clay content relations are very site specific and cannot be used in general for an interpretation of seismic velocities.

1.2 Electrical resistivity

1.2.1 Archie´s law – conductive pore fluid and resistive rock matrix

Since the electrical resistivity of most minerals is high (exception: clay, metal ores, and graphite), the electrical current flows mainly through thepore water. According to the famous Archie law, the resistivity of water-saturated clay-free material can be described as

$$\rho_{AQUIFER} = \rho_{WATER} \cdot F \tag{1.6}$$

$\rho_{AQUIFER}$ = specific resistivity of water saturated sand
ρ_{WATER} = specific resistivity of pore water

The formation factor F combines all properties of the material influencing electrical current flow like porosity ϕ, pore shape, and diagenetic cementation.

$$F = a \cdot \phi^{-m} \tag{1.7}$$

Different expressions for the material constant m are used like porosity exponent, shape factor, or (misleading for deposits) cementation degree. Factors influencing m are, e.g., the geometry of pores, the compaction, the mineral composition, and the insolating properties of cementation (Ransom 1984).

The constant a reflects the influence of mineral grains on current flow. If the mineral grains are perfect insulators (main condition for the validity of Archie´s law), then a = 1. If the mineral grains contribute to electrical conductivity to a certain degree, a is reduced accordingly.

Typical values for a and m are (after Schön 1996): loose sands, a = 1.0, m = 1.3, and sandstones, a = 0.7, m = 1.9. Further examples for a and m are given by Worthington (1993).

Fig. 1.10 shows the influence of the porosity and the porosity exponent m on the formation factor F. For sandy aquifers with porosities ranging from 20 – 30 % formation factors can be expected in the range of 4 - 8. However, as the porosity exponent m is normally unknown, it is difficult to predict the porosity from the measured resistivities of the aquifer, even if the resistivity of the pore water is known. Some values for formation factors in relation to grain size for loose sands are shown in Fig. 1.11.

As the constant m is influenced by pore geometry, the formation factor F is related to tortuosity T. Tortuosity describes how crooked the way of fluid flow through pore space is. Tortuosity depends on porosity, pore shape, and the shape of channels connecting the pores. Assuming that the electrical current flow follows the same path through the pore space as the fluid flow, a relation between formation factor and tortuosity can be found (TNO 1976).

$$F = T \cdot \phi^{-m^*} \tag{1.8}$$

m^* = modified porosity exponent.

A mean tortuosity of T= 1.26 was found by TNO (1976) for dune sands and deposits from the river Rhine. Since tortuosity is strongly related to the hydraulic conductivity, Eq. 1.8 gives a link between geophysical and hydraulic properties of the aquifer.

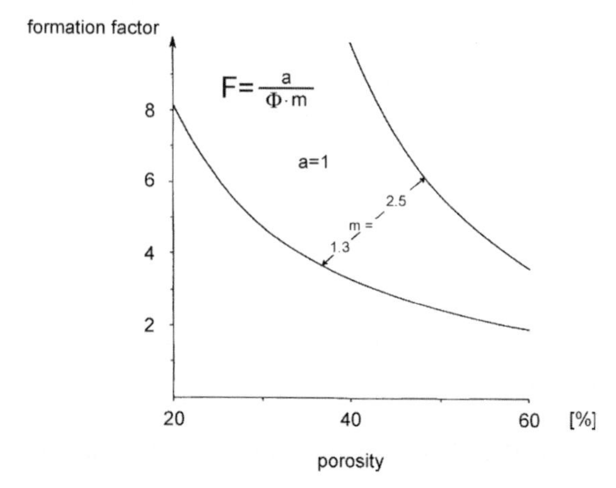

Fig. 1.10. Archie´s law: formation factor F vs. porosity for different porosity exponents

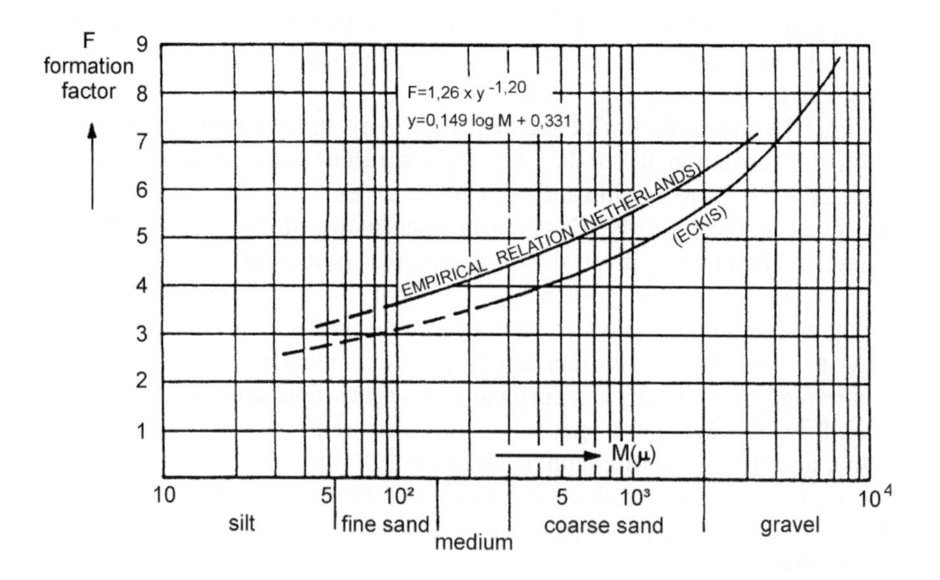

Fig. 1.11. Formation factor dependent on grain size for The Netherlands (TNO 1976, with permission from TNO) compared to results for California (Ecknis 1934), M(μ) = grainsize in micrometer

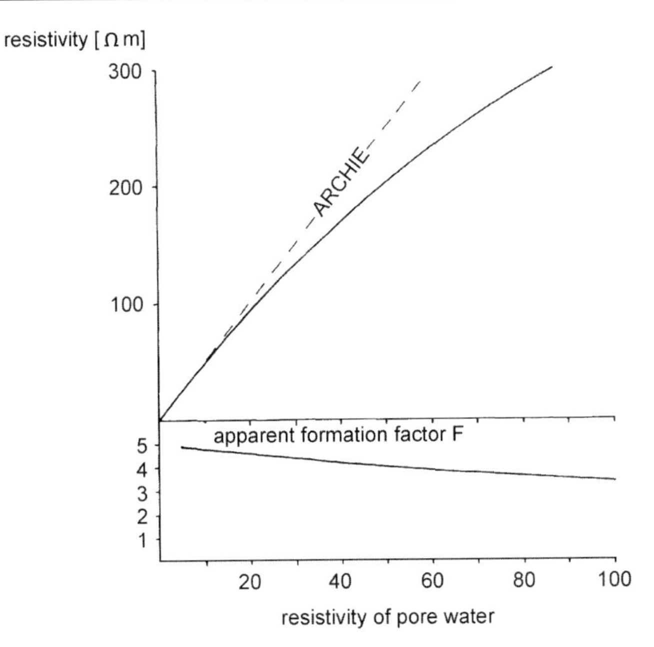

Fig. 1.12. Resistivity and apparent formation factor for high resistive pore water

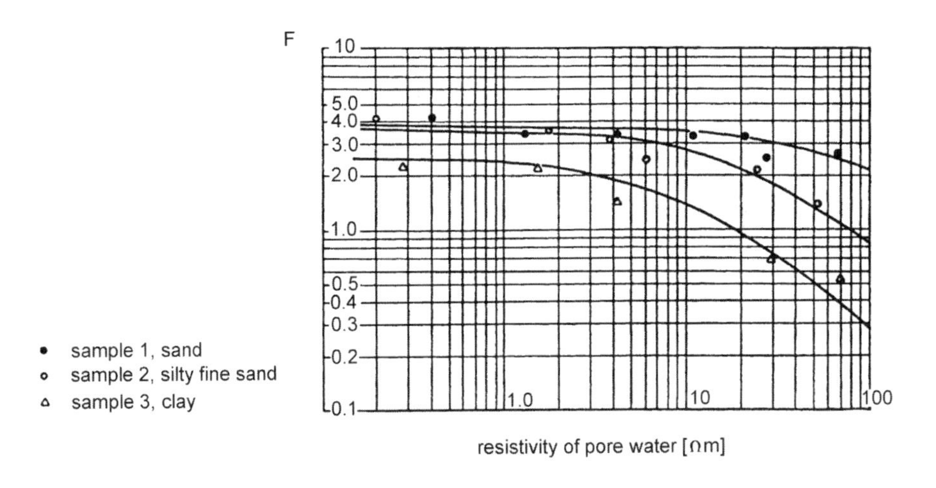

Fig. 1.13. Field examples measured in the Chaco of Paraguay (Repsold 1976, with permission from BGR) for formation factors depending on water resistivity

1.2.2 Limitations of Archie´s law – conducting mineral grains

The validity of Archie´s law and related formulae is restricted to materials with highly resistive mineral grains and conducting pore fluid. A minor contribution of the mineral grains to electrical conductivity can be taken into account by the constant a. However, when the resistivity of the pore water is sufficiently high that the electrical conductivity of the mineral grains is a substantial contribution to the electrical conductivity of the aquifer, the formulations of Archie are no longer valid. Modified formulations are also required for material with surface conductivity like clay.

High resistive pore water

The electrical resistivity of pore water is controlled by the mineral content (salinity) as described in chapter "Groundwater quality". If the mineral content of the groundwater is low resulting in a high bulk resistivity of the aquifer, current flow through the aquifer can be explained by parallel connection of rock matrix and pore fluid (Repsold 1976).

$$\frac{1}{\rho_{AQUIFER}} = \frac{1}{\rho_{MATRIX}} + \frac{1}{F \cdot \rho_{WATER}} \tag{1.9}$$

If we assume a matrix resistivity ρ_{MATRX} of 1000 Ωm and a formation factor of 5, then even for water resistivity of 20 Ωm aquifer resistivity is clearly lower than expected by Archie´s law. If a formation factor is calculated formally by $F=\rho_{AQUIFER}/\rho_{WATER}$, a decrease of the so obtained apparent formation factor is observed with increasing water resistivity (Fig. 1.12). Field examples for apparent formation factors depending on water resistivity are shown in Fig. 1.13.

Resistivity of clay and till

Clayey material is characterized by low electrical resistivity in the range of 5 - 60 Ωm and often a target in electrical or electromagnetic surveys. This low resistivity is caused by surface conductivity of clay minerals. As clay minerals are flat, water can diffuse between the minerals and so increase the specific surface area. The specific surface area of clays can be up to 1000 m^2/g, whereas for sands this area is less than 0.1 m^2/g (Scheffer and Schachtschabel 1984). The large specific surface area supports the surface conductivity. Because a number of cations in clay minerals is replaced by cations of higher valence, electrical charge of the clay mineral surface is negative. The negative charge is compensated by the concentration of cations in the pore water in the vicinity of the mineral surface. This process is quantified by the cation exchange capacity (CEC).

The calculation of the resistivity of clayey material is complicated, since the electrical current flow is possible through clay minerals as well as through pore fluid. A relatively easy approach is given by Frohlich and Parke (1989). They assume that the bulk conductivity of clayey material σ_0 can be explained by parallel connection of surface conductivity $\sigma_{SURFACE}$ and conductivity of pore water σ_{WATER} with volumetric water content Θ:

$$\sigma_0 = \frac{1}{a} \cdot \sigma_{WATER} \cdot \Theta^k + \sigma_{SURFACE} \tag{1.10}$$

or, expressed in terms of resistivity

$$\frac{1}{\rho_0} = \frac{\Theta^k}{a \cdot \rho_{WATER}} + \frac{1}{\rho_{SURFACE}} \tag{1.11}$$

The first part of Eqs. 1.10 and 1.11 is related to Archie´s law, when exponent k is defined by the saturation degree S_W

$$\Theta^k = S_W{}^n \cdot \phi^m \tag{1.12}$$

A special case of Eq. 1.10 is given by Mualem and Friedman (1991)

$$\sigma_0 = \sigma_{WATER} \cdot \frac{\Theta^{2.5}}{\phi} + \sigma_{SURFACE} \tag{1.13}$$

An expression of surface conductivity (in mS/cm) in terms of volumetric clay content C was found by Rhoades et al. (1989)

$$\sigma_{SURFACE} = 2.3 \cdot C - 0.021 \tag{1.14}$$

However, for the practical use of Eqs. 1.13 and 1.14, the validity of the empirically determined constants for the project area must be checked.

A more general approach to electrical conductivity of clayey material based on cation exchange capacity is given by Sen et al. (1988):

$$\sigma_0 = \frac{1}{F}\left(\sigma_w + \sigma_w \frac{AQ_v}{\sigma_w + BQ_v}\right) + EQ_v \tag{1.15}$$

Q_v can be expressed by cation exchange capacity CEC, matrix density ρ_{MAT}, and porosity ϕ:

$$Q_v = \frac{\rho_{MAT}(1-\phi)}{\phi} CEC \tag{1.16}$$

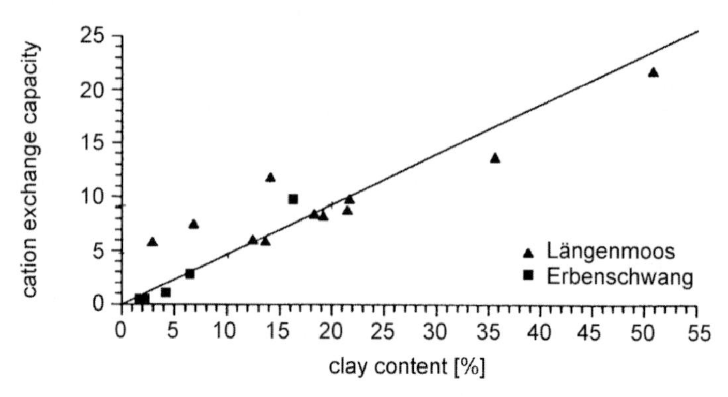

Fig. 1.14. Correlation between clay content and cation exchange capacity for two areas in Southern Germany (Günzel 1994)

According to Günzel (1994), constants A, B, and E are given by BQ_v=0.7, EQ_v=0, and $A=m\lambda^S_{na}$, with m=porosity exponent of Archie equation and λ^S_{na}= equivalence conductivity of Na^+-exchange cations, empirically derived as λ^S_{na}=1.94 (S/m)/(mol/l).

Sen et al. (1988) found an empirical relation between porosity exponent and cation exchange capacity for sandstone samples: $m=1.67+0.2\times CEC^{1/2}$. This can lead to an increase in resistivity with increasing clay content, a clear contradiction to the experience that increasing clay content of unconsolidated material leads to decreasing resistivity. The use of the empirical relation between m and CEC should be restricted to consolidated material. Sen et al. (1988) also mentioned that a good fit of measured data is possible using constant m=2.

Eq. 1.15 is valid for saturated material. For partly saturated material, Günzel (1994) replaced Q_V by $Q^* = Q_V/S_W$ (S_W = saturation degree), formation factor F is changed accordingly. Assuming clay free material with CEC = 0, Eq. 1.15 reduces to Archie´s law $\sigma = \sigma_W/F$.

As shown above, the critical parameter for conductivity of clayey material is not the clay content, but the cation exchange capacity. Cation exchange capacity strictly depends on the mineral composition of clay, which may differ from area to area. Günzel (1994) showed that for smaller areas, where a constant composition of clay minerals can be assumed, a linear relation CEC = i×C between clay content C and cation exchange capacity exists (Fig. 1.14). As a consequence, if in Eq. 1.16 CEC is replaced by i×C, Eq. 1.15 relates clay content to conductivity.

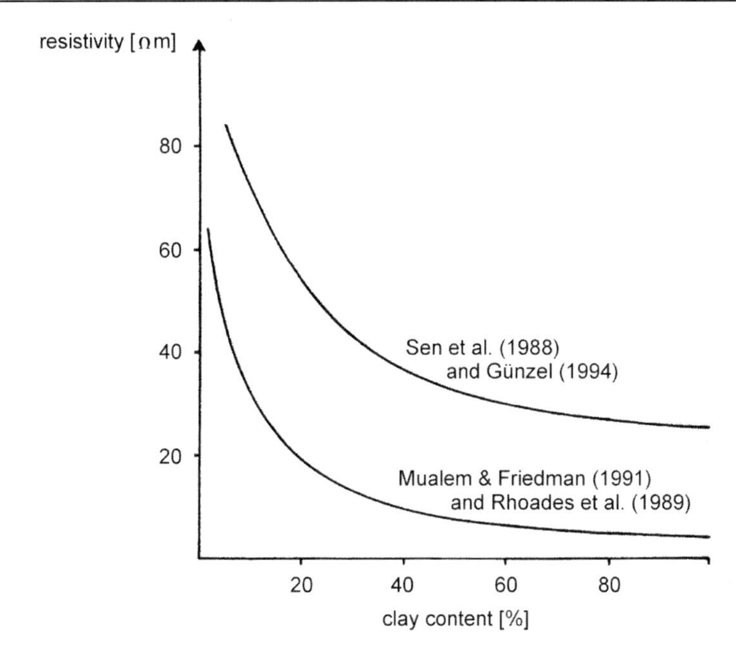

Fig. 1.15. Resistivity of clayey sediments related to clay content by Sen et al. (1988) and Mualem and Friedman (1991). For both relations, porosity of 30% and pore water resistivity of 15 Ωm were assumed

Using above formulae a comparison between the formalism given by Sen et al. (1988) and the easy formalism of Mualem and Friedman (1991) in Eq. 1.10 is possible. In Eq. 1.10 the surface conductivity σ_S is replaced by Eq. 1.14 (Rhoades et al. 1989), whereas in Eq. 1.16 CEC is replaced by the CEC/clay content ratio found by Günzel (1994) (Fig. 1.14). The results (conductivity converted to resistivity) are shown in Fig. 1.15. Both resistivity-clay content relations show similar shapes, but strong differences for the absolute values. This can be explained by local effects of clay mineral composition in the relations of Günzel (1994) and Rhoades et al. (1989).

Based on the formalism of Sen et al. (1988), an approach to determine clay content from resistivity data is given by Borús (2000). If the lateral distribution of clay content in the near surface subsoil has to be determined by electrical measurements, some reference points with known clay content and resistivity are required in this area. These clay content/resistivity values are displayed in a diagram showing resistivity/clay content curves for different CEC/clay content ratios i (Fig. 1.16). The curve which gives the best fit to the measured data can be used to determine the clay content

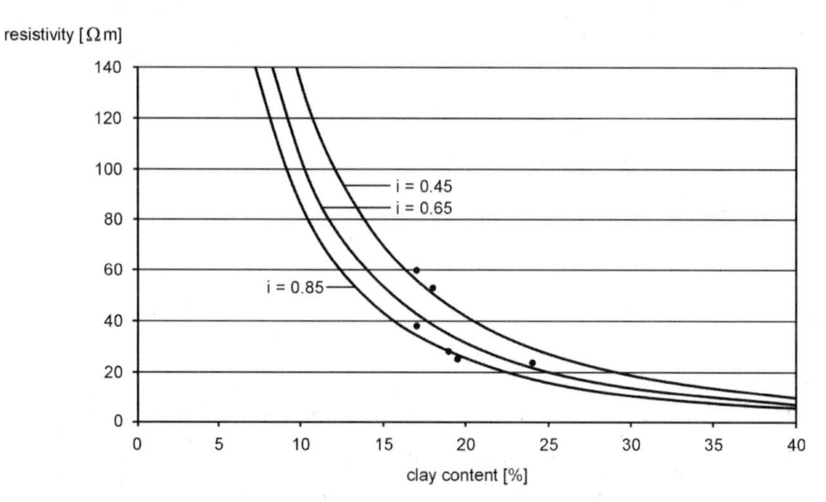

Fig. 1.16. Resistivity-clay content curves for different CEC-clay content ratios i after Sen et al (1988). Dots: measured clay content-resistivity values from the Baltic coast area near Kiel. The curve with i=0.65 gives the best approached to the measured data and can be used for an assessment of clay content from resistivity values in the project area (Borús 2000)

from the resistivity values measured in the project area apart from the reference points.

1.3 Electric Permittivity (Dielectricity)

Electric permittivity ε (more correct: relative permittivity ε_r) depends on the polarisation properties of material and is the dominating factor for the propagation speed of electromagnetic waves in the sub-surface which can be calculated from:

$$v = \frac{c}{\sqrt{\varepsilon}} \tag{1.17}$$

(c = speed of light in vacuum)

The propagation speed of electromagnetic waves is used for time/depth conversion of GPR sections. Because this speed is extremely high, e.g. 3×10^8 m/s in vacuum, normally the "easy to handle" unit cm/ns is used, 3×10^8 m/s then reduces to 30 cm/ns.

Typical values for permittivity ε are: water = 80, saturated sand = 20 – 30, and air = 1. High permittivity of water results in strong correlation

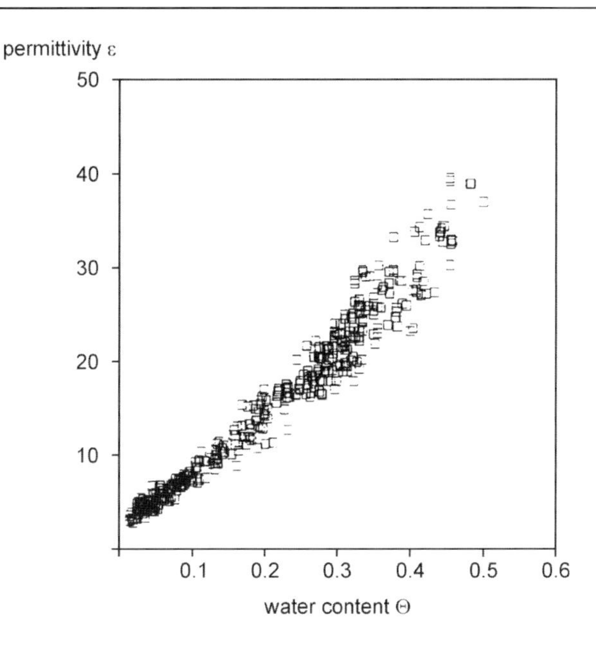

Fig. 1.17. Permittivity of glacial sediments from Finland and Wisconsin (USA) in relation to water content, the data can be fitted by $\varepsilon = 3.2 + 35.4\times\Theta + 101.7\times\Theta^2 - 63\times\Theta^3$ (Sutinen 1992, with permission from the Geological Survey of Finland)

between permittivity of material and water content, as shown in Fig. 1.17.

High permittivity of water results from dipole characteristic of water molecules leading to high polarisability. High polarisability is lost when water is frozen. In saltwater, permittivity is also reduced down to $\varepsilon = 35$ at total saturation (Kulenkampff 1988). This is caused by electrostatic grouping of dissociated anions and cations around the H^+ and O^2 ions of the water molecules reducing polarisability. For GPR measurements, this reduced permittivity and increased propagation speed in saltwater is not important, because due to high absorption of radar pulses in saltwater no sufficient penetration in salty soil can be achieved.

To quantify the influence of porosity ϕ and water content (quantified by the saturation degree S_W on the permittivity ε, a general mixing law for a multi-component rock system (Birschak et al. 1974) can be applied:

$$\varepsilon^{\varsigma} = \sum_{n} v_i \cdot \varepsilon_i^{\varsigma} \qquad (1.18)$$

v_i, ε_i volumetric content and permittivity of each component (rock matris, pore water, etc)

A special case is given by Schön (1996)

$$\varepsilon = \phi \cdot S_W \cdot \varepsilon_{WATER} + \phi \cdot (1 - S_W) \cdot \varepsilon_{AIR} + (1 - \phi) \cdot \varepsilon_{MATRIX} \quad (1.19)$$

ε_{MATRIX} = permittivity of rock matrix (e.g. quartz grains)
ε_{WATER} = permittivity of water
ε_{AIR} = permittivity of air

Another special case of the mixing law is the CRIM equation (complex refractive index method) (Schlumberger 1991)

$$\sqrt{\varepsilon} = \phi \cdot S_W \sqrt{\varepsilon_{WATER}} + (1 - \phi)\sqrt{\varepsilon_{MATRIX}} + \phi \cdot (1 - S_W)\sqrt{\varepsilon_{AIR}} \quad (1.20)$$

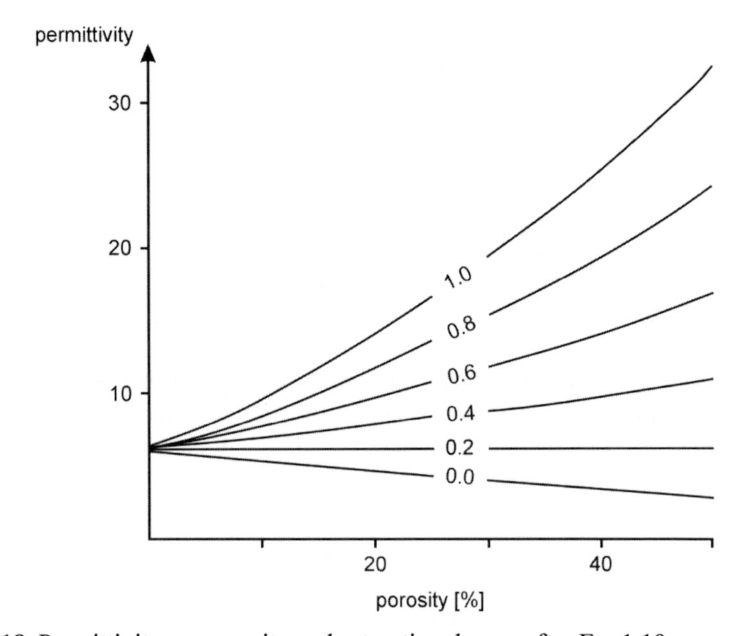

Fig. 1.18. Permittivity vs. porosity and saturation degree after Eq. 1.19

Based on Eq. 1.19 the influence of porosity on permittivity for different saturation degrees is shown in Fig. 1.18. Whereas for saturated pores permittivity increases with increasing porosity, a decrease of permittivity can be expected for air filled pores. For a saturation degree of about 30% which can be assumed for unsaturated sand, no influence of porosity variations on the permittivity can be expected.

A further formalism to calculate the permittivity for partial saturated sediments is the Hanai-Brüggemann mixing law (Graeves et al. 1996). First of all, for the pore filling water/air, an effective permittivity ε_{PORE} is calculated:

$$\varepsilon_{PORE} = \varepsilon_{WATER} \cdot S_W{}^{m_1} \cdot \left(\frac{1 - \frac{\varepsilon_{AIR}}{\varepsilon_{WATER}}}{1 - \frac{\varepsilon_{AIR}}{\varepsilon_{PORE}}} \right)^{m_1} \tag{1.21}$$

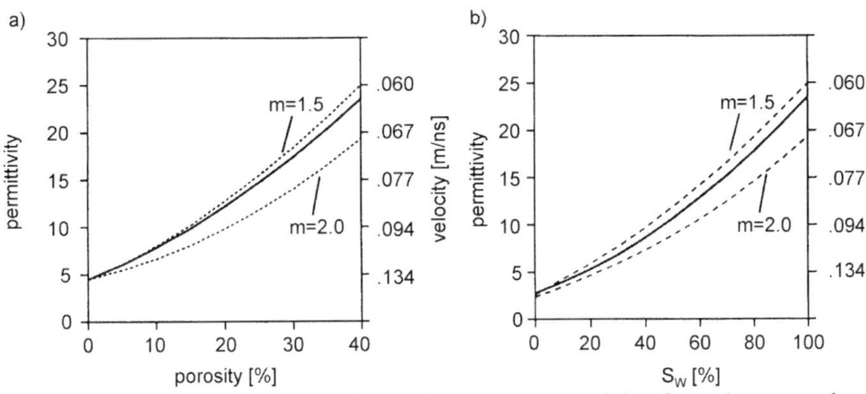

Fig. 1.19. Comparison of CRIM and Hanai-Brüggemann mixing law, a) saturated material, permittivity vs. porosity; b) partly saturated material, porosity 40%, permittivity vs. saturation degree (Graeves et al. 1996, with permission from SEG)

Using an analogue mixing law, the permittivity of partly saturated material can be obtained

$$\varepsilon = \varepsilon_{PORE} \cdot \phi^{m_2} \cdot \left(\frac{1 - \frac{\varepsilon_{MATRIX}}{\varepsilon_{PORE}}}{1 - \frac{\varepsilon_{MATRIX}}{\varepsilon}} \right)^{m_2} \tag{1.22}$$

Although constants m_1 and m_2 are different, for practical use both can be assumed to be equal (Graeves et al. 1996). A comparison of CRIM and Hanai-Brüggemann mixing law is given in Fig. 1.19. Identical values of 1.5 (unconsolidated sand) and 2 (cemented sandstone) were taken for constants m_1 and m_2. Assuming a constant m = 1.6, both formalisms lead to nearly identical results.

For clayey material with clay content C, the CRIM equation was extended by Wharton et al. (1980)

$$\sqrt{\varepsilon} = \phi \cdot S_W \cdot \sqrt{\varepsilon_{WATER}} + (1-\phi) \cdot (1-C) \cdot \sqrt{\varepsilon_{MATRIX}} + \phi \cdot (1-S_W)\sqrt{\varepsilon_{AIR}} + (1-\phi) \cdot C \cdot \sqrt{\varepsilon_{CLAY}} \tag{1.23}$$

Field values for permittivity of tills were reported by Sutinen (1992) as shown in Fig. 1.20.

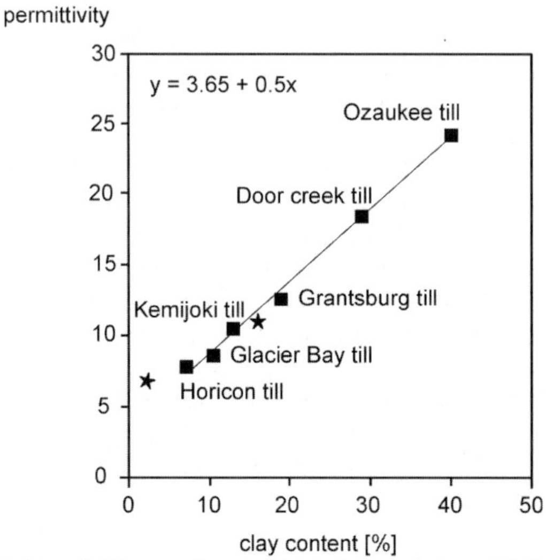

Fig. 1.20. Permittivity of till vs. clay content for samples of Wisconsin (USA), similar results were obtained in other parts of USA (Sutinen 1992, with permission from the Geological Survey of Finland)

1.4 Conclusions

As shown above, sufficient contrast of physical properties of saturated and unsaturated material can be expected leading to good conditions for geophysical prospecting. In Table 1.1 typical values for seismic velocities, resistivities, and permittivities for relevant materials are listed.

	seismics	geoelectrics, electromagnetics		GPR *)	
	V_P m/s	resistivity Ωm	conductivity mS/m	permittivity	wave velocity cm/ns (mean)
gravel, sand (dry)	300 – 800	500 – 2000	0.5 – 2	3 - 5	15
gravel, sand (saturated)	1500 – 2000	60 – 200	5 – 17	20 – 30	6
fractured rock	1500 – 3000	60 – 2000	0.5 – 17	20 – 30	6
solid rock	> 3000	> 2000	< 0.5	4 - 6	13
till	1500 – 2200	30 – 60	17 – 34	5 – 40	6
clay	1500 – 2500	10 – 30	34 - 100	5 – 40	6

Table 1.1. Physical properties of aquifers and impermeable layers
*) after Davis and Annan (1989)

1.5 References

Bachran R, Dvorkin J, Nur AM (2000) Seismic velocities and Poisson´s ratio of shallow unconsolidated sands. Geophysics 65:559-564

Baermann A, Hübner S (1984) Ingenieurgeologische und geophysikalische Untersuchungen an Geschiebemergeln im Norddeutschen Raum. Geologisches Jahrbuch C37:17-57

Birschak R, Gardener CG, Hipp JE, Victor JM (1974) High dielectric constant microwave probes for sensing soil moisture. Proc. IEEE, 62, 93-98

Borús H (1999) Einsatz geophysikalischer Meßverfahren zur Abschätzung der hydraulischen Durchlässigkeit tonhaltiger Sedimente als Beitrag zum Grundwasserschutz. PhD-Thesis. Christian-Albrechts-Universität Kiel

Davis, JL and Annan AP (1989) Ground-penetrating radar for high-resolution mapping of soil and rock stratigraphy. Geophysical Prospecting 37:531-551

Ecknis RP (1934) South Coastal Basin Investigation, Geology and Ground Water Storage Capacity of Valley Fill. State of California Bull. 45

Frohlich RK, Parke CD (1989) The electrical resistivity of the vadose zone - field study. Ground Water 27:525-530

Graeves RJ, Lesmes DP, Lee JM, Toksöz MN (1996) Velocity variations and water content estimated from multi-offset ground penetrating radar. Geophysics 61:683-695

Günzel F (1994) Geoelektrische Untersuchung von Grundwasserkontaminationen unter Berücksichtigung von Ton- und Wassergehalt auf die elektrische Leitfähigkeit des Untergrundes. PhD thesis Ludwig-Maximilians-Universität München

Han DH, Nur A, Morgan D (1986) Effects of porosity and clay content on wave velocities in sandstones. Geophysics 51:2093-2107

Hamilton EL (1971) Elastic properties of marine sediments. J Geoph Res 76:579-604

Klimentos T (1991) The effects of porosity-permeability-clay content on the velocity of compressional waves. Geophysics 56:1930-1939

Kowallis BJ, Jones LEA, Wang HF (1984) Velocity-porosity-clay content systematics of poorly consolidated sandstones. J Geophys Res 89:10355-10364

Kulenkampff J (1988) Untersuchung über die komplexe elektrische Leitfähigkeit von porösen Gesteinen. – Diploma Thesis, Institut für Geophysik, Technische Universität Clausthal

Marion D, Nur A, Yin H, Han D (1992) Compressional velocity and porosity in sand-clay mixtures. Geophysics 57:554-563

Mavko G, Mukerji T, Dvorkin J (1998) The rock physics handbook: tools for seismic analysis in pourous media. Cambridge University Press, Cambridge, New York, Melbourne

Morgan NA (1969) Physical properties of marine sediments as related to seismic velocities. Geophysics 34:529-545

Mualem Y, Friedman SP (1991) Theoretical prediction of electrical conductivity in saturated and unsaturated soil. Water Resources Research 27:2771-2777

Nur A, Mavko G, Dvorkin J, Galmudi D (1998) Critical porosity: A key to relating physical properties to porosity in rocks. The Leading Edge 17:357-362

Raymer LL, Hunt ER, Gardner JS (1980) An improved sonic transmit time - porosity transform. Trans. SPWLA, 21st Ann Log Symp:1-13

Ransom RC (1984) A contribution towards a better understanding of the modified Archie formation resistivity factor relationship. The Log Analyst:7-12

Repsold H (1976) Über das Verhalten des Formationsfaktors in Lockersedimenten bei schwach mineralisierten Porenwässern. Geologisches Jahrbuch, E9:19-34

Rhoades JD, Manteghi NA, Shouse PJ, Alves WJ (1989) Soil electrical conductivity and soil salinity: new formulations and calibrations. Soil Sci Soc Am J. 53:433-439

Scheffer F, Schachtschabel P (1984) Lehrbuch der Bodenkunde. Enke Verlag, Stuttgart

Schlumberger (1991) Log interpretation principles/applications. Schlumberger Educational Services, Houston

Schön, JH (1996) Physical properties of rocks: Fundamentals and Principles of Petrophysics. Pergamon Press, New York

Sen PN, Goode PA, Sibbit A (1988) Electrical conduction in clay bearing sandstones at low and high salinities. J. Appl. Phys. 63:4832-4840

Sutinen R (1992) Glacial deposits, their electrical properties and surveying by image interpretation and ground penetrating radar. Geological Survey of Finland Bulletin 359, Espoo

TNO (1976) Geophysical well logging for geohydrological purposes in unconsolidated formations. Groundwater Survey TNO, The Netherlands Organisation for Applied Scientific Research, Delft

Watkins, JS, Walters LA, Godson RH (1972) Dependence of in-situ compressional-wave velocity on porosity in unsaturated rocks. Geophysics 37:29-35

Wharton, RP, Hazen GA, Rau RN, Best DL (1980) Electromagnetic propagation logging: advances in technique and interpretation. Soc of Petr Eng, Paper 9267

Worthington PF(1993) The uses and abuses of the Archie equation, 1: The formation factor-porosity relationship. Journal of Applied Geophysics 30:215 - 228

Wyllie MRJ, Gregory AR & Gardner LW (1956) Elastic wave velocities in heterogenous and porous media. Geophysics 21:41-70

2 Seismic methods

Wolfgang Rabbel

2.1 Introduction

The following article provides an overview of principles and concepts of seismic prospecting on shore and its application to hydro-geological targets. In the introduction some questions regarding the role of seismic methods in near surface exploration are explained. The subsequent two paragraphs provide closer looks at seismic refraction and reflection measurements, respectively. Therein, basics of body wave propagation and its resolution potential for imaging shallow geological structure are treated.

2.1.1 What type of waves is applied in seismic exploration?

Seismic exploration is based on the propagation of elastic waves inside the earth. The propagation velocity and amplitude (signal strength) of these waves depend on the dynamic elastic constants of the rocks and on their density. There are two families of elastic waves: **body waves** which are capable to traverse and "probe" all depth levels of the subsurface, and **interface waves** existing only near the boundary of layers such as the earth surface. Except for some special applications, seismic prospecting is mainly based on acquiring and analysing body wave signals generated and recorded at the earth's surface.

There are two types of body waves: compressional and shear waves also called P- and S-waves, respectively. P- and S-waves exhibit particle movement parallel and orthogonal to the direction of wave propagation, respectively (Fig. 2.1 a and b, respectively). Their propagation velocities v_P and v_S are related with the compressional modulus k, the shear modulus μ and the density ρ of the subsoil following Eq. 2.1:

$$v_p = \sqrt{\frac{k + 4\mu/3}{\rho}} \qquad v_s = \sqrt{\mu/\rho} \qquad (2.1)$$

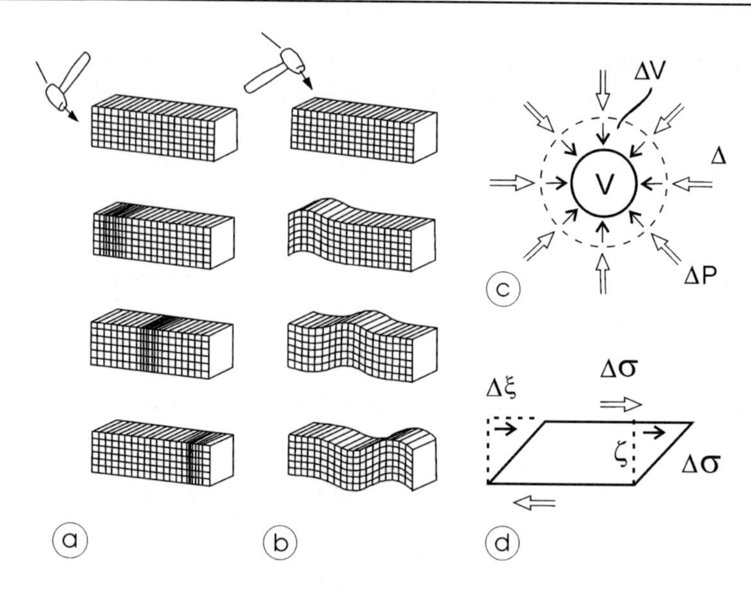

Fig. 2.1. (a) Particle movement of compressional waves (P-waves), (b) particle movement of shear waves (S-waves), (c) definition of compressional modulus k with respect the change ΔV of a volume V caused by a pressure increment ΔP (d) definition of shear modulus μ with respect to the shear displacement Δξ of a sample of width ζ caused by a shear stress Δσ

The compressional modulus k is defined as the ratio between pressure and volume deformation connected with the passage of a P-wave (Fig. 2.1c). The shear modulus μ is correspondingly defined as the ratio of shear stress to shear deformation (Fig. 2.1d). In contrast to P-wave movement, no volume change occurs in the rock during S-wave propagation (except at interfaces). Therefore, the S-wave velocity depends mainly on the properties of the rock matrix and is nearly independent of the pore fill whereas P-wave velocity depends on both matrix and pore fill. In particular, P-waves are sensitive to rock porosity and fluid saturation making them a suitable tool for groundwater exploration. For a more comprehensive lithological investigation of the subsurface advantage can be taken of combined P- and S-wave surveys.

2.1.2 How can seismic waves image geological structure?

Seismic waves travelling through a homogeneous layer with constant seismic velocity form circular wave fronts, the propagation of which can be described by straight rays. However, seismically homogeneous layers are rarely found near the earth's surface where both compositional and structural changes of the geological layers and increasing lithostatic pressure cause significant spatial variations of seismic velocity. Usually, these variations are gradual inside a layer and discontinuous at layer interfaces. Both kinds of velocity variation cause a bending of the propagating wave front and the corresponding rays (Fig. 2.2).

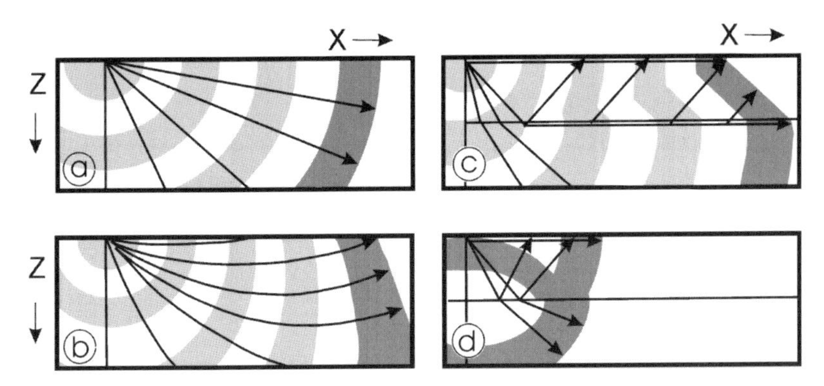

Fig. 2.2. (a) Wave fronts and rays of a seismic wave in a homogeneous layer, (b) wave fronts and rays of a seismic wave in a layer where seismic velocity increases continuously with depth such as caused by lithostatic pressure, (c) wave fronts and rays of direct and refracted (head) waves in a two-layer medium where the seismic velocity increases discontinuously with depth such as found at the water table for P-wave velocity, (d) wave fronts and rays of direct and refracted waves in a two-layer medium where the seismic velocity increases discontinuously with depth

Of particular importance is the case of a plane interface where the energy of an impinging seismic P- or S-wave is partly transmitted into the lying wall, partly reflected into the hanging wall, and partly converted into S- or P-waves, respectively (Fig. 2.2c). The angle directions i_1 and i_2 of the arriving and emerging waves, respectively, are related by Snell's law to the propagation velocities v_1 and v_2 applying to the considered wave type and half space (Fig. 2.3):

$$\frac{\sin(i_1)}{v_1} = \frac{\sin(i_2)}{v_2} \tag{2.2}$$

Refracted waves travelling parallel to the interfaces (so-called head waves, Fig. 2.2c and 2.3 left) are generated if the velocity increases in the lying medium and if the incident ray impinges under the critical angle $i_1 = i_C$ defined by:

$$i_c = \arcsin(v_1 / v_2) \tag{2.3}$$

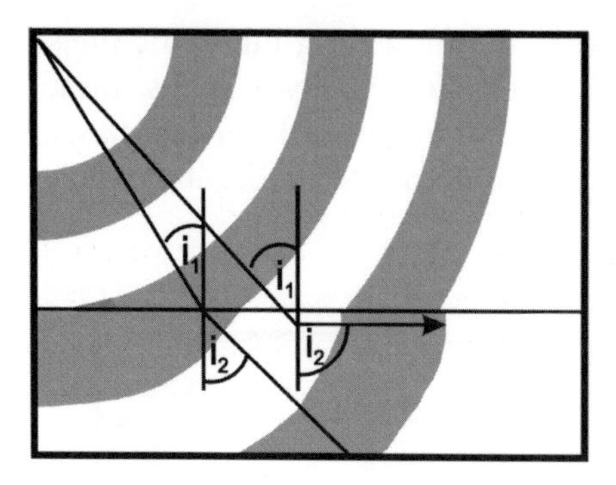

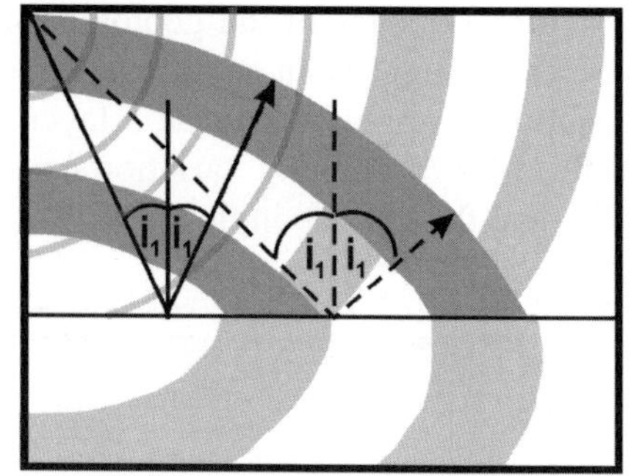

Fig. 2.3. Transmission (left) and reflection (right) of wave fronts at a plane seismic discontinuity illustrating Snell's law. Note that the incidence (arrival) and excidence (take-off) angles of rays are measured with respect to the normal of the interface. This normal may deviate from the vertical in case of dipping layers

Because layers are elastically coupled, head waves radiate back into the hanging wall so their propagation can be observed at the earth surface after a certain time delay. Transmitted refracted waves and head waves are the main observables in seismic refraction measurements.

In contrast to head waves which exist only in case of a velocity increase across an interface, reflected waves ("echos") are generated at any interface in the subsoil where the density or the velocity changes discontinuously, independently of whether an increase or decrease occurs. These waves form the basis of seismic reflection imaging. The strength of the reflected signals depends on the impedance contrast, which is the difference in the product of velocity and density, on the velocities and incidence angles of the considered waves.

Basic observables in seismic prospecting are the travel times and amplitudes of seismic waves reflected or refracted at the target horizons and at the layer interfaces in the hanging wall. The term travel time denotes the time a wave requires for propagating from the source point, where it was generated, down to the target horizon and back to a receiver. The travel time τ of a wave following a given ray path S can be computed by solving the integral:

$$\tau(S) = \int_S ds / v(s) \tag{2.4a}$$

where ds is the ray path increment and v(s) the seismic velocity along the ray. If the ray can be subdivided into a number of straight segments Δs_j along which the velocity v_j is constant, the integral can be replaced by the summation:

$$\tau(S) = \sum_j \Delta s_j / v_j \tag{2.4b}$$

Source points and receivers are usually located at the earth's surface – except in case of borehole seismics. Of course, a seismic receiver records refracted and reflected signals from all interfaces reached by the downward travelling wave train. In the resulting seismogram these differently delayed arrivals can only be distinguished from each other if the time duration of the seismic signal is short enough (Fig. 2.4).

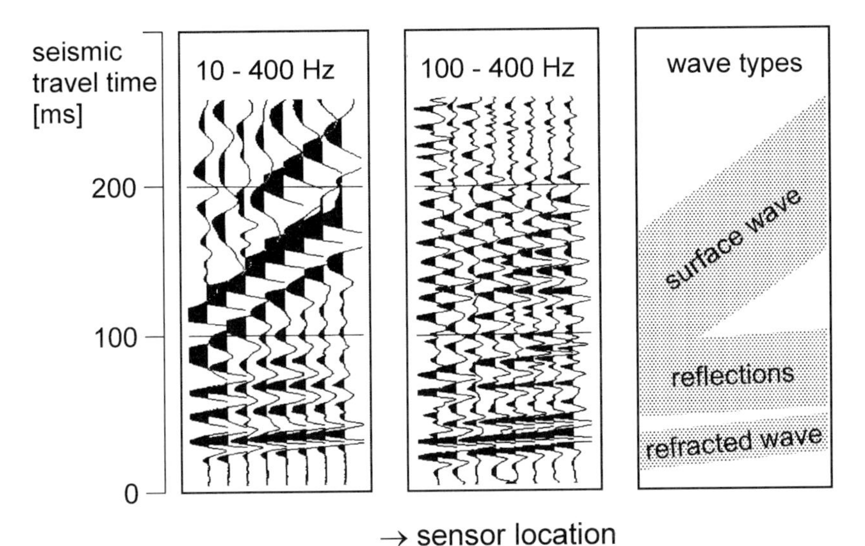

Fig. 2.4. Seismograms recorded at neighbouring sensor locations (positive and negative amplitudes in black and white, respectively. Arrivals of different wave types can be distinguished from each other because of their finite signal length and their differing signal shape and dip. Note that the appearance of the seismograms is strongly influenced by the width of their frequency spectra (compare left and middle figures). Signal spectra can be influenced by the choice of the seismic source and sensors and by analogue or digital filtering

2.1.3 How are seismic waves generated and recorded in the field?

For the investigation of near surface targets at less than a hundred meter maximum depth, seismic signals can be generated, for example, with hammer blows and weight drops. Accelerated weight drops and vibration sources (Fig. 2.5) can be used to explore targets up to 1 km depth. Small explosive charges are in use also. Compared to the lengths of the generated waves each of these sources can be regarded as a point force. Impact and explosion sources provide impulse type signals from the outset. In contrast, vibration sources stretch the signal input over up to twenty seconds and reconstruct an impulse type signal only after recording by digitally cross-correlating the source signal with the seismograms (so-called Vibroseis® technique). The sharpness of both impact and cross-correlated vibration signals depend (1) on the width of the frequency spectrum of the signal and (2) on its extent towards high frequencies (Fig. 2.4).

Fig. 2.5. Examples for seismic sources applied in near surface prospecting: Accelerated weight drop (THOR Geophysical, Kiel) (top left) and small Vibroseis® truck (GGA-Institut, Hannover) (bottom left). The piston and ground plate through which hydraulic movement of the Vibro-truck is transferred to the ground are shown on the right hand side

The seismic signals are usually recorded along linear geophone spreads where sensors (geophones) are placed at regular distances. The use of geophone spreads implies that a range of subsurface points can be investigated with each shot. Spread length, geophone spacing, and source-to-spread arrangement to be chosen depend on the target depth, source spectra, seismic velocities, wave type and on whether a 2D investigation along linear profiles or a 3D investigation is intended. These points will be discussed later in more detail. Typical for either type of investigation is that the source-geophone arrangements are moved along the earth surface with some overlap. This approach provides a multiple coverage of subsurface points with waves passing the same target range along different travel paths. It is required generally for determining the seismic velocity structure and specifically for improving the signal-to-noise ratio in reflection seismology.

2.1.4 What kind of seismic measurements can be performed?

From an interpretation point of view one may distinguish between seismic refraction and seismic reflection investigations. Seismic refraction studies are mainly based on analysing the travel times of waves refracted under the critical angle. Often, only first arrival times are picked from the seismograms which can be easily identified. A bunch of algorithms has been formulated in order to convert the observed travel time-distance functions into cross-sections of the subsurface. These cross-sections show the seismic velocity structure comprising the depth of seismic interfaces and the velocity inside layers (Fig. 2.6 top).

In contrast, the digital processing of seismic reflection data offers the opportunity to create cross-sectional images of the subsurface in a more direct way. The final outcome of these measurements usually is a seismic reflection section showing the amplitude of the reflected wave at the correct horizontal and depth coordinate of the reflection point (Fig. 2.6 bottom). Usually seismic reflection images are richer in structural detail than velocity sections based on refraction measurements.

Clearly, a seismic point source generates body waves which are turned into reflections and refractions regardless of which wave type the interpreter wishes to analyse. Therefore, from the acquisition point of view, both approaches differ mainly in the applied source-receiver distances and the source point spacing resulting in different multiplicity of subsurface point coverage. Modern approaches try to combine both reflection imaging and velocity tomography based on refraction analysis (Fig. 2.6).

2.1.5 What kind of hydro-geologically relevant information can be obtained from seismic prospecting?

The propagation velocity of P-waves depends strongly on the porosity and the water saturation of sediments. S-wave velocity or the shear modulus, respectively, is mainly determined by the stiffness of the rock matrix. In consequence, both P-and S-wave velocities depend significantly on the fracture density of rocks. Therefore seismic investigations can contribute basically in different regards to hydrogeological investigations: To find the groundwater table or define bed rock levels appear self-evident tasks. In a more general sense, seismic investigations can serve to determine the structural and lithological framework of hydrogeological studies and to quantify the heterogeneity of aquifers and aquitards off boreholes. Also porosity and fracture density can be investigated.

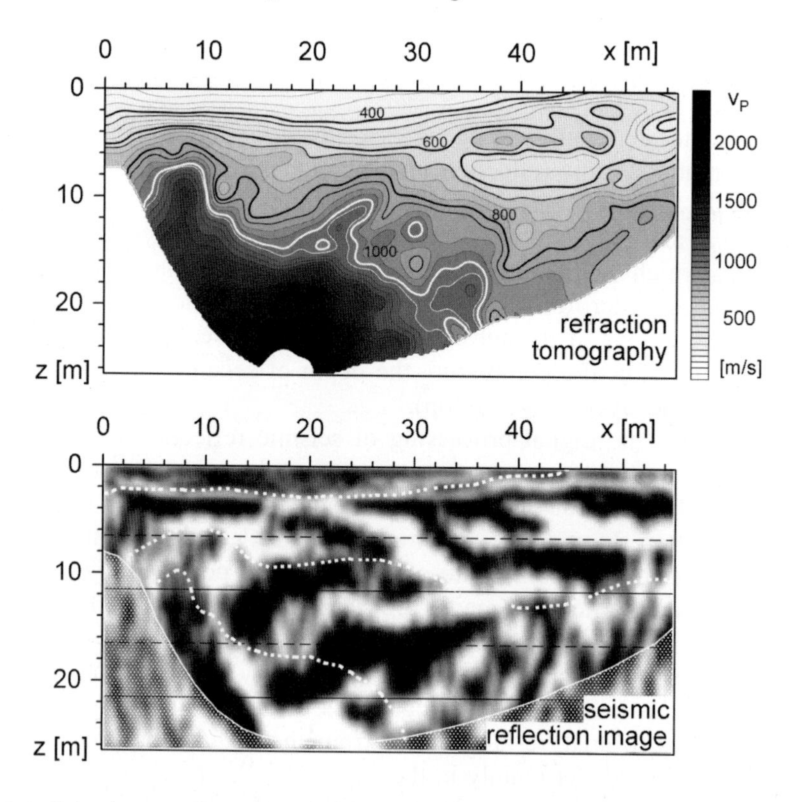

Fig. 2.6. Seismic structure of the weathering zone in a hard rock area based on the application of P-waves. Top: Tomographic interpretation of refracted arrivals The image shows a cross-section in terms of P-wave velocity v_P where unconsolidated material is indicated by low v_P values. Bottom: Corresponding seismic reflection section showing the complex layering within the vadoze zone. Velocity contours for $v_P = 400$, 800 and 1200 m/s are indicated by dashed lines for comparison (by courtesy of GeoExpert AG, Schwerzenbach, Switzerland)

2.1.6 What are the advantages and disadvantages of seismic measurements compared to other methods? How do seismics and other geophysical measurements complement each other?

Seismic prospecting provides reliable information on the depth of interfaces. In comparison to electromagnetic induction and DC-geoelectrical measurements, penetration depth and structural resolution of seismic measurements are usually higher and less ambiguous. These advantages are paid for with higher costs in acquisition and interpretation. The relation of v_P and v_S to rock parameters such as porosity and pore fill is not unique. Therefore, seismic measurements have to be combined with other types of geophysical methods if sedimentary parameters are to be determined in situ.

In hydrogeophysics it is most advantageous to combine geoelectrical and seismic methods to determine, for example, the porosity of sediments or the quality and depth distribution of groundwater horizons. In particular, seismic methods are useful to embed geoelectrical measurements in the framework of geological structure. The sequence “drill hole - seismics - DC-geoelectrics - EM-induction” can be regarded as a sequence of decreasing structural resolution and investigation costs. Seismics is a reliable means to extrapolate structural information gathered at boreholes into the area off the borehole.

2.2 Seismic refraction measurements

Seismic energy transmitted from a source point through a stack of layers can (partially) return to the earth’s surface in two ways: (1) by reflection and back scattering from the top of structural interfaces, the case which is treated in paragraph 3, or (2) if the transmitted waves are refracted into the bedding plane so the bended ray finds a turning point inside a layer. To observe these transmitted body waves is the basis of seismic refraction measurements.

A transmitted wave finds a turning point inside a particular layer only if it was radiated from the source point under the corresponding critical angle. Following Snell’s law (Eqs. 2.2 and 2.3) this critical angle exists if the seismic velocity of the considered layer is higher than any velocity in the hanging wall. Under the critical angle either head waves or diving waves are formed depending on whether the seismic velocity of the refracting layer is constant or gradually increasing, respectively (Fig. 2.7). For brevity the terms “refraction” or “refracted wave” will be used in the following for both types of critically refracted waves.

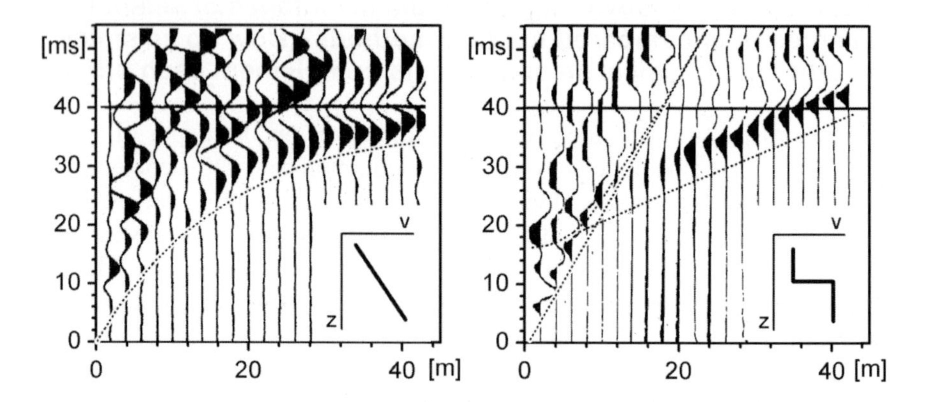

Fig. 2.7. Seismogram sections showing arrivals of refracted waves: diving waves (left) and head waves (right) appearing as first breaks. Corresponding wave fronts and ray diagrams are shown in Fig.2.2b and c, respectively. Travel time curves indicated by dotted curves. Inlays are symbolic velocity depth functions

Refractions form the first arrivals observed on P- or S-wave seismograms (so-called first breaks; cf. Fig. 2.7). They can be regarded as the most obvious realisation of Fermat's principle saying that a wave always chooses the fastest ray path to propagate from one point to the other. In near surface applications first breaks usually can be identified easily so they form a robust basis of interpretation.

In the following it is assumed that the investigated geological structure does not change significantly along the strike direction so the seismic investigation can be restricted to two dimensions. This means that the seismic measurements are thought to be performed along linear profiles perpendicular to the geological strike direction.

2.2.1 Targets for seismic refraction measurements

Seismic refraction surveys can be successfully applied to explore geological structures exhibiting a low number of interfaces where the seismic velocity increases with depth. In hydro-geological applications this situation is found, for P-waves, usually at the groundwater table, often at the aquifer-aquitard interface zone, and at the top of the basement or the bottom of the weathering zone in crystalline areas. Since refracted waves sample a layer interface horizontally over considerable distance their travel time-distance functions provide a more direct and reliable measure of seismic propagation velocity than those of reflections which show rather vertical or inclined ray paths. Therefore, seismic reflection imaging should always be combined with refraction measurements if there is special interest in precise velocity information. Lateral structural variation in the vadose zone as well as in the aquifer-aquitard sequence is often associated with gradual rather than abrupt changes of P- and S-wave velocity. This structural change can be imaged with seismic refraction tomography in terms of seismic velocity (cf. Fig. 2.6).

2.2.2 Body wave propagation in two-layer media with a plane interface

Basic aspects of the propagation of refracted and reflected body waves can be studied by considering a structure comprising two layers with constant seismic velocity and a plane dipping interface (Fig. 2.8). There are three primary types of body wave arrivals to be regarded: the direct wave showing a horizontal travel path through the uppermost layer, the reflection from the interface, and the critically refracted wave heading along the interface in the bottom medium and radiating back into the upper medium along its path. The basis of interpretation is the observation of the arrival times of these waves at increasing source-geophone distance along the seismic profile. These travel time-distance functions and their implications for seismic interpretation are described below. The following formulae can be applied to monotypic P- and S-waves by inserting the respective P- and S-wave velocities into the velocity variables. Note, however, that the equations cannot be applied to P-to-S or S-to-P converted waves without modification. To compute the respective travel time-distance functions the following variables are defined:

x	horizontal coordinate along a seismic profile,
x_A, x_B	coordinates of source points at A and B, respectively, with $x_A < x_B$,
v_1, v_2	constant seismic velocities of the upper and lower layers, respectively,
i_C	critical angle $i_C = \arcsin(v_1/v_2)$,
δ	dip angle of the interface, positive clockwise from horizontal to layer, so $\delta > 0$ represents a down oriented dip in positive x-direction,
z_A, z_B	depth of the interface beneath source points A and B,
z	depth of the interface defined by $z(x) = z_A + (x - x_A)/\cos(\delta)$
$\sigma(a)$	sign function =1 if $a \geq 0$, else –1

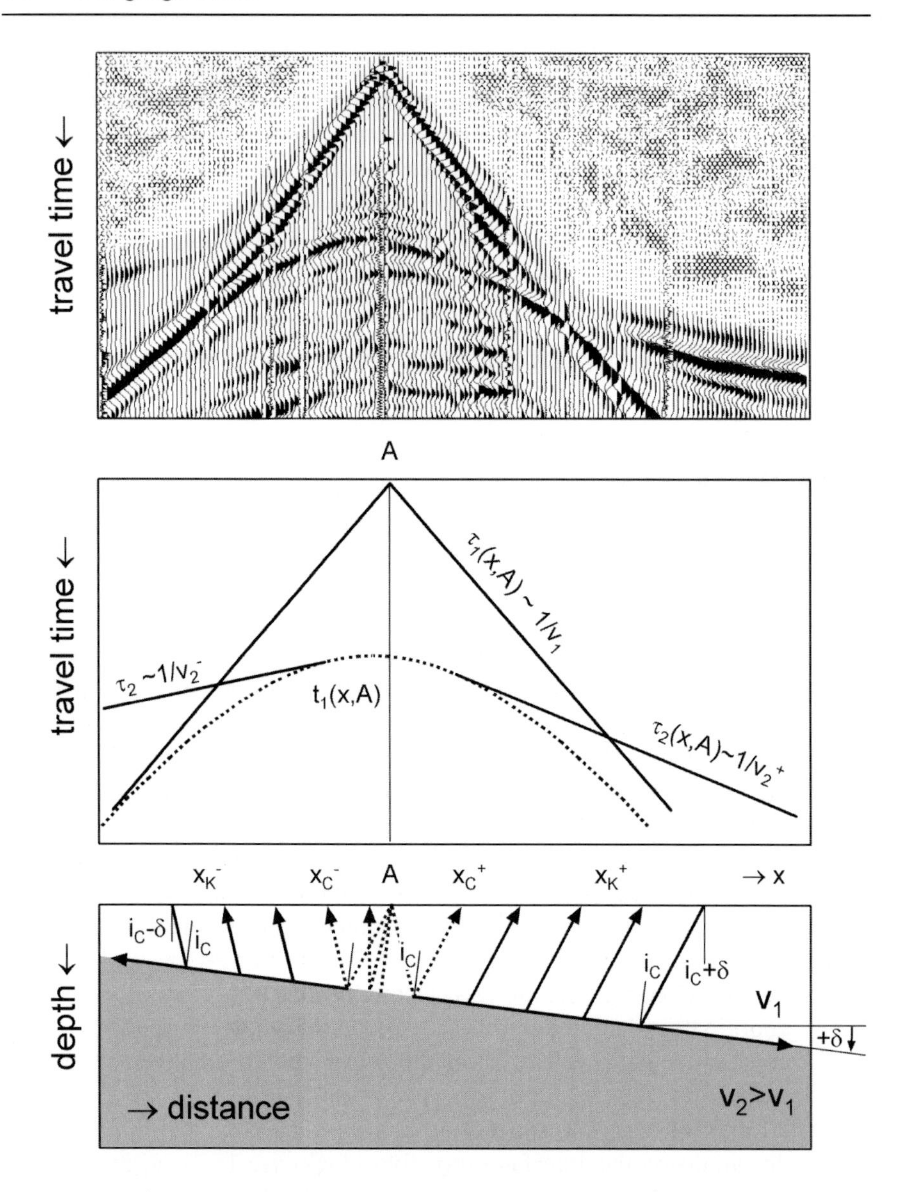

Fig. 2.8. Body waves in a two-layer medium with a dipping interface. Top: field seismogram. Middle: simplified travel time plan of direct, refracted and reflected waves (τ_1,τ_2 and t_1, respectively); not the asymmetry caused by layer dip. Bottom: ray paths of reflected and critically refracted waves (dashed and solid lines, respectively). $x_C^{\pm}$ and $x_K^{\pm}$ are critical and crossover distances, respectively; critical angle i_C and orientation of dip angle δ indicated

By applying Snell's law (Eqs. 2.2 and 2.3) and the travel time equation (Eq. 2.4b) the following expressions for the travel time-distance functions of the direct and refracted waves τ_1 and τ_2, respectively, can be derived for a receiver at the earth's surface (z = 0):

$$\tau_1(x, x_A) = (x - x_A)/v_1 \quad (2.5a)$$

$$\tau_2(x, x_A) = \tau_2(x_A, x_A) + (x - x_A) \cdot \sin(i_C + \sigma(x - x_A) \cdot \delta)/v_1 \quad (2.5b)$$

Equation (2.5b) is valid for $x \geq x_C^+$ or $x \leq x_C^-$. The term "$\tau(x,x_A)$" reads "traveltime τ between source point x_A and receiver point x". The so-called intercept time $\tau_2(x_A, x_A)$ and the critical points $x_C^{\pm}$ on either sides of a shot point at x_A are given by:

$$\tau_2(x_A, x_A) = 2 \cdot z_A \cdot \cos(\delta) \cdot \cos(i_C)/v_1 \quad (2.5c)$$

$$x_C^{\pm} = x_A \pm 2 \cdot z_A \cdot \cos(\delta) \cdot \sin(i_C)/\cos(i_C \pm \delta) \quad (2.5d)$$

There are some items to be noted at this point:

a) Both direct wave and refraction travel time curves are linear functions. The slope of the travel time curve depends on the incidence angle i_0 of the emerging ray at the earth's surface and on the velocity of the uppermost layer. The corresponding relation $d\tau_j/dx = \sin(i_0)/v_1$ is valid for all types of body wave arrivals. In case of the direct wave $i_0=90°$, in case of the refraction $i_0= i_C\pm\delta$. The expression can be applied locally also in heterogeneous media where i_0 and v_1 may vary with x.

b) The refracted arrival $\tau_2(x,x_A)$ can be observed only beyond the so-called critical points $x_C^{\pm}$ corresponding to the ray incidence at the critical angle. For further interpretation the intercept time $\tau_2(x_A,x_A)$ has to be determined by linear extrapolation of the observed travel time curve back to the source point coordinate.

c) The velocity v_1 of the top layer can be read directly from the slope of the arrival time curve because $d\tau_1/dx = 1/v_1$. However, the velocity v_2 of the bottom layer can be determined from the slope of function τ_2 only in case of horizontal layering where $\delta=0$. In this case the slope is $d\tau_2/dx = \sin(i_C)/v_1 = 1/v_2$.

d) In case of a dipping interface the velocity of the bottom layer can be determined if the refraction of a second source B placed at x_B is observed along the same segment of the seismic profile. The observations may be either "reverse" or "overlapping" (Fig. 2.9). In the following it is assumed that $x_B>x_A$. Then, in a "reverse shot" configuration the receivers are placed between source points A and B so $x_A<x<x_B$ and $\sigma(x-x_A)= -\sigma(x-x_B)$. For overlapping refraction observations, A and B are on the same side of the geophone spread, so $x<x_A<x_B$ or $x>x_B>x_A$, and $\sigma(x-x_A)=\sigma(x-x_B)$, respectively.

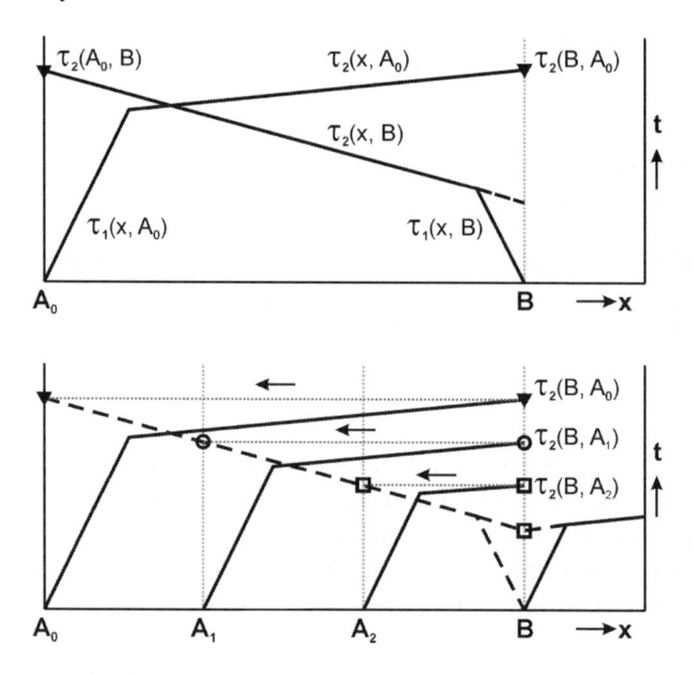

Fig. 2.9. Schematic travel time-distance curves of direct and refracted waves (τ_1 and τ_2, respectively) for a dipping layer and different source-receiver configurations. Top: Reverse shot configuration; signals from source points at A_0 and B observed along the same receiver spread. Bottom: Overlapping source-receiver configuration; equally spaced source points at A_0, A_1,... A_n ; signals observed along a moving receiver spread. If the sourcepoint spacing is dense enough, the reverse observation from a source at B can be reconstructed from the overlapping observations on the basis of travel time reciprocity (see dashed line)

In case of reversed observations the refraction travel time branches correspond to up-dip and down-dip travelling waves, respectively. The slopes of these travel time curves are:

$$d\tau_2(x, x_A)/dx = \sin(i_C + \delta)/v_1 = 1/v_2^+ \quad (2.6a)$$

$$d\tau_2(x, x_B)/dx = -\sin(i_C - \delta)/v_1 = -1/v_2^- \quad (2.6b)$$

for $x_A<x<x_B$. In contrast to the upper layer, where $d\tau_1/dx = 1/v_1$, these properties do not represent the seismic velocity of the refracting layer. Therefore, the inverse slopes v_2^+ or v_2^- are called *apparent* velocities. However, the slopes $1/v_2^+$ and $1/v_2^-$ can be averaged in order to eliminate the effect of the layer dip and to determine the *true* layer velocity v_2:

$$\begin{aligned} &1/v_2(x) = \sin(i_C)/v_1 \\ &\approx [1/v_2^+(x) + 1/v_2^-(x)]/2 = [\sin(i_C+\delta) + \sin(i_C-\delta)]/(2\cdot v_1) \end{aligned} \quad (2.6c)$$

With regard to field applications note that $1/v_2^+$ and $1/v_2^-$ have to be determined along corresponding segments $x_A<x<x_B$ of the refraction travel time branches in order to account for local variations of velocity and dip (more details see next section).

Regarding travel-time interpretation reverse and overlapping source-receiver configurations are equivalent to each other because a reverse travel time branch can be reconstructed from overlapping observations by applying the principle of travel time reciprocity (Fig. 2.9 bottom). It says that travel times do not change when source and receiver points are exchanged. Consider, for example, refraction travel times $\tau_2(x_R,x_A)$ and $\tau_2(x_R,x_B)$ from sources at A and B observed at the same receiver point $x=x_R$ ($x_R>x_B>x_A$). These time points can be used to construct the slope of a reverse refraction travel time branch attributed to a virtual source at x_R and to virtual receivers between A and B because $\tau_2(x_A,x_R)=\tau_2(x_R,x_A)$ and $\tau_2(x_B,x_R)=\tau_2(x_R,x_B)$ and, consequently:

$$d\tau_2(x,x_R)/dx \approx [\tau_2(x_R,x_A) - \tau_2(x_R,x_B)]/(x_A - x_B) \quad (2.7)$$

for $x_A<x<x_B$. The term of Eq. 2.7 may be substituted in Eq. 2.6b for $d\tau_2(x,x_B)/dx$.

The travel time-distance function of the reflected wave t_1 is given by:

$$t_1(x,x_A) = [(x-x_A)^2 + 4\cdot z_A^2\cdot\cos^2(\delta) + 4\cdot z_A\cdot(x-x_A)\cdot\sin(\delta)\cdot\cos(\delta)]^{1/2}/v_1 \quad (2.8)$$

It shows the shape of a hyperbola which is symmetrical with respect to a time axis placed at $x=x_A-2\cdot z_A\cdot\sin(\delta)\cdot\cos(\delta)$. Reflection and refraction time functions are tied to each other at the critical point x_C where they agree in both travel time $[t_1(x_C,x_A)=\tau_2(x_C,x_A)]$ and slope $[dt_1(x,x_A)/dx=d\tau_2(x,x_A)/dx$ for $x=x_C]$.

So, the critically refracted wave starts at x_C as a secondary arrival following the direct wave. Because of its faster advance it takes over as a first break at source-geophone distances larger than the "crossover distance" x_K leading to a sharp bend in the travel time curve of the first breaks. The crossover distances $x_K^{\pm}$ for the first refraction on either side of the a source point at coordinate x_A are given by:

$$x_K^{\pm} = x_A \pm 2 \cdot z_A \cdot \cos(\delta) \cdot \cos(i_C) / [1 - \sin(i_C \pm \delta)] \tag{2.9a}$$

Eq. 2.9a can be derived from Eq. 2.5a and b by applying the conditions that travel times $\tau_1(x_K,x_A)=\tau_2(x_K,x_A)$ at $x=x_K$. For horizontal layers, where $\delta=0$, Eq. 2.9b is the simplified equation:

$$x_K^{\pm} = x_K = x_A \pm 2 \cdot z_A \cdot [(v_2 + v_1)/(v_2 - v_1)]^{1/2} \tag{2.9b}$$

which can be applied for initial estimates of layer depth.

Regarding the interpretation of field data the first steps are to determine the slopes of the direct and reverse travel time branches from the observed seismograms and to compute the corresponding seismic layer velocities by applying Eqs. 2.5a, 2.6c, or 2.7, respectively. The next steps are to determine the critical angle $i_C=\arcsin(v_1/v_2)$, the layer dip δ via Eq. 2.6a,b and the interface depths below the source points from Eq. 2.5c or 2.9.

2.2.3 Seismic refraction in laterally heterogeneous two-layer media

In order to consider a more realistic subsurface situation the above treated two-layer case can be generalised by allowing lateral variation of both seismic velocity and interface depth. It is assumed that the seismic velocities of the upper and lower layers and the depth and dip of the interface are smoothly varying functions $v_1=v_1(x)$, $v_2=v_2(x)$, $z=z(x)$ and $\delta=\delta(x)$, respectively. As a consequence, both direct and refracted waves will show undulations in the slope of their travel time branches (Fig. 2.10).

In this situation $v_1(x)$ has to be measured by observing the direct waves of a number shot points placed along the seismic profile. In addition, reversed travel time branches $\tau_2(x,x_A)$ and $\tau_2(x,x_B)$ ($x_A<x<x_B$) of critically refracted waves have to be observed from which $v_2(x)$, $z(x)$ and $\delta(x)$ can be determined. For example, this can be achieved by applying the "Plus-minus method" of Hagedoorn (1959). In this approach it is assumed that $v_1=v_1(x)$, $v_2=v_2(x)$ and $\delta=\delta(x)$ can be approximated by average values within short segments of the geophone spread so the values are locally constant. The method is named after the functions:

$$t^{+}(x) = \tau_2(x,x_A) + \tau_2(x,x_B) - \tau_2(x_A,x_B) \tag{2.10a}$$

$$t^{-}(x) = \tau_2(x,x_A) - \tau_2(x,x_B) - \tau_2(x_A,x_B) \tag{2.10b}$$

where $\tau_2(x_A,x_B) = \tau_2(x_B,x_A)$ denotes the so-called reciprocal travel times between the source points A and B and reverse, respectively. A consideration of ray geometry, wave fronts and travel times (Fig. 2.11) shows that the depth, velocity and dip of the refracting layer beneath the geophone point x can be expressed by:

$$z(x) = v_1 \cdot t^+ /[2 \cdot \cos(\delta) \cdot \cos(i_C)] \tag{2.11a}$$

$$1/v_2(x) = [dt^- /dx]/2 \tag{2.11b}$$

$$\delta(x) = \arcsin[v_1 \cdot (dt^+ / dx)/(2 \cdot \cos(i_C)] \tag{2.11c}$$

where $\cos(i_C)=[1-(v_1/v_2)^2]^{1/2}$. Note that all quantities on the right hand side of Eqs. 2.11a-c are functions of x.

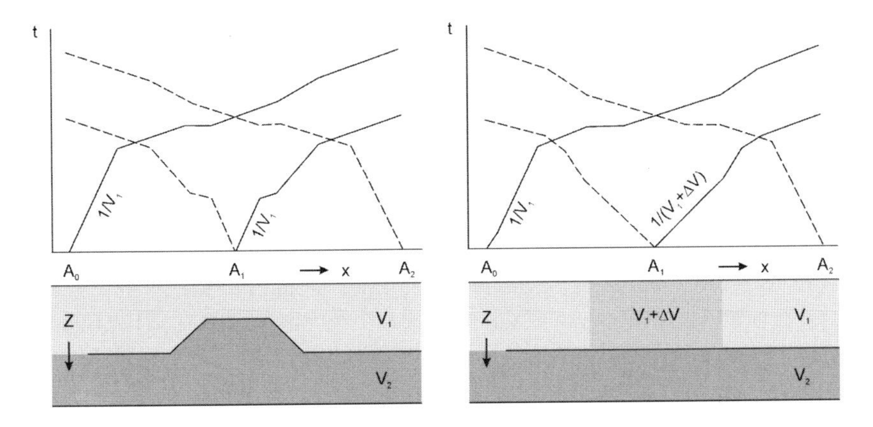

Fig. 2.10. Effect of lateral variation of interface depth (left) and seismic velocities (right) on reversed travel time branches. Note that in case of interface variation only the critically refracted waves are distorted, whereas in case of velocity variation both direct and refracted travel time branches are affected

In particular, Eq. 2.11b is equivalent to Eq. 2.6 where the velocity of the bottom layer is determined by averaging the slopes of up-dip and down-dip refractions. However, because of the different incidence directions, up-dip and down-dip slopes will belong to different sections of the interface if they are determined at the geophone point x (Fig. 2.12). The distance between these sections is:

$$2 \cdot r = 2 \cdot z \cdot \cos(\delta) \cdot \tan(i_C) = v_1 \cdot t^+ \cdot \sin(i_C)/\cos^2(i_C) \tag{2.12}$$

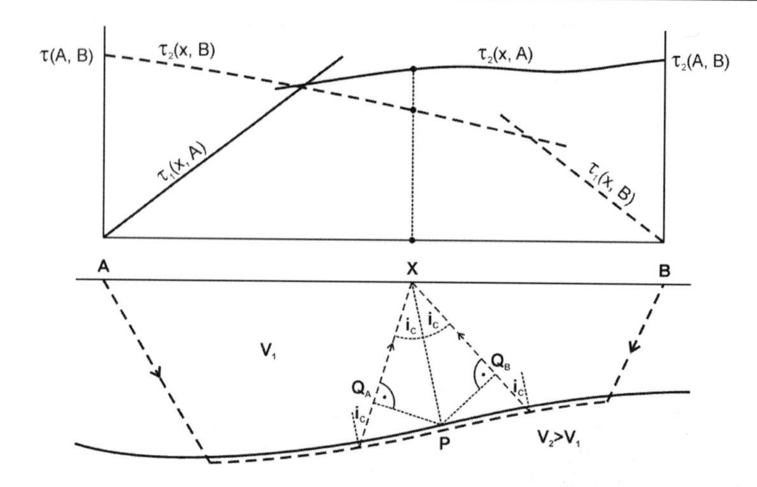

Fig. 2.11. Ray geometry to be considered to derive the Plus-minus method of Hagedoorn (1959). The meaning of the "plus-time" $t^+(x)$ can be understood by regarding (1) the refracted rays connecting the geophone at x with shot points A and B and (2) the positions of the refracted wave fronts when they reach the same point P beneath the geophone. P is the foot point of the interface normal pointing towards geophone position x. The refracted wave fronts emerging from shot points A and B arrive at P at the same travel time as they arrive at points Q_A and Q_B on the up-going parts of the respective refracted rays. Since, in addition, P-x-Q_A and P-x-Q_B form congruent rectangular triangles, the following travel time equations apply to respective segments of the refracted rays: $\tau_2(x,x_A)= \tau_2(x,Q_A)+\tau_2(Q_A,x_A)$, $\tau_2(x,x_B)= \tau_2(x,Q_B)+\tau_2(Q_B,x_B)$, $\tau_2(x_A,x_B)=\tau_2(P,x_A)+\tau_2(P,x_B)= \tau_2(Q_A,x_A)+\tau_2(Q_B,x_B)$, $\tau_2(x,Q_A)=\tau_2(x,Q_B)$. Therefore, $t^+(x)$ can be identified as $t^+(x) = 2\cdot\tau_2(x,Q_A) =2\cdot\tau_2(x,Q_B)$

This deviation will cause some inaccuracy in cases were the refracting interface is not locally plane. To image the interface segment vertically below x, the slopes of the refraction time derivatives would have to be computed at points $x^{\pm}=x\pm z\cdot\tan(i_C\pm\delta)$ for shot points A and B, respectively. Alternatively, the ray diagram (Fig. 2.12) shows that the same range of interface points will be covered by both shot observations if right and left hand segments of the respective travel time curves are considered, namely $\tau_2(x^+,x_A)$ for $x \le x^+ \le (x+s^+)$ and $\tau_2(x^-,x_B)$ for $(x-s^-) \le x^- \le x$ where:

$$s^{\pm} = 2\cdot r\cdot\cos(i_C \pm 2\cdot\delta)/\cos(i_C \pm \delta) \tag{2.13}$$

The velocity $v_2(x)$ of the bottom layer can be computed from averaging the slopes of these curve segments.

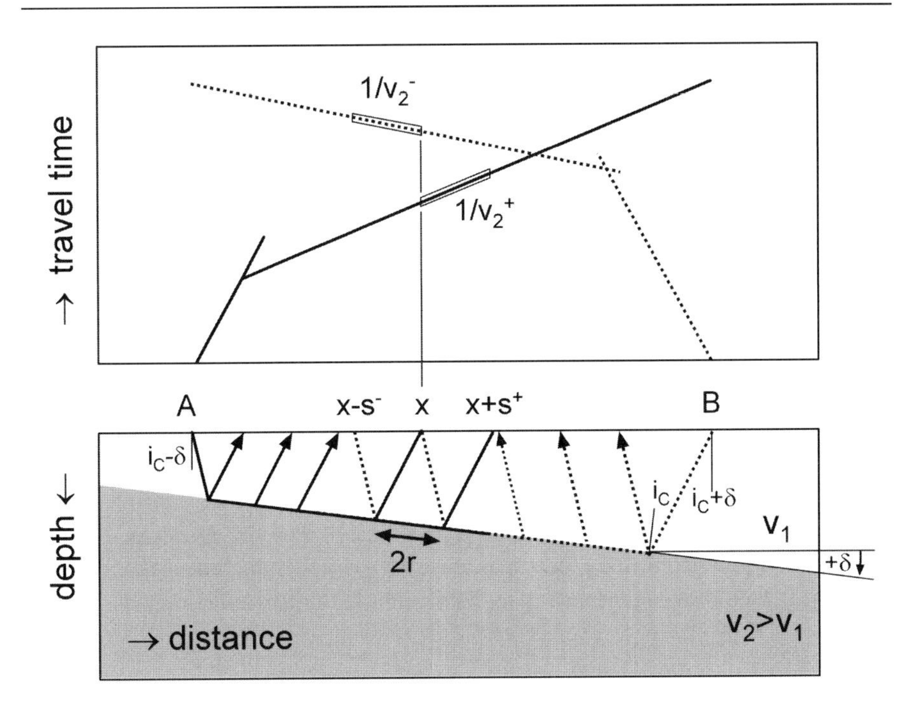

Fig. 2.12. Horizontal shift of segments of reversed refraction travel time curves belonging to the same interface segment (cf. Eqs. 2.12 and 2.13)

2.2.4 Consistency criteria of seismic refraction measurements

The above consideration of the two-layer case shows that there are a number of criteria of field layout and travel time analysis needed for interpretation of seismic refraction measurements. These requirements listed below are valid and equally important also for multi-layer situations with lateral velocity heterogeneity and more complicated interfaces:

1) The seismic profile has to be covered with multiple source points and continuous receiver lines so it is possible to observe or construct reversed branches of refracted arrivals (Fig. 2.13). Optimum is a combination of overlapping and reverse shots.
2) The seismic velocity of refracting layers can be determined by averaging the slopes of reversed corresponding travel time branches. This procedure can be applied not only to plane interfaces but also to smoothly curved interfaces if the reversed branches apply to the same underground segment.
3) The consistency of travel time observations from different shot points has to be checked by application of the principle of reciprocity

to each type of arrival [$\tau_j(x_P,x_0)=\tau_j(x_0,x_P)$]. Since the principle of reciprocity applies in a strict sense, field layout should provide adequate shot-geophone arrangements so the interpreter can take advantage from it.

4) The intercept times of refracted arrivals observed for left and right hand spread from the same shot point agree if the refracting interface is plane. This criterion of travel time consistency may be violated in case of curved or disrupted interfaces (Fig. 2.14).

5) Overlapping refracted travel time branches of the same interface should appear parallel. This criterion can be applied for consistency checks and for combining observations from different source points into one long travel time branch (Fig. 2.15). This latter procedure is required for some interpretation algorithms such the wave front method (see below). Note that this "parallelism of refraction travel time branches" applies only to plane homogeneous layers in a strict sense. Deviations may occur in case of curved interfaces and in case that the velocity increases with depth inside the refracting layer.

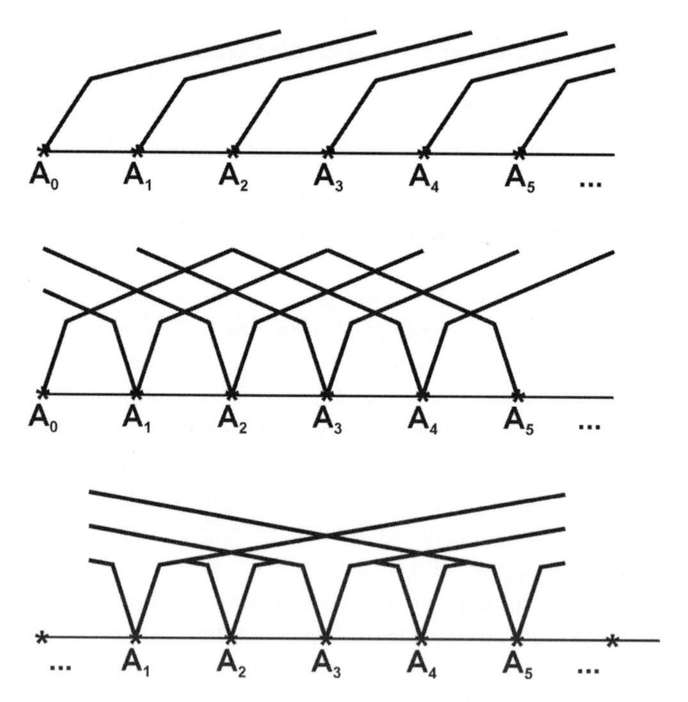

Fig. 2.13. Examples of typical overlapping acquisition schemes for seismic refraction measurements. Top: one-sided spread (end-on spread); middle and bottom: split spread

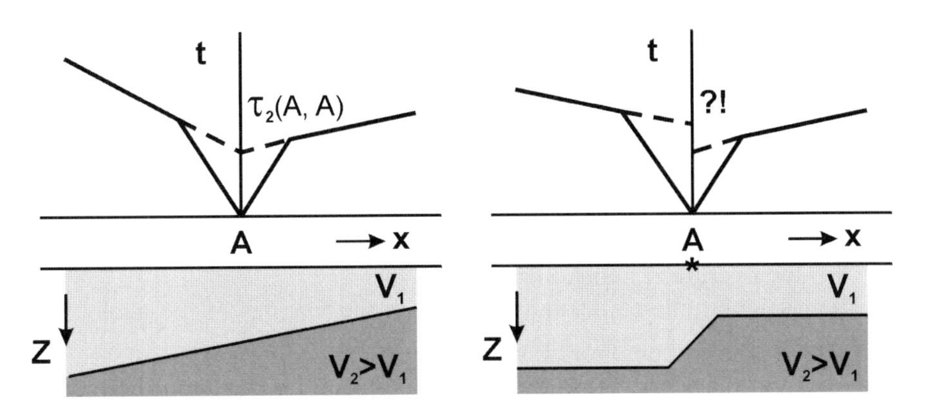

Fig. 2.14. Intercept times for plane and disrupted refracting horizons

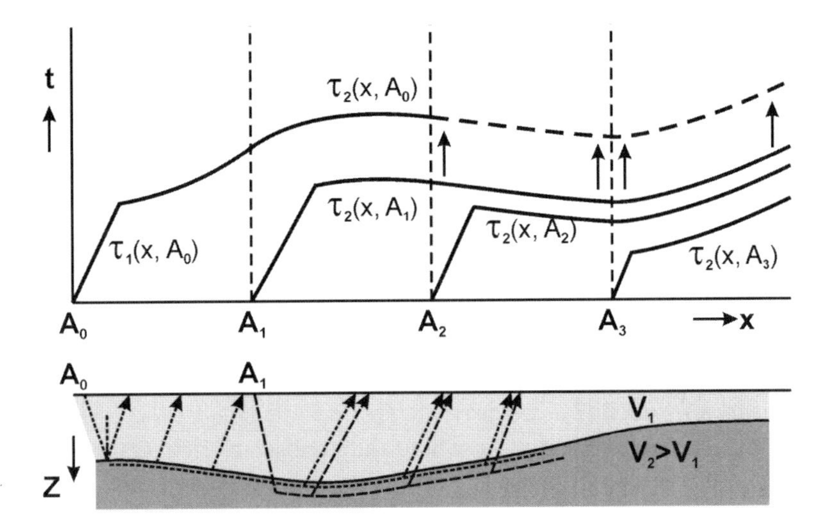

Fig. 2.15. Construction of a long refracted travel time branch (dashed line) from overlapping observations (solid lines). The underlying assumption of the "parallelism of travel time branches" (top figure) applies to smooth interfaces and negligible velocity increase inside the refracting layer. In this case refracted waves of adjacent source points travel along the same path (bottom figure)

2.2.5 Field layout of seismic refraction measurements

In order to perform seismic field measurements the geometrical arrangement of source and geophone points has to be defined. This involves the specification of

- the length of the geophone spread for each source point,
- the geophone spacing along the spread,
- the spacing of shot points, and
- the overlap of geophone spreads.

The particular choice depends mainly on the features of the geological structure to be investigated, namely:

- the seismic velocity of the overburden,
- the depth and seismic velocity of the refracting target layer, and
- the heterogeneity of the these properties.

The above considerations and formulae of the two-layer case provide the means to estimate the requested properties. They can be applied for multi-layer cases, too, if the upper layer is regarded as sort of an average of the hanging wall of the target layer.

The remarks below apply to the common situation that near surface seismic refraction measurements are performed by deploying a large number of equally spaced geophones along spreads compared to which the number and spacing of shot points are more sparse or wide, respectively. This assumption is clearly somewhat arbitrary because the role of shot and receiver points may be exchanged, which is a common situation for seismic measurements at the sea bottom, for example. For sea bottom measurements a number of seismometers is deployed and shots are fired at high rates from a moving boat.

Spread length. Refracted arrivals can be identified most securely beyond the crossover distance x_K where they appear as first breaks. In practice, as a rule of thumb, the slope or the corresponding apparent velocity of a refracted arrival can be determined if it is observed over a distance of one wave length or longer. For impulse type signals a wave length λ can be defined via:

$$\lambda = v \cdot T \tag{2.14}$$

where v is the propagation velocity and T is the apparent period or time duration of one oscillation cycle (extending over one positive plus one negative deflection). From a ray geometrical point of view the spread has to be long enough so the same underground segments can be imaged by shot and reverse shot observations. Applying Eqs. 2.12 or 2.13 leads to the conclusion that the spread S should extend longer than:

$$S > x_K + \max\{s^{\pm}, \lambda\} \tag{2.15a}$$

For planning the investigation of a multi-layer structure the following equation can be used to estimate the crossover distances x_{Kmn} of refracted arrivals from the m-th and n-th layer:

$$x_{Kmn} = x_A \pm v_m \cdot v_n \cdot [\tau_m(x_A, x_A) - \tau_n(x_A, x_A)]/(v_m - v_n) \tag{2.15b}$$

where v_j and $\tau_j(x_A,x_A)$ (j=m,n) are the seismic velocity and intercept time of the layers, respectively, and $x=x_A$ is the coordinate of the source point. For a given horizontal layer sequence the intercept time can be determined via Eq 2.18 (for dip angle δ_j=0). Usually, the objective of a survey is to determine the layer structure along a transect rather than to determine a depth structure at a certain point. Therefore, one will always try to extend the observational distance as far as the costs remain acceptable.

Geophone spacing. There are two criteria on the basis of which the geophone spacing Δx_G along a refraction spread can be determined. A strict criterion from filter theory is that spatial aliasing has to be avoided. Spatial aliasing occurs when short wave length signals are sampled at only sparse grid points so short wavelengths are imaged to apparently longer wavelengths in the registration. This criterion corresponds to the Nyquist criterion known from time sequence analysis saying that:

$$\Delta x_G < \lambda_{min}/2 = v_{min} \cdot T_{min} \tag{2.16}$$

Here, λ_{min}, v_{min} and T_{min} are the smallest wavelength, propagation velocity and signal period to be expected during the survey. A second criterion can be formulated with regard to the geometry of the travel time curve of the first breaks because one has to assure that its slope and crossover points can be identified clearly. Theoretically, the slope of a straight line can be defined by two points. In practice, however, a certain redundancy is required to reduce random errors. Therefore, under low noise conditions, each linear segment of the travel time curve between successive flexure points should be sampled at a minimum of five points. For a two-layer situation, for example, both joint criteria yield a maximum geophone spacing of $\Delta x_G < \min\{\lambda_{min}/2, x_K/4, (S-x_K)/4\}$.

Source point spacing. Source point spacing Δx_S to be chosen depends on the lateral heterogeneity of the seismic velocity of the overburden. If the uppermost layer is known to be laterally strongly variable, Δx_S should not exceed the distance along which the direct wave can be observed, so in many of these cases $\Delta x_S \approx x_K$ will be adequate. Clearly, if $v_1(x)$ varies only smoothly the shot density can be decreased and $v_1(x)$ can be approximated by interpolation based on more sparse sampling. Corresponding arguments apply to multi-layer cases where the lateral structure of intermediate layers placed between top and target horizons has to be determined continuously. The corresponding critical and crossover distances can be determined from the multi-layer formulae presented below.

Consideration of lateral velocity heterogeneity. From a more general point of view, lateral velocity heterogeneity can be thought of in terms of velocity perturbations revealing a certain spectrum of wavelengths. In this perception of heterogeneity, the sampling theorem (Eq. 2.16) can be adopted to formulate a criterion which can be applied to both shot and geophone spacing. High resolution of lateral velocity variation will be obtained if Δx_S and Δx_G are arranged in such a way that the travel time curves are sampled with a spacing of less than $\Lambda_{min}/2$. Λ_{min} represents the smallest significant wavelength component of the spectrum of the velocity fluctuation function $\delta v_1(x) = v_1(x) - v_1^{mean}$ (v_1^{mean} is the average of $v_1(x)$ along the seismic profile).

Overlapping geophone spreads. There are three aspects leading to apply overlapping geophone spreads in field surveys.

(1) Refracted arrivals exist only beyond the critical distance x_C and are often observed only beyond the crossover distance x_K. Therefore, a blind zone exists below each shot point in terms of refraction information. Therefore, in order to continuously cover a target horizon along a seismic profile, the spreads of adjacent shot points should overlap at least by respective crossover distances. Regarding the above two-layer case with shot points at x_A and x_B, for example, geophones should be placed at positions x within the interval $(x_A - s^-) < x < (x_B + s^+)$.

(2) Overlapping geophone spreads are needed for a consistency check of observations from different source points using the criterion of the parallelism of travel time branches (see above), and

(3) they are needed to reconstruct reverse travel-time branches in case of one-sided spreads. In this latter case, the multiplicity of overlapping determines the number of sample points available for the reverse shot reconstruction.

2.2.6 Near surface layering conditions and seismic implications

Because of the ongoing sediment transport and weathering at the earth's surface, both geological and seismic heterogeneity are usually strong in the uppermost some 10 meters compared to deeper, more consolidated levels. P- and S-wave velocities are usually some 100 m/s in the vadose zone where sediments are dry and mostly low cohesive. At the earth's surface the vertical as well as the horizontal variation of P- and S-wave velocities can easily exceed 50-100% within meter or ten-meter distances. This variability is caused by an exponential increase of compaction under the onset of sedimentary loading and by structural and compositional changes associated with variation in porosity and grain cementation.

A non-linear increase of seismic velocities with depth is caused by lithostatic pressure. This velocity increase is mainly a consequence of porosity reduction and improved grain contacts. The continuous increase of lithostatic pressure with depth leads to a pronounced vertical gradient of seismic velocities. Therefore, interpreters of near-surface seismic sections are often confronted with convexly curved first breaks instead of straight lines (Fig. 2.7 and 2.16). They are caused by "diving waves" the geometry and travel times of which are described in the formulae shown in Fig. 2.16. The ray bending associated with a continuous velocity increase can be explained with a successive application of Snell's law at infinitesimal depth increments. The diving leads to larger penetration depths of rays with increasing source-receiver offset. Therefore, the above noted parallelism of refracted travel time branches is not strictly valid in gradient type media. This criterion can be applied only to overlapping travel time branches of bended rays if they cover similar offset ranges.

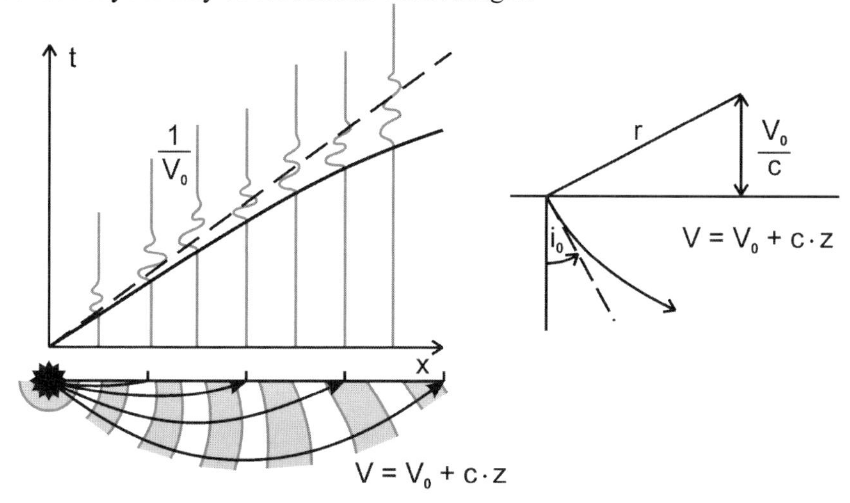

Fig. 2.16. Influence of a gradual increase of near-surface seismic velocity on wave propagation. Top left: schematic seismogram section and travel time curve (cf. field seismograms in Fig. 2.7). Bottom left: corresponding schematic wave fronts and rays. Right: Circular ray geometry applying to media with a constant velocity gradient c. The radius r of the ray depends on gradient c and velocity v_0 and take-off angle i_0 at the earth's surface according to $r = v_0/(c \cdot \sin(i_0))$ (e.g. Sheriff and Geldart 1995)

Of special importance is the ground water table where P-wave velocities step to more than 1500 m/s when water saturation is reached. The placement of water in the pore space increases the whole rock compressional modulus and diminishes the reduction of pore space with depth by increasing pressure. Therefore, water saturated sediments have higher P-wave velocities than partially saturated ones. In contrast, the vertical velocity gradient is comparably smaller leading to straight travel time lines of refracted P-wave arrivals. The presence of water in the still highly porous media has a tendency to decrease also P-wave velocity contrasts between lithologically different rock units. A consequence of decreasing velocity contrasts for P-wave refraction measurements is that the slopes of the travel time branches of successive refracting layers become more similar and that corresponding crossover points move to larger offset.

The increase of P-wave velocity at the groundwater table is also associated with an increase of the seismic wavelength (see Eq. 2.14) leading to a reduction of structural resolution (see discussion at the end of this paragraph). Usually, shear wave velocities are changed only little by water saturation so they are expected to represent lithological contrasts below the water table more clearly than P-waves. In ground water layers often a v_P/v_S ratio of 5-10 is found implying that S-wave lengths are shorter by a factor of 5-10 than P-wave lengths if signal periods are similar. Therefore, the complementary application of shear wave seismics to explore the lithological situation of ground water layers can be advantageous in comparison to pure P-wave surveys.

Typically, seismic sources for near surface surveys on land provide signals of 20-200 Hz main frequency, corresponding to signals length of 5-50 ms in one cycle. Often, seismic signals consist of more than one cycle increasing the signal duration accordingly. Depending on the seismic velocity, the corresponding wave lengths in near surface sediments are in the range of some meters to some tens of meters (Eq. 2.14) which is the same order of magnitude as the dimensions of typical shallow investigation targets. A consequence of these conditions for practical work is the interference of seismic arrivals which occurs where travel time differences are much shorter than the signal length. Not rarely, near surface reflections and the onset of refractions at the critical point are covered by direct and surface wave signals so effectively that they cannot be visualized even by digital filtering. Therefore, in contrast to deep seismic sounding, near surface seismic interpreters often have to rely on first break analysis only.

2.2.7 Seismic interpretation approaches for heterogeneous subsurface structures

Since the first commercial applications of refraction seismology in the 1920s a large number of seismic interpretation algorithms has been developed applying to geological situations of different complexity. Each algorithm has certain requirements regarding the shot-geophone configuration to be realised during the field measurements. Common to all methods is (1) that reversed-shot observation or equivalent overlapping spreads are required except in cases where geological layers are known to be horizontal, and (2) that, in the average trend, seismic velocities are assumed to increase with depth. As discussed above, this later condition is usually fulfilled in near surface sediments because of the strong influence of lithostatic loading. In the following, three examples selected from the wealth of interpretation methods will be discussed applying to increasingly complicated geological situations.

Intercept-time method

The intercept-time method (see e.g. Sheriff and Geldard, 1995, based on Adachi, 1954, and Johnson, 1976) is applicable to geological layer sequences showing smoothly varying interface depths and more or less constant velocity within each layer. The seismic profile is thought to be divided into segments along which the geological structure can be approximated by layers with plane dipping interfaces (Fig. 2.17). The travel time curves of refracted arrivals of all these layers have to be observed in reverse directions along each segment. They are approximated by straight lines corresponding to the assumption of constant velocity within each layer. Under these conditions the equations of the two-layer case discussed above (Eqs. 2.5, 2.6) can be extended to describe a multi-layer case. Below, the notation of the two-layer case is used extended by an additional index to label the refracting interface or the layers of the hanging wall, respectively (Fig. 2.17). Then, for two shot points A and B located at the respective endpoints $x=x_A$ and x_B of a profile segment, the reversed refraction travel time functions read:

$$\text{for } x \geq x_{Cj}^{+}: \quad \tau_j(x,x_A) = \tau_j(x_A,x_A) + (x - x_A)/v_j^{+} \tag{2.17a}$$

$$\text{for } x \leq x_{Cj}^{-}: \quad \tau_j(x,x_B) = \tau_j(x_B,x_B) + (x - x_B)/v_j^{-} \tag{2.17b}$$

respectively, where j=1,…, n is the index of the refracting interface (j=1 at the earth's surface), $x_{Cj}^{\pm}$ are the corresponding critical distances and $1/v_j^{\pm}$

are the slopes of the refraction travel time curves in up- and down-dip directions, respectively. For a layer sequence with interface dip angels δ_k<10° the intercept times at the shot points can be approximated by:

$$\tau_j(x_A,x_A) \approx \sum_{k=1}^{j-1}\{2\cdot\Delta z_{Ak}\cdot[1-v_k^2/v_j^2]^{1/2}\cdot\cos(\delta_k)/v_j\} \tag{2.18a}$$

$$\tau_j(x_B,x_B) \approx \sum_{k=1}^{j-1}\{2\cdot\Delta z_{Bk}\cdot[1-v_k^2/v_j^2]^{1/2}\cdot\cos(\delta_k)/v_j\} \tag{2.18b}$$

where Δz_{Ak} and Δz_{Bk} are the thickness of the k-th interface beneath shot points A and B, respectively (Sandmeier, 1997). As for the two-layer case (Eq. 2.6), the seismic velocity of each layer can be determined by averaging the slopes of the respective reverse shots observations:

$$1/v_j(x) \approx [1/v_j^+(x)+1/v_j^-(x)]/2 \tag{2.19}$$

In determining $1/v_j^{\pm}$ and the intercept times from field data attention has to be paid to comply with the criterion of reciprocal travel times $\tau_j(x_A,x_B)=\tau_j(x_B,x_A)$ for all pairs of refracted arrivals. Δz_{Ak}, Δz_{Bk} and δ_k are determined recursively by solving Eqs. 2.18a and b successively for the first until the n-th layer each time inserting the results of the previous steps. By rearranging of Eqs. 2.18 the recursion formula for layer thicknesses beneath shot points A and B are obtained:

$$\Delta z_{A(j-1)} \approx \tau_j(x_A,x_A)\cdot v_j/\{2\cdot\cos(\delta_j)\cdot[1-v_{j-1}^2/v_j^2]^{1/2}-\Psi_{Aj}\} \tag{2.20a}$$

$$\Delta z_{B(j-1)} \approx \tau_j(x_B,x_B)\cdot v_j/\{2\cdot\cos(\delta_j)\cdot[1-v_{j-1}^2/v_j^2]^{1/2}-\Psi_{Bj}\} \tag{2.20b}$$

where the terms:

$$\Psi_{Aj} = \sum_{k=1}^{j-2} 2\cdot\Delta z_{Ak}\cdot[1-v_k^2/v_j^2]^{1/2}\cdot\cos(\delta_k)/v_j\cdot \tag{2.21a}$$

$$\Psi_{Bj} = \sum_{k=1}^{j-2} 2\cdot\Delta z_{Bk}\cdot[1-v_k^2/v_j^2]^{1/2}\cdot\cos(\delta_k)/v_j\cdot \tag{2.21b}$$

depend only on the properties of the layers in the hanging wall of the j-th interface. Finally the dip and the depths of the j-th interface beneath shot points A and B are given by:

$$\delta_j = \arctan[(z_{Bj}-z_{Aj})/(x_B-x_A)] \tag{2.22a}$$

$$z_{Aj} = \sum_{k=1}^{j-1} 2 \cdot \Delta z_{Ak} \text{ and } z_{Bj} = \sum_{k=1}^{j-1} 2 \cdot \Delta z_{Bk} \tag{2.22b}$$

Along a seismic profile the above procedure can be applied to adjacent or overlapping segments in order to obtain a continuous coverage.

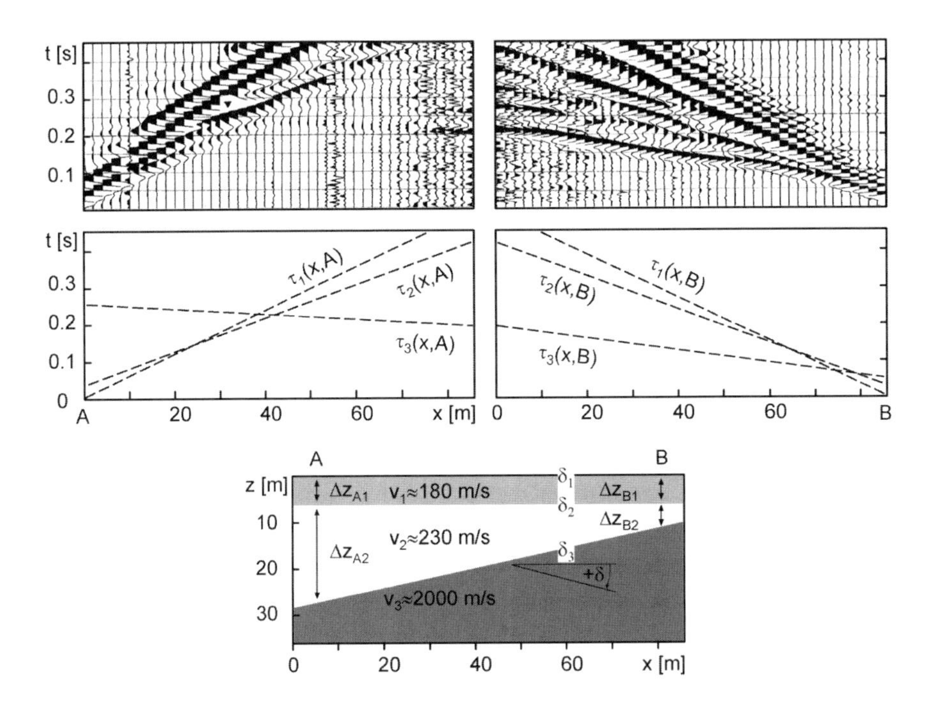

Fig. 2.17. Refraction investigation of the shore of a silted river bed comprising three-layers. Top: reversed shear wave seismogram sections. Middle: simplified travel time curves of the direct and two refracted waves extrapolated to zero-offset in order to gain corresponding intercept times. Bottom: subsurface model in terms of shear wave velocity. The upper interface is horizontal but the bed rock shows about 12° dip. Note that the up-dip travelling bed rock refraction of the seismogram section on the left shows a negative slope ("apparent velocity") because the strong velocity increase compared to the sediments

Wave-front method

The wave-front method (Fig. 2.18) is a most elegant approach which can be applied to layered geological structures with widely irregular interfaces. The velocity variation within the hanging wall of the refracting layer under consideration is allowed to be almost arbitrary (but has to be known, of course). The seismic velocity of the refracting layer itself may vary laterally as well. To investigate multi-layer situations, the wave-front method has to be applied successively to each interface from top to bottom, just like the intercept-time method. The observational basis for determining the depth contour and the seismic velocity of a particular refracting layer is a continuous mapping of the respective reversed travel time branches between two shot locations (Fig. 2.18 top). To construct these travel time branches for source-geophone distances less than the crossover distance, adjacent shot observations have to be combined. In chaining the refracted travel time branches the above mentioned consistency criteria, especially the parallelism of overlapping branches, have to be applied (Fig. 2.15). The wave-front method works best if there is a pronounced contrast in the seismic velocities at the interface, and if the depth gradient of velocity is rather low in the refracting layer.

In order to find the depth contour of the j-th layer the refracted travel time fields $\tau_j(x,z;x_A)$ and $\tau_j(x,z;x_B)$ of two respective source points A and B are considered not only at the earths surface but also at depth $z \geq 0$. Points A and B form the end points of the mutually reverse travel paths along the refracting interface (Fig. 2.18 middle). Therefore, the following travel time condition applies to a point $P=(x_P,z_P)$ of this interface (Thornburgh 1930):

$$\tau_i(x_P, z_P; x_A) + \tau_i(x_P, z_P; x_B) = \tau_i(x_A, x_B) \tag{2.23}$$

This relation indicates that the reciprocal travel time $\tau_j(x_A,x_B)$ observed at the earth's surface can be used to identify the interface points if the travel time fields $\tau_j(x,z;x_A)$ and $\tau_j(x,z;x_B)$ of shot points A and B, respectively, can be reconstructed below the earths surface (Fig. 2.18 bottom). If the velocity function of the overburden is known the wave field continuation from the surface back to depth is straightforward. A simple approach, which can be applied in case of smooth velocity variation in the overburden, is based on the so-called eikonal equation for wave propagation in two-dimensional structures. The eikonal equation:

$$(\partial\tau_i / \partial x)^2 + (\partial\tau_i / \partial z)^2 = 1 / v^2 \tag{2.24}$$

connects ray geometry with local velocity. At the earth's surface $\partial\tau_j/\partial x$ is represented by the slope of the refracted travel time branch, namely, $\partial\tau_j/\partial x=\sin[i(x)]/v(x)$ for $z=0$ where i is the incidence angle of the ray with respect to the vertical. The vertical derivative $\partial\tau_j/\partial z$ of the travel time field is given by $\partial\tau_j/\partial z = \cos[i(x)]/v(x) = [1/v^2-(\partial\tau_j/\partial x)^2]^{1/2}$ because v(x,z) is (assumed to be) known. Based on these properties, the travel time field observed at z=0 can be traced back to the slightly deeper level z=Δz by applying the linear approximation:

$$\tau_i(x-\Delta x,\Delta z;x_A)=\tau_i(x,0;x_A)-(\partial\tau_i/\partial x)\cdot\Delta x-(\partial\tau_i/\partial z)\cdot\Delta z \tag{2.25}$$

with $\Delta x=\Delta z\cdot\sin(i)/\cos(i)=\Delta z\cdot(\partial\tau_j/\partial x)/[1/v^2-(\partial\tau_j/\partial x)^2]^{1/2}$. In repeated application of this procedure the travel time field $\tau_j(x,\ q\cdot\Delta z;\ x_A)$ is reconstructed for successive depth levels $z=q\cdot\Delta z$ (q=1,2,…). Note that $\partial\tau_j/\partial x$ has to be newly calculated at each depth level. For the determination of the interface depth, also the travel time field $\tau_j(x,\ q\cdot\Delta z;\ x_B)$ of the reverse shot has to be continued to depth. At each depth level $q\cdot\Delta z$, points $(x_P,\ q\cdot\Delta z)$ of the refracting interface are identified by checking whether the imaging condition is satisfied according to:

$$|\,\tau_i(x_P,q\cdot\Delta z;x_A)+\tau_i(x_P,q\cdot\Delta z;x_B)-\tau_i(x_A,x_B)\,|<\varepsilon \tag{2.26}$$

where ε>0 denotes a numerical threshold. The local seismic velocity of the refracting layer can be determined from the travel time values of either shot A or B or both along the interface. At a point $P=(x_P,z_P)$ located between two neighbouring points $P1=(x_{P1},z_{P1})$ and $P2=(x_{P2},z_{P2})$ of the interface the velocity of the refracting layer can be determined approximately by the following formulae:

$$1/v_i(x_P,z_P;x_A)\approx\frac{\tau_i(x_{P2},z_{P2};x_A)-\tau_i(x_{P1},z_{P1};x_A)}{[(x_{P2}-x_{P1})^2+(z_{P2}-z_{P1})^2]^{1/2}} \tag{2.27a}$$

$$1/v_i(x_P,z_P;x_B)\approx\frac{\tau_i(x_{P2},z_{P2};x_B)-\tau_i(x_{P1},z_{P1};x_B)}{[(x_{P2}-x_{P1})^2+(z_{P2}-z_{P1})^2]^{1/2}} \tag{2.27b}$$

$$1/v_i(x_P,z_P)\approx[1/v_i(x_P,z_P;x_B)+1/v_i(x_P,z_P;x_A)]/2 \tag{2.27c}$$

Note that the above outlined schedule of wave field continuation can be replaced by graphical solutions, ray-tracing or by more elaborate (and accurate) finite-difference implementations of the wave- and eikonal equations (e.g. Meissner and Stegena 1977, Sandmeier 1997, and the references therein).

The so-called generalised reciprocal method (GRM-method) (Palmer 1980) is another widely-used interpretation approach based on similar arguments as the wave-front method. However, it requires the assumption of some processing parameters making it less strict than the wave-front method.

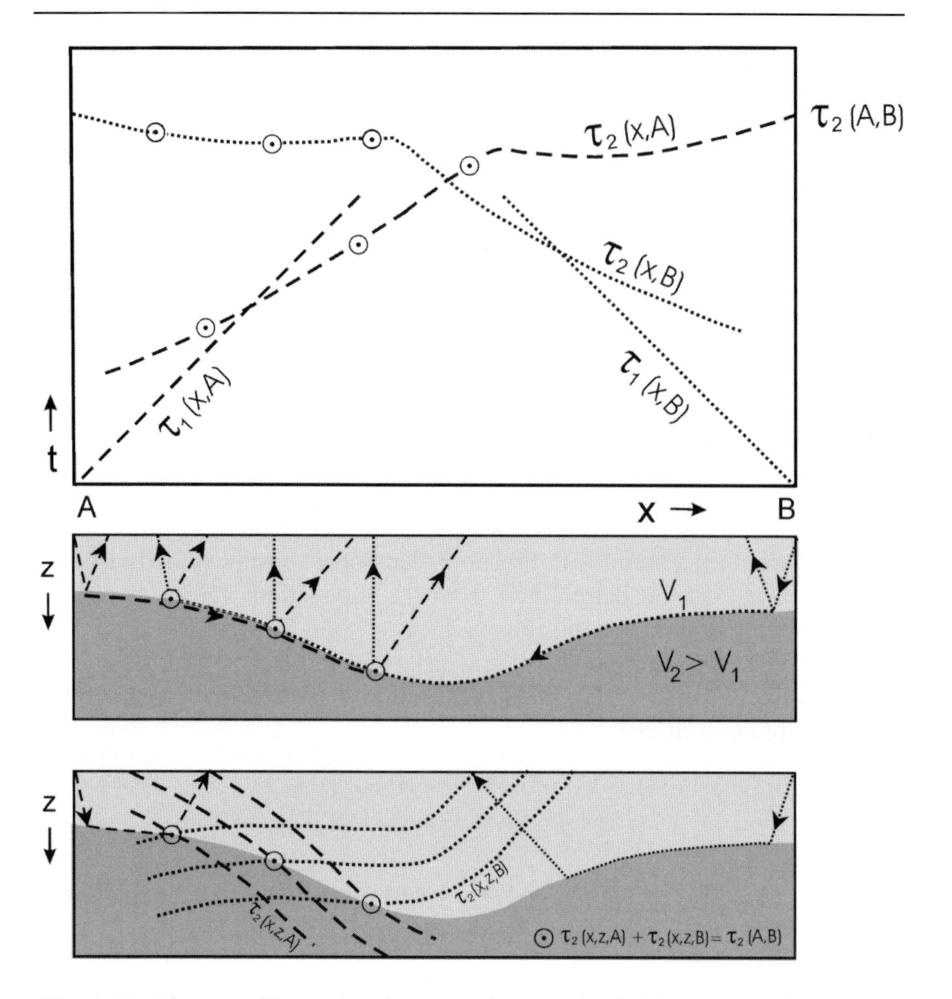

Fig. 2.18. Diagrams illustrating the wave-front method. Top: Reversed direct and refracted travel-time branches for source points at A and B. Middle: Ray paths of the refracted waves of source points A and B (dashed and dotted lines, respectively). Circles denote points on the interface and on the travel time curves corresponding to each other. The coordinates of travel time points and the interface points are connected through the up-diving parts of the refracted rays. Bottom: To determine the interface points the travel time field is traced back from the surface into the earth along the up-diving part of the ray. The resulting wave-front positions can be determined from the slope of travel time curve and the velocity of the upper layer via eq. 2.24-25. Interface points are found by applying eq. 2.23 or 2.26 to the reversed travel time fields

Seismic refraction tomography

Seismic refraction tomography (Fig. 2.6 and 2.19) is the most general approach to determine vertically and laterally heterogeneous velocity fields of the subsurface (see, e.g., Sharma 1997, for a review of tomographic inversion methods). Its application requires a dense multi-fold coverage of subsurface segments following the sampling theorem as it was discussed above in the seismic field layout section. In order to guarantee high structural resolution, the seismic field configuration has to ensure that each relevant underground segment is sounded under different ray directions. Mathematically the relation between seismic travel times and heterogeneous seismic velocity structure is a non-linear problem which cannot be solved directly. Therefore, seismic tomography is usually based on iterative algorithms comprising four elements:

1) the formulation of an initial simplified subsurface model of seismic velocities (Fig.2.19a-c),
2) the application of a modelling algorithm to simulate seismic travel times for a given subsurface model (Fig. 2.19d-e),
3) a validation of the agreement of simulated and observed travel times (Fig. 2.19f),
4) the application of a computing scheme to convert the observed differences between simulated and observed travel times into an improvement of the initial underground model (Fig. 2.19g),
5) an update of the initial underground model (Fig. 2.19h).

Steps 2) to 5) are repeated until simulated and observed travel times agree satisfactorily or until no further significant improvement is found between successive iteration steps.

In step 1), a starting model of the subsurface can be determined by applying one of the above outlined algorithms. Alternatively, it may also be based on drilling information or the analogy to structurally similar areas which were investigated previously. Usually, both convergence rate and reliability of tomographic results improve with the quality of the starting model.

For the simulation of travel times in step 2) two types of algorithms are widely applied: Ray-tracing and finite-difference (FD) implementations of the eikonal equation (e.g. Červený 2001, Vidale 1988, Zelt and Smith 1992, Hole and Zelt 1995). Since travel times have to be computed for all source-receiver combinations the operating expense during iteration may be considerable for large survey areas. In terms of computation time and memory requirement, ray-tracing is more efficient than FD-eikonal solvers, but it can be applied to smoothly heterogeneous media only. For

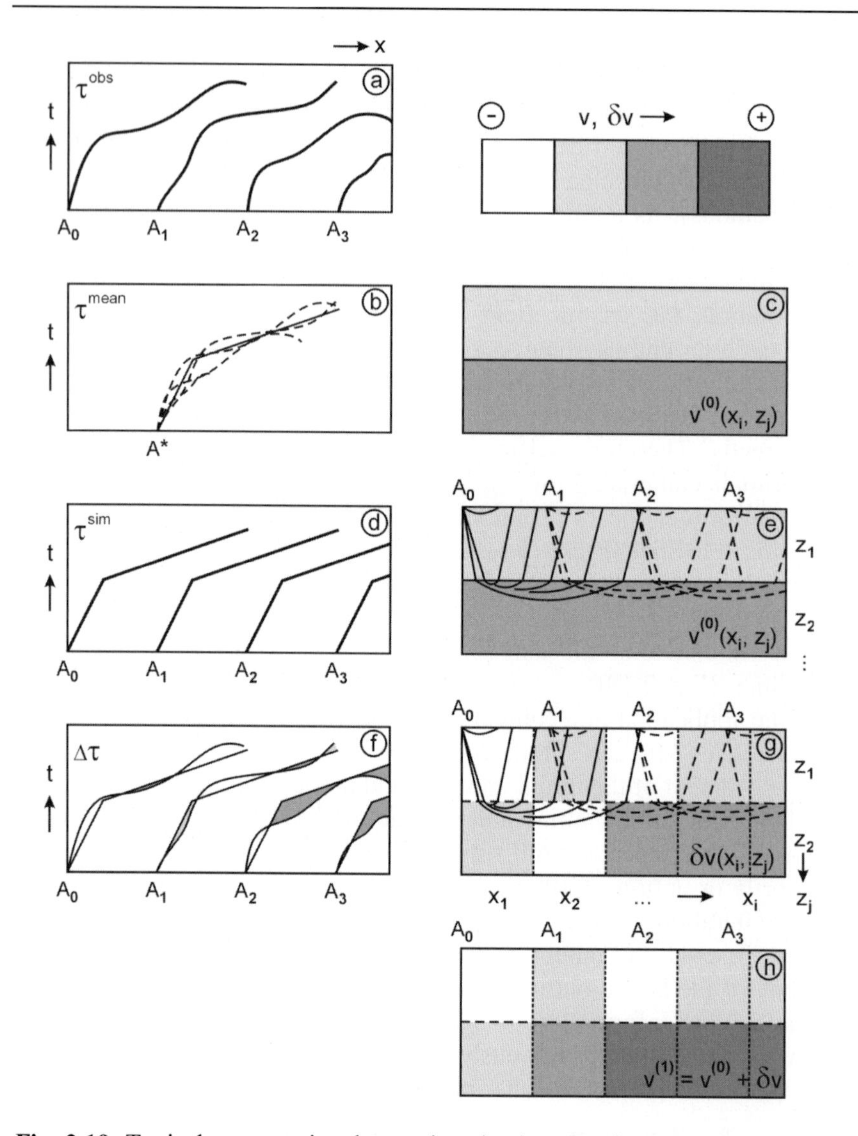

Fig. 2.19. Typical computational steps in seismic refraction tomography: (a) Observational basis: Refraction travel time curves τ^{obs} for overlapping spreads and a dense sequence of source points, (b) computation of simplified average travel time-distance curve τ^{mean}, (c) derivation of a one-dimensional starting model $v^{(0)}$ based on b), (d) computation of simulated travel-time curves τ^{sim} for the actual model, (e) computation of rays for the actual model, (f) determination of travel-time residuals $\Delta\tau=\tau^{obs}-\tau^{sim}$ (and checking of data fit), (g) application of an inversion technique to convert travel-time residuals $\Delta\tau$ into corrections δv of the actual velocity model, (h) updating of the actual velocity model $v^{(1)}= v^{(0)}+\delta v$, or more generally, $v^{(n+1)}= v^{(n)}+\delta v$ for the next iteration starting again with d)

ray-tracing these smooth subsurface models can be parameterised very effectively with piecewise analytical functions whereas FD-eikonal solvers work on a regular grid of points at which the seismic velocity is specified. To avoid discretisation errors the grid spacing in both horizontal and vertical directions should be significantly less than the geophone spacing ($>x_G/4$).

In most algorithms the validation step 3) is based on a least squares approach such as to minimise an objective function

$$L = \sum_i \sum_k \{\tau_{ik}^{obs} - \tau_{ik}^{sim}\}^2 / (n \cdot \sigma_{ik}^2) \tag{2.28}$$

during the iteration. Here, $\tau_{ik}^{obs}=\tau^{obs}(x_i,x_k)$ and $\tau_{ik}^{sim}=\tau^{sim}(x_i,x_k)$ are observed and simulated travel times respectively, applying to source and receiver points at x_k and x_i, respectively. σ_{ik} is a weight factor which can be related, for example, to data quality, and n is the number of data points along the considered travel time branches. The term L corresponds to the normalised variance of the travel time data after subtraction to the portion explained by the actual subsurface model. $L^{1/2}$ can be interpreted as some kind of standard deviation of the fit. The iteration procedure can be terminated, for example, when this value reaches the accuracy of the travel time picks from the field data or when no further significant improvement of the fit is achieved.

For the improving of the subsurface model during iteration (step 4) different approaches are in use including Monte Carlo simulation and genetic algorithms which are based on selecting seismic velocity updates from random numbers. Another most widely spread family of tomographic implementations is based on a linearization of the travel time integral equation (Eq. 2.4), such as:

$$\Delta\tau_{ik} = -\sum_i \Delta s_i^{(0)} \cdot \Delta v_i / v_i^{(0)2} \tag{2.29}$$

i=1,..,n; k=1,..,m.

Here S_{ik} stands for a ray connecting source k with receiver i; the Δs_j terms are ray segments along which seismic velocity values v_j are found. The velocity values $v_j^{(0)}$ represent the actual velocity model based on which the v_j values are to be updated for the next iteration step according to $v_j=v_j^{(0)}+\Delta v_j$. Here, Δv_j are velocity corrections to be determined from the travel time portion which remained unexplained after the actual iteration step. The first term on the right hand side of Eq. 2.32 is the simulated travel time corresponding to the actual velocity model and respective ray path segments $\Delta s_j^{(0)}$. By definition of the travel time residuals $\Delta\tau_{ik}=\tau_{ik}^{obs}-\{\sum_j \Delta s_j^{(0)}/v_j^{(0)}\}$ for all source-receiver combinations a system of linear equations is obtained from which the velocity updates Δv_j can be estimated in a least-squares sense:

$$\Delta\tau_{ik} = -\sum_i \Delta s_i^{(0)} \cdot \Delta v_i / v_i^{(0)2} \tag{2.30}$$

Note that the velocity updates Δv_j are computed on the assumption that the ray paths remain approximately unchanged. This is an inaccuracy which is corrected in the subsequent modelling step.

Since tomographic computations formally require a starting model, a number of tomographic runs with differing starting models should be performed in order to determine up to which degree the solutions depend on the initial conditions of the iteration. The accuracy of results depends also on the geometric distribution of active sources and geophones in the field. Therefore, it is important to complement the tomographic study by an investigation of its resolution. This can be performed, for example, by a "chequerboard test" where the tomographic algorithm is applied to travel time data simulated numerically for the real acquisition geometry and a specially designed seismic velocity model. This subsurface model shall be composed of a smooth back ground velocity function similar to the real conditions which is overprinted by chequerboard type velocity variations. The length of the edges of the chequerboard segments can be varied in order to determine the limits of tomographic resolution. Resolution is fine if the chequerboard pattern can be recovered from tomography at all relevant depth levels.

2.2.8 Structural resolution of seismic refraction measurements

As discussed in the field layout and tomography sections, the resolution of seismic refraction measurements may suffer from deficiencies of the source-geophone configuration under unfavourable field conditions. Numerical simulations such as the chequerboard test described above can serve to quantifying the corresponding limits of resolution.

In addition, the resolution of seismic refraction measurements is limited by a number of circumstances which are partly fundamental and partly solvable by an extension of the interpretation approach. These are: the existence of Fresnel zones and the possible occurrence of thin refracting layers and non-refracting low-velocity zones.

Fresnel zones of refracted waves

Following Huygens' principle, the propagation of a wave front can be explained by the interference of elementary waves emanating from subsurface points which were subsequently hit by an incident wave (Fig. 2.20). That this assumption applies to seismic waves can be observed, for example, at faulted layers where incident refracted or reflected wave fronts are disrupted. These disrupted refractions (and reflections as well, cf. Fig. 2.26 right) continue into curved diffracted waves apparently emanating from the edges of the faulted layers, so the fault edges can be identified as Huygens' source points. A consequence of this understanding of wave propagation is that each seismic signal recorded at a geophone is influenced not only by the geological structure found along the ray which connects source and geophone, but it is also influenced by the structure within a certain surrounding of the ray. This surrounding from which constructively interfering elementary waves arrive at the geophone is called (first) Fresnel volume. The surface of the Fresnel volume is the envelope of all possible ray paths between source and geophone along which the travel time is delayed by half a signal period or less compared to the fastest one.

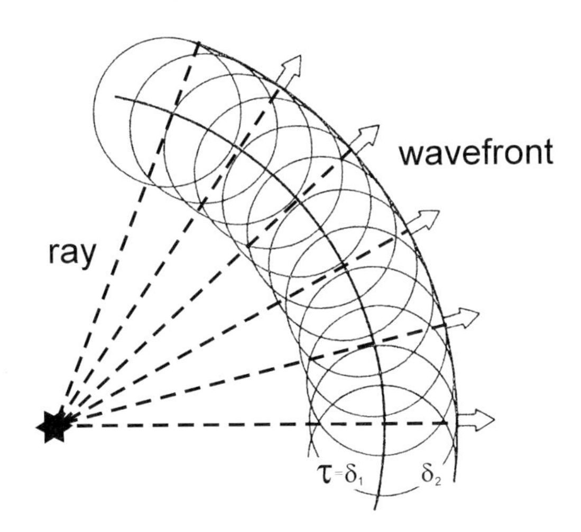

Fig.2.20. Huygens' principle: Forming of a wave front by the interference of elementary waves

The radius r_F of the Fresnel zone is of practical importance because it provides a measure for the resolution of seismic measurements. Subsurface structure is well resolved on a scale which is of the order of r_F or larger. Signals from segments or objects the distance of which is smaller than the radius of one Fresnel zone interfere so they cannot be observed separately from each other. At the interface of a two-layer structure the Fresnel zone radius of the refracted wave is given by:

$$r_F^{refr} = [v_1 \cdot T \cdot z_1 + (v_1 \cdot T/2)^2]^{1/2} /(1 - v_1^2 / v_2^2) \approx [v_1 \cdot T \cdot z_1]^{1/2} \quad (2.31a)$$

corresponding to a geophone located at the earth's surface (Fig. 2.21). Here, T is the period of the seismic signal, z_1 is the layer thickness and v_1 and v_2 are the seismic velocities of the upper and lower layers, respectively. A groundwater table located at 20 m depth, for example, is sampled by a refracted P-wave at 100 Hz with Fresnel zone radius of 19 m if the P-wave velocity of the vadose zone and the aquifer are assumed to be 750 and 1500 m/s, respectively.

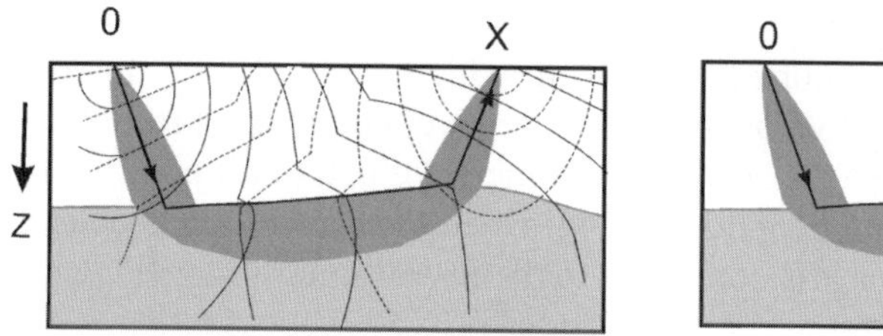

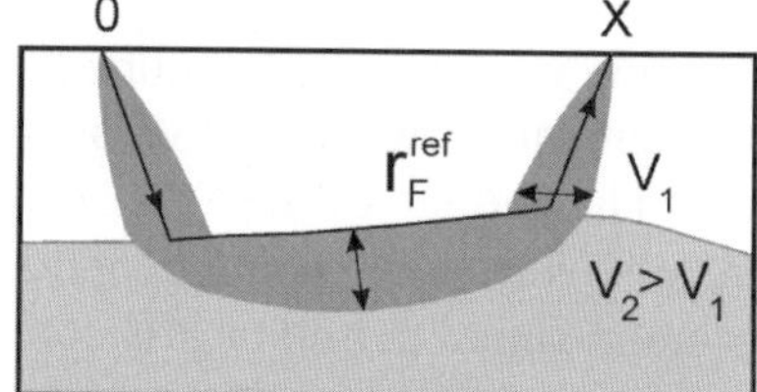

Fig. 2.21. Fresnel volume of refracted rays (after Hagedoorn and Diephuis 2001). Left: The Fresnel volume of a ray conncting a source a 0 and a receiver at x can be determined by identifying all subsurface points $P=(x_P,z_P)$ around a central ray for which the travel time condition $\tau(x_P,z_P;0)+\tau(x_P,z_P;x)-\tau(x_A,x_B)<T/2$ (T = signal period). Right: Fresnel zone width and thickness referring to Eq. 31

The maximum thickness of the Fresnel volume below the refracting interface is given by

$$r_F^{refr} \approx 0.5 \cdot \{v_2 \cdot T \cdot [x - \frac{2 \cdot z_1 \cdot v_1}{\sqrt{v_1^2 - v_2^2}}]\}^{1/2} \quad (2.31b)$$

if the source-geophone distance $x>>v_2 \cdot T/4$. Eq. (2.31b) shows that a refracted arrival actually "looks" much deeper into the bottom layer when it travels along an interface than a ray path based on Fermat's principle suggests. To detect the groundwater table in the above example, the geophone spread should have a minimum length of about 90 m (Eq. 2.15a). Eq. 2.31b implies that the corresponding maximum thickness of the Fresnel volume would amount to approximately 16 m below interface.

Hidden refracting layers

In near surface prospecting the interpretation of seismic refraction measurements is often restricted to the analysis of first breaks. This is the case especially if later body wave arrivals are obscured by interfering surface waves, comparably long source signals, and reverberations caused by multiply reflected and refracted arrivals. If the travel time interpretation is

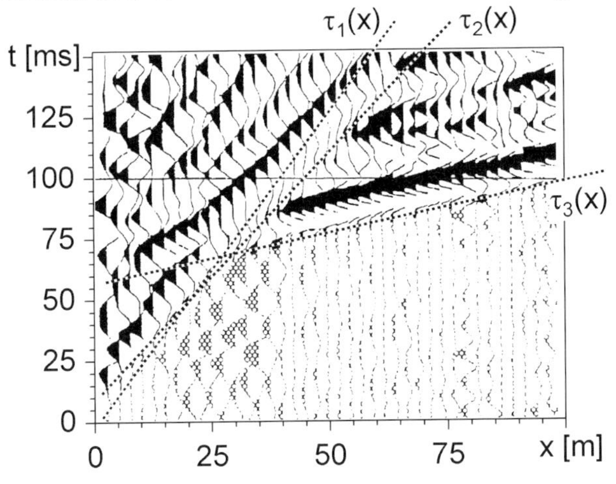

Fig. 2.22. Seismogram example showing direct and refracted arrivals for a three-layer case. τ_1, τ_2 and τ_3 indicate direct, first and second refracted waves, respectively. In the first breaks the first refraction (τ_2) cannot be identified clearly, but it appears in the slope of the later arrivals at x>30 m. (cf. Fig. 2.17, right column)

based on first breaks only, a thin layer can remain "hidden" even if its seismic velocity increases with respect to the hanging wall because a minimum thickness is required to make refractions show up as a first break (e.g. Gebrande and Miller 1985). An example is shown in Fig. 2.22. Based on the formulae for crossover points and intercept times (Eq. 2.15 and 2.18, respectively) it can be shown, for example, that the second of three layers will be visible as a first break only if the criterion of inequality (Eq. 2.32) is fulfilled. Denoting thickness and seismic velocities of the three layers by Δz_1, Δz_2, Δz_3 and $v_1<v_2<v_3$, respectively, the criterion reads:

$$\Delta z_2 / \Delta z_1 > (V_{12} - V_{13}) \cdot V_{23} \cdot (v_2 / v_1) \qquad (2.32)$$

where

$$V_{12}=[(1+v_1/v_2)/(1-v_1/v_2)]^{1/2}$$
$$V_{13}=[(1+v_1/v_3)/(1-v_1/v_3)]^{1/2}$$
$$V_{23}=(1-v_1/v_3)/(1-v_2^2/v_3^2)]^{1/2}.$$

For example, if the P-wave velocities of a vadose zone, an aquifer and an aquitard are 750 m/s, 1500 m/s and 2000 m/s, respectively, the aquifer will become visible as a first break if its thickness exceeds one third of the thickness of the vadose zone.

The hidden layer problem clearly diminishes if the direct and refracted travel time branches can be followed up beyond the first breaks (τ_2 for x>30 m in Fig. 2.22). If this is not possible because of wave interference, the readability of the seismogram sections can (sometimes) be improved by digital signal processing such as deconvolution and dip filtering. Also, the application of spatial Fourier and Radon transforms can help to identify otherwise hidden travel time branches.

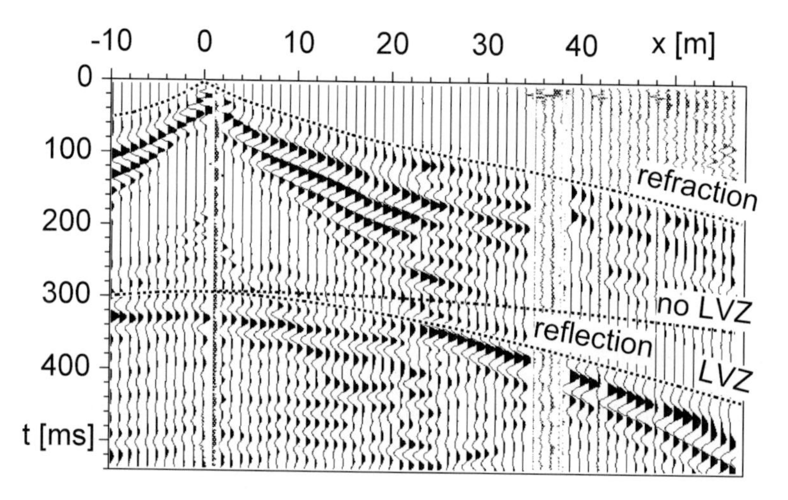

Fig. 2.23. Seismogram example for a low velocity zone (LVZ) beneath refracting layers. Travel-time curves are shown for waves refracted along the surface layers (upper most curve) and reflections from the bottom of a deeper layer. The strong curvature of the observed reflection indicates that a low-velocity layer is present (lower most curve). If there were no LVZ the reflection hyperbola would appear more flat (middle curve)

Low velocity zones

In layered media, the thickness and velocity of embedded low velocity zones cannot be determined by seismic refraction measurements from the earth's surface because, according to Snell's law, they do not return refracted rays to the surface. In hydro-geological prospecting this situation can occur where cohesive soil overlays lowly cohesive dry sediments in the vadose zone, for example, in a sequence consisting of marl or boulder clay overlaying dry sand. Sometimes, the existence of low velocity layers is indicated by out-dying phases and discontinuous time-shifts in the first break sequence. Note, that this basic problem cannot be solved by applying refraction tomography. Refraction tomography is capable to detect and quantify low velocity zones only if they are, on the one hand, of limited lateral extent and if, on the other hand, their minimum extent is in the order of the radius of the first Fresnel zone. For the investigation of embedded zones of low S- wave velocity up to a few meters depth the analysis of surface waves dispersion sometimes helps. However, the situation improves considerably if reflections from top and bottom of an embedded low velocity zone can be identified over a sufficient range of source-geophone offsets so their travel times and curvatures can be included in the analysis (Fig. 2.23).

2.3 Seismic reflection imaging

Seismic reflections can be regarded as echoes generated at geological interfaces where the seismic impedance shows a contrast. Seismic impedance is defined as the product of seismic velocity and density. At every interface the energy of an incoming wave is split into reflected and transmitted portions. Therefore, the structure of layered media can be imaged as a sequence of reflected seismic impulses arriving at a geophone spread with time delays in between. The time delays between two successive reflections are the differences between the respective "two-way travel times" (TWT) from the earth's surface down to the reflecting interfaces and reverse. Reflection seismograms can be mounted to form a reflection time section showing a distorted image of the layer sequences of the subsurface (Fig. 2.24, 2.25). The distortion can be removed by transforming the reflection time section into a depth section. This so-called migration process requires in advance the determination (or assumption) of seismic velocity-depth functions. In contrast to seismic refraction interpretation, reflection surveys provide a more direct way to image geological structure which can be applied even in case of low velocity zones (cf. Fig. 2.6). A disadvantage of reflection imaging is that it requires comparably more efforts in data acquisition and that its applicability to very shallow structure is often limited by the interference of surface waves and direct and refracted body waves.

2.3.1 Targets for seismic reflection measurements

The strength of seismic reflection imaging is its structural resolution. Therefore, it should be applied to investigate lithological sequences involving thin and complex layering or embedded low velocity zones. In contrast to refracted waves, which can be observed only beyond the critical distance, seismic reflections can be received, in principle, at all shot-geophone offsets including "zero-offset" where source and receiver points coincide (Fig. 2.8, 2.23). Therefore, from an acquisition point of view, reflection seismics is best suited to image deep layers the investigation of which would require very long spreads if refraction measurements were applied (compare the discussion of minimum spread lengths in Sect. 2.2.5).

Regarding hydro-geological investigations, typical targets for seismic reflection imaging are found in glacially formed sedimentary areas, for example, buried, complexly refilled valleys cutting through elsewhere sealed aquifer and opening new hydraulic pathways, or compressively deformed moraine sequences with vertical and horizontal alternations of high- and low-permeable material. Other examples are the mapping of deep bedrock and of near surface fault zones where dip and displacement can be determined by analysing the time- or depth shifts of reflections on either sides of the fault plane. Reflection imaging can also be applied to localise small scale heterogeneities such as sand lenses in clayish layers or lateral compositional changes.

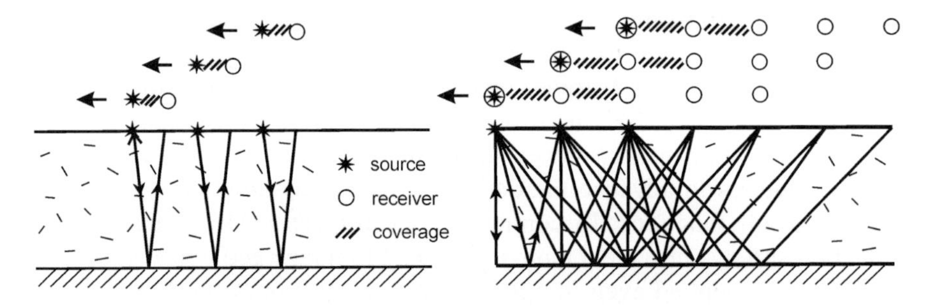

Fig. 2.24. Acquisition scheme and ray paths for seismic reflection measurements with single-fold (left) und multi-fold coverage (right) of subsurface points. Single-fold measurements require a source and just one sensor which are moved along the profile. Multi-fold coverage can be realised by moving a group of sensors. The underground segment covered at each position of the source-sensor group is indicated by a hatched line

2.3.2 Seismic reflection amplitudes

The signal strength of reflected seismic waves depends on the contrast in seismic velocity and density found at geological interfaces. For a seismic wave impinging vertically or nearly vertically on a layer interface (so-called near-vertical incidence) the reflection strength is expressed by the reflection coefficient R_0

$$R_0 = \frac{\rho_2 \cdot v_2 - \rho_1 \cdot v_1}{\rho_2 \cdot v_2 + \rho_1 \cdot v_1} \tag{2.33a}$$

where ρ_1, v_1 and ρ_2, v_2 are densities and seismic velocities of the hanging and laying layers with respect to the ray incidence, respectively. R_0 is defined as the amplitude ratio of the reflected and incident waves at the interface. The case of near-vertical incidence is of special practical importance because it can be applied as an approximation to most seismic reflection surveys after respective data processing (Figs. 2.24 left, and 2.25). In near surface sediments reflection coefficients between a few and a few ten percent are found for typical contrasts of seismic velocity and density.

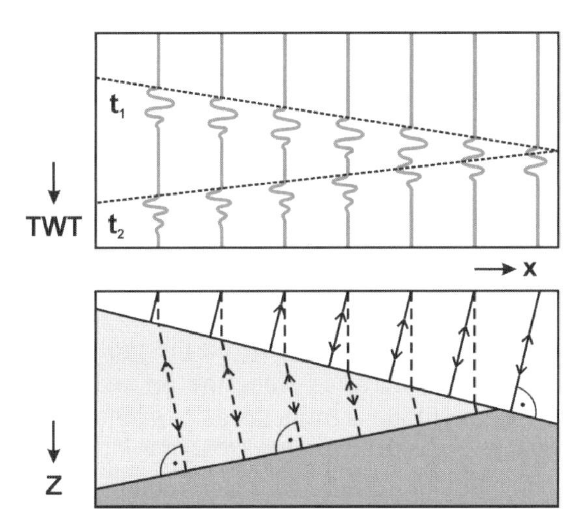

Fig. 2.25. Zero-offset reflection time-section (top) of a simple subsurface model (bottom). The time section corresponds to the acquisition scheme of Fig. 2.24 (left) if source and sensor are placed at the same positions for each shot ("zero-offset"). In this case the recorded reflection signals correspond ray paths arriving under normal incidence at the reflecting interface irrespectively of the layer dip (see rays in the bottom figure). Therefore, regarding the horizontal position, the reflection time section shows aberration which needs to be corrected. This effect is even more obvious for non-plane layering (Fig. 2.26)

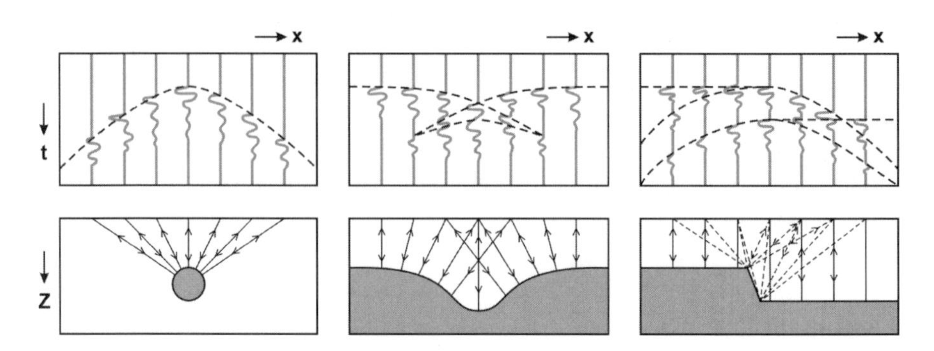

Fig. 2.26. Image aberration in zero-offset reflection time sections (top) for different subsurface models (bottom). Ray paths and travel time curves indicated. Left: A small isolated body is imaged as diffraction hyperbola. Middle: A syncline appears as a travel time loop. Right: Fault-related displacement is connected with the appearance of diffracted waves originating at the edges of interfaces

The corresponding relative amplitude of the wave portion transmitted vertically through the interface is described by the transmission coefficient T_0

$$T_0 = \frac{2 \cdot \rho_1 \cdot v_1}{\rho_2 \cdot v_2 + \rho_1 \cdot v_1} \tag{2.33b}$$

Note that reflection and transmission coefficients generally depend on the incidence angle and have to be represented by complex numbers for inclined incidence beyond the critical angles to include shape changes of the seismic signals (for details see, e.g., Aki and Richards 2002).

The amplitude of a reflected wave observed at the receiver depends not only on the reflection coefficient but also on the product of transmission coefficients related to the interfaces along the up and down directed ray paths. In addition the amplitude is influenced by geometrical spreading and attenuation.

Geometrical spreading describes the decrease of the energy density along a propagating wave front caused by the widening of the wave front area with increasing travel distance. In case of nearly plane layering and observations at near-vertical incidence, the amplitude decay by geometrical spreading is approximately proportional to the length of the ray path or to the two-way travel time, respectively. The decay of the energy density is proportional to the square of the ray path length.

When a downward travelling wave reaches a reflecting interface at 20 m depth, for example, its amplitude will have decreased to only 5% of the value it had at 1 m depth because of geometrical spreading. The value of

5% corresponds to the ratio of wave front radii at the considered depths which is 1m/20m. Back to the earth's surface the spreading factor will have further decreased to 2.5% (1m/40m). To obtain the relative reflection amplitude which would be observed at the earth's surface, this value has to be multiplied with the reflection coefficient and the product of the transmission coefficients along the ray path. The example shows that reflection amplitudes are relatively small compared to the direct wave, for instance, which decays only with the inverse of the shot-geophone distance.

Note that the proportionality of spreading factor and ray length cannot generally be applied. In special structural situations, for example, geometrical spreading can lead even to a local amplitude increase near focussing zones (so-called caustics) (see, e.g., Červený 2001 for details).

The attenuation of seismic waves comprises two different aspects: Scattering and absorption. Scattering is the conversion from coherent into incoherent energy caused by wave diffraction at small-scale irregular heterogeneity. Here, "small" means dimensions in the order of half a wave length and below. Absorption is the conversion of seismic into non-seismic energy caused, for example, by seismically induced flow of pore fluids or by slip at grain boundaries in low-cohesive sediments. Attenuation is quantified in terms of the attenuation factor 1/Q defined as the ratio of energy loss and maximum energy stored in a volume during one cycle of deformation. For near surface sediments typical values of Q are between 5 and 100.

2.3.3 Concepts of seismic reflection measurements

In performing seismic reflection measurements and interpretation different concepts can be followed. In the order of increasing complexity, these are, on the one hand, zero-offset or constant-offset measurements, and, on the other hand, the common-midpoint method (CMP method) and pre-stack migration. The first approach provides a single-fold coverage of the subsurface meaning that each segment of a geological interface is illuminated by just one shot-geophone pair out of the survey. The latter approaches require a multi-fold coverage of subsurface points where the reflections of each interface segment are gathered by different shot-geophone pairs at different offsets (Fig. 2.24). In the following, the zero-offset and common midpoint configurations are discussed as representatives of single and multi-fold methods, respectively.

Zero-offset image of subsurface structures

A straight way to generate a seismic reflection image of the subsurface is to perform measurements where source and receiver points coincide or nearly coincide. The resulting seismograms represent a zero- or near-offset section, respectively. For example, such kind of a survey can be realised at sea by a boat pulling just one sensor and a marine source firing at a constant rate, or on land by moving along a profile with just one source and one geophone planted near the source each time a new location is reached.

The arrivals recorded in zero-offset configuration are reflected at the geological interfaces at vertical incidence angles irrespectively of the dip and seismic velocity of the layers (Fig. 2.25). Since the wave starts and ends at the same point, downwards and upwards directed ray paths are identical. Therefore, zero-offset is the case to which the formula of the reflection coefficient (Eq. 2.33a) applies in a strict sense. Despite its conceptual straightness there are some complications connected with zero-offset imaging which deserve a closer view.

The primary outcome of a seismic reflection survey is a seismogram section where the amplitudes of the reflected waves are plotted as a function of the receiver location and of the arrival time at the receiver. However, the goal of imaging is a depth section where reflection amplitudes appear at the spatial coordinates of the corresponding reflecting elements. Compared to a depth section, zero-offset time sections suffer from aberrations. Except for the case of near horizontal layering and low velocity variation, these aberrations cannot satisfactorily be compensated by a linear rescaling of the vertical axis from TWT to depth using, for instance, an average velocity (Fig. 2.25). In case of horizontal layers, for example, the two-way travel time of the normal-incidence reflection from the n-th interface below the surface is

$$t_n(0) = 2 \cdot \sum v_i \cdot \Delta z_i \qquad (2.34)$$

where v_j and Δz_j are seismic velocity and thickness of the layers in between. Successive reflections of the hanging wall are separated by a time delay of $\Delta t_j = t_j - t_{j-1} = 2 \cdot v_j \cdot \Delta z_j$ where v_j is usually different for each layer. Therefore, a correct depth scale can only be achieved if the velocity-depth function is known. Moreover, in the more general case of non-horizontal layering, reflection points are not located directly beneath the receiver position but laterally displaced depending on the dip and seismic velocity of the hanging wall (Fig. 2.26). This lateral displacement can be compensated by applying a so-called migration algorithm which is treated below. Again, such kind of operation requires the knowledge of seismic velocities of the subsurface which cannot be determined solely on the basis of a zero-offset section.

Another complication is the relative weakness of reflection amplitudes the visibility of which may be affected severely by ambient noise, in particular with increasing travel-time. A solution to both seismic velocity and noise problems is provided by multi-fold coverage schemes of seismic data acquisition, for example, the common-midpoint method.

In summary, zero- or near-offset surveys are advantageous to be applied in situations where seismic velocities are known at least approximately, where sufficient reflection strength can be expected, and where the budget is low.

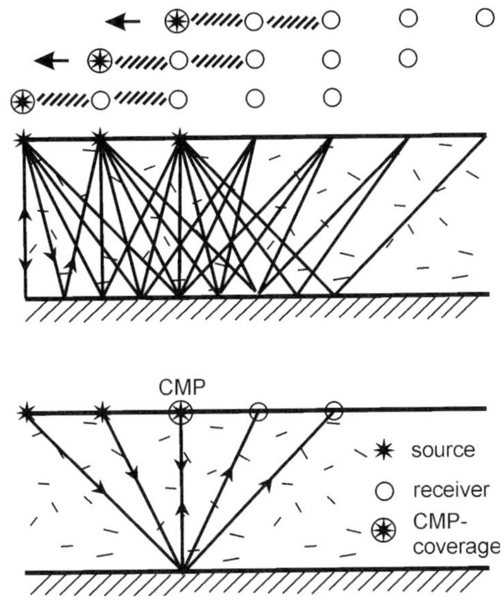

Fig. 2.27. Records of a multi-fold covered profile (top, cf. Fig. 2.24) can be combined into so-called common-midpoint (CMP) gathers (bottom). A CMP-gather comprises all records of a survey revealing the same coordinate for the midpoint between source and receiver. In case of near horizontal layering the records included in each CMP-gather cover the same subsurface points (common reflection points) which are located vertically beneath the CMP

The common-midpoint (CMP) method

Seismic velocities can be determined from the travel times of reflections only if they are observed at a number of different source-geophone offsets. This can be realised, for example, by continuously moving source points and receiver spreads along a seismic profile (Fig. 2.27). The overlapping spreads provide a multi-fold coverage of the segments of the reflecting interfaces in the sense that each segment is illuminated several times under different angles of wave incidence. The multi-fold reflection records belonging to the interface segments beneath a point of a profile can be found by selecting the records of those source-receiver pairs which are placed symmetrically to this point. The seismograms assigned to a "common midpoint" (CMP) are called a CMP-gather. In a strict sense, the reflection points of the seismograms of a CMP-gather coincide only if the layering is horizontal and if the lateral variation of seismic velocity is smooth.

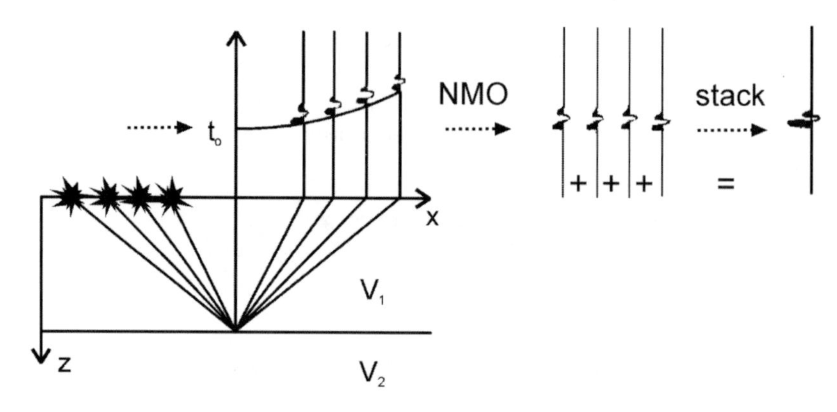

Fig. 2.28. Normal move-out (NMO) correction. Left: Reflection signals of different traces in a CMP-gather show different travel-times caused by the different lengths of the corresponding ray paths or source-sensor offsets, respectively. Right: The NMO-correction serves to compensate these differences so the signals can be stacked (summed) in order to improve the signal-to-noise ratio. In terms of travel time the CMP-stacked trace corresponds to a zero-offset seismogram. NMO analysis is also used for determining average seismic velocity

In a CMP-gather the travel time-offset curve of a reflection forms a hyperbola symmetrical to zero-offset from the curvature of which the seismic velocity can be determined. Moreover, the signal-to-noise ratio of the records can be improved by summing or "stacking" the seismograms of a CMP-gather. Since reflection travel times increase with source-geophone offset, a corresponding time shift has to be applied to the reflection signals before they can be stacked (Fig. 2.28). This time shift, called normal move-out correction (NMO-correction), reduces the travel times observed at non-zero offset to those observed at zero-offset. Therefore, the outcome of CMP-stacking is a seismic time-section corresponding to a zero-offset section with an improved signal-to-noise ratio and backed by seismic velocity information.

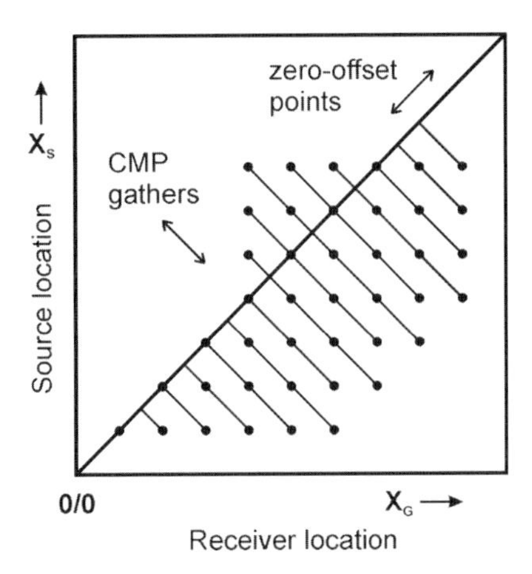

Fig. 2.29. Acquisition plan for seismic reflection measurements along linear profiles. Source and receiver coordinates realised in a survey are plotted against each other. Common midpoints are found along the +45° diagonal whereas the shot-geophone pairs of each CMP-gather are found along respective lines of -45° tilt showing the coverage of the respective CMP. In the example shown in the figure, a spread of six geophones was thought to be moved first from left to right with the source at its left-hand side (one-sided or end-on spread; see dotted zones). Then geophone spread was left in its position, and only the source was moved (split spread)

In the following, the most important formulae involved in CMP-reflection analysis and stacking are listed (see, e.g., Sheriff and Geldart, 1995, for more details). The coordinates x_{CMP}, x_S and x_G of CMP, source and receiver locations, respectively, and the source-receiver offset $\Delta x = x_G - x_S$ are related to each other by:

$$x_{CMP} = (x_S - x_G)/2 = x_S + \Delta x/2 = x_G - \Delta x/2 \quad (2.35)$$

The spacing and coverage of common midpoints can be visualised by an acquisition plan where source and receiver coordinates realised in a survey are plotted against each other (Fig. 2.29). In this diagram common midpoints are found along the +45° diagonal whereas the shot-geophone pairs of each CMP-gather are found along respective lines of -45° tilt. The spacing of common midpoints along the diagonal is:

$$\Delta x_{CMP} = \min\{\Delta x_S/2, \Delta x_G/2\} \quad (2.36)$$

where Δx_S and Δx_G are source and receiver spacing, respectively. Usually, seismic reflection surveys are recorded with source points and receiver-spread advancing continuously along the profile (Figs. 2.27, 2.29). For a receiver-spread of length $N_G \cdot \Delta x_G$ maximum coverage is given by:

$$N_{CMP} = N_G \cdot \Delta x_G /(2 \cdot \Delta x_S) \tag{2.37}$$

if both source and receiver-spread are moved by Δx_S after each shot. Maximum coverage is reached at a distance of $N_G \cdot \Delta x_G/2$ from the start and endpoints of the profile, respectively.

In a CMP-gather the travel time-offset curve of the reflection can be approximated by:

$$t_n(\Delta x) \approx [t_n^2(0) + \Delta x^2 / V_n^2]^{1/2} \tag{2.38}$$

for $\Delta x < V_n \cdot t_n(0)/2$ where n is the index of the reflecting interface, $t_n(0)$ is the travel time at zero-offset and V_n is the so-called normal move-out velocity (NMO-velocity). In case of plane horizontal layering and small offsets V_n can be expressed by the seismic velocities v_j and the interval travel times $\Delta t_j = 2 \cdot \Delta z_j / v_j$ at zero-offset of the n layers above the reflecting interface (layer thickness Δz_j, j=1,..,n). The corresponding equation reads:

$$V_n^2 = \Sigma v_i^2 \cdot \Delta t_i / t_n(0) \tag{2.39}$$

V_n as defined in Eq. 2.39 is also called root mean square (RMS) velocity. Since the V_j, Δt_j and $t_j(0)$ can be determined from the seismograms of the CMP-gather for each layer interface, Eq. 2.39 can be rearranged in a recursion formula from which the layer velocities v_j can be derived successively from top to bottom:

$$v_n^2 = [V_n^2 \cdot t_n(0) - \Sigma v_i^2 \cdot \Delta t_i] / \Delta t_n(0) \tag{2.40}$$

The sequence of layer velocities v_j can be applied to determine the depth of the reflecting interfaces.

CMP-geometry and Eq. 2.38 can be applied to describe reflection travel times even in case of dipping layers, but the formula of the NMO-velocity (Eq. 2.39) has to be modified. Formulae for dipping multi-layer cases can be found, for example, in Hubral and Krey (1980). If the uppermost interface shows a dip angle of δ_1, the corresponding NMO-velocity reads:

$$V_1 = v_1 / \cos(\delta_1) \tag{2.41}$$

Note that $V_1 = \infty$ for a vertical interface (δ_1=90°). Along a dipping interface the reflection points of a CMP-gather disperse more and more up-dip with increasing shot-geophone offset Δx (Fig. 2.30). The distance d_1 between two reflection points of arrivals at zero and non-zero offsets is given by:

$$d_1 = \Delta x^2 \cdot \sin(\delta)_1 /(2 \cdot h_1) \tag{2.42}$$

where $h_l=v_l \cdot t_l(0)/2$ is the distance of the reflecting element from the CMP measured along the zero-offset ray. The reflection point dispersal has a maximum for dip angles of 45° and gets smaller with increasing interface depth.

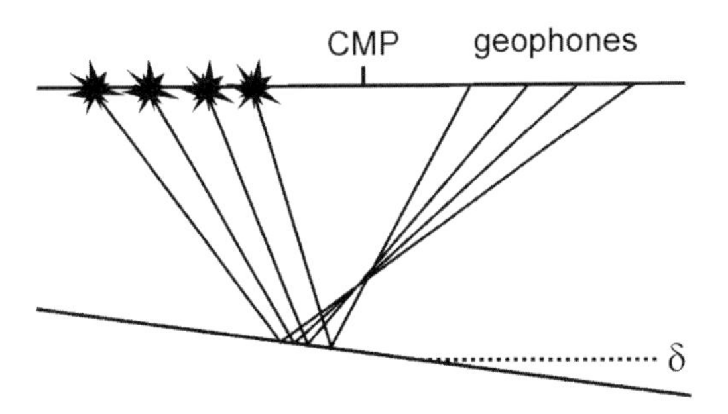

Fig. 2.30. Reflection point dispersal in a CMP gather in case of a dipping interface

The digital processing of CMP-data resulting in a stacked time-section usually comprises the following steps:

a) The application of digital filters to remove direct, refracted and surface wave arrivals from the data and to improve the signal-to-noise ratio,
b) a velocity analysis where both NMO- and interval velocities V_n and v_n are determined as a function of zero-offset travel time. The analysis is performed for a number of sample points along the profile. The spacing of the sample points depends on the lateral heterogeneity of the seismic velocity field,
c) a normal-moveout correction is applied to each CMP-gather where the travel times of non-zero offset seismograms are shifted so the arrivals align at equal times with the zero-offset seismogram. Let $u_j(x_{CMP}, \Delta x, t)$ represent the j-th seismogram of a CMP-gather and $V(x_{CMP}, t_0)$ be the NMO-velocities found in the analysis (t_0 = travel time for Δx=0). Then the NMO-corrected seismogram $U_j(x_{CMP}, \Delta x, t_0)$ can be expressed by:

$$U_i(x_{CMP}, \Delta x, t_0) = u_i(x_{CMP}, \Delta x, [t_0^2 + \frac{\Delta x^2}{V(x_{CMP}, t_0)^2}]^{1/2}) \tag{2.43}$$

d) Finally, the NMO-corrected seismograms of each CMP-gather are averaged ("stacked") and the resulting stacked seismograms compiled in a CMP-stacked section:

$$U(x_{CMP},t_0) = \Sigma U_i(x_{CMP},\Delta x,t_0)/N(x_{CMP},t_0) \tag{2.44}$$

where N is the CMP-coverage which may be variable along the profile and with reflection time. CMP-coverage gets time dependent if shallow reflections cannot be observed over the full offset range. Note that NMO-corrections cause a signal stretching in the transformed data which increases with offset and decreases with travel time. The relative frequency shift $\Delta f/f$ caused by NMO-stretching corresponds to the relative travel time shift of the NMO-correction: $\Delta f/f=[t_n(\Delta x)-t_n(0)]/t_n(0)$ (cf. Eq. 2.38). Since arrivals stretched by more than 10% should not be added to the stack the coverage can be additionally reduced in particular for shallow arrivals.

2.3.4 Seismic migration

The final objective of seismic reflection imaging is to create a seismic depth section where reflection amplitudes are plotted at the spatial coordinates of the interface segments from which they were returned to the earth's surface. Seismic migration is the (digital) processing step by which zero-offset and CMP-stacked time sections can be transformed into depth sections.

This conversion clearly requires the definition of a seismic velocity model of the subsurface. The determination of an accurate migration velocity model can be implemented as an iterative process in which the interval velocities derived from normal-moveout velocity analysis can be used as a starting model. A consistency criterion which can be applied to improve the initial velocity model is that the structural boundaries of the seismic velocity model and of the seismic reflection section should coincide after the migration has been performed.

Out of the many different migration algorithms one basic scheme shall be outlined briefly. It is based on the idea to determine the locations of all subsurface points from which an observed seismic arrival could have been reflected (Fig. 2.31). Starting with just one seismogram out of a zero-offset or CMP-stacked section, the possible reflection points of a particular arrival will form a circle around the CMP position if the seismic velocity is constant. For a non-constant velocity model this circle will be deformed accordingly. The location of the reflecting element can be identified precisely only if more than one seismogram is considered. In this case the coordinates of the reflecting interface can be traced by the envelope of neighbouring circles. In digital implementations of this migration approach the (deformed) circles are realised as 2D-filter operators along which the observed seismic amplitudes are "smeared". The smeared arrivals of adjacent traces interfere constructively at the coordinates of the reflecting interface and interfere destructively elsewhere. The filter operators can be obtained by applying finite-difference computations or ray-tracing. This migration scheme can be adopted also to three-dimensional problems where circles have to be replaced by (deformed) spherical surfaces.

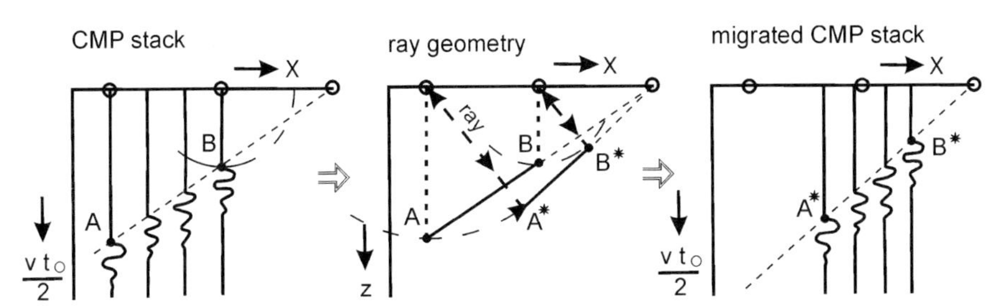

Fig. 2.31. Graphical migration scheme applying to constant velocity media. Left: The possible reflection points of an arrival appearing on one seismogram of a zero-offset or CMP-stacked section form a circle around the CMP position. Here the time scale (t_0) is replaced by a pseudo-depth scale ($vt_0/2$). Middle: The location of the reflecting element can be identified as the envelope of circles belonging to neighbouring seismograms. Reflection signals A and B of the CMP-stack "migrate" to the new positions A* and B* which correspond to the real position of the reflection points. Right: Migration is usually applied directly to the wave field in form of a digital filter process. Note that reflection elements get steeper and move up-dip during migration

The horizontal and vertical shift of a dipping reflection element during migration can be estimated by considering corresponding ray paths assuming constant seismic velocity v_1 (Fig. 2.31 middle). In a zero-offset or CMP-stacked time section, the dip p_0 of a reflected arrival corresponds to the incidence angle i_0 of the reflected wave front at the earth's surface:

$$p_0 = \frac{dt_0}{dx_{CMP}} = 2 \cdot \sin(i_0) / v_1 \tag{2.45}$$

In the migrated section, incidence angle i_0 and the dip angle δ_1 of the considered reflection element are identical so δ_1 follows from $\sin(\delta_1)= v_1 \cdot dt_0/(2 \cdot dx_{CMP})$. This result can be compared with the dip appearing in CMP-stacked section if its time axis is multiplied by v_1. In this case the considered reflection element would show an apparent dip angle δ_1^{app} according to $\tan(\delta_1^{app})= v_1 \cdot dt_0/(2 \cdot dx_{CMP})$. Since $\arcsin(\delta) \geq \arctan(\delta)$, reflection images appear at steeper dip after migration. The horizontal and vertical movement of a reflection element can be described similarly: In the CMP-stacked section a reflection at position x_{CMP} shows an apparent depth of

$z_1^{app}=v_1 \cdot t_0/2$. After migration it is found at the endpoint of the zero-offset ray corresponding to coordinates $x_{mig}=x_{CMP}-z_1^{app} \cdot \sin(\delta_1)$ and $z_{mig}=z_1^{app} \cdot \cos(\delta_1)$ so reflection elements migrate up-dip. Inserting Eq. 2.45 shows that the migration of reflection points can be completely described in terms of tilt and travel time of arrivals in the CMP-stacked section provided the seismic velocity v_1 is known:

$$x_{mig} = x_{CMP} - v_1^2 \cdot p_0 \cdot t_0 / 4 \tag{2.46}$$

$$z_{mig} = 0.5 \cdot v_1 \cdot t_0 \cdot (1 - v_1^2 \cdot p_0^2 / 4)^{1/2}$$

The effect of migration on imaging shallow layering near the top of a salt dome can be studied in Fig. 2.32. Note the steepening of dipping interfaces and the correction of the aberration effects of the CMP stack such as diffractions and travel time loops (cf. Fig. 2.26).

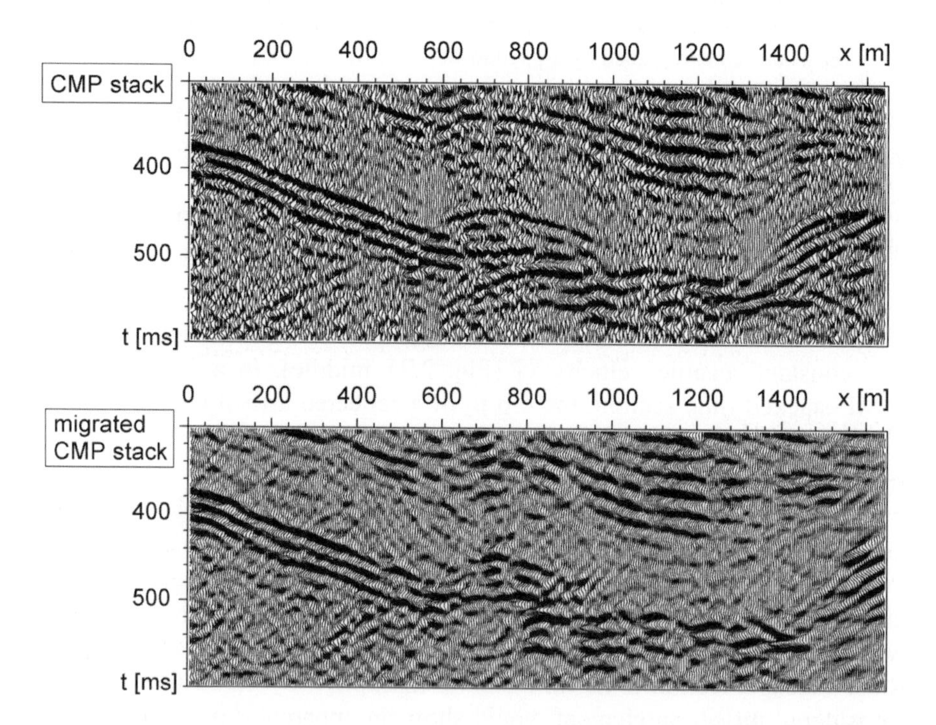

Fig. 2.32. Top: CMP-stacked section of tertiary sediments and cap rock of salt dome with a pseudo depth scale. Bottom: Migrated version of top figure

Whereas the application of Eq. 2.46 is restricted to estimating the effect of migration on zero-offset type time sections, the above outlined filter concept of migration can be transferred easily to a concept of pre-stack migration in order to create depth images more or less directly from single shot gathers. In case of non-coincident source and receiver points an (deformed) ellipse has to be considered instead of a circle to describe the possible locations of a reflection point. Pre-stack migration should always be applied if the subsurface structure is so complicated that the reflection point dispersal inherent in CMP-stacking of dipping reflections leads to blurred images. Like CMP-stacking, also pre-stack migration requires multi-fold covered data for velocity analysis and improvement of the signal-to-noise ratio. The major difference between both processes is the order in which stacking and imaging are performed and the procedures applied to determine the velocity model.

Details of velocity analysis and all aspects of digital processing of seismic reflection data are outlined, for example, in the monographs of Hubral and Krey (1980), Keary et al. (2002), Sheriff and Geldart (1995) and Yilmaz (2001).

2.3.5 Field layout of seismic reflection measurements

Many aspects to be considered in selecting seismic field configurations have already been discussed in the seismic refraction section. Especially the sampling criteria can be transferred almost directly to reflection measurements. A new aspect is the multi-fold coverage connected with reflection velocity analysis and improvement of the signal-to-noise ratio.

Spread length

In reflection seismology, seismic velocity analysis is based of measuring the curvature of reflection hyperbolas. Therefore, the spread length to be chosen depends on the depth level up to which seismic velocities are to be determined reliably. A rule of thumb is that reflection hyperbolas are well constrained if the source-receiver offsets are in the order of at least half the reflection depth. Another rule of thumb can be formulated in analogy to the velocity determination along refracted branches: The move-out of the hyperbola can be determined well if the move-out along the observed travel-time branch is in the order of one signal period T or more. Based on Eq. 2.38 this condition can be formulated as the following inequality:

$$\Delta x_{max} > V_{max} \cdot T \cdot [1 + 2 \cdot t_{max}(0)/T]^{1/2} \approx V_{max} \cdot [2 \cdot t_{max}(0) \cdot T]^{1/2} \qquad (2.47)$$

where Δx_{max} is the spread length, and V_{max} and $t_{max}(\Delta x = 0)$ are estimates of the NMO-velocity and two-way travel time for the deepest reflection under consideration. The right hand expression applies if $t_{max}(0) >> T$.

It should be emphasized that the spread length influences mainly the quality of the velocity analysis and respective reflection depth determination whereas the quality of the stacked reflection image increases simply with the number of stacked traces. Since NMO-stretching should not exceed 10% large offset seismograms even have to be excluded from stacking sometimes.

Geophone spacing

In selecting the geophone spacing Δx_G, criterion (2.16) should be strictly followed saying that $\Delta x_G < \lambda_{min}/2 = v_{min} \cdot T_{min}$ where λ_{min}, v_{min} and T_{min} are the smallest wavelength, propagation velocity and signal period to be expected during the survey. This criterion is of particular importance because the digital filters applied to suppress non-reflection arrivals will be affected by spatial aliasing if the criterion is violated. In addition, it has to be considered that the spacing Δx_{CMP} of the traces in the CMP-stacked section as well as in the migrated section is related to the geophone spacing by $\Delta x_{CMP} = \Delta x_G/2$ (Eq. 4.36). Since Δx_{CMP} determines the structural sampling of the investigation target, the geophone spacing has to be chosen according to the required resolution.

Source point spacing and advance of the receiver spread

For both stacking and velocity analysis it is advantageous to provide CMP-gathers with an almost uniform coverage of subsurface points in terms of offsets and number of traces. Therefore, the source point spacing should be as regular as possible where $\Delta x_S = \Delta x_G$ can be regarded as an optimum. For the same reasons receiver spreads should be advanced together with the shot positions.

Acquisition plan

Recommendations concerning the uniformity of coverage are sometimes difficult to follow because of logistical or budget restrictions. The consequence of irregularities in the shot-receiver configuration or reduced shot density, for example, can be analysed by considering an acquisition plan such as shown in Fig. 2.29.

2.3.6 Problems of near surface reflection seismics

The implications of near surface geological conditions with respect to the seismic velocity field have been outlined already in Sect. 2.2.6. Strong contrasts and gradient zones in seismic velocity can also cause complications regarding seismic reflection. Some major points of these are discussed below.

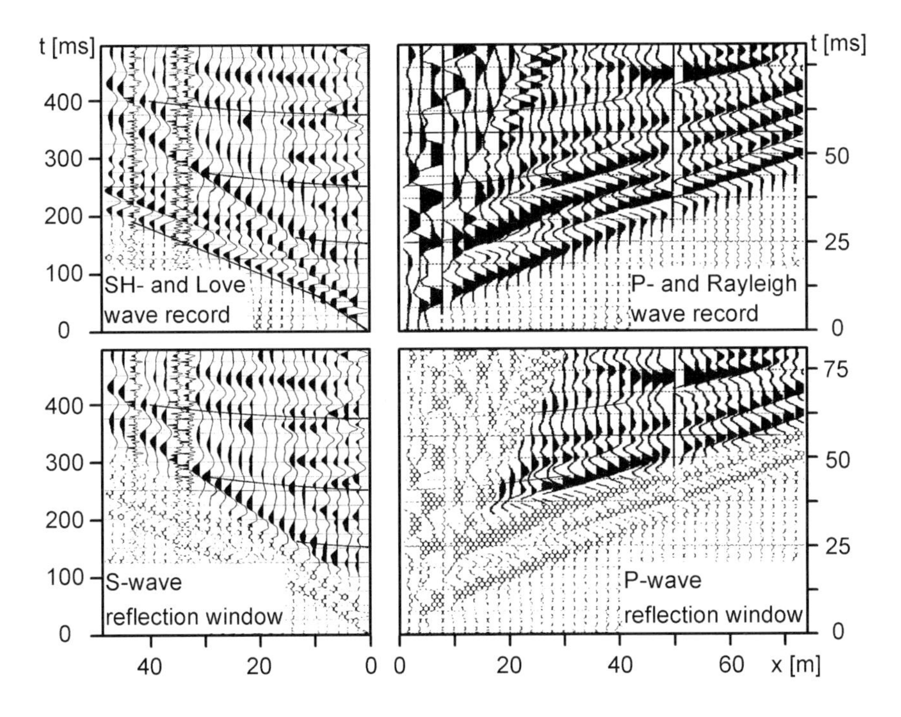

Fig. 2.33. Top: S- (left) and P- (right) seismogram sections covering the same segment of a boulder clay-sand layer sequence. Note the different time scales of P- and S-seismograms corresponding to a v_P/v_S-ratio of 6. Bottom: Reflection windows of S- and P-waves. Outside the reflection windows strong interference with refracted and surface waves occurs

Reflection window

Near the earth's surface reflections often interfere with surface waves and direct and refracted body waves (Fig. 2.33). Since the amplitudes of these wave types can be larger than reflections by orders of magnitudes the potential of digital filtering to suppress them can be very limited. In this case reflection processing and interpretation have to be restricted to space-time windows where reflected arrivals show up clearly. In P-wave surveys this window is often found in an offset range between the onsets of Rayleigh and direct waves, or between the direct wave and the first refractions. Unfortunately, these reflections are usually exposed to significant NMO-stretching. Sometimes, a certain improvements can be achieved if the NMO-correction is not carried out with respect to zero-offset but with respect to a common non-zero offset. In S-wave surveys the situation is gradually better because surface waves travel with a velocity higher or equal the one of the direct wave. So, at least in principle, S-wave reflections find a cone free of disturbances symmetrical to zero-offset.

Seismic velocity gradients

A special problem, which can affect S-waves even more than P-waves, is the occurrence of strong vertical gradients of the seismic velocity fields. Strong velocity gradients can cause even near-vertical rays to bend towards the horizontal and to propagate in a multiple garland parallel to the earth's surface. In pathological cases this process can lead to the vanishing of reflections or to reducing the observation window almost to zero-offset.

Multiple reflections

If a near surface interface shows a strong contrast in seismic impedance it creates multiply reflected arrivals travelling back and forth between the earth's surface and the interface. Corresponding reflection coefficients of 30% and more can be found at the groundwater table or at soft-to-hard rock interfaces. Multiple reflections can efficiently cover primary reflections from deeper levels and are often hard to overcome by applying digital filters.

2.3.7 Structural resolution of seismic reflection measurements

The Fresnel zone concept outlined in Sect. 2.2.8.can be applied to reflected waves, too, if respective differences in the ray paths of refractions and reflections are considered. In analogy to the refraction case, it can be stated that reflected signals from different interface segments can be observed separately from each other if the minimum distance between these segments corresponds to the radius of one Fresnel zone (see Sheriff and Geldart, 1995, and Hagedoorn and Diephuis, 2001, for a detailed treatment). Here, the Fresnel zone is defined as the envelope of all possible ray paths between source, interface and back to the geophone along which the travel time is delayed by half a signal period or less compared to the zero-offset reflection. Two cases have to be distinguished: the lateral resolution of structure along one interface and the interference of signals from two follow-up interfaces (Fig. 2.34).

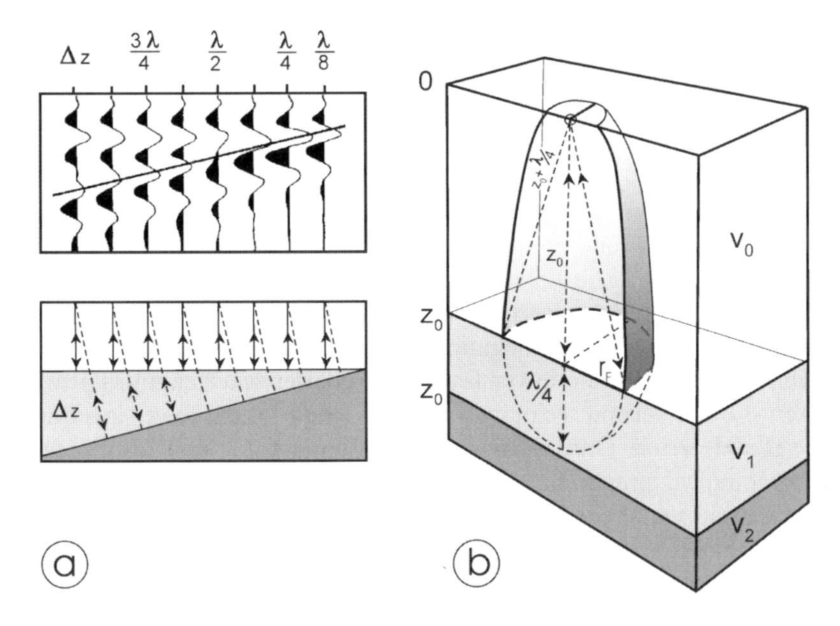

Fig. 2.34. a) Tuning effect: interference of reflections from top and bottom of a thin layer, b) Fresnel volume of a reflected ray as a measure of structural resolution of seismic reflection sections

Lateral resolution

On a plane interface the radius of the Fresnel zone r_F^{refl} of a reflected wave can be expressed by:

$$r_F^{refl} = (v/2)\cdot[T\cdot t + T^2/4]^{1/2} \approx (v/2)\cdot[T\cdot t]^{1/2} \tag{2.48}$$

where T is the signal period and v and t are the average propagation velocity and two-way travel time. The equation applies to both zero- and non-zero offsets. The approximation on the right hand side of Eq. 2.48 requires that the signal period is much shorter than the travel time ($t >> T/4$) which should be usually the case even in near surface prospecting. A comparison with Eq. 2.31a shows that the Fresnel zone of a zero-offset reflection $r_F^{refl} \approx v\cdot[T\cdot t_0/4]^{1/2}$ is smaller by a factor of 0.7 than the Fresnel zone of the corresponding refraction $r_F^{refr} \approx v\cdot[T\cdot t_0/2]^{1/2}$.

The radius of the reflection Fresnel zone can be reduced if a migration algorithm is applied to the CMP-stacked time section. The decrease of the lateral extent of the Fresnel volume during migration can be understood if seismic migration is regarded as a computational shift of the seismic acquisition plane from the earth's surface to the depth level of the reflecting interface (see e.g. Yilmaz, 2001). The theoretical minimum to which the radius of the Fresnel zone can be reduced is $v \cdot T/4$ corresponding to a quarter wavelength.

Vertical resolution

"Vertical resolution" can be regarded as the ability to separate the images of two interfaces found at subsequent depth levels or two-way reflection times, respectively. The Fresnel volume concept suggests that the deeper interface can be imaged clearly if it is placed outside the Fresnel volume of the upper reflection. In prolongation of the zero-offset ray the upper Fresnel zone extents into the lower layer by a quarter wavelength $\lambda_2/4=v_2 \cdot T/4$ where v_2, λ_2 are seismic velocity and wave length of the lower layer and T is the signal period. On the basis of this argument it is generally accepted that a geological structure can be resolved if its thickness exceeds $\lambda/4$. Lateral structural variations of a layer packet can often be recognized even if the layer thickness falls below $\lambda/4$ but it is more difficult to quantify. If a layer thins out amplitude tuning effects can occur leading to lateral increase or decrease of the interference amplitude from which the actual thickness can be reconstructed under favourable conditions (Fig. 2.34).

2.4 Further reading

Further information on the role of seismics in hydro-geological prospecting and, more generally, environmental applications can be found, for example, in the following review articles and monographs: Rubin and Hubbard (2005), Steeples (2005), Knödel et al. (1997), Kirsch and Rabbel (1997).

2.5 References

Adachi R (1954) On a proof of fundamental formula concerning refraction method of geophysical prospecting and some remarks. Kumamoto J. Sci., Ser. A, 2:18-23

Aki K, Richards PG (2002) Quantitative Seismology. University Science Books, Sausalito CA

Červený V (2001) Seismic ray theory. Cambridge University Press, Cambridge

Gebrande H, Miller H (1985) Refraktionsseismik. In: Bender F (ed) Angewandte Geowissenschaften Band II. Ferdinand Enke Verlag, Stuttgart, pp 226-260
Hagedoorn JG (1959) The plus-minus method of interpreting seismic refraction sections. Geophysical Prospecting 7:158-182
Hagedoorn JG, Diephuis G (2001) The seismic transmission volume. Geophysical Prospecting 49:697-707
Hole JA, Zelt B C (1995) 3-D finite-difference reflection traveltimes. Geophys. J. Int. 121:427-434
Hubral P, Krey Th (1980) Interval velocities from seismic reflection measurements. Society of Exploration Geophysicists, Tulsa OK
Johnson SH (1976) Interpretation of split spread refraction data in terms of dipping layers. Geophysics 41:418-24
Kearey P, Brooks M, Hill I (2002) Introduction to Geophysical Exploration (3rd ed.). Blackwell, Oxford
Kirsch R, Rabbel W (1997) Seismische Verfahren in der Umweltgeophysik. In Beblo M. (ed), Umweltgeophysik, Ernst und Sohn Verlag, pp 243-311
Knödel K, Krummel H, Lange G (eds) (1997) Geophysik. Handbuch zur Erkundung des Untergrundes von Deponien und Altlasten,vol 3. Springer, Berlin
Meissner R, Stegena L (1977) Praxis der seismischen Feldmessungen und Auswertung. Gebrüder Bornträger, Berlin
Palmer D (1980) An introduction to the generalized reciprocal method of seismic refraction interpretation. Geophysics, 46:1508-1518
Rubin Y, Hubbard S (eds) (2005) Hydrogeophysics. Springer, Berlin
Sandmeier K-J (1997) Standard-Inversionsverfahren. In Knödel K, Krummel H, Lange G (eds) (1997), Geophysik. Handbuch zur Erkundung des Untergrundes von Deponien und Altlasten, vol 3. Springer, Berlin, pp 533-545
Sharma PV (1997) Environmental and engineering geophysics. Cambridge University Press, Cambridge
Sheriff RE, Geldart LP (1995) Exploration Seismology (2nd. ed.). Cambridge University Press, Cambridge
Steeples DW (2005) Shallow seismic methods. In Rubin Y, Hubbard S (eds) Hydrogeophysics. Springer, Berlin, pp 215-251
Thornburgh HP (1930) Wave-front diagram in seismic interpretation. Bull. Amer. Ass. Petrol. Geol. 14:185-200
Vidale JE (1988) Finite-difference calculation of travel times. Bulletin of the Seismological Society of America 78:2062-2076
Yilmaz Ö (2001) Seismic Data Analysis (vol 1 and 2, 2nd ed.). Society of Exploration Geophysicists, Tulsa OK
Zelt CA, Smith RB (1992) Seismic travel-time inversion for 2-D crustal velocity structure. Geophys. J. Int. 108:16-34

3 Geoelectrical methods

Geoelectrical methods are applied to map the resistivity structure of the underground. Rock resistivity is of special interest for hydrogeological purposes: it allows, e.g., to discriminate between fresh water and salt water, between soft-rock sandy aquifers and clayey material, between hard-rock porous/fractured aquifers and low-permeable claystones and marlstones, and between water-bearing fractured rock and its solid host rock. These applications are discussed in Chaps. 12 – 17. In this chapter, the basic principles, the field techniques, and the data evaluation and interpretation procedures of resistivity measurements are shown.

3.1 Basic principles

Kord Ernstson, Reinhard Kirsch

Resistivity of the ground is measured by injected currents and the resulting potential differences at the surface. The general field layout is sketched in Fig. 3.1. Two pairs of electrodes are required: electrodes A and B are used for current injections, while electrodes M and N are for potential difference measurements.

For a homogeneous ground and an arbitrary electrode arrangement (Fig. 3.1A) the resistivity ρ (unit: Ohm*meter, Ωm) as the relevant petrophysical parameter can be calculated from the current I and the potential difference U by

$$\rho_A = K \cdot \frac{U}{I} \tag{3.1}$$

K is called geometric factor (unit: meter) and can be calculated from the electrode spacing by

$$K = \frac{1}{2\pi} \cdot \left[\left(\frac{1}{AM} - \frac{1}{BM} \right) - \left(\frac{1}{AN} - \frac{1}{BN} \right) \right] \tag{3.2}$$

Current flow-lines and equipotentials for a homogeneous underground are sketched in Fig. 3.2.

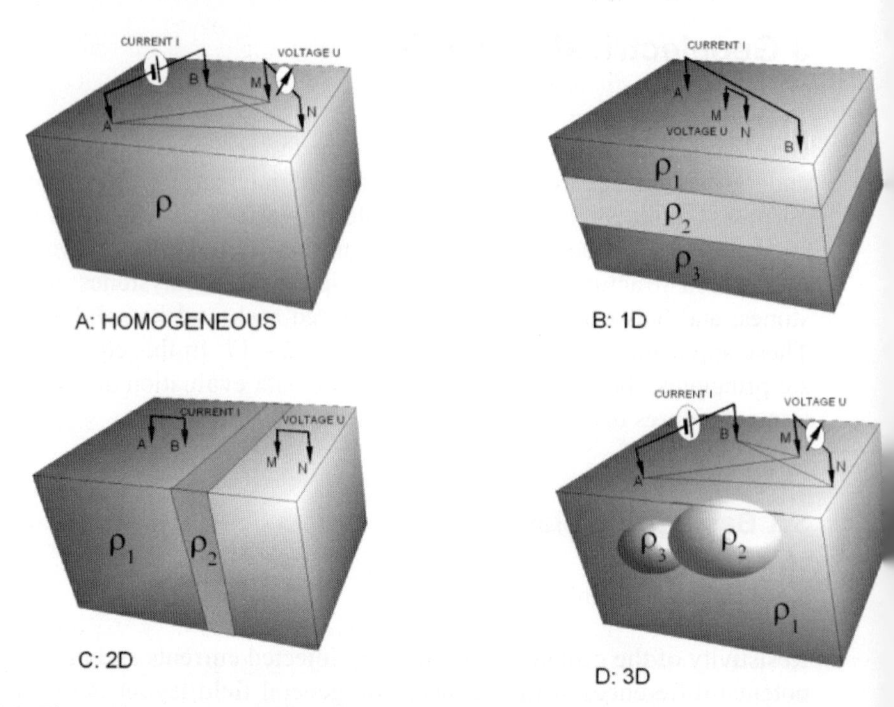

Fig. 3.1. Electrode arrangement for apparent resistivity measurements, A: homogeneous ground, B: layered ground, C: 2D resistivity distribution in the ground, D: 3D resistivity distribution in the ground

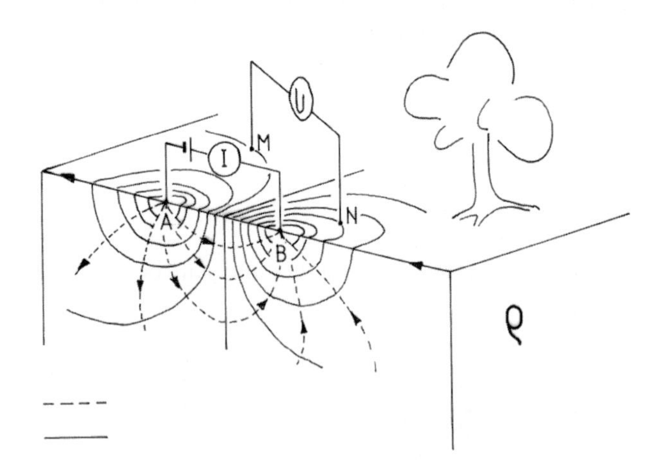

Fig 3.2. Current flow lines and equipotentials in a homogeneous ground

A homogeneous ground is in general not verified with real geological conditions, which leads to the basic concept of *apparent resistivity* formally measured by

$$\rho_A = K \cdot \frac{U}{I}$$

over inhomogeneous ground (Figs. 3.1B – 3.1D). From Figs. 3.1B –3.1D it is obvious that the resistivities and the shapes of the different geological units contribute to ρ_A. Different from the homogeneous ground (Fig. 3.1A) and the true resistivity ρ, the apparent resistivity ρ_A depends also on the location of the electrodes with respect to the geological units. Resistivity soundings are done by selecting an appropriate electrode configuration (see, e.g., Figs. 3.1B – 3.1D), by systematically changing and/or moving its configuration, and by sampling the related apparent resistivities ρ_A. This data set is computer-processed with the aim to get the underground true resistivity distribution, which has to be interpreted in terms of geological structures.

In general, a distinction is made whether one has to do with a horizontally layered earth (commonly in sedimentary rocks), with elongated, so-called two-dimensional geological structures (e.g., dikes, fracture zones), or with arbitrarily shaped structures (e.g., lenses, karst caves). Accordingly, the terms 1D (vertical electrical sounding, VES), 2D (electrical imaging) and 3D (electrical mapping, horizontal electrical sounding, HES, resistivity tomography) geoelectrics are frequently used.

3.2 Vertical electrical soundings (VES)

Vertical (1D) electrical soundings are applied to a horizontally or approximately horizontally layered earth. Geological targets may be, e.g., sedimentary rocks of different lithologies, layered aquifers of different properties, sedimentary rocks overlying igneous rocks, or the weathering zone of igneous rocks. In the most favorable case, the number of layers, their thicknesses and resistivities are the outcome of a VES survey.

The basic idea of resolving the vertical resistivity layering is to stepwise increase the current-injecting electrodes AB spacing, which leads to an increasing penetration of the current lines and in this way to an increasing influence of the deep-seated layers on the apparent resistivity ρ_A (Fig. 3.3). The step-wise measured apparent resistivities are plotted against the current electrode spacing in a log/log scale and interpolated to a continuous

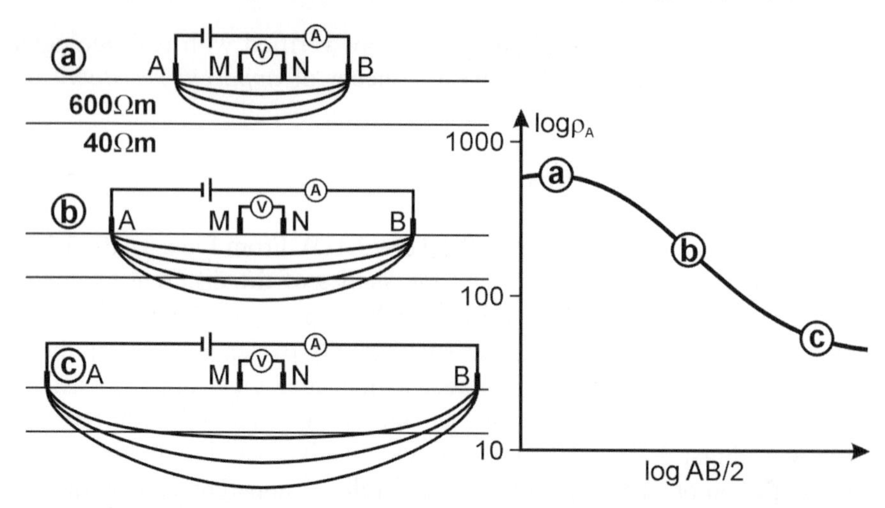

Fig. 3.3. Apparent resistivity measurements with increased current electrode spacing leading to increased penetration depths of the injected current. Results are compiled in the sounding curve

curve. This plot is called sounding curve, that is the base of all data inversion to obtain the resistivity/depth structure of the ground.

In general, linear electrode configurations are used for resistivity measurements. Common configurations are the Schlumberger, Wenner, and Dipole-Dipole spreads (Fig 3.4). Because of practical and methodical advantages, vertical electrical soundings mostly use the symmetrical Schlumberger configuration where the voltage electrodes M, N are closely spaced and fixed to the center of the array and the current electrodes A, B move outwards. The geometrical factor is

$$K_{\text{SCHLUMBERGER}} = \frac{\pi}{MN} \cdot \left(\frac{AB}{2} \right)^2 \tag{3.3}$$

Frequently, the Half-Schlumberger (or pole-dipole) configuration (Fig. 3.4) proves advantageous when topographic conditions prevent a full Schlumberger spread. Theoretically, in the pole-dipole array the current electrode B is placed to infinity with zero contribution to the potential electrodes. Then, the geometrical factor is

$$K_{\text{Halfschlumberger}} = \frac{2\pi}{MN} \cdot \left(\frac{A0}{2} \right)^2 \tag{3.4}$$

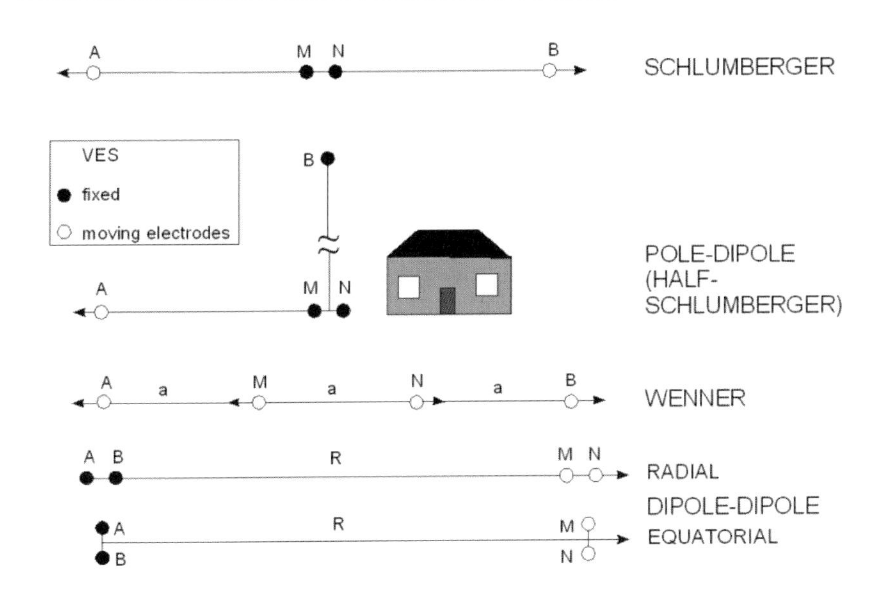

Fig. 3.4. Electrode configurations: Schlumberger, Half-Schlumberger, Wenner, and dipole-dipole spread

In a Wenner configuration VES, all four electrodes have to be moved on increasing the spread. The electrode spacing AM = MN = NB = a remain identical, and the geometrical factor is thus computed to be

$$K_{WENNER} = 2 \cdot \pi \cdot a \tag{3.5}$$

In the dipole-dipole configurations (Fig. 3.5), electrodes A, B and M, N are in each case closely spaced to form a current dipole and a potential dipole. A VES is in general done by the current dipole fixed to the ground and moving potential dipole. The apparent resistivity is plotted in a log/log scale as a function of dipole separation R. In principle, the dipoles may have arbitrary azimuthal orientation, but the radial and equatorial arrays are most common. The geometric factor must be calculated from Eq. (3.2). Although in the dipole arrays the current electrodes remain fixed, a depth penetration is likewise achieved: As can be seen from Fig. 3.2, on increasing separation, the potential dipole will increasingly be influenced by equipotential lines of larger depth extent.

3.2.1 Field equipment

Although in principle resistivity measurements are simple and reduced to current, voltage and electrode-spacing records, highly sophisticated electronics withstanding hard field conditions is required. Although VES and its theoretical background are based on direct current propagation, alternating rectangular DC pulses or even low-frequency AC are used. Depth of investigation and surface resistivity conditions are important for selecting an appropriate instrumentation. A broad range including equipments of very high power as well as small, hand-held instruments are available on the market. Increasingly, VES are performed in connection with or as part of Induced Polarization (IP) which is addressed also by instrument manufacturers. Most versatile are geoelectric equipments allowing the complete frequency spectrum to be applied to resistivity and IP soundings (Chap. 4). Standard of modern resistivity instrumentation is signal stacking as well as analogue-digital conversion of data and their storage in a memory.

Field equipment is completed by cable wheels and electrodes for current injection and voltage record. Good insulation of the cables to prevent leakage and to survive in rugged terrain is required. Stainless steel rods are normally used for the current electrodes and may be appropriate also for the potential electrodes if AC is applied. For highly accurate DC or DC pulse measurements, unpolarizable electrodes as used in self-potential surveys are strongly recommended.

3.2.2 Field measurements

Planning of resistivity measurements for solving a hydrogeological problem requires both the so far known geological setting and the terrain conditions. An assessment of the required length of electrode spreads can be made by model calculations (next section), if some information about the resistivity/depth structure exist. AB ~ 5*DI (DI = depth of investigation) for Wenner and Schlumberger arrays and 2R ~ 5*DI for dipole-dipole arrays is a rough estimate. Basically, a resistivity survey comprises a set of depth soundings which depending on field conditions and the survey problem may be arranged on profiles or in a grid. The distance between sounding locations also depends on the survey problem and should be chosen in the range of some ten meters rather than in the range of some hundred meters.

To avoid topographic effects affecting the sounding curve, the direction of the electrode spread should be chosen to enable a more or less horizontal electrode line. Therefore, in hilly or generally difficult terrain, a Schlumberger half spread (pole-dipole array) may be preferable.

Interpretation of sounding curves is done under the assumption of horizontal layering. For dipping layers, the direction of the electrode spread should follow the general strike to minimize disturbing effects.

In a VES, the increasing spacing AB/2 (Schlumberger), a (Wenner) and R (dipole) should be chosen as equidistant in a logarithmic scale and therefore also in the field. Correspondingly, in the range from 1 m to 10 m there are as many apparent resistivity measurements (preferably not less than 10) as in the range from 10 m to 100 m.

Increasing AB/2 leads to rapidly reduced potential difference to be measured at electrodes M,N. This can be compensated by increased voltage for current injection or increased number of signal stacks. If necessary, MN distance can also be increased to get a better signal. When MN is increased, e.g., from 1 m to 5 m, a static shift in the sounding curve due to near-surface resistivity inhomogeneities is frequently observed. This shift can be interpolated by overlapping measurements with both MN spacing (Fig. 3.5).

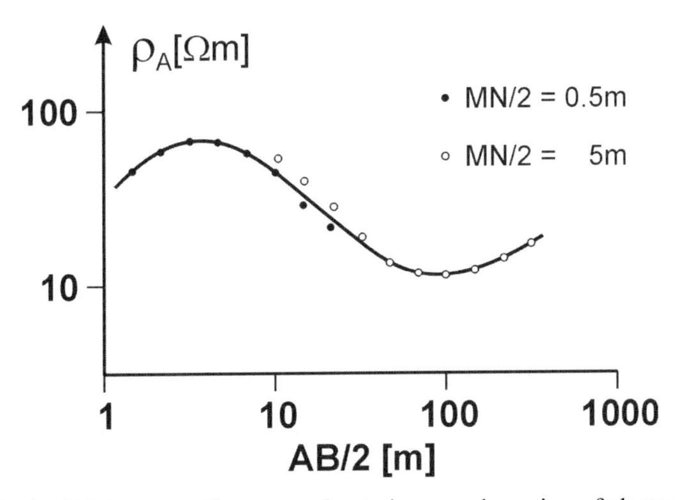

Fig. 3.5. Static shift in a sounding curve due to increased spacing of electrodes M and N

Sounding curves over horizontally layered ground are generally smooth. In case the operator observes strong scattering of the measured apparent resistivity values and even clear shifts along the sounding curve, strong deviations of the horizontal layering, man-made conductivity anomalies (power lines, pipes etc) or instrumental failure are evident. Geological causes like traversing a fault in the course of the electrode spread may be responsible which can advantageously be used to establish the existence of tectonic structures. Nearby metallic power lines and pipes may considerably disturb a VES not allowing a reliable processing of the data. In the case of a suspected metallic pipe, a short profile with a VLF receiver will rapidly locate it. Instrumental causes for bad data sampling may be current leakage from electrode cables, electromagnetic coupling between current and potential cable layouts, electromagnetic coupling with the ground, and simply a faulty resistivity meter. Strong electromagnetic coupling may occur even at very low frequencies in an AC equipment, which requires a very careful cable layout. And on suspicion of a defective instrument, a VES at a test site may be advisable. Electric noise from industry and power lines may be overcome by sufficient signal stacking, but in unfavorable cases it may prevent any survey.

3.2.3 Sounding curve processing

For a horizontally layered earth model with defined resistivities and thicknesses of the layers, apparent resistivity as a function of electrode AB spacing can be calculated. The theoretical background is discussed in detail by Mundry and Dennert (1980). In practice, the reverse process has to be done: to deduce the number of layers, their resistivities and thicknesses from a measured sounding curve. Different from theory, the solution is not unambiguous within the frame of data scattering.

Basically, the inversion proceeds from the comparison of the measured sounding curve with calculated sounding curves for given model layering. Originally, formalized master curves and graphic procedures were used. Now, a variety of PC-based programs is available, which allow a rapid computation for an arbitrary number of layers. Purely automatic inversion programs without any assumptions of the layering exist as well as programs that require the handling of the interpreter. In the latter case, a starting model is defined incorporating known geological parameters (e.g., standard thicknesses of stratigraphical units, depth to the groundwater table), available borehole data, or the results of other geophysical measurements. Based on the starting model, the computer conducts an iteration process by trying to adjust the theoretical model and its sounding curve to the measured curve, which can be controlled on the computer's monitor. The convergence may interactively be speeded up by, e.g., introducing additional layers. A "best fit" (Fig. 3.6) to stop the iteration may be defined by the computer calculating a root mean square (RMS) error or, preferentially, by the interpreter. In both cases, the interpreter should be aware that the obtained solution is only one out of a couple of physically equivalent models and that he has to consider the complete range of equivalence with respect to geological reliability. Equivalence of physical solutions is a basic principle in geophysics, and especially with vertical electrical soundings, the principle may turn out to be a problem, which is discussed in the following section.

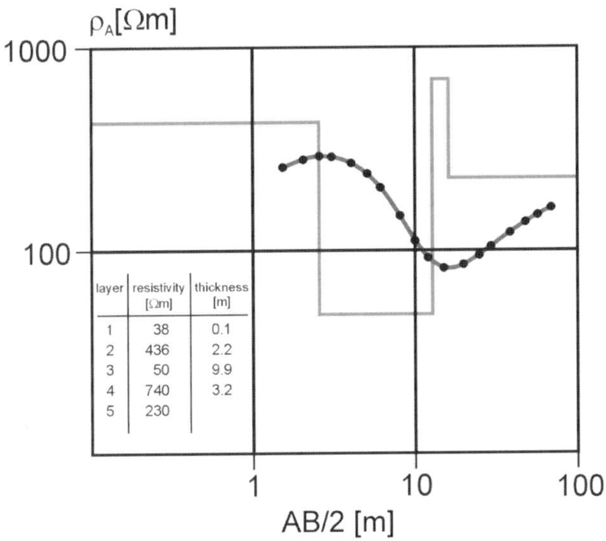

Fig. 3.6. Modeling of a sounding curve: measured apparent resistivities (dots), best fit layer model (tabulated and as a graph in the logarithmic diagram) and related model curve (full line) are shown. For theoretical reasons the thickness of the lowermost layer is always assumed to be infinite

For the planning of field surveys, the interpretation programs can be used to calculate sounding curves for interpreter-defined underground models. If an assumption on the expected resistivity conditions in the project area can be made, this forward calculation of sounding curves allows an assessment of the length of electrode spread required to get information about the target horizon, or, for thinner horizons, can show if they are really resolved in the sounding curve.

3.2.4 Ambiguities of sounding curve interpretation

In VES, sounding curve modeling and interpretation is inextricably linked with the *principle of equivalence* The principle expresses that a measured sounding curve is basically related with many physically equivalent models that may differ considerably.

A simple example is shown in Fig. 3.7. The sounding curve matches two four-layer models with alternating high and low resistivities. It is evident that for the second, high-resistivity layer a doubling of the resistivity corresponds approximately with a halving of the thickness. For the third, low-resistivity layer, a halving of the resistivity can be compensated by halving of the thickness. According to this rule of equivalence, the maximum of a sounding curve can be modeled by the product of $T = m*\rho$ only and the minimum of a sounding curve by the ratio $S = m/\rho$ only. T (unit ohm*m^2) and S (unit S or mho) are termed transverse resistance and longitudinal conductivity, respectively. Consequently, the thickness of a layer cannot unambiguously be determined from VES without knowing its resistivity, and vice versa. Accordingly, the top of the 600 Ωm-layer (say the top of an aquifer) in the example of Fig. 3.7 may exist at depths of 31 m (model I) or 17 m (model II) depending on the resistivities of the overlying layers.

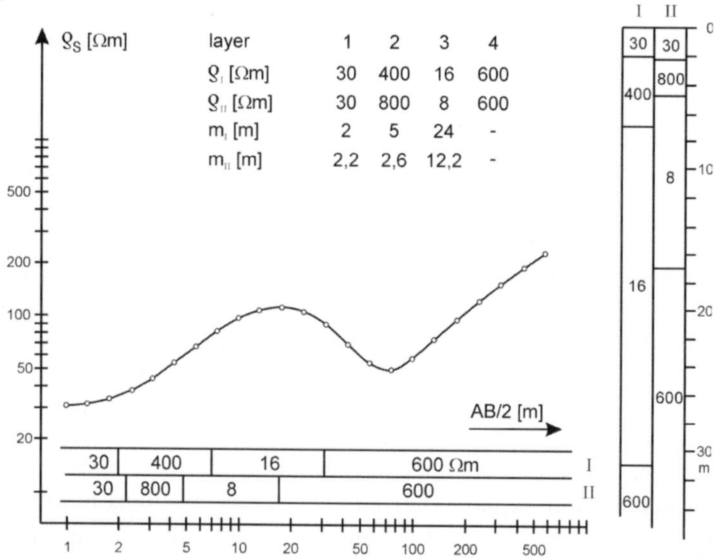

Fig. 3.7. Equivalence in VES (see text)

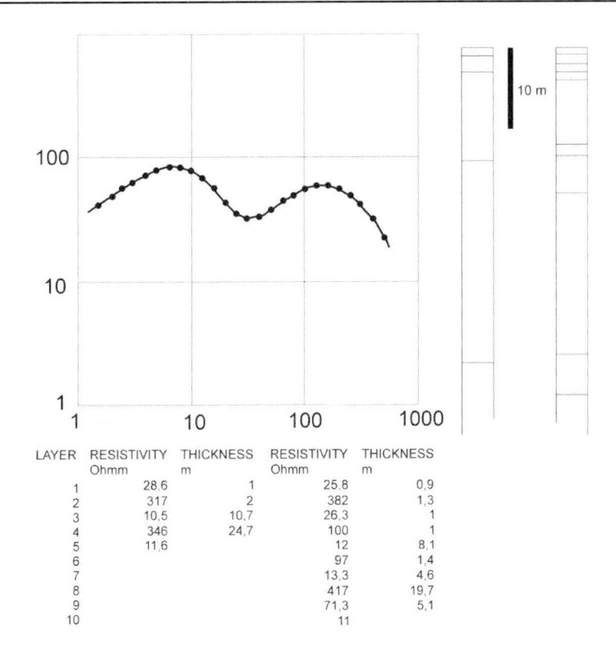

LAYER	RESISTIVITY Ohmm	THICKNESS m	RESISTIVITY Ohmm	THICKNESS m
1	28,6	1	25,8	0,9
2	317	2	382	1,3
3	10,5	10,7	26,3	1
4	346	24,7	100	1
5	11,6		12	8,1
6			97	1,4
7			13,3	4,6
8			417	19,7
9			71,3	5,1
10			11	

Fig. 3.8. Equivalence of a five-layer and a ten-layer underground

Equivalence is not restricted to this easily understandable situation. The sounding curve in Fig. 3.8 can be modeled by a five-layer as well as by a ten-layer earth without, from the physical point of view, any preference.

From Fig. 3.8, another important property of vertical electrical sounding curves is evident. By all appearances, only the minimum number of layers can be deduced implying that thin layers are preferentially suppressed. As shown by the logarithmic scale of Fig. 3.8, this effect rapidly increases with increasing sounding depth. For the five-layer case, the comparable amplitudes of both sounding curve maxima are attributed to a thickness of 2 m (layer 2) at one meter depth but to about 25 m (layer 4) at about 14 m depth. A rule of thumb indicates that a layer becomes clearly visible in a sounding curve if its thickness is comparable to its depth of deposition. The decreasing resolution with increasing depth is a basic drawback of the VES method since at the same time the problem of macroanisotropy as a special case of equivalence increases.

Electrical anisotropy describes the effect that generally the resistivity of rocks is smaller parallel to bedding or schistosity compared with the resistivity transverse to these textures. Correspondingly, a longitudinal resistivity ρ_L and a transverse resistivity ρ_T are defined. For electrical soundings, current flow-lines are curved, so that transverse as well as longitudinal resistivities are encountered. From theory, modeling of an anisotropic bed (layer i) with thickness h_i in a sounding curve supplies a substitute resistivity ρ^*_i and substitute thickness h^*_i:

$$\rho_i^* = \sqrt{\rho_T \cdot \rho_L} \tag{3.6}$$

$$h_i^* = h_i \cdot \sqrt{\frac{\rho_T}{\rho_L}} \tag{3.7}$$

While for the substitute resistivity $\rho_L < \rho^*_i < \rho_T$, the substitute thickness h^*_i is larger than the true thickness h_i by a multiplier of $\lambda=(\rho_T/\rho_L)^{1/2}$, λ being the coefficient of anisotropy (Fig. 3.9).

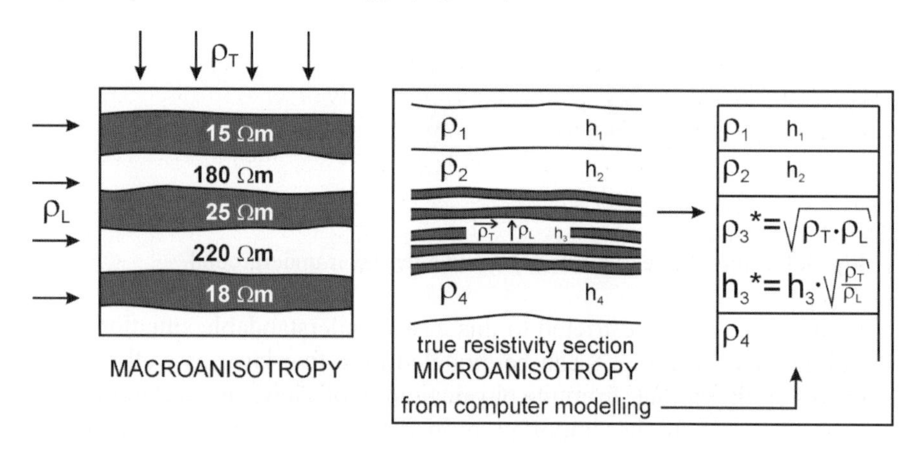

Fig. 3.9. Electrical anisotropy on different scales: a macroanisotropic resistivity section with average transverse and longitudinal resistivities ρ_T and ρ_L (left) and a single microanisotropic layer with transverse and longitudinal resistivities ρ_T and ρ_L (right). Interpretation of the sounding curve leads to higher thickness of the microanisotropic layer (h_3^*) than in reality

Geoelectric anisotropy is a matter of scale and of the resolution of a vertical resistivity section. While graphite, a slate or a lake sediment may be termed microanisotropic, a sequence of well-defined electrically isotropic beds can behave macroanisotropic if they are not resolved in a geoelectric sounding curve (see, e.g., the ten-layer case in Fig. 3.8).

For a macroanisotropic section, average longitudinal and average transverse resistivities can be calculated from the parameters of the single beds (Fig. 3.9):

$$\rho_L = \Sigma h_i^* / \Sigma h_i / \rho_i \tag{3.8}$$

$$\rho_T = \Sigma h_i^* \rho_i / \Sigma h_i \tag{3.9}$$

Substitute resistivity and thickness as well as coefficient of anisotropy are the same as for microanisotropic behavior (3.6, 3.7).

Equivalence between true resistivity sections and sections with substitute layers due to anisotropy may cause serious modeling and interpretation errors if the anisotropy (both macro and micro) is not recognized. As can be calculated from Eqs. 3.6 - 3.9, intermittently occurring high- and low-resistivity beds may lead to large coefficients of anisotropy and, correspondingly, to large errors in modeled thicknesses.

Anisotropy also leads to discrepancies between results of vertical electrical soundings and electromagnetic induction measurements (horizontal current flow-lines). Comparing VES date with data from resistivity borehole logging (mostly horizontal flow lines), anisotropy must also be taken into consideration.

3.2.5 Geological and hydrogeological interpretation

The discussion of the principle of equivalence shows that singular depth soundings are in general little meaningful. Likewise, the sometimes used term "electrical drilling" should basically be avoided, because VES is not intended to and cannot replace boreholes but is methodically a different complex. VES interpretation comprises the more or less synchronous handling of measured sounding curves in the survey area and their modeling results. Continuity of layers in the area should be checked as well as the reality of obvious breaks in the geologic layering.

With regard to equivalence, reinterpretation of some soundings can be necessary, and additional field measurements may be helpful. In areas of young Cenozoic unconsolidated deposits (molasse, glacial sediments) with rapidly changing thicknesses (Fig. 3.10A), data of a key borehole may be required to fix modeling parameters and thus to get absolute depths independent of equivalence. In hard sedimentary rocks where stratigraphic standard thicknesses and rock resistivities are frequently well known and constant over large areas, VES modeling and interpretation may be easier leading to a detailed knowledge of the tectonics in many cases (Fig. 3.10B). As the final result, a resistivity model of the project area which is geologically and hydrogeologically reasonable and without discrepancies with drilling or other geophysical results should be obtained.

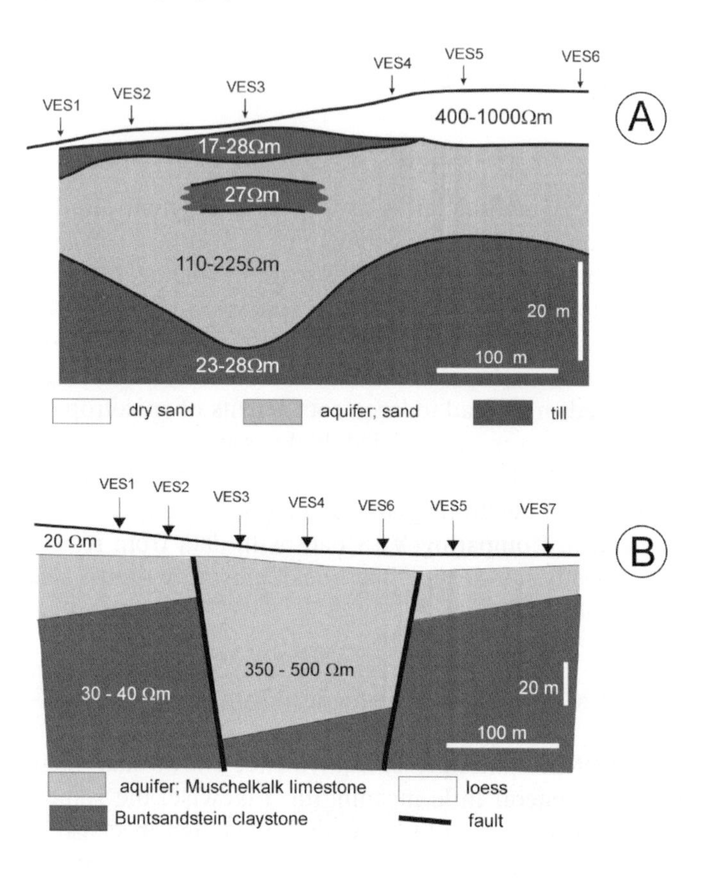

Fig. 3.10. Resistivity depth profiles from vertical electrical soundings. A: Quaternary sandy aquifer partly covered with till, B: tectonic graben as a fractured and partially karstified limestone

3.3 Resistivity mapping

Targets of resistivity mapping (or profiling) are near surface resistivity anomalies, caused by, e.g., fracture zones, cavities or waste deposits. Any common electrode configuration (e.g., Wenner or Dipole-Dipole) can be used for mapping purposes. In general, the chosen four-point configuration is kept constant and moved along profiles, while apparent resistivity is recorded (Fig. 3.11). Prior to the field works, optimum electrode spacing of the configuration can be determined by model calculations, if assumptions on resistivity and depth of the target and on resistivity of the surrounding material are possible.

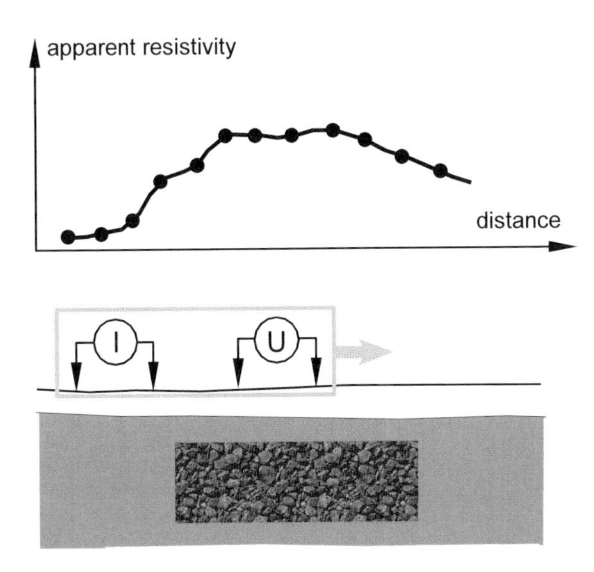

Fig. 3.11. Resistivity mapping with a dipole-dipole configuration

Another common array is the gradient array (Fig. 3.12). Here electrodes A and B are fixed and only electrodes M and N are moved, and a rectangular area between the electrodes is mapped. The apparent resistivities are calculated from (3.1, 3.2) and plotted as a map of isoohms (Fig. 3.13). Instead of point electrodes line electrodes may be used for current injection (e.g. grounded cables or a number of lined-up connected steel rods).

Although the mapping response of an arbitrary resistivity distribution can be calculated, interpretation is in general done qualitatively by locating structures of interest and outlining their extension and strike. Nevertheless, a study of resistivity mapping model curves (see, e.g., Keller & Frischknecht 1970, Schulz 1985) may be very useful to learn that even simple geometries may produce complex apparent-resistivity profiles and that anomalies may be quite different when measured with different electrode configurations.

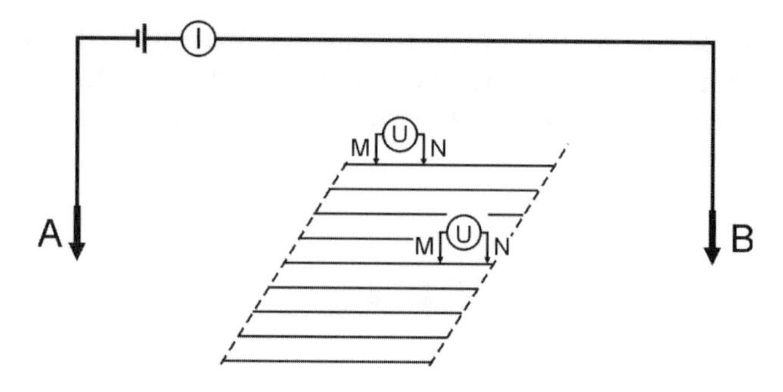

Fig. 3.12. Gradient array for resistivity mapping

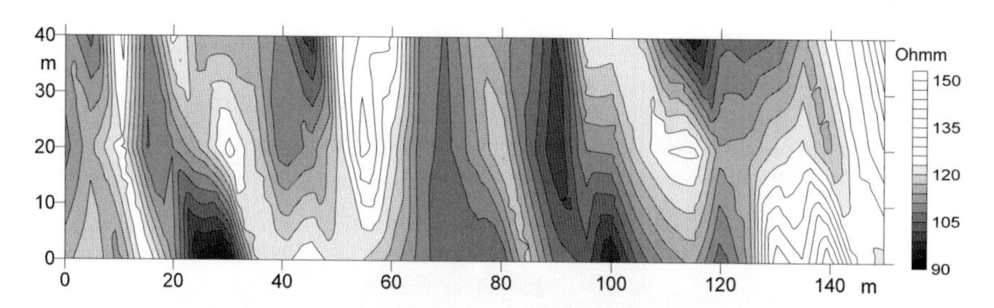

Fig. 3.13. Apparent resistivities over fracture zones in limestone mapped by gradient array

3.3.1 Square array configuration

The square array configuration is especially designed for the mapping of resistivity anisotropy, caused by e.g. fracture zones. Fracture zones may behave electrically anisotropic, because the resistivity parallel to strike is in general lower than perpendicular. The electrodes are arranged to form a square (Fig. 3.14) whose side length is a, and the apparent resistivity assigned to the midpoint is computed from

$$\rho_A = K \cdot \frac{U}{I} \tag{3.10}$$

with the geometric factor of the square array defined by

$$K = \frac{2\pi a}{2-\sqrt{2}} \tag{3.11}$$

At each location, the square is rotated by 45°, and four apparent resistivity values ρ_{A1} ... ρ_{A4} are measured (Fig. 3.14). They depend on the resistivity anisotropy and on the strike of the fracture zone which can be deduced graphically or analytically (Lane et al. 1995, Habberjam 1975). Using anisotropy as defined by Habberjam (1975), secondary porosity (porosity of the fractures) can also be estimated from the measured apparent resistivities (Taylor 1984).

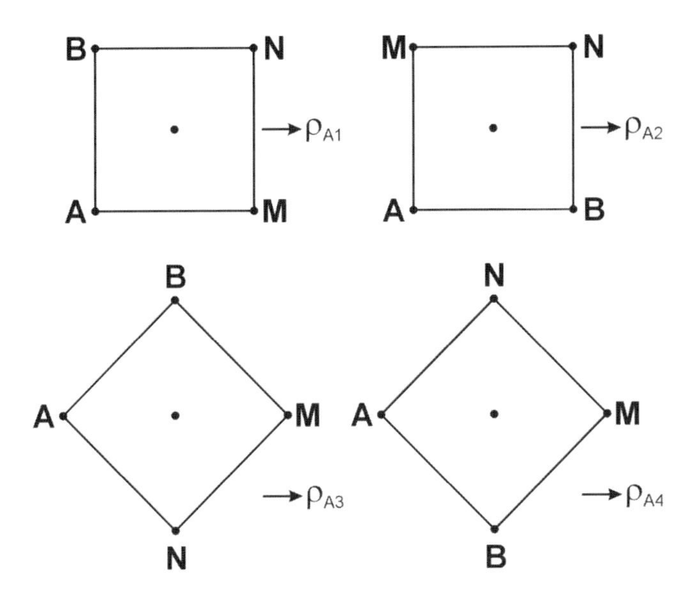

Fig. 3.14. Field layout of square array configuration

For mapping purposes, square arrays are positioned to form a continuous row (Fig. 3.15). Side length a of the squares depends on the penetration required, a typical value is a=5m.

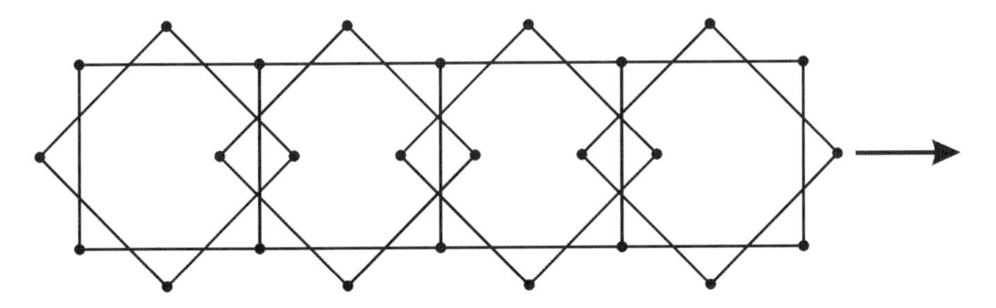

Fig. 3.15. Square array mapping

For the assessment of secondary porosity the following quantities are calculated from the measured apparent resistivities:

$$A = \left[(\rho_{A3} + 3\rho_{A1})/2 + (\rho_{A4} + \rho_{A2})/(2)^{1/2}\right] \cdot \left[2 + (2)^{1/2}\right] \tag{3.12}$$

$$B = \left[(\rho_{A1} + 3\rho_{A3})/2 + (\rho_{A2} + \rho_{A4})/(2)^{1/2}\right] \cdot \left[2 + (2)^{1/2}\right]$$

$$C = \left[(\rho_{A4} + 3\rho_{A2})/2 + (\rho_{A1} + \rho_{A3})/(2)^{1/2}\right] \cdot \left[2 + (2)^{1/2}\right]$$

$$D = \left[(\rho_{A2} + 3\rho_{A4})/2 + (\rho_{A3} + \rho_{A1})/(2)^{1/2}\right] \cdot \left[2 + (2)^{1/2}\right]$$

$$T = A^{-2} + B^{-2} + C^{-2} + D^{-2}$$

$$S = 2\left[(A^{-2} - B^{-2})^2 + (D^{-2} - C^{-2})^2\right]^{1/2}$$

$$N = \left[(T + S)/(T - S)\right]^{1/2}$$

Secondary porosity Φ can then be calculated after

$$\phi = \frac{3.41 \cdot 10^4 (N-1)(N^2-1)}{N^2 C(\rho_{max} - \rho_{min})} \tag{3.13}$$

with: C conductivity of groundwater (microsiemens/cm)
ρ_{max}, ρ_{min} maximum and minimum apparent resistivity.

However, it should be kept in mind that only a rough estimate of fracture pore space can be obtained. Fracture fillings of clay or clay/water can lead to incorrect results.

3.3.2 Mobile electrode arrays

Mobile electrode arrays are useful for mapping of large areas, and in some circumstances they may be an alternative to electromagnetic soundings with moving transmitter - moving receiver arrays. A number of ingenious instruments has been developed, e.g., with electrodes mounted on wheels or pushed into the ground pneumatically. Two instrumentations especially designed for groundwater exploration and brownfield mappings are presented here: the pulled array by the Hydrogeophysics Group, University of Aarhus, and the OhmMapper by Geometrics Inc.

Pulled array configurations

To cover large areas with spatially dense measurements, a towed electrode array was developed by Hydrogeophysics Group, Aarhus University. Originally designed for resistivity mapping (pulled array continuous electrical profiling, PACEP) in Wenner configuration with electrode spacing of 10m, 20m, and 30m, the electrode array has been extended to allow resistivity soundings with 8 electrode spacing ranging from 2m to 30m in Wenner and dipole-dipole configuration (pulled array continuous electrical sounding, PACES). Electrodes are steel cylinders coupled by weight to the ground.

Measurements are typically made at 1m intervals along the survey lines, with the distance between lines being 50-300m. With the PACEP method, two people can complete 10 to 15km of profile in one day.

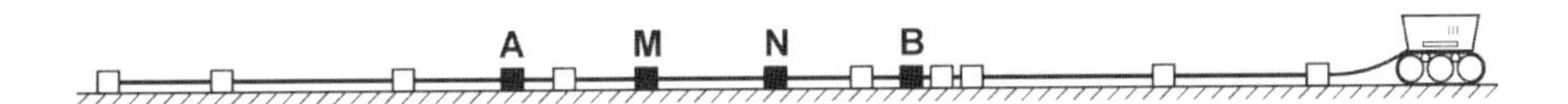

Fig. 3.16. PACES electrode configuration, activated electrodes for Wenner measurements are in black

OhmMapper

OhmMapper is also a towed instrument, but without galvanic ground coupling. The basic principle is shown in Fig. 3.17. Two conducting plates on the ground are electrically charged like a capacitor by AC current. In the ground, the opposite charge occur leading to balancing currents. As in classical resistivity methods, the resulting potential difference is measured. For practical use, the capacitors for energizing the ground are replaced by electrified cables acting as transmitter dipole. The instrument is operated in dipole-dipole configuration, penetration depth depends on dipole spacing. Instrument frequency is 16.5 Hz.

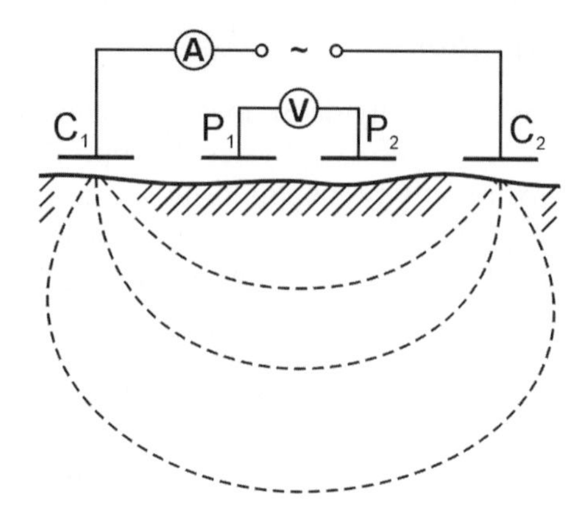

Fig. 3.17. Basic principle of capacitively coupled OhmMapper system

3.3.3 Mise-à-la-masse method

The mise-à-la-masse method originally used as a geophysical technique in mining only, has enjoyed some revival in environmental geophysics and hydrogeology. It is a kind of cross between resistivity and self-potential measurements. The technique uses a four-point electrode configuration as shown in Fig. 3.18. One current electrode (B) and one potential electrode (N) are (theoretically) set to infinity, and the potential electrode M serves to measure the potential field related with the current that is injected at electrode A. In an electrically homogeneous ground, the equipotentials are hemispherical (Fig. 3.18A). If, however, a buried conductive body is contacted by the current electrode A (Fig. 3.18B), then the potential distribution at the surface will reflect the shape of the body and its spatial position which is the objective of a mise-à-la-masse survey.

Different from other geophysical methods, the target must be known at one point at least in order to contact it, e.g., by a borehole. While in mining the electrically charged body is an ore body, groundwater is normally mise-à-la-masse-contacted for hydrogeological and environmental geophysical purposes. Mise-à-la-masse is used to track groundwater flow paths, to evaluate contaminant plumes near waste disposal sites, and to delineate the migration of conductive tracers (Nimmer & Osiensky 2002).

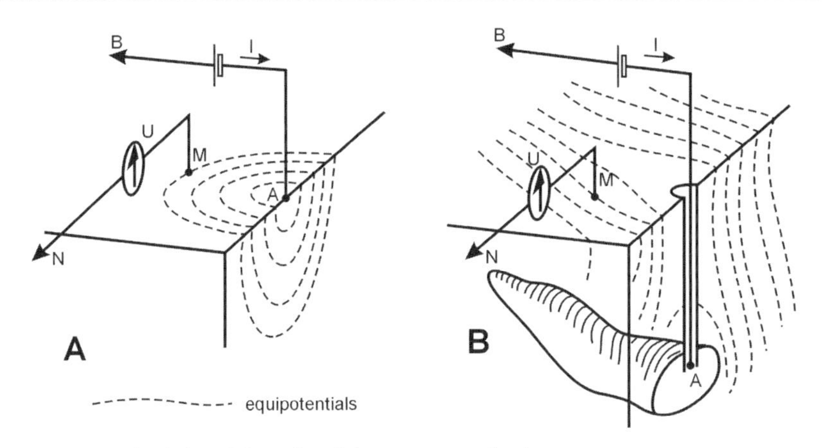

Fig. 3.18. Principle of the mise-à-la-masse method

3.4 Self- potential measurements

Self-potential (or spontaneous potential, SP) measurements belong to the earliest methods used in applied geophysics. Originally applied in mining to ore body exploration, SP became a standard tool with borehole logging and is now of increasing interest in environmental geophysics, geothermal application, and hydrogeology.

Self potentials describe natural electrical direct currents originating from various electrochemical, electrophysical, and bioelectrical processes in the ground. Reduction and oxidation processes above and below the ground-water table define mineralization potentials of some 100 millivolts related with highly conductive ore bodies or graphite deposits. Electrochemical potentials in the order of ten millivolts result from ion flow in connection with variable electrolytic concentrations of the ground water and with clay mineral membrane effects. Heat flow may produce SP anomalies, and hydrogeologically most relevant are streaming (or electrokinetic or filtration) potentials from ground-water flow in porous rocks.

3.4.1 Basic principles of streaming potential measurements

In porous rocks, the contact between rock matrix and pore fluid is characterized by an electric double layer (Helmholtz double layer). Electrically charged, this double layer usually fixes pore fluid anions while cations remain mobile. On water flow, the cations are transported, synonymous with an electric current and the setting-up of a positive potential in the direction of the flow.

For a capillary, the amplitude E_K of the streaming potential is given by

$$E_K = \frac{\zeta \cdot \Delta P \cdot \rho \cdot \varepsilon}{4\pi\eta} \quad (3.14)$$

where ξ is the potential of the double layer (zeta potential), and ρ, ε, μ are the resistivity, dielectric constant and dynamic viscosity of the pore fluid. ΔP is the hydraulic pressure drop across the capillary which may be replaced by hydraulic conductivity and flow velocity. Experience shows that streaming potentials can reach amplitudes of the order of several 100 millivolts.

Although the relation of (3.14) has widely been confirmed experimentally (e.g., Bogoslowsky & Ogilvy 1972, Maineult et al. 2004), it is of only limited importance for quantitative considerations, because the zeta potential of rocks is a poorly defined property and water movements in rocks are far from being like in simple capillaries. However, the similarity between electrokinetic and hydraulic potentials is the base of many attempts to describe hydraulic conditions from self-potentials (e.g., Rizzo et al. 2004), and the dependence of E_K of resistivity and hydraulic conductivity has initiated SP measurements also for the study of groundwater contamination (e.g., Buselli & Lu 2001, Baker & Cull 2004, Vichabian et al. 1999).

3.4.2 Field procedures

Technically, the measurement of self-potentials is simple. The voltage between two unpolarizable electrodes, a fixed base station and a moving probe, is measured by a high-impedance voltmeter and a cable layout as shown in Fig. 3.19A. SP stations are arranged in a grid or - in the case of two-dimensional geologic structures - on profiles perpendicular to the strike. Station spacing depend on the project conditions and may be of the order of meters or tens of meters. In large survey areas, a gradient layout may preferentially be used (Fig. 3.19B). Modern equipments use multichannel digital voltmeters or large PC-controlled electrode arrays which allow also a monitoring of SP fields.

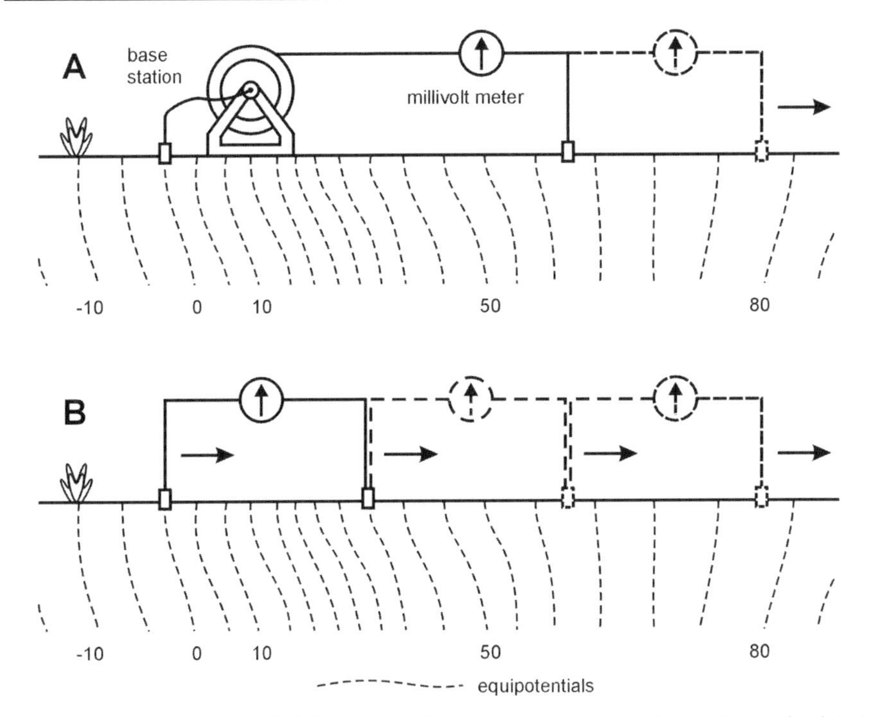

Fig. 3.19. Self-potential field layouts: Fixed base - moving electrode method (A) and gradient method (B). Because of the lack of absolute potential values on the earth's surface the potential of the base station may arbitrarily be set to zero

Despite the simple measuring technique, SP noise sources like telluric currents, industrial currents, electric railway systems, electrode drifts, and inhomogeneous soils may considerably affect a survey making reliable data acquisition a difficult task (Corvin 1990). High-amplitude bioelectric potentials from vegetation can seriously overprint the geologic SP signature, and time variations of SP fields are a matter of further complexity (Ernstson and Scherer 1986).

3.4.3 Data processing and interpretation

Self-potential data are plotted on profiles (Fig. 3.20) or in the form of isopotentials (Fig. 3.21). As with other geophysical potential field measurements (e.g., gravity and geomagnetic surveys), SP data may be processed by filtering and computing of gradient fields.

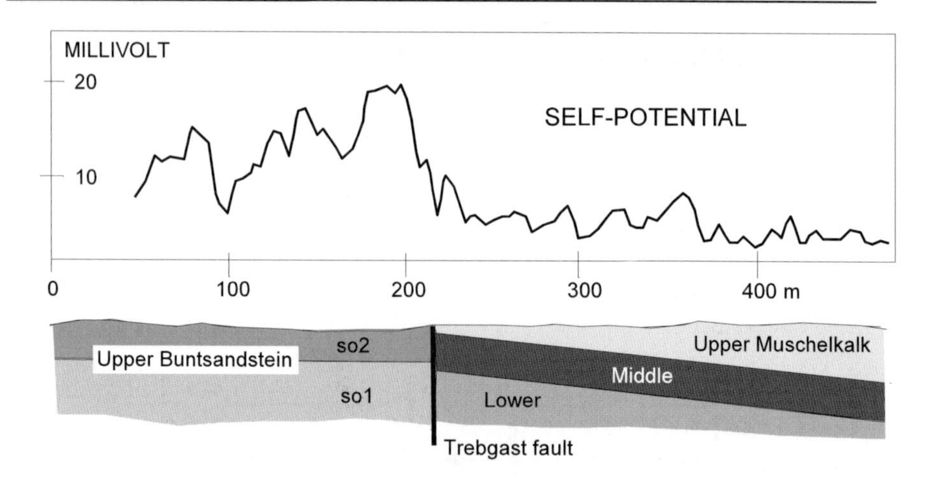

Fig. 3.20. Profile of self-potential anomalies across a fault

SP interpretation is mostly done qualitatively with the aim to, e.g., delineate ground-water flow and to locate zones of ground-water infiltration and ascent. Computer programs for the modeling of SP anomalies over causative bodies of simple shape are available which allow to estimate depth, strike and dip of the source of SP anomalies. New approaches use mathematical methodologies to directly relate SP signals with hydraulic conductivity distribution in the ground (Rizzo et al. 2004) and with the contours of the ground-water table (Birch 1998, Revil et al. 2003).

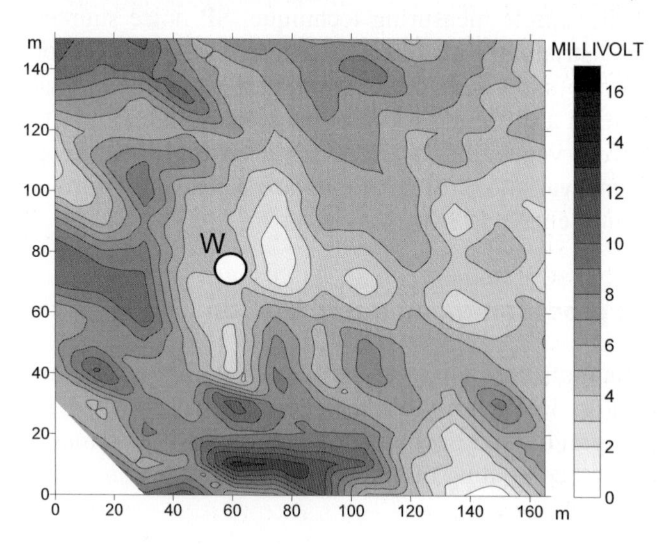

Fig. 3.21. Self-potential anomalies around a well (w) during a pumping test

3.5 2D measurements

Markus Janik, Heinrich Krummel

Sounding and profiling can be combined in a single process (2D resistivity imaging) to investigate complicated geological structures with lateral resistivity changes. This combination provides detailed information both laterally and vertically along the profile. In the 2D case it is assumed that the resistivity of the ground varies only in the vertical and one horizontal direction along the profile. There is no resistivity variation perpendicular to it (strike direction). 2D resistivity surveys (2D imaging) have played an increasingly important role in the last few years. The advantages of 2D measurements are their high vertical and lateral resolution along the profile, comparatively low cost due to computer-driven data acquisition, which means only a small field crew is needed (one operator and, depending on terrain roughness and profile lengths, one or two assistants). The following section describes briefly the field measurements, data processing and interpretation of this method, partly following Lange (1997).

3.5.1 Field equipment

The basic equipment for all dc resistivity measurements consists of a transmitter, receiver, the power supply, electrodes and cables. An additional switching unit is necessary for 2D and 3D measurements to activate the electrodes individually. 3D measurements will not be further discussed in this section. For some instruments a laptop is required for data acquisition, display and storage. Many different instruments have been developed for 2D measurements, usually multi-electrode systems, sometimes multichannel ones.

Stainless steel electrodes seem to be the best choice for the galvanic coupling of the electrodes with the ground, as they are durable and have fairly low self-potentials, when placed in the ground.

Two types of operation are common for 2D resistivity measurements. The first one uses "passive" electrodes connected to a switching box via a cable containing 20 or more cores. The other type of operation uses "active" electrodes: A cable containing only a few cores is connected to an addressable switch on each electrode. The switch is used to activate or deactivate the electrode either for current injection or for voltage measurements. Cross-talk between cores can influence data acquisition. The advantage of the second type is that the cross-talk is reduced. Moreover, it is easier to increase the number of electrodes (up to 256 and more) and to use basic electrode spacing of 10 m and more, even 50 m and 200 m have been applied. (Friedel 2000).

3.5.2 Field measurements

Before a survey is started, it is necessary to select the optimal kind of measurement. Such a decision is influenced by geology and topography in the study area, by the size of the area, and last but not least by economic considerations. For data processing and interpretation it is necessary to collect all available information about drilling results, the local geology and hydro-geology as well as potential sources of cultural noise (metal pipes, power lines, industrial objects).

The investigation depth of commonly used arrays is, as a rule of thumb, in the range of L/6 to L/4, where L is the spacing between the two outer active electrodes. The dipole-dipole array, which offers the best depth of investigation, has the disadvantage of a comparatively low signal/noise ratio (for more details, see Friedel 2000). Therefore, to obtain good quality data, the electrode must be adequately coupled to the ground and the equipment must have a high sensitivity.

It is fundamental to select the most appropriate electrode array for the given problem. Each array has advantages and disadvantages. Experience, confirmed by theoretical studies (e.g., Roy and Apparao 1971, Edwards 1977, Barker 1989), suggests that the dipole-dipole array has the best resolution with regard to the detection of single objects, and is the best choice for multi-electrode measurements. At regions with high electrical noise, other configurations like Schlumberger or Wenner arrays should be used, to increase the signal strength.

Another factor which affects the results is topography (Fox et al., 1980). The terrain relief may have a strong influence on shaping the equipotential surfaces and, therefore, the *K*-factor has to be modified for the terrain geometry.

Fig. 3.22 shows an example of the electrodes arrangement and measurement sequence that can be used for a 2D electrical imaging survey, here for a Wenner electrode array. The electrodes should be positioned with equal spacing along the profile. A multi-core cable with equidistant "takeouts" for connecting the electrodes is placed along the profile. It is important that the electrodes make good contact to the ground. Data acquisition is begun after the cables are connected to the switching box and the electrode contacts have been checked. The measurements are controlled by the microprocessor-driven resistivity meter or by a computer.

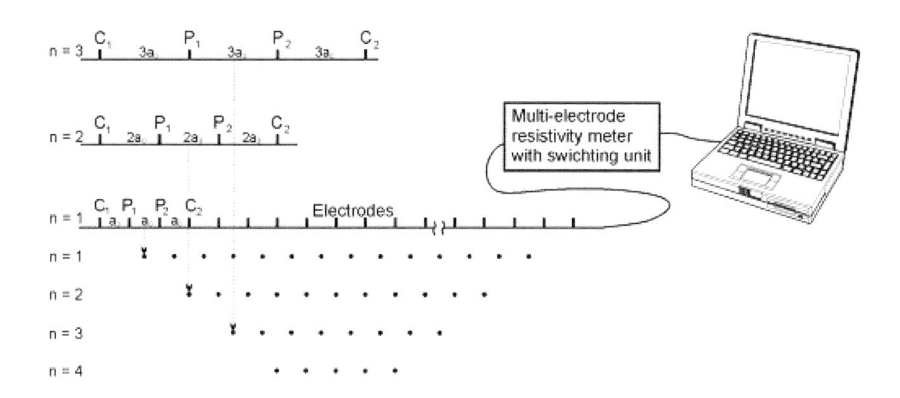

Fig. 3.22. Setup for a 2D resistivity measurement with a Wenner electrode array

The electrodes are switched to take measurements of the ground resistivity as shown in Fig. 3.22. Several depth levels may be entered by increasing the electrode spacing. The apparent resistivities are plotted as a function of location along the profile and electrode separation. For the Wenner electrode array, the location on the profile is given by the center of the array, and the depth ("pseudodepth") is given by the spacing of the current electrodes. Pseudodepth and location depend on the used array geometry. The points in Fig. 3.22 mark the position of the measured data values within the pseudosection.

Normally, two cables are used for each measurement. The resistivity meter and the switching unit are connected in the middle of the cable layout. When a profile is longer than the electrode layout, the profile can be measured using the two cables in a leap-frogging fashion. The first cable is placed at the end of the second one and a new measurement is made. This technique is also known as the roll-along method and can be used to measure the full length of the profile. Strong profile bends should be avoided. In rough terrain the elevation along the profile should be determined.

3.5.3 Data Processing and Interpretation

For many groundwater-related problems it is preferable to carry out 2D resistivity measurements instead of 1D soundings and/or profiling. The aim of a geoelectrical survey at the Earth surface is to provide detailed information about the lateral *and* vertical resistivity distribution in the ground. Unfortunately 3D measurements still are too time consuming and expensive. As the principles applied are similar to those of computer tomography in medicine, such geoelectrical investigations are also called impedance tomography or 2D resistivity imaging.

The observed data of a 2D resistivity survey are displayed as a pseudosection along the profile. Pseudosections provide an initial picture of the subsurface geology. In recent years, several computer programs have been developed to carry out such inversions (Barker 1992, Loke and Barker 1996, Advanced Geosciences 2002). Depending on the algorithm used, the result is a smoothed layer model or a block model showing sharp layer boundaries.

The commonly used 2D-model divides the subsurface into a number of rectangular blocks that will produce an apparent resistivity pseudosection that agrees with the actual measurements. The model used for inversion consists of a number of rectangular cells. The size of these cells is determined either automatically as a function of the electrode spacing, or manually by the user. In general, the size of the cells increases with increasing depth and towards the beginning and the end of a profile, due to decreasing sensitivities in these areas. The optimization method tries to reduce the difference between the calculated and measured apparent resistivity values by adjusting the resistivity of the model blocks. There are still problems with the determination of the confidence intervals and parameter limits of models for the 2D cases. In 2D inversion, ambiguity is influenced by several factors: the structure of the grid used to approximate the geological structures, limited data density, and errors in the data. As in the 1D inversion of vertical soundings it is advantageous if *a priori* information can be included.

In spite of the various capabilities of different programs, the final assessment of inversion results has to be done by the user, i.e., the geophysicist. Due to the limited number of values and precision of the data, all results are ambiguous. This ambiguity of a 2D survey is fairly reduced compared to a 1D survey, and it can be further reduced by including all available information from bore holes and other geophysical surveys.

The final result of a 2D resistivity survey is a cross-section of the calculated rock resistivities along the profile line. This cross-section should include the structural interpretation of the resistivity data (e.g., the boundaries of fracture zones and caves). If the spacing between parallel profiles is not too large, horizontal resistivity sections for different depths can also be derived from the data.

3.5.4 Examples

The following example describes the investigation of a dyke. The objective was to map a marl layer, which is the first low permeable layer and was used as the base of the dyke. Depth profile of this layer and sandy intrusions within the marl were targets of this measurement.

Parameters of the dc resistivity survey example:	
Profile length:	135 m
Electrode spacing:	5 m
Array configuration:	2D-Schlumberger
Total number of stations:	28
Personnel:	1 technician, 1 assistant
Duration of the survey:	2½ hours
Duration of data acquisition:	45 minutes
Number of readings:	215
Equipment:	Sting R1/IP multi-electrode resistivity meter (Advanced Geosciences Inc.)

The inversion procedure is demonstrated in Fig. 3.23. At the top the measured pseudosection is shown (part A of the figure). Depth values are calculated using the survey dimensions and the geometry factor. The scale shows the decreasing resolution of the survey with increasing depth. Correctly spoken, the array configuration is not pure Schlumberger, but a combination of Wenner and Schlumberger geometry. The inversion is done in that way, that for all model blocks resistivities (shown in part C) are adjusted to fit the measured pseudosection. To control this, the pseudosection is calculated from the block model. Part B of the figure shows the calculated pseudosection of the adjusted model. As an optional final step, the resistivity values assigned to the model blocks are taken as point values at the center at each block and displayed as a colour or greyshaded contour image (part D of the figure). Here, the margins of the block model are clipped to the part of the model which is sufficiently covered by the data points.

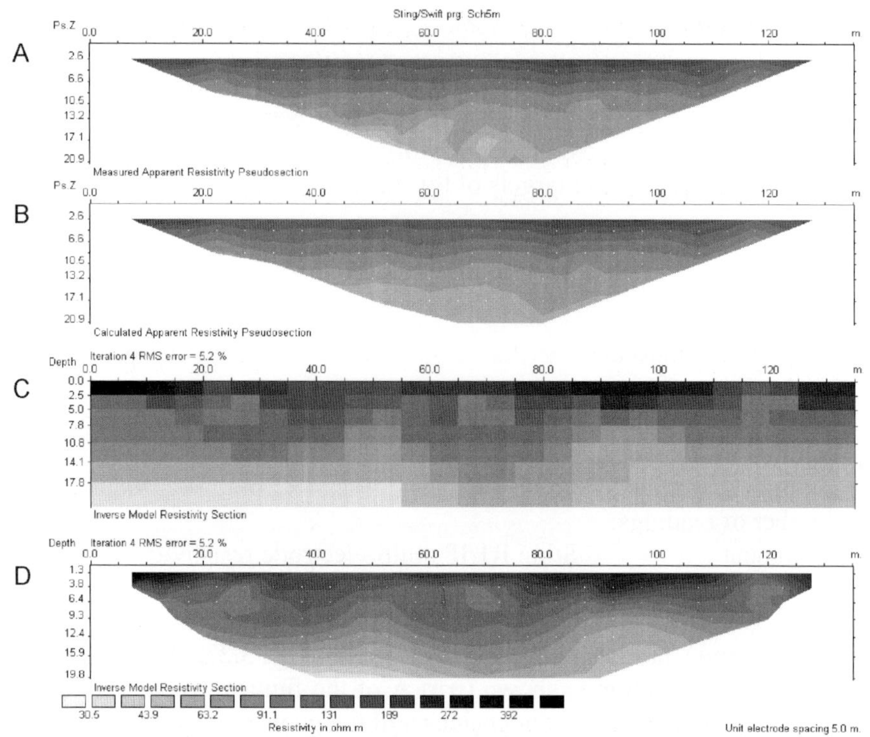

Fig. 3.23. Procedure of 2D resistivity inversion.

The final result is shown in Fig. 3.24 as a contour map of the resistivity distribution. The maximum investigation depth is about 20 m. The result clearly indicates the depth of the marl layer and its undulating shape. The estimated depth of the marl layer was verified by drilling results. The two borehole locations are indicated in the profile. The resistivity map shows that the marl layer is continuous, sandy intrusions within the marl are not indicated in this profile section.

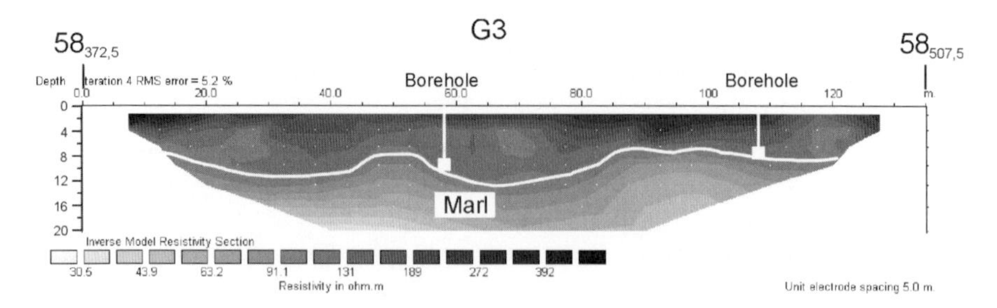

Fig. 3.24. 2D resistivity inversion result of a Schlumberger pseudosection along a dyke.

Another example shows the mapping of a fracture zone within sandstone. Drilling results showed variations of depth to the groundwater table of up to 4 m within a distance of 100 m. One possible explanation was the existence of a fracture zone within the sandstone formation. The exact location of the (geologically known) fracture zone had to be determined. The example shows the result of one resistivity profile close to the two drill sites.

Parameters of the dc resistivity survey example:

Profile length:	275 m
Electrode spacing:	5 m
Array configuration:	2D-dipole-dipole
Total number of stations:	56
Personnel:	1 technician, 1 assistant
Duration of the survey:	4 hours
Duration of data acquisition:	2½ hours
Number of data points:	694
Equipment:	Sting R1/IP multi-electrode resistivity meter (Advanced Geosciences Inc.)

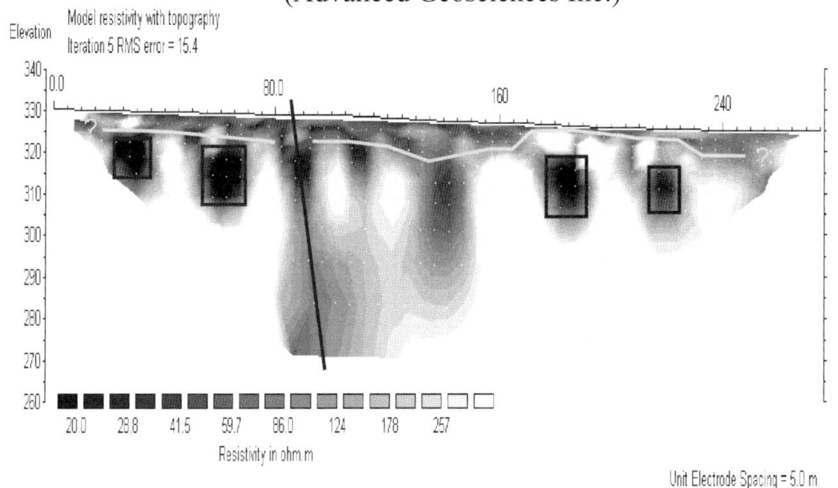

Fig. 3.25. 2D resistivity inversion result of a dipole-dipole pseudosection for the localisation of a fracture Zone in sandstone

The inversion result is shown as a contour map of the resistivity distribution in Fig. 3.25. The maximum investigation depth is about 50 m. The top of the sandstone layer is indicated by a grey line. Boxes show dark spots with low resistivity interpreted as clay lenses. The fracture zone is indicated by low resistivities between 80 m and 150 m.

3.6 References

Advanced Geosciences Inc (2002) EarthImager, 2D Resistivity and IP Inversion Software

Baker SS, Cull JP (2004) Streaming potential and groundwater contamination. Exploration Geophys 35:41-44

Barker RD (1989) Depth of investigation of collinear symmetrical four-electrode arrays. Geophysics 54:1031-1037

Barker RD (1992) A simple algorithm for electrical imaging of the subsurface. First Break 10:52-62

Birch FS (1998) Imaging the water table by filtering self-potential profiles. Ground Water 36:779-782

Bogoslowsky VA, Ogilvy AA (1972) The study of streaming potentials on fissured media models. Geophys Prosp 20:109-117

Buselli G, Lu K (2001) Groundwater contamination monitoring with multichannel electrical and electromagnetic methods. J Appl Geophys 48:11-23

Corvin RF (1990) The self-potential method for environmental and engineering applications. Geotechnical and Environmental Geophysics, vol 1. Soc Expl Geophys, Tulsa, pp 127-145

Edwards LS (1977) A modified pseudosection for resistivity and induced polarisation. Geophysics 42:1020-1036

Ernstson K, Scherer HU (1986) Self-potential variations with time and their relation to hydrogeologic and meteorological parameters. Geophysics 51:1967-1977

Fox RC, Hohmann GW, Killpack TJ, Rijo L (1980) Topographic effects in resistivity and induced-polarization surveys. Geophysics 45:75-93

Friedel S (2000) Über die Abbildungseigenschaften der geoelektrischen Impedanztomographie unter Berücksichtigung von endlicher Anzahl und endlicher Genauigkeit der Messdaten. PhD thesis Universität Leipzig. Berichte aus der Wissenschaft, Shaker, Aachen

Habberjam GM (1975) Apparent resistivity, anisotropy, and strike measurements. Geoph Prosp 23:211-247

Keller GV, Frischknecht FC (1970) Electrical Methods in Geophysical Prospecting. Pergamon Press, Oxford etc, 517 pp

Lane JW, Haeni FP, Watson WM (1995) Use of square-array direct-current resistivity method to detect fractures in crystalline bedrock in New Hampshire. Ground Water 33:476-485

Lange G (1997) Geoelektrik. In: Knödel K, Krummel H, Lange G (eds) Handbuch zur Erkundung des Untergrundes von Deponien und Altlasten, Bd. 3 Geophysik. Springer, Heidelberg, pp 122-165

Loke MH, Barker R D (1996) Rapid least-squares inversion of apparent resistivity pseudosections using a quasi-Newton method. Geophys Prosp 44:131–152

Maineult A, Barnabé Y, Ackerer P (2004) Electrical response of flow, diffusion, and advection in a laboratory sand box. Vadose Zone J 3:1180-1192

Mundry E (1985) Gleichstromverfahren. In: Bender F (ed) Angewandte Geowissenschaften Bd. 2 Enke Verlag, Stuttgart, pp 299-338
Mundry E, Dennert U (1980) Das Umkehrproblem in der Geophysik. Geologisches Jahrbuch, Reihe E19:19-39
Nimmer RE, Osiensky JL (2002) Using mise-a-la-masse to delineate the migration of a conductive tracer in partially saturated basalt. Environmental Geosciences 9:81-87
Revil A, Naudet V, Nouzaret J and Pessel M (2003) Principles of electrography applied to self-potential sources and hydrogeological applications. Water Resources Res 39: 1114,doi:10.1029/2001WR000916
Rizzo E, Suski B, Revil A, Straface S, Troisi S (2004) Self-potential signals associated with pumping tests experiments. J Geophys Res 109:B10203, doi:10.1029/2004JB003049
Roy A, Apparao A (1971) Depth of investigation in direct current methods. Geophysics 36:943-959
Schulz R (1985) Interpretation and depth of investigation of gradient measurements in direct current geoelectrics. Geoph Prosp 33:1240-1253
Taylor RW (1984) The determination of joint orientation and porosity from azimuthal resistivity measurements. In: Nielsen DM, Curl M (eds) National Water Well Association/U.S. Environmental Protection Agency Conference on Surface and Borehole Geophysical Methods in Goundwater Investigations, pp 37-49
Vichabian Y, Reppert PM, Morgan FD (1999) Self potential mapping of contaminants. Proc Symp Appl Geophys Engin Envir Probl (SAGEEP), pp 657-662
Ward SH (1990) Resistivity and Induced Polarization Methods. In: Ward SH (ed) Geotechnical and Environmental Geophysics, vol 1. Soc Expl Geophys, Tulsa, pp 127 - 145

4 Complex Conductivity Measurements

Frank Börner

4.1 Introduction

Geohydraulic properties of aquifers and soils are of general interest for environmental and geotechnical applications. Aquifer and soil characterizing parameters are used for the modelling the water flow and the migration of hazardous substances. Compared to the application of conventional direct methods, non-invasive geophysical measurements have the potential to predict parameter distributions more realistically. Additionally, such measurements are more cost-effective.

Geoelectrical surveys have become an increasingly important tool in subsurface hydrogeological applications. The spatial distribution of electrical parameters of the subsurface can provide valuable information for characterizing the heterogeneity of the groundwater and the soil zone. However, due to various petrophysical influences on electrical rock properties, the attempt to convert electrical conductivity variations to variations of geohydraulic parameters brings often ambiguous results. Geoelectrical measurements can contribute to, e.g.,

- the assessment of aquifer vulnerability and depth to watertable (Kalinsky et al. 1993, Kirsch 2000),
- the determination of catchment areas and aquifer characteristics (hydraulic conductivity, sorption capacity, dominant flow regime, Mazac et al. 1985, Boerner et al. 1996, Kemna 2000, Weihnacht 2005),
- the monitoring of water content and water movement (e.g. Daily et al. 1992, Gruhne 1999, Berger et al. 2001, Berthold et al. 2004, Liu and Yeh 2004),
- the monitoring of changes of water quality and mineral alteration connected with active remedial measures on contaminated sites as well as with natural attenuation processes (Grissemann and Rammlmair 2000, Atekwana et al. 2004).

The complex conductivity measurement is an innovative geoelectrical method which is sensitive to physico-chemical mineral-water-interaction at the grain surfaces. In comparison to conventional geoelectrics a complex electrical measurement can be provide besides conductivity also information on the electrical capacity and the relaxation processes in the frequency range below some kHz. At higher frequencies the electrical behaviour of rock is increasingly determined by physical interactions alone.

Complex measurements can reduce the petrophysical ambiguity caused by the influence of textural and state properties. This advantage compensates for the at times expensive and time-consuming measurements.

In the context of the general topic the following section focuses on the utilizable petrophysical potential of complex conductivity measurements to provide non-invasively

- geohydraulic parameters like the hydraulic conductivity or the sorption capacity, and
- information to the state and distribution of contaminants in the pore space.

Complex conductivity as well as IP phenomena and measuring techniques are reviewed by, e.g. Wait (1959), Sumner (1976), Klein et al. (1984) or Ward (1990). Olhoeft (1985) presented an overview of the properties and state conditions which influence complex conductivity spectra. The complex electrical properties of porous rocks were investigated by Buchheim and Irmer (1979), Vinegar and Waxman (1984) or Boerner (1992). The effect of elevated pressure on the broad band complex conductivity was discussed by Lockner and Byerlee (1985). Recently Slater and Lesmes (2002a), Klitsch (2004), Ulrich and Slater (2004) focused their research on unconsolidated material as well as on the distributed geometry of rock texture and related relaxation phenomena.

4.2 Complex conductivity and transfer function of water-wet rocks

The electrical transfer function of a defined rock volume can be characterized by the dependence of conductivity and dielectric permittivity on frequency as well as on the thermodynamic state parameters like temperature, pressure and water content. Based on Maxwell´s equations of electrodynamics, an effective current density J* is defined as the response of a time-varying electrical field E. For a sinusoidal time dependence of E the process is described by

$$J^*(\omega) = (\sigma_{eff} + i\omega\varepsilon_{eff})E(\omega) \tag{4.1}$$

where ω is the angular frequency. The term in brackets is the electrical transfer function. The quantities that describe the effective electrical or dielectrical properties are the electrical conductivity σ_{eff} and the dielectric permittivity ε_{eff}. The transfer function in Eq. 4.1 can be formulated in terms of a complex conductivity σ* (Olhoeft 1985):

$$\sigma_{eff} + i\omega\varepsilon_{eff} = \sigma^* = \sigma'(\omega) + i\sigma''(\omega) \tag{4.2}$$

A complex electrical measurement of rock material provides information on ohmic and capacitive electrical processes and its frequency dependence. This information can be used to characterize the rock texture and state. Complex conductivity or permittivity data for a broad frequency range were published e.g. by Lockner and Byerlee (1985), Dissado and Hill (1984), and Kulenkampff et al. (1993).

There are already several basic formulae that describe the frequency dependence of a more or less complicated material. The Debye-model describes a single relaxation process of polarized dipols in a nonpolar viscose medium. However, single relaxation processes are seldom observed in natural rock. The Cole-Cole-model (Cole and Cole, 1941), which considers a distribution of relaxation times, is a more realistic model for heterogeneous materials:

$$\varepsilon^*(\omega) = \varepsilon_\infty + \frac{\varepsilon_0 - \varepsilon_\infty}{1 + (i\omega\tau)^{1-\alpha}} \tag{4.3}$$

The exponent 1-α describes a symmetric distribution of relaxation times. Kulenkampff (1994) developed a Cole-Cole-like power law model for the frequency range up to 1 MHz. It considers a low frequency limit of conductivity and a high frequency limit of dielectric permittivity:

$$\sigma^*(\omega) = \sigma_O\left[1 + (i\omega/\omega_N)^{1-\alpha}\right] + i\omega\varepsilon_\infty\varepsilon_0 \tag{4.4}$$

A similar model was successfully used by Fechner et al. (2004) to analyze the frequency dependence of clay-limestone-mixtures.

The so-called ‘constant phase angle response’ (CPA) has been identified in many spectra of systems with interface related electrochemical or biological processes (Jonscher 1981). The transfer function has a proportionality according to

$$\sigma^*(\omega) \propto (i\omega)^{1-\eta} \tag{4.5}$$

This behaviour has been attributed to the self-similar structure of internal rough surfaces. Liu (1985) could explain the CPA-behaviour at very low frequencies by using an equivalent circuit model with a hierarchical structure (Fig. 4.1). The CPA frequency exponent η predicts a dynamic scaling relationship between static capacitors and conductors. Pape et al. (1992) used a CPA-like-model to analyze IP decay curves measured in

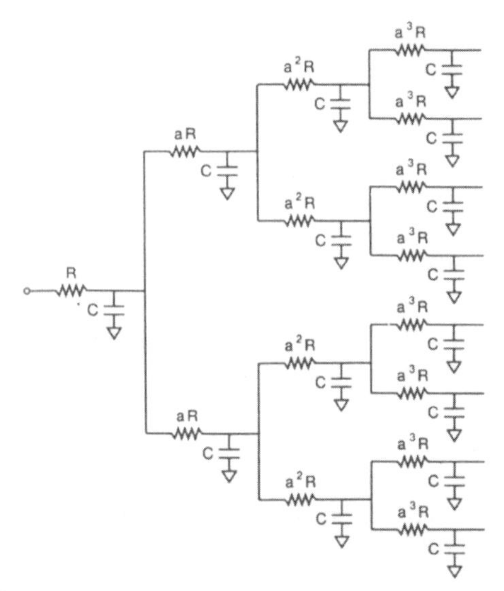

Fig. 4.1. Self similar equivalent circuit of a rough model interface from Liu et al. (1986)

the super deep KTB-borehole in northern Bavaria/ Germany. Its algorithm relates the CPA-exponent to the fractal 'pigeon hole model' of rock pore space (Pape et al. 1987). Recently Spangenberg (1996) used successfully the fractal concept to model other petrophysical properties and the interrelation between various properties.

Jonscher (1981) analyzed a large number of spectra of different water-wet materials and found that most of them can be explained with relatively simple power laws. He formulated an empirical law of universal relaxation that is characterized by the overlap of two ranges with different frequency exponents caused by different relaxation processes. On this basis Dissado and Hill (1984) developed the model of anomalous low frequency dispersion consisting of hierarchical clusters of relaxing elements. The exponents describe on the one hand the coupling of elements (e.g. charges) within the cluster and on the other hand the coupling between clusters (Kulenkampff, 1994). Generally, the low frequency part is dominated by electrical conduction processes whereas the high frequency part is dominated by dielectrical polarization processes. Comparable with the model of anomalous low frequency dispersion by Dissado and Hill (1984) an empirical broad band model is used for an interpretation combining electrical and dielectrical measurements that describes the low as well as the high frequency portion by different power law frequency dependences (Kulenkampff et al 1993, Boerner 2001). The result is the Combined Conductivity and Permittivity Model (CCPM). It is in terms of complex conductivity:

$$\sigma^*(\omega) = \sigma_{DC} + \sigma_{O,NF}(i\omega/\omega_N)^{1-p} + \omega_N \varepsilon_{O,HF}(i\omega/\omega_N)^n + i\omega\varepsilon_\infty \tag{4.6}$$

where σ_{DC} is the DC-conductivity, $\sigma_{0,NF}$ the amplitude factor and p the frequency exponent of interface conductivity, $\varepsilon_{0,HF}$ the amplitude factor and n the exponent of interface permittivity, ε_∞ the high frequency dielectric permittivity and ω_N a normalizing angular frequency. All parameters used can be obtained from a broad band electrical and dielectrical measurement. The CCPM takes into consideration the fact that the low frequency constant phase angle behaviour changes to a Cole-Cole-behaviour for the conductivity at frequencies higher than 10 kHz. (Kulenkampff and Schopper 1988, Ruffet et al. 1991). Some experimentally measured broad band conductivity spectra are shown in Fig. 4.2.

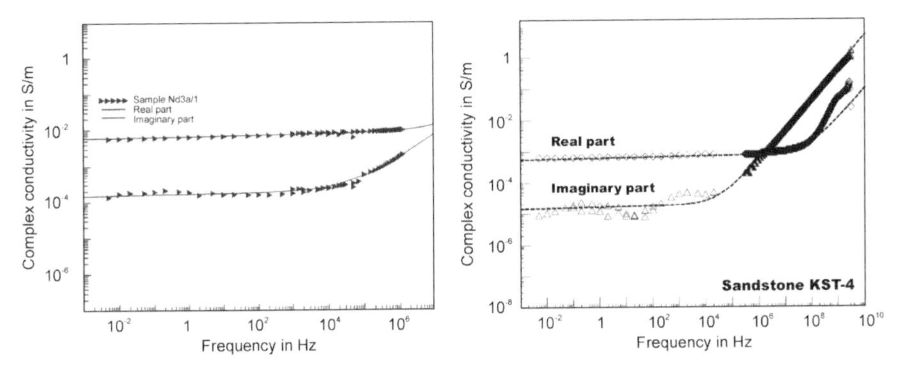

Fig. 4.2. Broad band Complex conductivity spectra of rocks. Dashed lines are fits with the CCPM (left: Kulenkampff et al. 1993, right: Boerner 2001)

4.3 Quantitative interpretation of Complex conductivity measurements

4.3.1 Low Frequency conductivity model

The capacitive behaviour of aquifer materials is caused by the accumulation and diffusion of ions in a pore network with a distributed geometry and an electrochemical double layer at the mineral grain-water interface. The pore scale polarization is interpreted mainly by the processes of membrane polarization and space charge polarization (Ward and Fraser 1967, Waxman and Smits 1968, Schön 1996).

The close relation between the observed polarization of water-wet rocks and the pore space structure is reflected in physical models e.g. from Marshall and Madden (1959), Buchheim and Irmer (1979), Pape et al. (1992), Titov et al. (2002) or Klitsch (2003).

A constant phase angle behaviour was identified by different authors for a varity of water-wet rock material, e.g. for dispersed copper ores (Van Voorhis et al. 1973), for shaly sands (Vinegar and Waxman 1984, Boerner 1992), for sandstone and crystalline rock (Lockner and Byerlee 1985, Pape et al. 1991, Börner and Schön 1993) and for unconsolidated sand (Gruhne 1999).

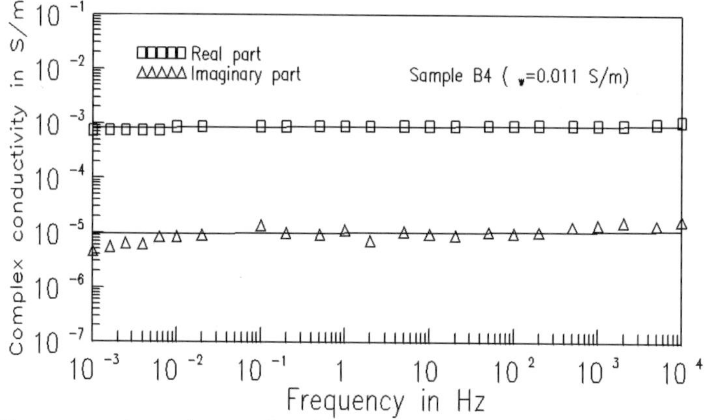

Fig. 4.3. Frequency dependence of real and imaginary part of complex conductivity and fit according to cpa-model in Eq.4.7

Aquifer materials like clean and shaly sandstones as well as unconsolidated sand and silt generally show an almost identical power law dependence of the real as well as the imaginary part on frequency as the main feature observed (see Fig. 4.3). Based on the constant phase angle behaviour (Jonscher 1981) the general dependence of complex conductivity σ^* on frequency in the range from 10^{-3} to 10^{2} Hz can be expressed by:

$$\sigma^*(\omega) = \sigma_O (i\omega/\omega_N)^{1-p} \tag{4.7}$$

where σ_0 is an amplitude factor, ω the angular frequency (normalized at $\omega = 1\ s^{-1}$ and (1-p) is an exponent describing the frequency dependence. The separation of a true DC-conduction σ_{DC} component in the low frequency range according to Eq. 4.6 is impossible if exponent p>0,95. Generally this is true for high porous water saturated rocks with low interface conductivity. For this simple case σ_O in Eq. 4.7 is the sum of σ_{DC} and the interface conductivity $\sigma_{0,NF}$ in Eq. 4.6. The advantages of such an uncomplicated transfer function model are

- it has only two free parameters σ_0 and p, and
- all parameters may be obtained from one single complex electrical measurement.

Examples of measured cpa-spectra are shown in Figs. 4.3 and 4.4.

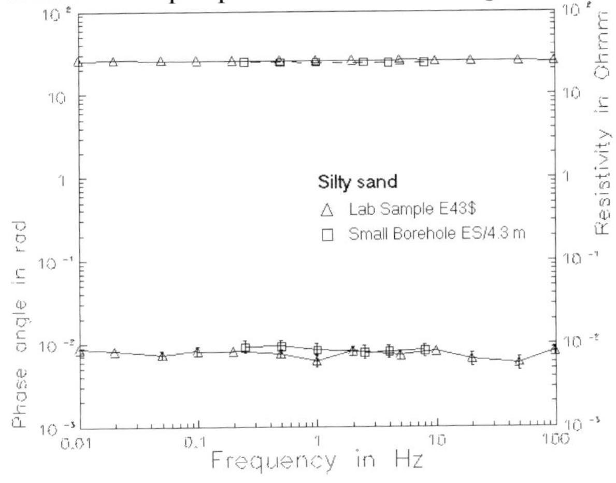

Fig. 4.4. Comparison between field and lab data: resistivity and phase angle vs. frequency (Boerner et al. 1996, permission from Blackwell Publishing)

The cpa-behaviour can be superimposed by phase maxima between 0.01 and 10 Hz . Some authors relate this maxima to deviations from an 'ideal' fractal pore space structure, sized sands or high clay contents. Phase maxima with a small amplitude were modelled for rock with water filled pores by Buchheim and Irmer (1979) or Titov et al. (2002). Klitsch (2003) used a capillar network model to reflect experimental data. Center frequency and amplitude of the anomaly are related to mean grain or pore size. It should be noted that similar effects can be caused by small amounts of conductive minerals (Grissemann et al. 2000), organic material like peat or coal (Comas and Slater 2004) or an insufficient working measurement system.

At frequencies below 10^{-2} Hz an onset of a true DC-conductivity accompanied by a vanishing imaginary part is often observed (see Fig. 4.3).

4.3.2 Complex conductivity measurements

A complex conductivity measurement (similar to a spectral IP-measurement) determines the conductivity amplitude (or resistivity) and the phase shift in the low frequency range for a conductive material (Fig. 4.6). Typical is the measurement of discrete frequencies in logarithmic e-qual spaced steps over a frequency range from 0.001 to 10 Hz (in lab scale to 10 kHz). The same information contains a voltage decay curves obtained from time domain IP measurement. Fourier transformation relates frequency domain and time domain data.

The commonly measured quantities resistivity magnitude $|\rho^*|$ and phase angle φ, can be expressed in terms of complex conductivity thus

$$|\sigma^*| = \frac{1}{|\rho^*|} = \sqrt{(\sigma')^2 + (\sigma'')^2} \tag{4.8}$$

and

$$\tan(\varphi) = \frac{\sigma''}{\sigma'} \tag{4.9}$$

By using the cpa-model two parameters can be obtained from a single frequency scan: conductivity amplitude and frequency dependence. The correlation between conductivity and increasing frequency is described by the frequency exponent *1-p* in Eq. 4.7, which is related to the constant phase angle φ. The frequency exponent *1-p* can be obtained from a regression of log[$\sigma^*(\omega)$] vs. log[ω]. Matching the frequency effect *FE* (Parasnis 1966) *1-p* was named "logarithmic frequency effect" *LFE* (Börner 1992):

$$\mathrm{LFE} = 1 - \mathrm{p} = \frac{\partial \lg|\sigma^*(\omega)|}{\partial \lg(\omega)} \tag{4.10}$$

LFE is similar to *FE* but more clearly related to the constant phase angle. Additionally, it is valid over a broad frequency range.

The same information can be obtained from the conductivity or resistivity amplitude and the phase angle measured at one fixed frequency. The relationship between the frequency exponent (*1-p*) and the frequency independent phase angle φ results from Eq. 4.7:

$$\tan\varphi = \frac{\sigma''(\omega)}{\sigma'(\omega)} = \frac{\sigma_0''}{\sigma_0'} = \tan\left[\frac{\pi}{2}(1-\mathrm{p})\right] \tag{4.11a}$$

$$\varphi = \frac{\pi}{2}(1-\mathrm{p}) \tag{4.11b}$$

Splitting Eq. 4.7 into a real and an imaginary part leads to

$$\sigma_0' = \sigma_0 \cos\left[\frac{\pi}{2}(1-p)\right] \tag{4.12a}$$

$$\sigma_0'' = \sigma_0 \sin\left[\frac{\pi}{2}(1-p)\right] \tag{4.12b}$$

where σ_0' and σ_0'' are frequency independent prefactors of the real and imaginary component.

The key question is then to find an applicable parameter model to relate σ_0' and σ_0'' to parameters like hydraulic conductivity, water content, water composition or contamination indicators.

Further parameters may be obtained when the deviation of the frequency dependence on that of an ideal cpa-model is quantified.

The validity of the Eq. (4.11b) was tested by plotting p, calculated from φ vs. p, obtained by regression (see Fig. 4.5).

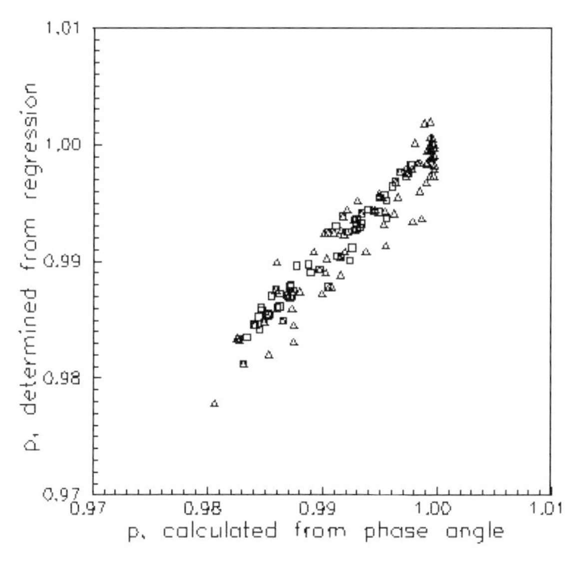

Fig. 4.5. p from regression vs. p calculated from φ by Eq. 4.11b

The complex conductivity measurement can be made in laboratory scale, in pilot scale in boreholes, at the surface or in artificial objects like buildings or foundations. In order to prevent noise and electrode polarization special electrodes are used for current injection and potential measurement. Electrode configurations are principally the same which are used in geoelectrics or resistivity logging (Figs. 4.7 and 4.8).

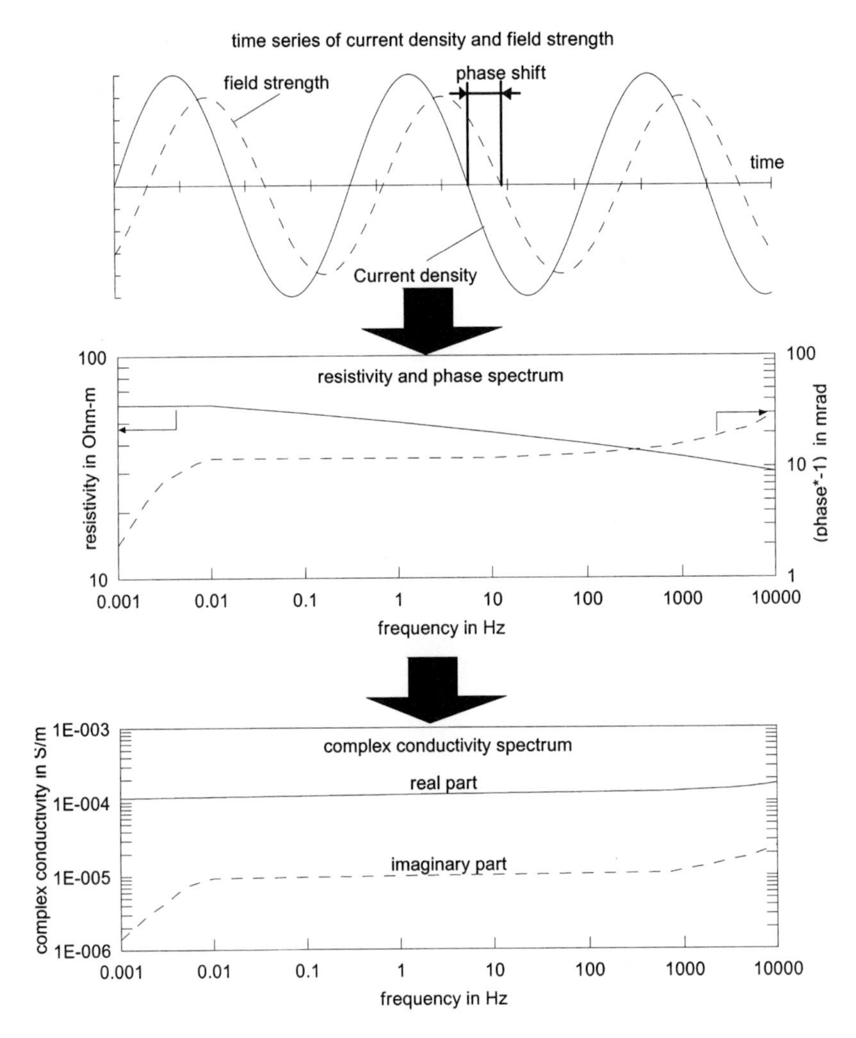

Fig. 4.6. Schema of complex conductivity measurement

The major problem of complex electric measurements in the field is the determination of exact frequency dependencies and phase angles respectively. The phase angles to be measured are of the size of a few milliradian (1 mrad=0.057°). Their numerical values are in clay-free sands or gravels only slightly greater than the measurement error. There are two possibilities to measure the phase angle under field conditions: (1) the direct measurement of the phase shift between injected current and registered voltage and (2) the determination of the logarithmic frequency effect according to Eq. 4.8. In the first case only one frequency, and in the second case, at least two frequencies have to be measured.

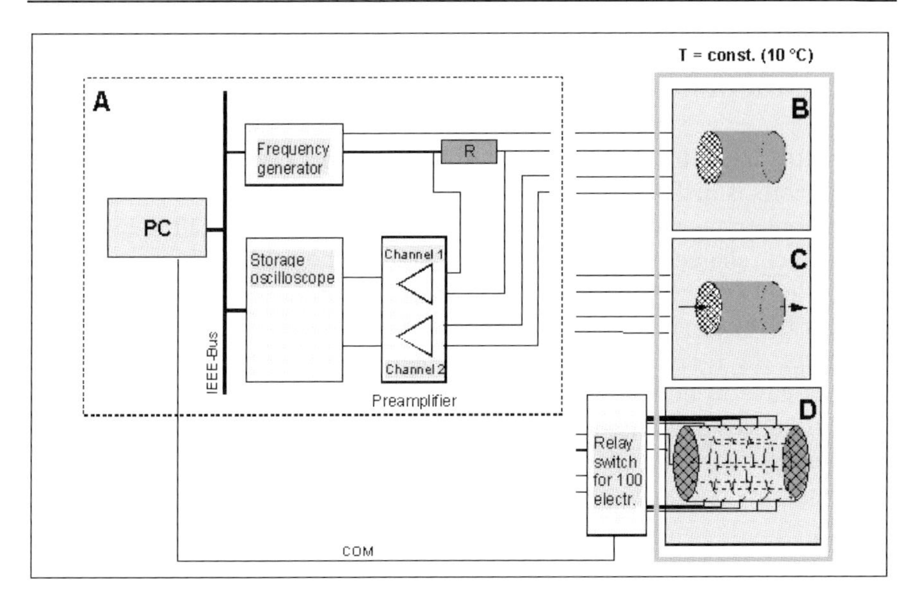

Fig. 4.7. Laboratory complex conductivity measurement system for near surface applications: A – data acquisition system, B – sample holder for measuring electrical parameters in equilibrium state, C – sample holder for monitoring integral electrical parameters during fluid exchange, D – multi electrode sample holder for monitoring the spatial distribution of electrical parameters during fluid exchange experiments

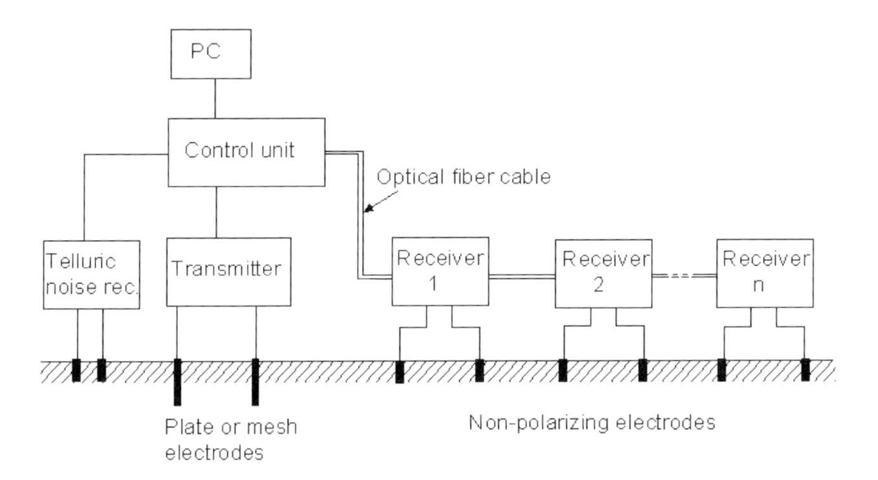

Fig. 4.8. Equipment and configuration for field measurements

(see e.g. www.radic-research.de)

4.4 Relations between complex electrical parameters and mean parameters of rock state and texture

The electrical conductivity of water-wet porous rocks without any conductive minerals is mainly related to the properties of the pore fluids, the pore space geometry as well as to interactions between mineral matrix and pore water. Rock characterizing parameters which are measurable on samples independently in the laboratory are of great importance. The knowledge of relations between single conductivity components and textural parameters constitutes the basis for a quantitative interpretation. These petrophysical relations have to be simple and physically plausible. To separate the different conductivity components, their dependence on, e.g., pore water salinity, temperature and saturation (water content) has to be analysed.

The salinity dependence of the complex conductivity has shown that σ_0 in Eq. 4.7 consists of two components (see Fig 4.9): real electrolytical volume conductivity σ_{el} and complex interface conductivity $\sigma_i^* = \sigma_i' + i\sigma_i''$.

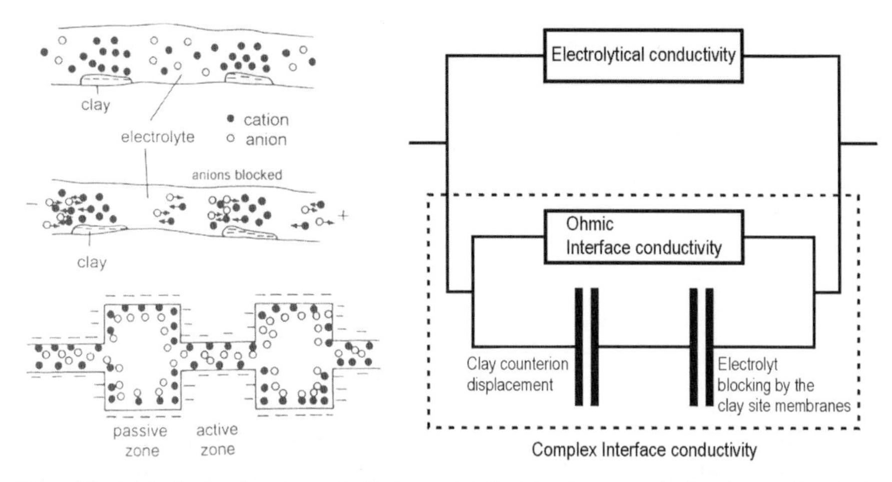

Fig. 4.9. (a) Polarization in rock (Schön 1986): Membrane polarization and polarization in clay-free rocks; (b) Equivalent circuit of the mean behaviour of water wet rock (after Buchheim and Irmer 1979, Vinegar and Waxman 1984)

Consequently, frequency independent prefactors of the real and imaginary part in a low frequency range have the general structure:

$$\sigma_0' = \sigma_{el} + \sigma_i' \tag{4.14a}$$

$$\sigma_0'' = \sigma_i'' \tag{4.14b}$$

It is a parallel connection of the real electrolytical and the complex interface conductivity. Experimental data confirmed this approach (e.g. Vinegar and Waxman (1984), Börner (1992) and Kulenkampff (1994). σ_{el} corresponds to Archies law (Archie 1942):

$$\sigma_{el} = \sigma_W / F \tag{4.15}$$

and σ_i^* is the product of parameters considering, on one hand, equivalent surface conductance and, on the other hand, geometry of current paths along the surface area.

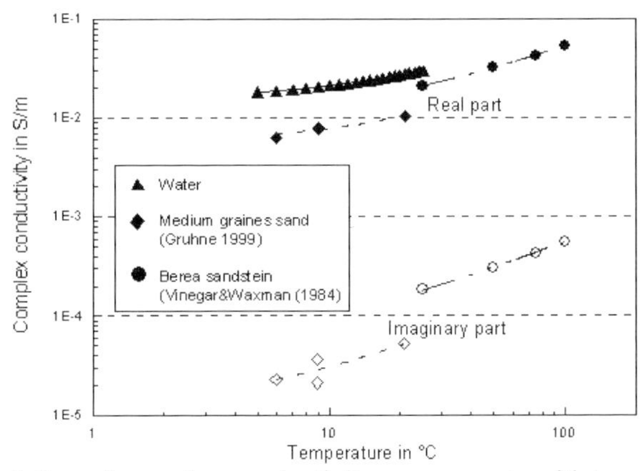

Fig. 4.10. Variation of complex conductivity components with temperature (after Gruhne 1999 and Vinegar and Waxman 1984)

The experimentally measured salinity dependence of complex conductivity components for different rock types is given in Fig. 4.11. The linear increase of the real rock conductivity with increasing water conductivity is typical for Archie-behaviour with slope 1/F. At a low water conductivity the curve corresponds with the level of the interface conductivity.

The weak salinity dependence of the imaginary part is shaped like an adsorption isotherm (Waxman and Thomas 1974). This effect is caused by a salinity dependent charge mobility and/or surface charge density. Models and experimental data for salinity dependant interface conductivity can be found in Waxman and Smits (1968), Vinegar and Waxman (1984), Kulenkampff and Schopper, 1988, Sen et al. (1988), Stenson and Sharma (1989) or Boerner (1992). The parameters in these models are related to the counter ion mobility, permittivity, temperature, surface potential and others. Experimental data of temperature dependence are shown in Fig. 4.10.

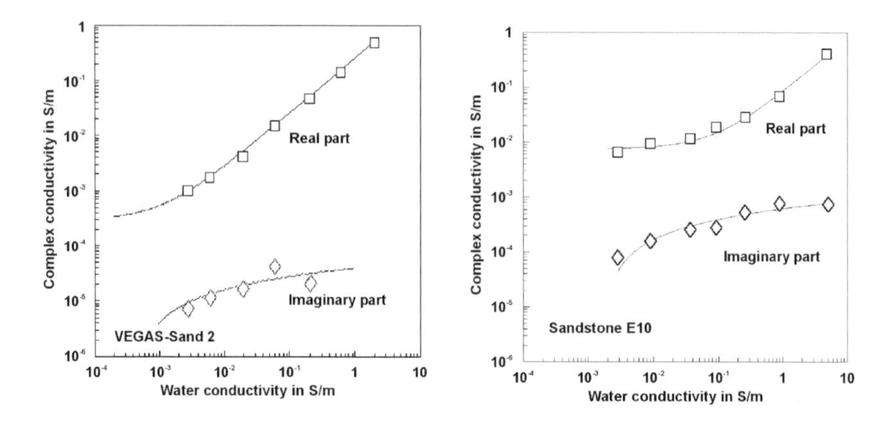

Fig. 4.11. Variation of complex conductivity components with electrolyte conductivity σ_w : (a) medium grained sand, (b) shaly sandstone

For practical purposes the same salinity function $f(\sigma_W)$ is used in Eqs. 4.14 a,b for the real and the imaginary part of the complex interface conductivity. Therefore the model ignores slight differences in the salinity dependence between the ohmic and capacitive interface contribution (Vinegar and Waxman 1984). In case of pore water with high salinity the function $f(\sigma_W)$ converges to a constant value of about 6 nS (range 1 to 30 nS) (Kulenkampff 1994) that is equal to the product of the mean surface charge density and cation mobility.

Water saturated rock

Various methods are used to express the real component of interface conductivity σ_i-term in Eq. 4.14a for a water saturated rock. The models have a very similar structure: Interface conductivity is formulated as the product of charge density in pore space and parameters that describe the charge mobility. The difference lies in the property used to quantify the size of the internal interface and the accompanying electrical process (see Table 4.2).

Similar to the real part of interface conductivity the imaginary part of conductivity shows a significant dependence on the size of grain-water-interface. It can be assumed that the capacitive behaviour is caused solely by interaction effects between the non-conductive matrix and pore water. An equivalent electrical circuit with the main conductivity components is shown in Fig. 4.9. Experimental results in Fig. 4.12 show the nearly linear dependence of conductivity on S_{POR} or Q_V . This dependence has been discovered by different authors.

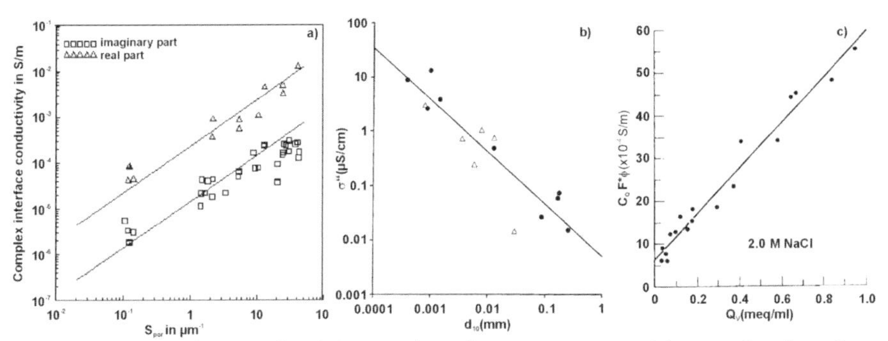

Fig. 4.12. Complex conductivity vs. interface parameters: (a) complex interface conductivity vs. S_{POR} for sandstones (Boerner 1992); (b) imaginary part vs. d_{10} for unconsolidated rocks (Slater and Lesmes 2002, Copyright American Geophysical Union); (c) imaginary part vs. Q_V (Vinegar and Waxman 1984)

Vinegar and Waxman (1984) developed on the basis of the Waxman-Smits-model a complex conductivity model with cation exchange capacity per unit pore volume Q_V as a shalyness parameter.

$$\sigma_0'' = \lambda Q_V / F_q \tag{4.16}$$

Eq. 4.16 is suitable for water saturated shaly sands (see Fig. 4.12c). λ is a function of pore water salinity. The shalyness parameter Q_V is defined as the cation exchange capacity CEC per unit pore volume:

$$Q_V = \mathrm{CEC} d_M (1 - \Phi) / \Phi \tag{4.17}$$

where d_M is the matrix density. CEC is an easily measurable quantity and hydrogeologists often use it to quantify the adsorption and retardation in the aquifer and the soil zone. However, chemical Q_V-estimation is a destructive method with a limited resolution for clay-free unconsolidated sand and gravel with very low Q_V-values.

A model that is additionally applicable for clay-free as well as for unconsolidated material is based on Rink and Schoppers (1974) S_{POR}-concept (Börner 1992). The size of the electrical active interface is described by using the surface-area-to-porosity ratio S_{POR} (Fig. 4.12a), :

$$S_{POR} = S_M d_M (1 - \Phi) / \Phi \tag{4.18}$$

The surface area per matrix mass S_M is easily measurable on almost any material using e.g. the nitrogen adsorption method. In case of water saturated rock the imaginary part σ_0'' is then described as

$$\sigma_0'' = \frac{l f(\sigma_w) S_{por}}{F} \tag{4.19}$$

where F, for purposes of simplicity, is the same formation factor for all conductivity components, $f(\sigma_W)$ is a general function concerning salinity dependence of interface conductivity and depending on surface charge density and the ion mobility, and l is the ratio between real and imaginary component of interface conductivity that is assumed to be nearly independent of salinity. Lesmes and Frye (2001) used a similar model in which the magnitude of interface conductivity is primarily determined by a weighted surface-to-volume-ratio S_0 and a geometric factor f (see also Table 4.1). Slater and Lesmes (2002b) found a linear relation between σ'' and the grain size related parameter d_{10} (Fig. 4.12b).

Sometimes the internal surface area of sand is related to the content of fine grained iron minerals. The effect was analyzed by using natural sands with different contents of iron hydroxide (Fig. 4.13).

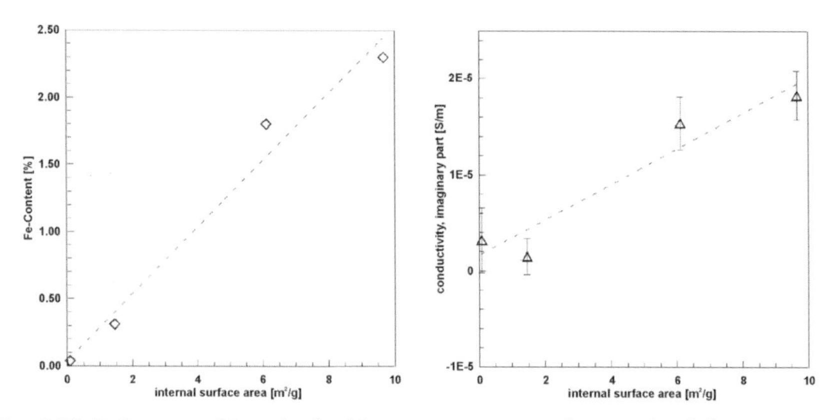

Fig. 4.13. Influence of iron-hydoxide content on complex conductivity of unconsolidated sands: (a) Fe-content vs. internal surface area, (b) imaginary part vs. internal surface area for the same samples

Partially water saturated rock

Both conductivity components in Eq. (4.14) are specifically dependant on water saturation S_W (see Fig. 4.14). According to Waxman and Smits (1968), two petrophysical effects result in a specific dependence of the interface conductivity on decreasing water saturation:

- reduction of cross section area for conduction paths in the water phase
- increasing concentration of counter ions near the matrix-water-interface.

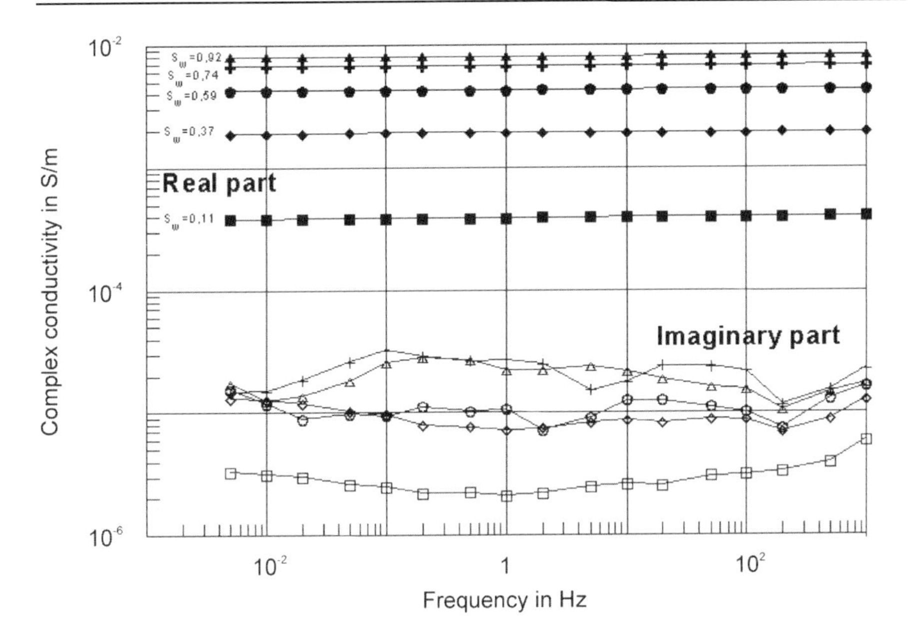

Fig. 4.14. Dependence of complex conductivity on water saturation for unconsolidated sand (Gruhne 1999)

For this reason the interface-terms have lower saturation exponents n* and n** than the ARCHIE term. The real part of an unsaturated rock $\sigma'_{0,t}$ and the accompanying imaginary part $\sigma''_{0,t}$ have the general equations

$$\sigma'_{0,t} = \sigma_{el} S_W^n + \sigma'_i f(\sigma_W) S_W^{n^*} \tag{4.20a}$$

$$\sigma''_{0,t} = \sigma''_i f(\sigma_W) S_W^{n^{**}} \tag{4.20b}$$

Laboratory measurements (DC-conductivity and complex conductivity) show that n* takes values between 0.5 and 1.5 (e.g. Waxman and Smits 1968, Vinegar and Waxman 1984, Boerner 1992, Gruhne 1999, Ulrich and Slater 2004). The determination of the exponent n* is very complicated and expensive from an experimental point of view. Practical applications sometimes necessitate two simplifications:

- the exponent of the interface term is related to the exponent of the ARCHIE-term e.g. by n*=n-1 (Waxman and Smits 1968) and
- the real and the imaginary part of interface conductivity have the same saturation exponent n*=n**.

The advantage is that only the exponent n has to be determined experimentally. The other exponents can be deduced from n.

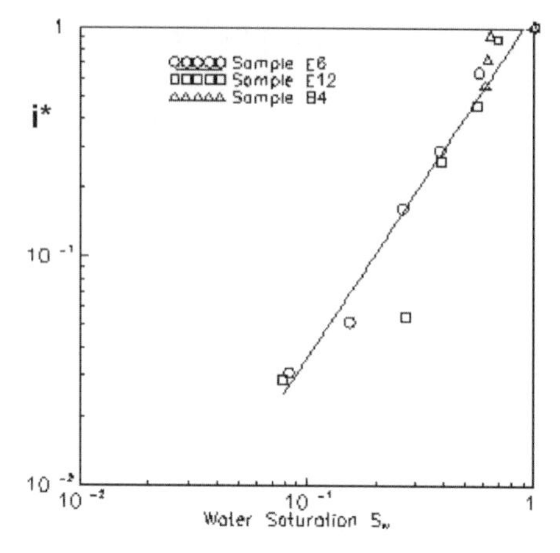

Fig. 4.15. Variation of i* with water saturation for three different sandstones (Boerner 1992)

Another simplification is the use of an apparent saturation exponent n_a for the real part of conductivity and an independent exponent n** for the imaginary part:

$$\sigma'_{0,t} = S_W^{n_a}\left[\sigma_{el} + \sigma'_i f(\sigma_W)\right] \tag{4.21a}$$

$$\sigma''_{0,t} = \sigma''_i f(\sigma_W) S_W^{n^{**}} \tag{4.21b}$$

n_a and n* may be directly obtained from multi phase experiments with a complex electrical monitoring (see Eq. 4.18).

The disadvantage is that n_a in Eq. (4.21a) depends on water salinity, interface conductivity, and water saturation. This effect is plausible, because electrolytical and interface conductivity have different types of dependency on salinity and water saturation. The conductivity is dominated by the interface component at low salinity and low water saturation. The electrolytical conductivity is the dominating component at high salinity and high water saturation. In case of material with high internal interface or cation exchange capacity the effect is more significant.

A "complex" resistivity index *i** was derived from the imaginary part of complex rock conductivity (Vinegar and Waxman 1984):

$$i^*(S_w) = \frac{\sigma''_{0,t}(S_w)}{\sigma''_0(S_w = 1)} \tag{4.22}$$

i* depends on the water saturation but it is independent of the rock texture. Figs. 4.15 and 4.16 show the decrease of *i** with decreasing water saturation for sandstone samples which are different in their S_{POR}- or Q_V-values respectively.

The dependence of both conductivity components on water saturation is plotted in Fig. 4.17 for an unconsolidated sand (Gruhne 1999). The slope of the two curves is characterized by the exponents n_a and n^{**}.

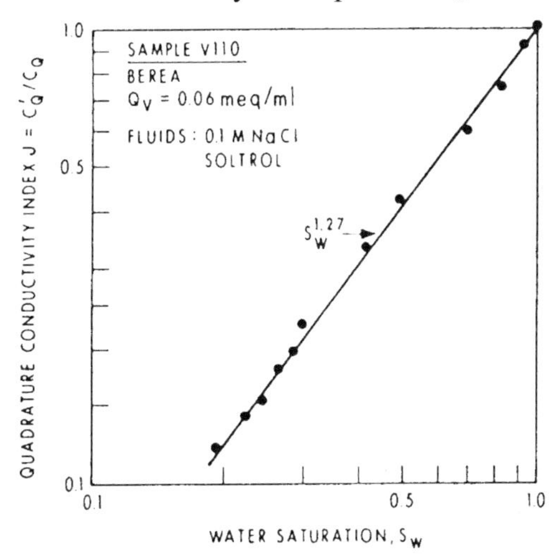

Fig. 4.16. Variation of i* with water saturation for a shaly sandstone (Vinegar and Waxman 1984)

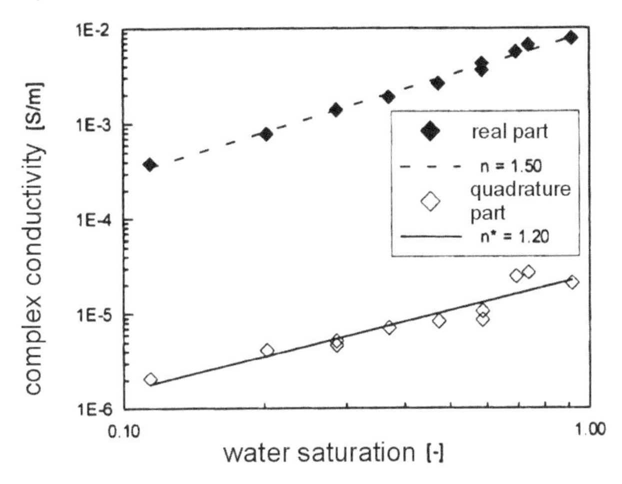

Fig. 4.17. Real and imaginary part of conductivity vs. water saturation measured during a three-fluid-phase-experiment (Gruhne 1999)

4.5 The potential of complex conductivity for environmental applications

4.5.1 Organic and inorganic contaminants

Changes of rock components in the vicinity of waste disposals or other contamination sources can be caused by migration of hazardous substances resulting from gradients in hydraulic pressure, chemical potential, or temperature. The electrical conductivity of water-bearing rock or soil is changed due to an increasing content of various substances in the pore space. The contrast between e.g. the leachate and the native groundwater is reflected in the contrast between contaminated and uncontaminated rock or soil. In this case a simple interpretation can be given on the basis of the Archie-equation (4.15). However, in cases of complex inorganic or organic contamination the conditions are much more difficult. Various substances influence electrical processes differently and may have opposite effects. Generally, the complex conductivity is influenced by changes of

- composition of the pore fluid (solute substances),
- volumetric content of nonaquaous phase liquids and
- grain surface-water-interphase microstructure.

The interface region of a rock or soil can be expected to be very sensitive to changes in material composition caused by organic and inorganic contamination. To quantify this effect independently on rock or soil texture a so-called normalized interface conductivity B_n

$$B_n = \frac{\sigma''_{0,\mathrm{cont.}}}{\sigma''_{0,\mathrm{clean}}} \tag{4.23}$$

was defined (Boerner et al. 1993). B_n depends only on fluid properties and resulting changes of interface microstructure. Therefore, B_n seems to be an indicator for contamination induced changes in physical properties of pore water like dielectric permittivity, type of dissolved salts but also larger concentrations of organic compounds.

The first experimental example in Fig. 4.18 shows the variation of B_n with varying amounts of a polar organic substance in pore water. The sample was contaminated with methanol (CH_3OH, $\varepsilon_r = 34$) of different volumetric content of the water phase. A similar behaviour has been found for contaminants like hexane, dichlormethane, benzene, toluoene. The frequency dependence decreases with increasing content and decreasing dielectric constant of the contaminant.

The effect of different salt types on the magnitude of interface conductivity is shown in Fig. 4.19. Different mobility and cation radii cause deviations in the frequency dependence of conductivity. A similar behaviour was found by Kulenkampff (1994) for the real part of interface conductivity by using the multiple salinity method.

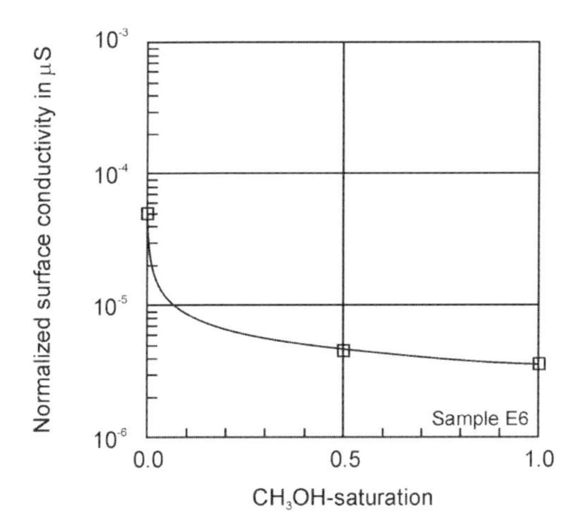

Fig. 4.18. Variation of the imaginary part of conductivity with polar organic substance content (Boerner et al. 1993)

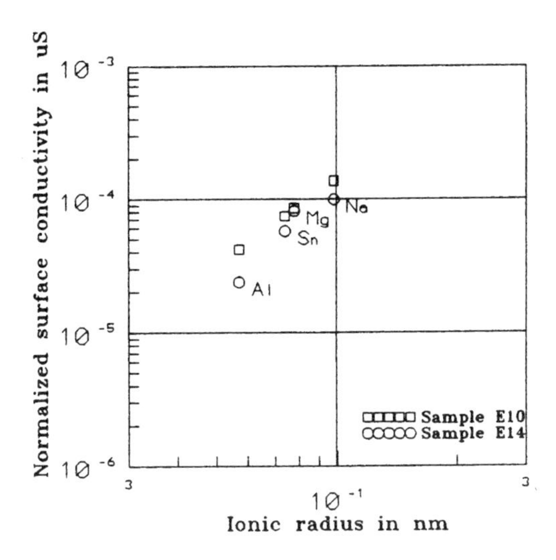

Fig. 4.19. Variation of imaginary part of conductivity with salt type dissolved in pore water, experimental results for NaCl, $MgCl_2$, $AlCl_3$ and $SnCl_4$

Another objective is the detection of nonaqueous phase liquids (NAPL) like kerosene, diesel fuel, oil, xylene, or TCE (see Figs. 4.20 and 4.21). Vanhala et al. (1992) showed that a NAPL-contamination influences the conductivity spectra of glacial till. The same effect was found in an experiment by Boerner et al. (1993) on sand and clay samples. In contrast to these results Gruhne (1999) showed that a NAPL-contamination does not significantly influence the complex conductivity in the soil zone when the water content is constant.

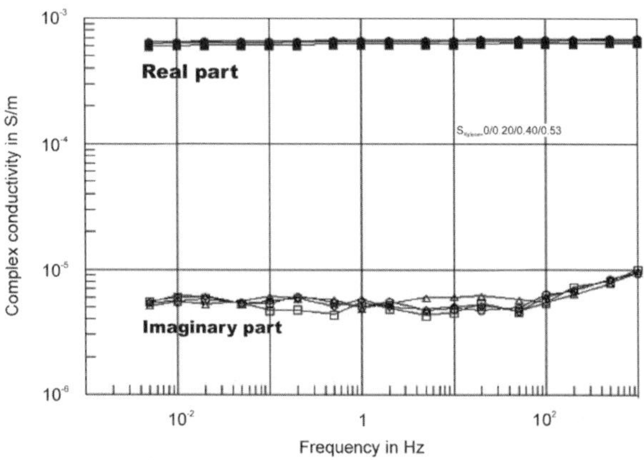

Fig. 4.20. Complex rock conductivity of LNAPL (xylene)-contaminated sand (Gruhne 1999)

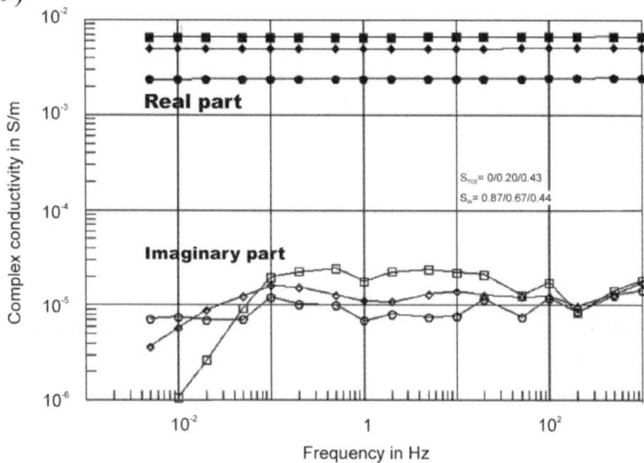

Fig. 4.21. Complex rock conductivity of DNAPL (TCE)-contaminated sand (Gruhne 1999)

In case of a NAPL-contamination different effects influence the complex electrical properties of the rock:

- Changing the water content (aquifer as well as soil zone),
- Solution of components of the NAPL-phase in the pore water,

- Microbiological degradation and its solution in the water phase (Atekwana et al. 2004).

Weller et al. (1996) present the result of a complex electrical Schlumberger sounding in a NAPL-contaminated brown field (Fig. 4.22). The contaminated layer was identified by its low conductivity and low phase shift. In some cases the phase shift changed its sign.

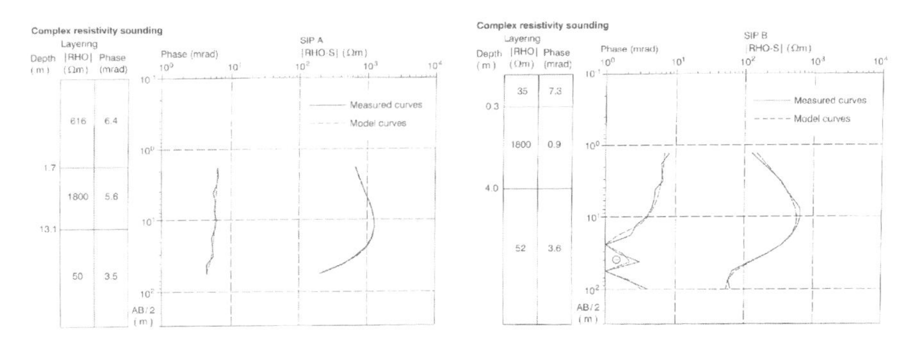

Fig. 4.22. Contamination indication in a complex conductivity sounding (Weller et al. 1996, permission from Springer Verlag GmbH)

4.5.2 Monitoring subsurface hydraulic and migration processes

Controlled large-scale experiments are necessary to bridge the gap between small scale laboratory investigations and field scale applications. In contrast to field investigations, experiments in large containers offer controlled and reproducible conditions which are the basis for petrophysical modelling and the testing of geophysical equipment.

The detectability of organic contaminations with complex conductivity measurements was investigated by Gruhne (1999) under controlled hydraulic and chemical conditions. He used multielectrode complex conductivity measurements to monitor the movement and the fluid-distribution of NAPL-contaminants during large-scale hydraulic experiments at the VEGAS-test site of Stuttgart University/Germany. To prevent electrode polarization and noise all electrodes were made of platinized platinum mesh and then connected with screened wire. Fig. 4.23 shows the 80 m^3-container used for 3D- experiments. It was filled with medium grained sand. The upper part of the container was divided into two hydraulically connected parts: One part was used as a reference where the second part was contaminated with NAPL.

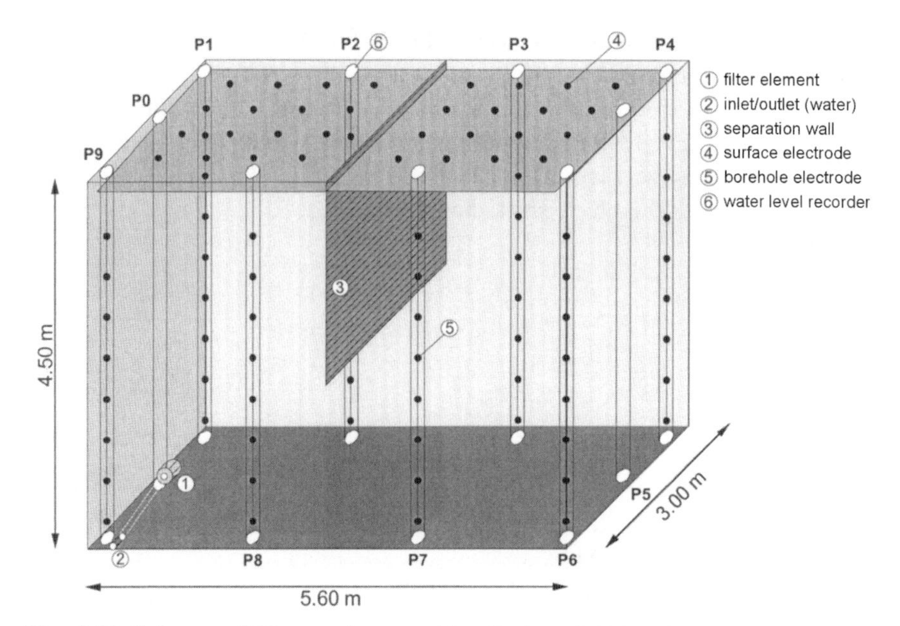

Fig. 4.23. Schema of 3D-container used by Gruhne (1999) with monitoring wells, electrode array and a wall to separate contaminated and uncontaminated part

The experiment includes monitoring of vertical NAPL-distribution as a result of a fluctuating water table (Fig. 4.24) and the monitoring of tracer infiltration (Fig. 4.25).

In a 15 m long and 3 m deep sand-filled channel two-dimensional electrode arrays were installed perpendicular to the flow direction. The typical break through indication of a NAPL is shown in Fig. 4.26. The relatively large NAPL-volume produces a significant effect in the potential and phase shift of a dipol-dipol-configuration. Between indications in potential and phase shift a delay of about 10 minutes was observed.

The dependence of the fluid distribution of a xylene-contamination on the depth of water table was monitored with complex electrical measurements during a hydraulic experiment (Fig. 4.27). Gruhne (1999) found that

- the residual NAPL-contamination below the water table can be detected relatively easy and may be quantified.
- a quantitative detection of the NAPL-phase in capillary and soil zone was possible only with additional hydraulic information like vertical distribution of air content in pore space.

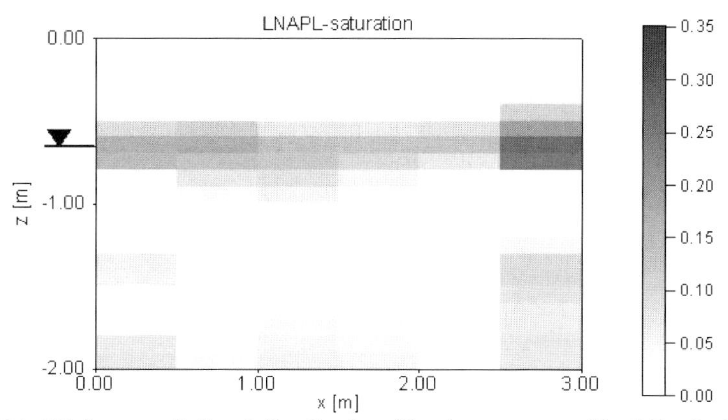

Fig. 4.24. 2D-image of diesel-distribution (depth to water table 0.7 m) obtained from an integrated electrical and hydraulic measurement (Giese 2001) in the container showed in Fig. 4.23

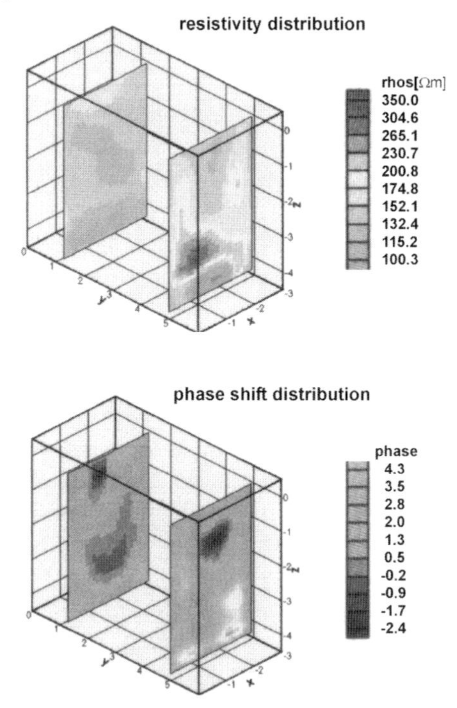

Fig. 4.25. Resistivity and phase shift in two sectional planes of 3D-container approx. 11 h after infiltration of deionised water at point (x=-1.5 m, y=6 m, z=-4 m). From Gruhne (1999) with 3D-inversion after Weller et al. (1996)

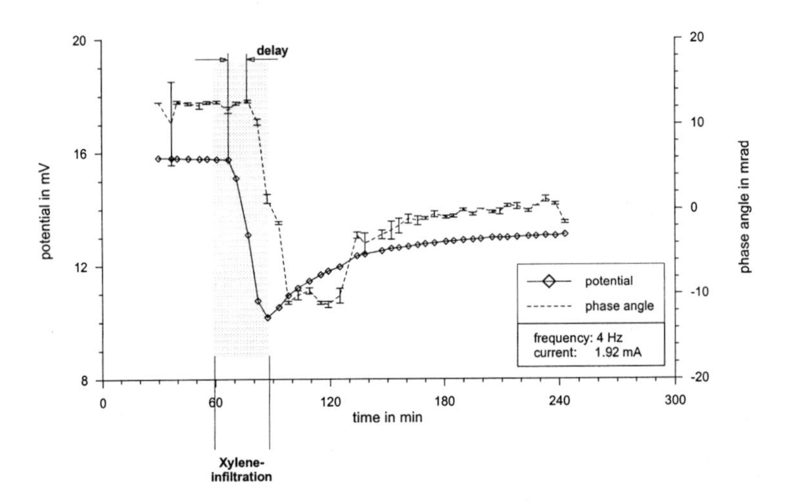

Fig. 4.26. Delay between break through indications of potential and phase shift during a Xylol-infiltration in sand

4.5.3 Geohydraulic parameters

Fluid flow, migration of substances and electrical conduction are transport processes governed by the geometry of the pore network of the rock. Additionally, these processes are significantly related to the microstructure of the mineral grain surfaces.

Many different models have been developed to relate hydraulic conductivity to electrical parameters (e.g. Brace 1977, Wong et al. 1984, Katz and Thompson 1986). Walsh and Brace (1984) used the direct correlation between formation factor F and hydraulic conductivity. Mazac et al. (1985) developed a model for relations between electrical and hydraulic properties based also on the formation factor. A review of many hydraulic conductivity models used in petrophysics is given in Schoen (1996). Most of the relations are valid for only one formation because the hydraulic conductivity estimation is based on the formation factor F alone. The very different exponents found for various rock types are due to significant differences of the internal surface area.

Starting point of a pragmatic way of geophysical hydraulic conductivity estimation is the Kozeny-Carman-equation,

$$k = \frac{\Phi}{c} \frac{m^2}{T} \tag{4.24}$$

a generalized Hagen-Poiseulle-equation for arbitrary capillary cross sections. It relates the hydraulic conductivity of a porous medium amongst the porosity Φ with its hydraulic radius m, the tortuosity T and further structural parameters (Wyllie and Gardner 1958, Engelhardt 1960). The use of KC-type equations is therefore a promising way to estimate hydraulic conductivity from geophysical proxies.

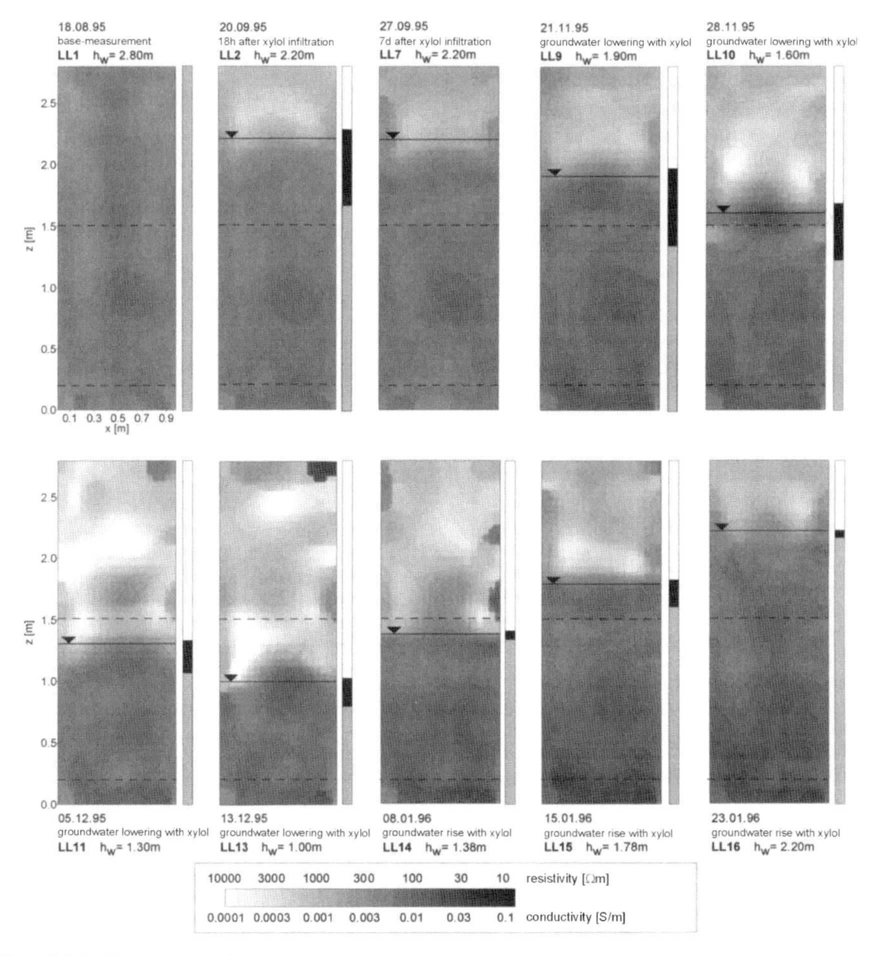

Fig. 4.27. Tomographic image of real part of conductivity, variation of depth to water table in a sand-filled channel contaminated with xylene

The hydraulic radius m is proportional to the reciprocal value of the surface-area-to-porosity-ratio $1/S_{POR}$. Sen et al. (1981) analysed the relationship between hydraulic conductivity, surface area-to porosity ratio and porosity by using a self-similar sandstone model. Pape et al. (1981, 1987) showed that the effective hydraulic radius m can be deduced from the surface area-to-porosity ratio measured with the laboratory BET-method for a given fractal dimension of porous material. Porosity and tortuosity in Eq. 4.24 can be replaced by the true formation factor F. The resulting PaRiS-equation (Pape et al. 1987) for consolidated rock

$$k = 475 \frac{1}{F S_{POR}^{3.1}} \tag{4.25}$$

is a modified Kozeny-Carman-equation, which relates the hydraulic conductivity to the electrical determined formation factor and a "non-geophysical" determined internal surface area.

The calculation of the hydraulic conductivity from geophysical "proxies" alone is possible on the basis of complex electrical measurements and the assumption of a cpa-response (Boerner et al. 1996). The determination of the real and imaginary part of conductivity allows the separation of the electrolytic bulk conductivity. The true formation factor F can be calculated from the ARCHIE-equation when the brine conductivity σ_w is known. The surface area-to-porosity ratio S_{POR}^{el} is directly related to the imaginary part of conductivity (Boerner and Schön 1991):

$$k = \mathrm{const}_1 \frac{1}{F\left(S_{POR}^{el}\right)^{\beta}} \approx \mathrm{const}_2 \frac{1}{F\left(\sigma_0''\right)^{\beta_2}} \tag{4.26}$$

The model enables an estimation of the hydraulic conductivity from F and S_{POR} obtained from one single complex electrical measurement (Fig. 4.28). The model relies on the constant phase angle model in the electrical response of the aquifer material at different frequencies. Slater and Lesmes (2002b) used the linear relationship between the imaginary part of conductivity and the grain size diameter that is larger than 10 percent of the sample for the hydraulic conductivity estimation (see Fig. 4.12b).

Some of the first geophysical hydraulic conductivity estimations in field scale were made on a test site in the catchment area of a drinking water supply near Elsnig in Northern Saxony (Germany). In order to get representative inverted data for hydraulic conductivity calculation simple one-dimensional complex resistivity soundings in Schlumberger configuration with spacing from 1.8 m to 125 m were used (Boerner et al. 1996).

The data including amplitude and phase values of five frequencies were processed as shown in Fig. 4.28. The complex inversion procedure results in two sounding curves: one for the real and one for the imaginary part of the apparent conductivity (see Fig. 4.29). The complex sounding curve was

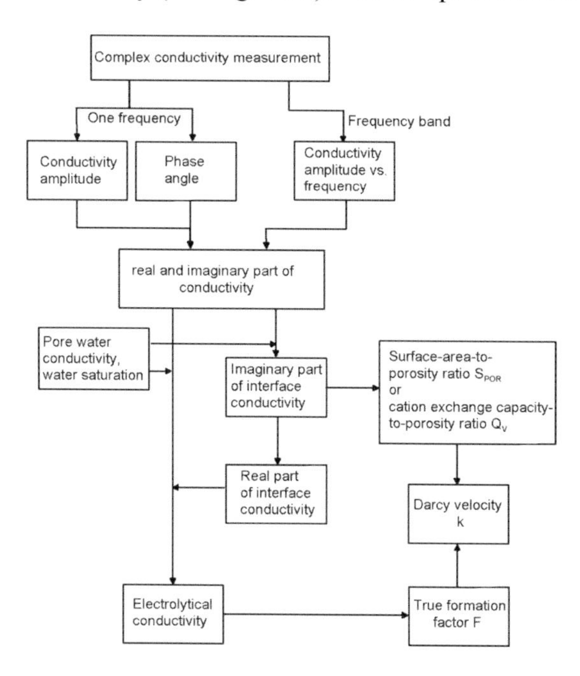

Fig. 4.28. Calculation scheme of geohydraulic parameters from complex conductivity measurements

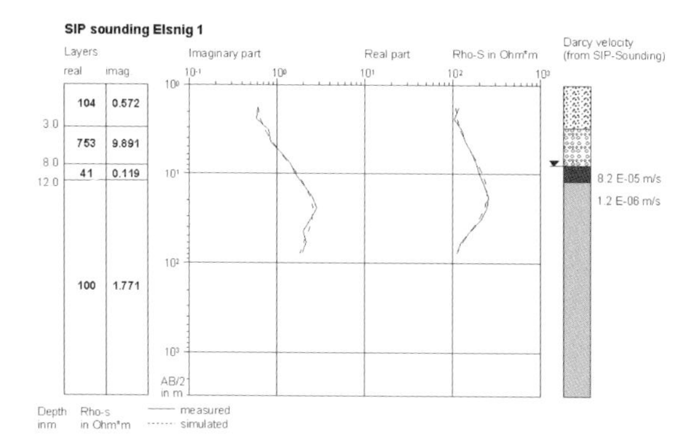

Fig. 4.29. Example of a 1-dimensional complex resistivity-sounding (SIP) at the Elsnig test site in Saxony/Germany and calculation of darcy-velocity (Boerner et al. 1996)

inverted by software VES-K (Weller et al. 1996). The result of the one-dimensional inversion is a model consisting of a certain number of layers, giving for each layer a thickness and a complex conductivity. The integration of the cpa-model according to Eq. 4.7 into the procedure provides the basis for the calculation of the conductivity components. Both the real and the imaginary part of the layer conductivity are used for further interpretation in order to find the hydraulic conductivity k of the layer.

Fig. 4.30 shows a comparison of field scale estimations of hydraulic conductivity from complex conductivity soundings with data obtained from grain size distribution. The data were collected at the Elsnig test site and the Zeithain brownfield area both in Saxony. Each data point represents one "electrical" layer which is nearly identical with an aquifer. The electrical k-value is therefore an integrated value for the whole thickness of the aquifer. The lab data are obtained from depth-specific sampling within the same aquifer. The result is a broad variation in k-values for the layer in question.

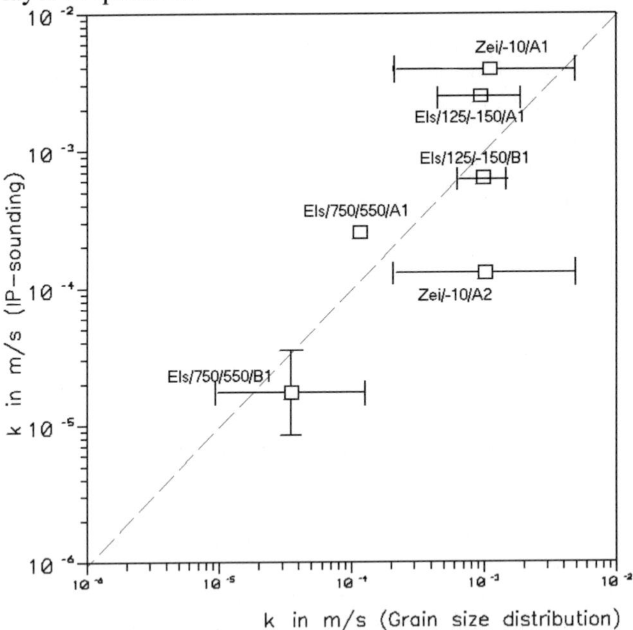

Fig. 4.30. Hydraulic conductivity estimated from field data (Spectral IP-Measurements) vs. hydraulic conductivity from grain size parameters (Boerner et al. 1996, permission from Blackwell Publishing)

Recent examples of parameter estimation based on complex conductivity tomography are presented by Andreas Hoerdt in Chap. 15.

4.6 References

Archie GE (1942) The electrical resistivity log as an aid in determining some reservoir characteristics. Transactions of the American Institute of Mining, Metallurgical and Petroleum Engineers 146:54-62

Atekwana EA, Werkema DD, Duris JW, Rossbach S, Atekwana EA, Sauck WA, Cassidy DP (2004) In-situ apparent conductivity measurements and microbial population distribution at a hydrocarbon contaminated site. Geophysics 69:56-63

Berger W, Börner F, Petzold H (2001) Consecutive geoelectric measurements reveal the downward movement of an oxidation zone. Waste Management 21:117-125

Berthold S, Bentley LR, Hayashi M (2004) Integrated hydrogeological and geophysical study of depression-focused groundwater recharge in the Canadian prairies. Water Resour Res 40: 1029-1039

Boerner FD (1991) Investigation of the complex conductivity between 1 MHz and 10 kHz. PhD-Thesis, Mining Academy Freiberg

Boerner FD (1992) Complex conductivity measurements of reservoir properties. Proc Third European Core Analysis Symposium, Paris, pp 359-386

Boerner FD (2001) A novel study of the broad band complex conductivity of various porous rocks. Proc 2001 International Symposium of the SCA, Edinburgh, UK, SCA-2001-38

Boerner FD, Schön JH (1991) A relation between the quadrature component of electrical conductivity and the specific surface area of sedimentary rocks. The Log Analyst 32:612-613

Boerner FD, Schön JH (1995): Low Frequency Complex Conductivity Measurements of microcrack Properties. Surveys in Geophysics, Kluwer Academic Publishers

Boerner FD, Gruhne M, Schön JH (1993) Contamination indications derived from electrical properties in the low frequency range. Geophysical Prospecting 41:83-98

Boerner FD, Schopper W, Weller A (1996) Evaluation of Transport and storage properties in the soil and groundwater zone from induced polarization measurements. Geophysical Prospecting 44:583-601

Brace WF (1977) Permeability from resistivity and pore shape. Journal of Geophysical Research 82:334-339

Buchheim W, Irmer G (1979) Zur Theorie der induzierten galvanischen Polarisation in Festkörpern mit elektrolytischer Porenfüllung. Gerlands Beiträge Geophysik 88:53-72

Busch K-F, Luckner L, Tiemer K (1993) Geohydraulik. Gebrüder Bornträger, Berlin-Stuttgart

Clavier C, Coates G, Dumanoir J (1977) Theoretical and experimental basis for the "Dual water" model for interpretation of shaly sands. Proceedings of the 52nd Annual Technical Conference and Exhibition of the Society of Petroleum Engineers, Denver, SPE 6859

Cole KS, Cole RH (1941) Dispersion and absorption in dielectrics. J Chem Phys 9:341

Comas X, Slater L (2004) Low frequency electrical properties of peat. Water Ressources Research 40 W12414

De Lima, OAL (1995) Water saturation and permeability from resistivity, dielectric, and porosity logs. Geophysics 60:1751-1764

Daily WD, Ramirez AL, LaBrecque DJ, Nitao J (1992) Electrical resistivity tomography of vadose water movement. Water Ressources Research 28:1429-1442

Dissado L, Hill RM (1984) Anomalous low frequency dispersion. J Chem Soc Faraday Trans 80:291-319

Engelhardt W v (1960) Der Porenraum der Sedimente. Springer Publishers Berlin, Göttingen, Heidelberg

Fechner T, Boerner FD, Richter T, Yaramanci U, Weihnacht B (2004) Lithological interpretation of the spectral dielectric properties of limestone. Near Surface Geophysics:150-159

Giese R (2001) Zur Hydraulik dreier nichtmischbarer Fluide in porösen Medien., Proceedings des DGFZ 22, Dresden

Grissemann C, Rammlmair D, Siegwart C, Foullet N (2000) Spectral induced polarization linked to image analyses: A new approach. In : Rammlmair et al. (eds) Applied mineralogy. Balkena, Rotterdam pp 561-564

Gruhne M (1999) Überwachung von Untergrundkontaminationen mit Messungen der komplexen elektrischen Leitfähigkeit. Proceedings des DGFZ 16, Dresden

Jonscher AK (1981) A new understanding of the dielectric relaxation of solids. Journal of Material Sciences 16:2037-2060

Kalinsky RJ, Kelly WE, Bogardi I, Pesti G (1993) Electrical resistivity measurements to estimate travel times through unsaturated ground water protective layers. Journal of Applied Geophysics 30:161-173

Katz AJ, Thomson AH (1986) Quantitative prediction of permeability in porous rock. Physical Review B 34:8179-8181

Kemna A (2000) Tomographic inversion of complex resistivity. PhD-Thesis. Berichte des Inst. Geophysik der Ruhr-Universität Bochum, Reihe A, 56

Kemna A, Binley A, Slater L (2004) Crosshole IP imaging for engineering and environmental applications. Geophysics 69: 97-107

Klein J, Biegler T, Horne M (1984) Mineral interfacial processes in the method of induced polarization. Geophysics 49:1105-1114

Klitsch N (2003) Ableitung von Gesteinseigenschaften aus Messungen der spektralen induzierten Polarisation an Sedimentgesteinen. PhD-Thesis, Inst Geophysik and Geol Univ Leipzig und Inst Angewandte Geophysik RWTH Aachen

Korvin G (1992) Fractal Models in the Earth Sciences. Elsevier, Amsterdam, 396p

Kirsch R (2000) Geophysikalische Oberflächenmethoden. In: Balke KD (ed) Grundwassererschließung. Gebrüder Borntraeger, Berlin, Stuttgart

Knödel K, Krummel H, Lange G. (1997) Handbuch zur Erkundung des Untergrundes von Deponien und Altlasten, Bd. 3 Geophysik. Springer Berlin-Heidelberg

Kulenkampff J (1994) Die komplexe elektrische Leitfähigkeit poröser Gesteine im Frequenzbereich von 10 Hz bis 1 MHz – Einflüsse von Porenstruktur und Porenfüllung. PhD-Thesis, Technical University Clausthal
Kulenkampff J, Schopper JR (1988) Low frequency complex conductivity - a means for separating volume and interlayer conductivity. In Transactions of the 12th European Formation Evaluation Symposium, Oslo
Kulenkampff J, Boerner FD, Schopper JR (1993) Broad band complex conductivity laboratory measurements enhancing the evaluation of reservoir properties. Trans of the 15th Europ Form Evaluation Symp, Stavanger
Lesmes DP, Frye KM (2001) Influence of pore fluid chemistry on the complex conductivity and induced polarization responses of Berea sandstaone. Journal of Geophysical Research 106:4079-4090
Liu SH (1985) Fractal model for the ac response of a rough interface. Phys Rev Letters 55:529-532
Liu S, Yeh T-CJ (2004) An integrativ Approach for monitoring water Movement in the vadose zone. Vadose Zone Journal 3:681-692
Lockner DA, Byerlee JD (1985) Complex resistivity measurements of confined rock. Journal of Geophysical Research 90:7837-7847
Marshall DJ, Madden TR (1959) Induced Polarization, a study of its causes. Geophysics 24:790-816
Mazac O, Kelly WE, Landa I (1985) A hydrogeophysical model for relations between electrical and hydraulic properties of aquifers. Journal of Hydrology 79:1-19
Niederleithinger E, Grissemann C, Rammlmair D (2000) SIP geophysical measurements on slag heaps: A new way to get information about subsurface structures and petrophysical parameters. In: Rammlmair et al. (eds) Applied mineralogy, Balkena, Rotterdam
Olhoeft GR (1985) Low frequency electrical properties. Geophysics 50:2492-2503
Pape H, Grinat M, Vogelsang D (1992) Logging of induced polarization in the KTB-Oberpfalz VB interpreted by a fractal model. Scientific Drilling 3:105-114
Pape H, Riepe L, Schopper JR (1981) Calculating Permeability from surface area Measurements. Trans 7th Europ Form Evaluation Symposium, Paris
Pape H, Worthington PF (1983) A surface-structure model for the electrical conductivity of reservoir rocks. Transactions of the 8th European Formation Evaluation Symposium, London
Pape H, Riepe L, Schopper JR (1987) Theory of self-similar network structures in sedimentary and igneous rocks and their investigation with microscopical and physical methods. Journal of Microscopy 148:121-147
Parasnis DS (1966) Mining Geophysics. Elsevier, Amsterdam, New York
Riepe L, Rink M, Schopper JR (1979) Relations between specific surface dependent rock properties. Transactions of the 6th European Logging Symposium, London
Rink M, Schopper JR (1974) Interface conductivity and its implication to electric logging. Transactions of the 15th Annual Logging Symposium, London

Ruffet C, Gueguen Y, Darot M (1991) Complex conductivity measurements and fractal nature of porosity. Geophysics:56 758-768

Schön J, Boerner F (1985) Untersuchungen zur elektrischen Leitfähigkeit von Lockergesteinen - der Einfluß matrixbedingter Leitfähigkeitsanteile. Neue Bergbautechnik 15:220-224

Schön J (1996) Physical Properties of Rocks: fundamentals and principles of petrophysics. In: Helbig, K., Treitel, S., (eds) Handbook of geophysical exploration. Section I, Seismic exploration. Elsevier, Oxford

Sen PN, Scala C, Cohen MH (1981) A self similar model for sedimentary rocks with application to the dielectric constant of fused glass beads. Geophysics 46:781-795

Sen PN, Goode PA, Sibbit A (1988) Electrical conduction in clay bearing sandstones at low and high salinities. Journal of Applied Physics 63:832-4840

Simandoux P (1963) Dielectric measurements on on porous media: application to the measurement of water saturation: study of the behaviour of argillaceous formations. Revue de l'Institut Francais du Petrol 18, supplementary issue:93-215

Slater LD, Lesmes DP (2002a) Electric-hydraulic relationships observed for unconsolidated sediments. Water ressources research 8:-13

Slater LD, Lesmes DL (2002b) IP interpretation in environmental investigations. Geophysics 7:7-88

Spangenberg E (1995) Ein fraktales Modellkonzept zur Berechnung physikalischer Gesteinseigenschaften und dessen Anwendung auf die elastischen Eigenschaften poröser Gesteine. PhD thesis. Scientific Technical Report STR95/23, Geoforschungszentrum Potsdam

Stenson JD, Sharma MM (1989) A petrophysical model for shaly sands. In Proceedings of the 64th Annual Technical Conference and Exhibition of the Society of Petroleum Engineers, San Antonio, paper SPE 19574

Sumner JS (1976) Principles of induced polarization for geophysical exploration. Elsevier, Amsterdam, Oxfort, New York

Titov K, Komarov V, Tarasov A, Levitski A (2002) Theoretical and experimental study of time-domain induced polarization in water-saturated sands. Journal Applied Geophysics 50:417-433

Titov K, Kemna A, Tarasov A, Vereecken H (2004) Induced polarization of unsaturated sands determined through time domain measurements. Vadose Zone Journal 3:1160-1168

Ulrich C, Slater LD (2004): Induced Polarization on unsaturated, unconsolidated sands. Geophysics 68:762-771

Vanhala H, Oininen H, Kukkonen I (1992): Detecting organic chemical contaminants by spectral-induced polarization method in glacial till environment. Geophysics 57:1014

Van Voorhis GD, Nelson PH, Drake TL (1973) Complex resistivity spectra of porphyry copper mineralization. Geophysics 38:49-60

Vinegar HJ, Waxman MH (1984) Induced polarization of shaly sand. Geophysics 49:1267-1287

Wait JR (1959) Overvoltage Research and Geophysical Applications. Pergamin Press, London, New York, Paris

Walsh JB, Brace WF(1984) The nature of pressure on porosity and the transport properties of rocks. J. Geophysical Research 89:9425-9431

Ward SH (1990) Resistivity and Induced Polarization Methods. In: Ward SH (ed) Geotechnical and Environmental Geophysics. SEG Series Investigations in Geophysics, Vol. 5

Ward SH, Fraser DC (1967) Conduction of electricity in rocks. In: Mining Geophysics, Vol. II, Tulsa, SEG

Waxman MH, Smits LJM (1968) Electrical Conductivities in oil-bearing Shaly Sands. Society of Petroleum Engineers Journal 243:107-122

Waxman ML, Thomas EC (1974) Electrical conductivities in shaly sands. Trans AIME 257:213–225

Weihnacht B, Boerner F (2005) Ermittlung geohydraulischer Parameter aus kombinierten geophysikalischen Messungen im Technikumsmaßstab. Proc. 65. JT der DGG, Graz, p. 39

Weller A, Boerner F (1996) Measurements of Spectral Induced Polarization for Environmental Purposes. Environmental Geology 27:329-334

Weller A, Gruhne M, Seichter M, Börner F (1996) Monitoring hydraulic experiments by complex conductivity tomography. European Journal of Environmental and Engineering Geophysics 1:209-228

Wong PZ, Koplik J, Tomanie JP (1984) Conductivity and Permeability of rocks. Physical Review B 30:6606-6614

Wyllie MR, Gardner GHF (1958) The generalized Kozeny-Carman equation. World Oil 3:210

5 Electromagnetic methods – frequency domain

5.1 Airborne techniques

Bernhard Siemon

5.1.1 Introduction

Modern frequency domain airborne electromagnetic (AEM) systems utilise small transmitter and receiver coils having a diameter of about half a metre. The transmitter signal, the primary magnetic field, is generated by sinusoidal current flow through the transmitter coil at a discrete frequency. As the primary magnetic field is very close to a dipole field at some distance from the transmitter coil, it can be regarded as a field of a magnetic dipole sitting in the centre of the transmitter coil and having an axis perpendicular to the area of the coil. The oscillating primary magnetic field induces eddy currents in the subsurface. These currents, in turn, generate the secondary magnetic field which is dependent on the underground conductivity distribution. The secondary magnetic field is picked up by the receiver coil and related to the primary magnetic field expected at the centre of the receiver coil. As the secondary field is very small with respect to the primary field, the primary field is generally bucked out and the relative secondary field is measured in parts per million (ppm). Due to the induction process within the earth, there is a small phase shift between the primary and secondary field, i.e., the relative secondary magnetic field is a complex quantity. The orientation of the transmitter coil is horizontal (VMD: vertical magnetic dipole) or vertical (HMD: horizontal magnetic dipole) and the receiver coil is oriented in a maximum coupled position, resulting in horizontal coplanar, vertical coplanar, or vertical coaxial coil systems.

5.1.2 Theory

Calculation of secondary field values

The secondary magnetic field for a stratified subsurface caused by an oscillating (frequency f) magnetic dipole source in the air is calculated using well-known formulae (e.g. Wait 1982, Ward and Hohmann 1988). They are based on Maxwell's equations and solve the homogeneous induction equation in the earth for the electromagnetic field vector **F**

$$\frac{d^2F}{dz^2} = \nu^2 F, \quad \nu^2 = \lambda^2 + i\omega\mu\sigma - \omega^2\mu\varepsilon \tag{5.1}$$

assuming a homogeneous and isotropic resistivity ρ, which is the reciprocal of the conductivity σ, $\omega = 2\pi f$ is the angular frequency, λ is the wave number, and $i = (-1)^{1/2}$ is the imaginary unit. Magnetic effects and displacement currents are normally neglected, i.e., the magnetic permeability μ is set to that of free space, $\mu = \mu_0 = 4\pi \times 10^{-7}$ Vs/Am, and the dielectric permittivity ε is assumed to be far less than σ/ω yielding a propagation factor $\nu^2 = \lambda^2 - i\omega\mu_0\sigma$. The inhomogeneous induction equation containing the source term has to be solved in a non-conductive environment (air) and both solutions are combined at the earth's surface.

For a horizontal-coplanar coil pair with a coil separation r and at an altitude h above the surface, the relative secondary magnetic field Z is given by (e.g. Wait 1982)

$$Z = r^3 \int_0^\infty R_1(f, \lambda, \rho(z)) \lambda^2 e^{-2\lambda h} J_0(\lambda r) d\lambda \tag{5.2}$$

where R_1 is the complex reflection factor containing the underground vertical resistivity distribution ρ(z) with z pointing vertically downwards, and J_0 and J_1 are Bessel functions of first kind and zero or first order, respectively, which can be approximated by

$$J_0(x) \approx 1 - \frac{x^2}{4} + \frac{x^4}{64} - \dots \tag{5.3}$$

$$J_1(x) \approx \frac{x}{2} - \frac{x^3}{16} + \frac{x^5}{384} - \dots$$

Similar formulae are valid for vertical-coplanar

$$Y = r^2 \int_0^\infty R_1(f,\lambda,\rho(z))\lambda e^{-2\lambda h} J_1(\lambda r) d\lambda \tag{5.4}$$

and vertical-coaxial

$$X = r^2 \int_0^\infty R_1(f,\lambda,\rho(z))\lambda e^{-2\lambda h} \frac{J_1(\lambda r) - \lambda J_0(\lambda r)}{2} d\lambda \tag{5.5}$$

coil configurations. As long as r < 0.3h, the horizontal secondary field values perpendicular (Eq. 5.4) and along (Eq. 5.5) the transmitter-receiver direction can be approximated by $Y \approx Z/2$ and $X \approx -Z/4$, respectively (Mundry 1984). Thus, only Z is regarded in the following.

The reflection factor R_1 is derived for the model of a n-layered half-space by a recurrence formula, see Frischknecht (1967) or Mundry (1984):

$$R_{j-1} = \frac{K_{j-1} + R_j e^{-2t_j v_j}}{1 + K_{j-1} R_j u_j}, \quad R_{n-1} = K_{n-1}, \quad j = n-1,\ldots,2 \tag{5.6}$$

$$K_{j-1} = \frac{v_{j-1} - v_j}{v_{j-1} + v_j}, \quad v_j = \sqrt{\lambda^2 + \frac{i\omega\mu_0}{\rho_j}}$$

where ρ_j and t_j are the model parameters resistivity and thickness of the jth layer (j = 1: air layer, i.e., ρ_1 infinite, and t_1 = h, sensor altitude; j = n: substratum, i.e., t_n infinite). Substituting $k = \lambda h$ and setting $\gamma = r/h$ in Eq. 5.2 yield to

$$Z = \gamma^3 \int_0^\infty R_1(f,k,\rho(z)) k^2 e^{-2k} J_0(k\gamma) dk \tag{5.7}$$

The complex integrals in Eqs. 5.2, 5.4, 5.5, and 5.7 can be evaluated numerically using fast Hankel transforms (Anderson 1989, Johanson and Sørenson 1979). A very fast calculation of the integral in Eq. 5.7 is realised by using a Laplace transform (Fluche and Sengpiel 1997). In this case, the coil separation has to be sufficiently smaller than the sensor altitude in order to approximate the Bessel function J_0 by the first term(s) of Eq. 5.3.

The reflection factor for a homogeneous half-space model (n = 2, $\rho_2 = \rho$, t_2 infinite) can be derived from Eqs. 5.6 and 5.7 using the substitution $k = \lambda h$ (cf. Mundry 1984)

$$R_1 = \frac{v_1 - v_2}{v_1 + v_2}, \quad v_1 = k, \quad v_2 = \sqrt{k^2 + i2\delta^2} \equiv v, \tag{5.8}$$

$$\delta = \frac{h}{p}, \quad p = \sqrt{\frac{2\rho}{\omega\mu_0}}$$

and Eq. 5.7 reduces for ratios $\gamma < 0.3$, i.e., $J_0 \approx 1$ (cf. Eq. 5.3), to

$$Z = \gamma^3 \int_0^\infty \frac{k - \nu}{k + \nu} k^2 e^{-2k} dk = \gamma^3 Z' \tag{5.9}$$

It is obvious from Eqs. 5.8 and 5.9 that the transformed secondary field Z′ depends only on the ratio $\delta = h/p$. Therefore, two curves are sufficient to describe the transformed secondary field Z' for all combinations of half-space resistivity, system frequency, and sensor altitude. Fig. 5.1 shows the components of the complex function Z' as in-phase R' (real part of Z') and quadrature Q' (imaginary part of Z') as well as amplitude $A' = (R'^2+Q'^2)^{1/2} = A/\gamma^3$ and ratio $\varepsilon = Q'/R' = Q/R$.

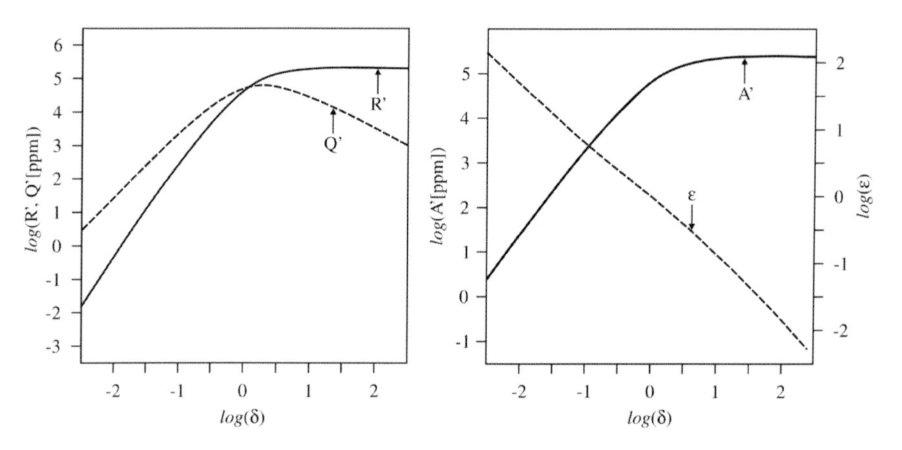

Fig. 5.1. In-phase R' and quadrature Q' (left) and Amplitude A' and phase ratio ε (right) of the transformed secondary field Z' for arbitrary half-space resistivity, system frequency, and sensor altitude on a log-log scale

If the ratio $\gamma = r/h$ is greater than 0.3, the integral of Eq. 5.7 including J_0 has to be evaluated in order to calculate Z', because the argument ($k\gamma$) of the Bessel function J_0 gains a non-negligible importance. Instead of a single pair of curves, two arrays have to be calculated for the various ratios of γ (Fluche et al. 1998).

In case of lateral resistivity changes, a numerical calculation of the secondary field is necessary, e.g. Avdeev et al. (1998), Newmann and Alumbaugh (1995), Stuntebeck (2003), or Xiong and Tripp (1995). Analytical solutions only exist for simple geometries, e.g. a conducting sphere or cylinder. An overview is given by Ward and Hohmann (1988).

Inversion of secondary field values

Generally, the (measured) secondary field data (R, Q) is inverted into resistivity using two principal models: the homogeneous half-space model and the layered half-space model (Fig. 5.2). While the homogeneous half-space inversion uses single frequency data, i.e., the inversion is done individually for each of the frequencies used, the multi-layer (or one dimensional, 1D) inversion is able to take the data of all frequencies available into account.

The resulting parameter of the half-space inversion is the apparent resistivity (or half-space resistivity) ρ_a which is the inverse of the apparent conductivity. Due to the skin-effect (high frequency currents are flowing on top of a perfect conductor) the plane-wave apparent skin depth

$$p_a = \sqrt{\frac{2\rho_a}{\omega\mu_0}} \approx 503.3\sqrt{\frac{\rho_a}{f}} \tag{5.10}$$

of the AEM fields increase with decreasing frequency f and increasing half-space resistivity ρ_a. Therefore, the apparent resistivities derived from high-frequency AEM data describe the shallower parts of the conducting subsurface and the low-frequency ones the deeper parts.

Any of the field components shown in Fig. 5.1 can be used to calculate the apparent resistivity, if the distance between the AEM sensor and the top of the half-space is known. Unfortunately, the dependency of the secondary field on the half-space resistivity is highly non-linear. Thus, the inversion is not straightforward and the apparent resistivities have to be derived by the use of look-up tables, curve fitting or iterative inversion procedures. A single-component inversion, however, has the disadvantage that the inphase component R' is very small for low frequencies and high resistivities, i.e., $\delta < 0.1$ (low induction mode), and nearly constant for high frequencies and low resistivities, i.e., $\delta > 10$ (strong induction mode), and the quadrature component Q' is not unique (cf. Fig. 5.1). On the other hand, the utilisation of the amplitude A' or the ratio ε yields to inaccurate half-space resistivities for a half-space with overburden (Siemon 2001). Using the sensor altitude as an input parameter for the inversion, a further strong disadvantage occurs: The sensor altitude, which is measured in field surveys by laser or radar altimeters, may be affected by trees or buildings (see Fig. 5.3). Therefore, the calculation of the apparent resistivity from both the amplitude A' and the ratio ε (or inphase R' and quadrature Q') is not only more accurate but yields also the apparent distance D_a of the AEM system to the top of the conducting half-space, i.e., a pseudo-layer half-space model is taken into account (Fraser 1978). Siemon (2001) has published a very fast and accurate approach for the calculation of both half-space parameters (ρ_a and D_a) from the curves A'(δ) and ε(δ) (cf. Fig. 5.1) approximated by polynomials.

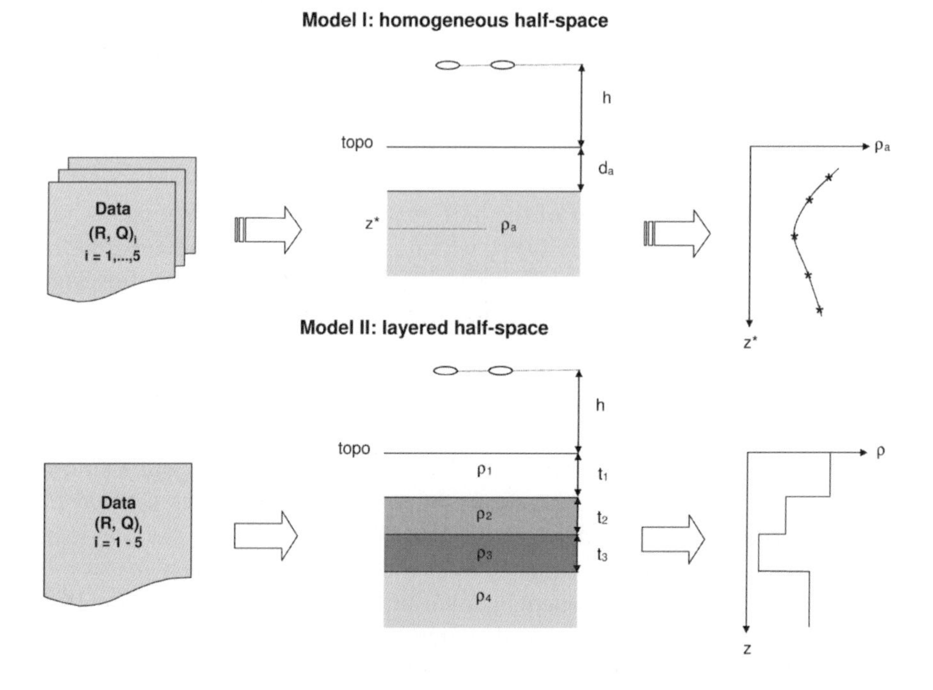

Fig. 5.2. AEM data inversion: I) homogeneous half-space model, II) layered half-space model (for a five-frequency data set)

The apparent distance D_a can be greater or smaller than the measured sensor altitude h: The difference of both, which is called apparent depth

$$d_a = D_a - h \tag{5.11}$$

is positive in case of a resistive cover (including air) as it is the case in Figs. 5.3 and 5.4; otherwise a conductive cover exists above a more resistive substratum. If the apparent distance equals the measured sensor altitude, no resistivity change with depth is supposed. This is the only case where all approaches for calculating the apparent resistivity will yield identical results.

From the apparent resistivity and the apparent depth, a third parameter can be derived: The centroid depth (Fig. 5.3)

$$z^* = d_a + p_a/2 \tag{5.12}$$

is a measure of the mean penetration of the induced underground currents (Siemon 2001). Each set of half-space parameters is obtained individually for each of the AEM frequencies at each of the measured sites.

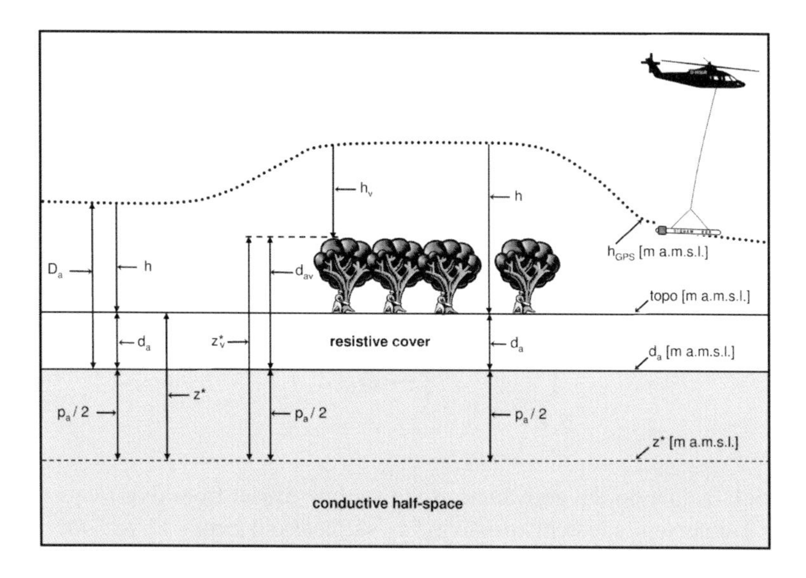

Fig. 5.3. Graphical display of apparent distance D_a, apparent depth d_a, centroid depth z^*, sensor altitude h, sensor elevation h_{GPS} and topographic elevation topo. In case of buildings or trees, the sensor altitude h_v, and thus, the apparent depth d_{av} and the centroid depth z^*_v differ from their correct values, but their associated elevations in m a.m.s.l. (metre above mean sea level) are correct

The model parameters of the 1D inversion (cf. Fig. 5.2) are the resistivities ρ and thicknesses t of the model layers (the thickness of the underlying half-space is assumed to be infinite). There are several procedures for the inversion of AEM data available (e.g. Qian et al. 1997, Fluche and Sengpiel 1997, Beard and Nyquist 1998, Ahl 2003, or Huang and Fraser 2003) which are often adapted from algorithms developed for ground EM data. We use a Marquardt inversion procedure which requires a starting model. Due to the huge number of inversion models to be calculated in an airborne survey, it is not feasible to optimise the starting model at each of the models sites. Therefore, an automatic generation of starting models is necessary, e.g. on the basis of apparent resistivity vs. centroid depth values (Fig. 5.4). The standard model contains as many layers as frequencies used plus a highly resistive cover layer. The layer resistivities are set equal to the apparent resistivities, the layer boundaries are chosen as the logarithmic mean of each two neighbouring centroid depth values. The thickness of the top layer is derived from the apparent depth d_a of the highest frequency used for the inversion. If this apparent depth value is less than a given minimum depth value, the minimum depth value (e.g. 1 m) is used.

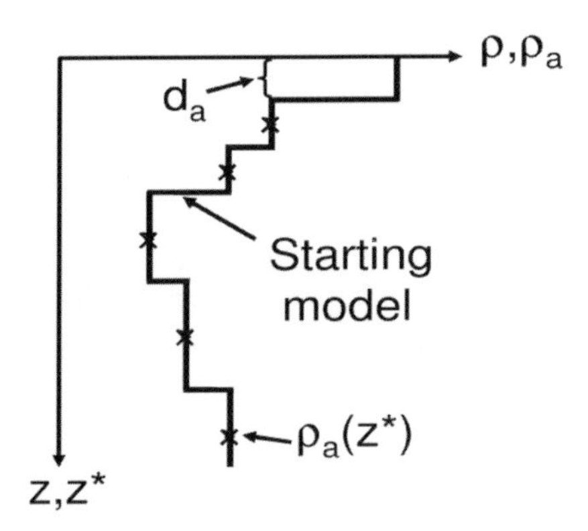

Fig. 5.4. Sketch on the derivation of the starting model from five-frequency half-space parameters apparent resistivity ρ_a vs. centroid depth z^* and the apparent depth d_a of the highest frequency

The inversion procedure is stopped when a given threshold is reached. This threshold is defined as the differential fit of the modelled data to the measured HEM data. We normally use a 10% threshold; i.e., the inversion stops when the enhancement of the fit is less than 10%. This standard starting model requires more model parameters than data values are available, i.e., an underdetermined equation system has to be solved. This is feasible because the inversion procedure is constraint and searches the smoothest model fitting the data (Sengpiel and Siemon 2000).

3D AEM inversion procedures (e.g. Liu et al. 1991, Sasaki 2001) are not only scarce but also very intensive in computing time and storage. In practice, 3D modelling is only necessary when strong lateral resistivity changes occur on a local scale. Due to the limited footprint of AEM systems (Beamish 2003), it is mostly adequate to invert the AEM data using an 1D inversion procedure (Sengpiel and Siemon 1998).

5.1.3 Systems

Far more helicopter than fixed-wing systems are used in airborne frequency-domain surveys. A summary of AEM systems is listed in Table 5.1.

Helicopter systems use a towed rigid-boom AEM system. On a fixed-wing system, the transmitter and receiver coils are mounted at the wing tips. While the helicopter-borne electromagnetic (HEM) system is towed at a sufficiently long distance below the helicopter, the fixed-wing electromagnetic (FEM) system has to cope with interactions with the aircraft.

Table 5.1. Helicopter (HEM) and fixed-wing (FEM) frequency-domain systems (#F: no. of frequencies f, #×: no. of coils, coil orientation: hor: horizontal, vert: vertical, copl: coplanar, coax: coaxial, r: coil separation)

Method	System	Properties	www link
HEM	AWI	2F, 2× hor copl, r = 2.1 / 2.8 m, f = 3.7 / 112 kHz	awi-bremerhaven.de
HEM	Impulse	6F, 3× hor copl / 3× vert coax, r = 6.5 m, f = 870 Hz – 23 kHz	aeroquestsur-veys.com
HEM	Humming-bird	5F, 3× hor copl / 2× vert coax, r = 4.7 m, f = 880 Hz - 35 kHz	geotechairborne-surveys.com
HEM	GEM2-A	6F, 1× hor copl, r = 5.1 m, f = 300 Hz - 48 kHz	geophex.com
HEM	Dighem^V	5F, 5× hor copl, r = 6.3 / 8 m, f = 400 Hz - 56 kHz	fugroairborne.com
HEM	RESOLVE	6F, 5× hor copl / 1× vert coax, r = 7.9-9 m, f = 380 Hz - 101 kHz	fugroairborne.com
HEM	Dighem-BGR	5F, 5× hor copl, r = 6,7 m, f = 385 Hz - 195 kHz	bgr.de
FEM	GSF-95	2F, 2× vert copl, S = 21.4 m, F = 3.1 / 14.4 kHz	gsf.fi\aerogeo
FEM	Hawk	1 - 10F, 1× hor copl / 1× vert copl, r = wing span, f = 200 Hz – 12.5 / 25 kHz	geotechairborne-surveys.com

Commonly, several geophysical methods are used simultaneously in an airborne survey. A typical helicopter-borne geophysical system operated by the German Federal Institute for Geosciences and Natural Resources (BGR) is shown in Fig. 5.5. It includes geophysical sensors that collect five-frequency electromagnetic, magnetic, and gamma-ray spectrometry data, as well as altimeters and positioning systems.

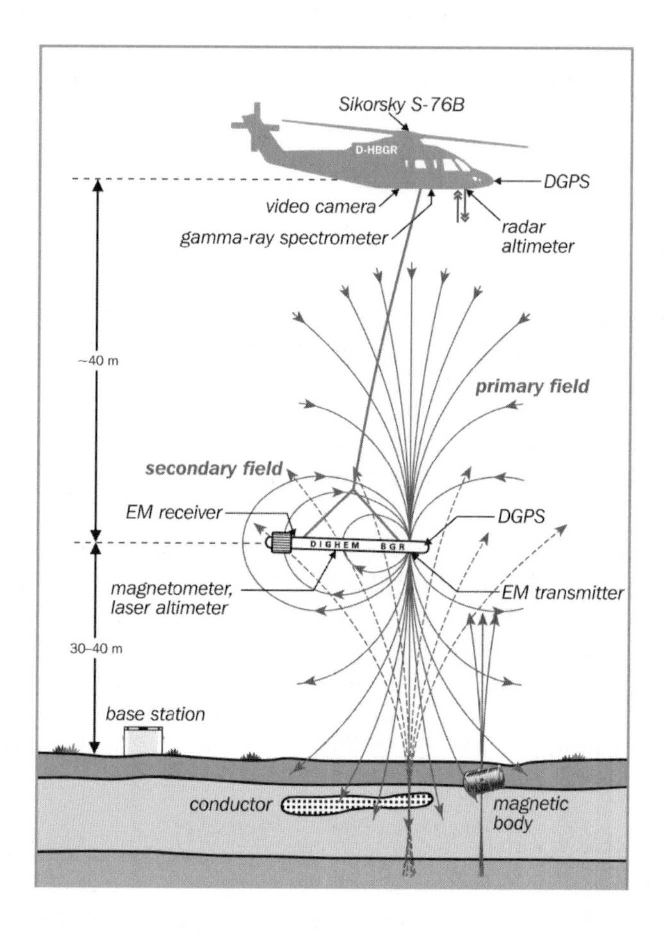

Fig. 5.5. BGR's helicopter-borne geophysical system: Electromagnetic, magnetic, GPS and laser altimeter sensors are housed by a "bird", a cigar-shaped 9 m long tube, which is kept at about 30–40 m above ground level. The gamma-ray spectrometer, additional altimeters and the navigation system are installed into the helicopter. The base station records the time varying parameters diurnal magnetic variations and air pressure history. The sampling rate is 10 Hz except for the spectrometer (1 Hz), which provides sampling distances of about 4 m and 40 m, respectively, taking an average flight velocity of 140 km/h into account

5.1.4 Data Processing

The receiver coils of a frequency-domain AEM system measure the induced voltages of the secondary magnetic fields at specific frequencies. These voltages have to be converted to relative values with respect to the primary fields at the receivers. These conversions are done using special calibration coils which produce definite signals in the measured AEM data. Based on these well-known calibration signals, the measured secondary field voltages are transformed into ppm (parts per million) values. The unit ppm is adequate to display the tiny secondary AEM fields.

Due to the induction process within the conducting earth, phase shifts occur between primary and secondary fields, i.e., the secondary field is a complex quantity having inphase and quadrature components. As a consequence, a phase adjustment has to take place at the beginning of each survey flight, e.g. using the well-defined signals of the calibration coils.

Calibration and phase adjustment are best performed above a highly resistive subsurface or in the air at high flight altitude. From the formulae for calculating the secondary fields, it is evident that these fields are strongly dependant on the sensor altitude, even for a homogeneous subsurface (cf. Eq. 5.9). Flight altitudes of several hundred meters (e.g. 350 m for a common HEM system) are sufficient to drop down the signal of the secondary field below the system noise level. This effect is not only used for accurate calibrations and phase adjustments but also for determining the zero levels of the AEM data. Remaining signals due to insufficiently bucked-out primary fields, coupling effects with the aircraft, or (thermal) system drift are generally detected at high flight altitude several times during a survey flight.

These basic values measured at reference points are used to shift the AEM data with respects to their zero levels (cf. Valleau 2000). This procedure enables the elimination of a long-term, quasi-linear drift; short-term variations caused by e.g. varying air temperatures due to alternating sensor elevations, however, cannot be determined successfully by this procedure.

Therefore, additional reference points – also along the profiles at normal survey flight altitude – may be determined where the secondary fields are small but not negligible. At these locations, the estimated half-space parameters are used to calculate the expected secondary field values which then serve as local reference levels.

A standard airborne survey consists of a number of parallel profile lines covering the entire survey area and several tie-lines which should be flown perpendicular to the profile lines. At the cross-over points, the AEM data from profile and tie-lines are compared and statistically analysed to correct the AEM line data for remaining zero-level errors. Due to the altitude dependency of the AEM data, the tie-line levelling and further AEM levelling procedures (e.g. Huang and Fraser 1999) are normally applied to half-space parameters (apparent resistivity and apparent depth) which are less affected by the changes of the sensor altitude.

Noise from external sources (e.g. from radio transmitters or power lines) should be eliminated from the AEM data by appropriate filtering or interpolation procedures. Induction effects from buildings and other electrical installations or effects from strongly magnetised underground sources should not be erased from the data during the first step of data processing. These effects appear particularly on a low-frequency resistivity map as conductive or resistive features outlining the locations of man-made installations or strongly magnetised sources, respectively. If necessary, these effects can be cancelled out after a thorough inspection, and the data may be interpolated in case of small data gaps and smoothly varying resistivities.

5.1.5 Presentation

The AEM results are generally presented as maps and vertical resistivity sections (VRS). The maps display, e.g. the half-space parameters apparent resistivity and centroid depth or the resistivities derived from the 1D inversion results for certain model layers or at several depth or elevation levels. An example is shown in Fig. 5.6.

The VRS - also based on the 1D inversion results - are produced along the survey lines. The vertical sections are constructed by placing the resistivity models for every sounding point along a survey profile next to each other using the topographic relief as base line in metre above mean sea level (m a.m.s.l.). The altitude of the HEM bird, the misfit of the inversion, and the HEM data are plotted above the resistivity models (Fig. 5.7).

Example

The area between the coastal towns of Cuxhaven and Bremerhaven, Germany, was surveyed in 2000/2001 using HEM (Siemon et al. 2004) in order to map glacial meltwater channels and saltwater intrusions from the estuaries of the Elbe and Weser rivers. The HEM data set serves as a base to revise and upgrade the groundwater model of the entire survey area (Fig. 5.6) and to assess the groundwater potential of the area in view of the increasing water consumption used by industry and tourism.

The morphology of the survey area is described by wet marshlands just above sea level and smooth sand ridges called "Geest" with elevations of ten to thirty metres above sea level. Marshlands run more or less parallel to the banks of the Elbe and Weser rivers and along the shore line, whereas the Geest ridge forms the centre part of the area under consideration. Minor settlements are spread all over the survey area with major centres in the north and south (cities of Cuxhaven and Bremerhaven). Roads, railway tracks, power lines, fences and many other infrastructural networks are existent. All these are potential sources of noise resulting in reduced quality of the airborne geophysical signals.

Figure 5.6 depicts the lateral variation of apparent resistivity for the frequency of 1830 Hz. The central area between the marshlands in the west and north-east is characterised by elevated resistivity values ($\rho_a > 65\ \Omega$m, light grey) associated with freshwater saturated sands. A few linear, north-east to north-west striking conductive features ($\rho_a = 10$-$65\ \Omega$m, grey) can be ascribed to channels incised by glacial meltwater during Pleistocene glacial regression epochs. The channels were subsequently filled with coarse sands and gravels in the bottom and mostly silt and clayish materials in the upper parts forming ideal freshwater aquifers. The total thickness of the channel fillings is approximately 300 m (Kuster and Meyer 1979) covered by several ten metres thick clay layers, referred to as cover or lid clays which are covered by about 30-60 m thick Pleistocene sediments.

The conductive response of the lid clays indicates the existence of buried meltwater channels. The lowest apparent resistivity values $\rho_a < 2\ \Omega$m, very dark grey) occur offshore clearly due to the presence of seawater and in the north-east and west $\rho_a = 2$-$7\ \Omega$m, dark grey) where saltwater intrudes several kilometres inland. Noteworthy, elevated resistivity values $\rho_a = 20$-$50\ \Omega$m, grey) are mapped offshore close to the north-western tip of the mainland. A freshwater aquifer extending seawards across the shore line is the hydrogeological source of this resistivity high.

The resistivity section in Fig. 5.7 shows the results of the 1D inversion of four-frequency HEM data collected along the survey line 219.1. This line runs WNW–ESE and is approximately half way between the cities of Cuxhaven and Bremerhaven (cf. Fig 5.6). From the top to the bottom, Fig. 5.7 displays the inphase (R) and quadrature (Q) HEM data, the resistivity section, and the relative misfit of the 1D inversion. The line above the resistivity models indicates the sensor (bird) elevation in metres above mean sea level which is the difference between the barometric altimeter record and the effective cable length of about 40 m.

The ground elevation is obtained as the difference between the sensor elevation and the radar/laser altitude of the bird above ground level. As radar/laser signals are frequently distorted by the tree canopy, the ground elevation may be too high in wooded areas (cf. Fig. 5.3).

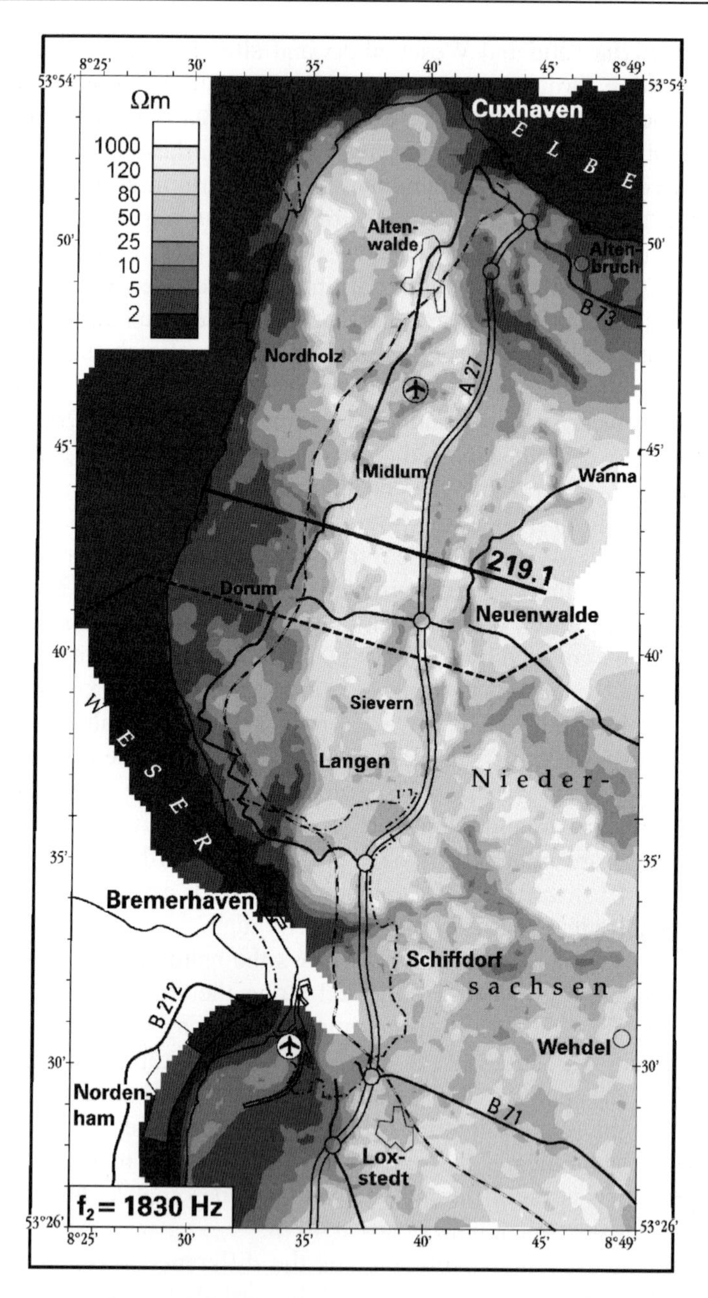

Fig. 5.6. Example for an apparent resistivity map of a coastal area in NW Germany. The survey was flown in two parts: the dashed line marks the boundary of the Cuxhaven (2000) and Bremerhaven (2001) survey areas. The solid line shows the location of profile 219.1 (see Fig. 5.7)

The resistivity models are plotted downward from the ground elevation line. Therefore, the top layer of the inversion models includes both trees and resistive cover layers. The clear drop in resistivity from the highly resistive top layer to the medium resistive underlying model layer is obviously caused by the groundwater table. Between profile-km 10 and 14, a pair of meltwater channels (buried valleys) has been identified due to the conductive properties of the lid clays (cf. B2.1). It can also be noted that the penetration depth of the HEM system is reduced within these meltwater channels because of the lid clays, i.e., the induced eddy currents may not have reached the bottom of the buried valley. The occurrence and the lateral extent of the buried valleys have reliably been located by the HEM system (see Fig. 5.6) as was confirmed by comparison with existing boreholes (Siemon and Binot 2002) and results of ground geophysics (Gabriel et al. 2003).

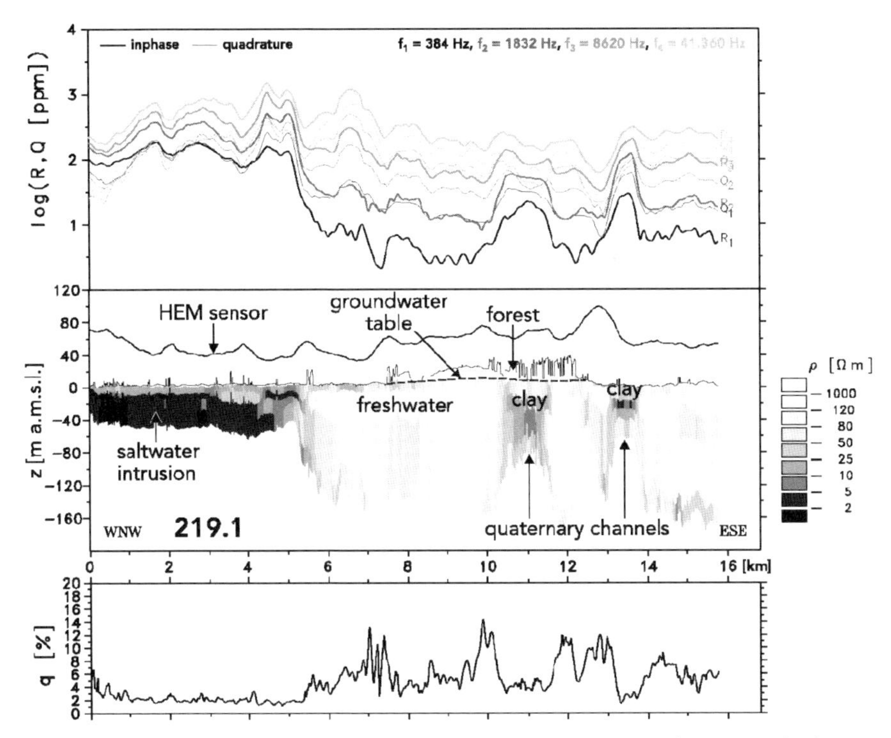

Fig. 5.7. Example for a vertical resistivity section (VRS): From the top to the bottom, the inphase and out-of-phase (quadrature) values of the HEM data (in ppm) for the four frequencies (f_1-f_4), the 1D resistivity models (in Ωm) using the topographic relief as base line in meters above mean sea level (m a.m.s.l.), and the misfit of the inversion q (in %) are displayed. The altitude of the HEM bird is plotted above the resistivity models

5.1.6 Discussion and Recommendations

Airborne electromagnetics is a very useful method for surveying large areas in order to support hydrogeological investigations. Due to the dependency of the geophysical parameter electrical conductivity from both the mineralization of the groundwater and the clay content, information about water quality and aquifer characteristics, respectively, can be derived from AEM data. The results, however, are sometimes ambiguous: A clayey aquitard in a freshwater environment and a brackish, sandy aquifer are associated with similar conductivities. As a consequence, additional information, e.g. from drillings, are required for a solid hydrogeological interpretation of the AEM data.

Frequency-domain electromagnetic measurements are suitable for high-resolution surveys as long as the targets are seated not deeper than 100 m. For deeper targets ground-based or airborne time-domain measurements are more suitable. Helicopter-borne multi-frequency systems are widely used in groundwater explorations due to their high-resolving properties and their applicability even in rough terrain. Fixed-wing systems are applicable for reconnaissance surveys in a flat terrain because these systems outrange helicopter-borne systems and they are less expensive, but they are less flexible and have less-resolving properties. Frequency-domain systems using natural (e.g. audio-frequency magnetic (AFMAG) systems) sources are not very practicable for detailed groundwater surveys.

5.2 Ground based techniques

Reinhard Kirsch

Electromagnetic methods enable the fast mapping of highly conducting underground structures. Comparable to the electromagnetic airborne methods, soundings are also possible when several frequencies are used for the measurements. The theory of induction and the use of secondary fields for resistivity determination are discussed in details in the previous section. The application of electromagnetic methods for the detection of fracture zone aquifers is illustrated in Chap. 13.

5.2.1 Slingram and ground conductivity meters

Originally, transmitter and receiver coils with horizontal orientations are used for exploration purposes leading to the name Horizontal Loop Electromagnetic (HLEM) systems or the Swedish name slingram. Both coils are operated at a fixed distance during the survey. Depth penetration can be controlled by the frequency or the coil separation, both are often coupled. High frequency systems specially designed for mapping of shallow underground structures (e.g. GEONICS EM 31 and EM 38) often combine transmitter and receiver coil in one instrument, so one-man operation is possible. Operation is also possible with vertical coils. Some instruments, e.g. MAXMIN (APEX) are not restricted to coplanar coil orientations. Unlike geoelectrical methods, no galvanic contact to the ground is required; therefore measurements on sealed terrains are possible.

Normally, the magnetic component of the superposition of primary and secondary field is measured. The measured field is split into the inphase and outphase (=quadrature, 90° phase shift) component with respect to the primary field. Both components are recorded. A typical response of the inphase and quadrature signal to a steeply dipping and highly conducting fracture zone is shown in Fig. 5.8. Examples for the use of HLEM measurements for fracture zone detection are given in Chap. 13, Figs. 13.11, 13.12, 13.16, 13.18, and 13.19.

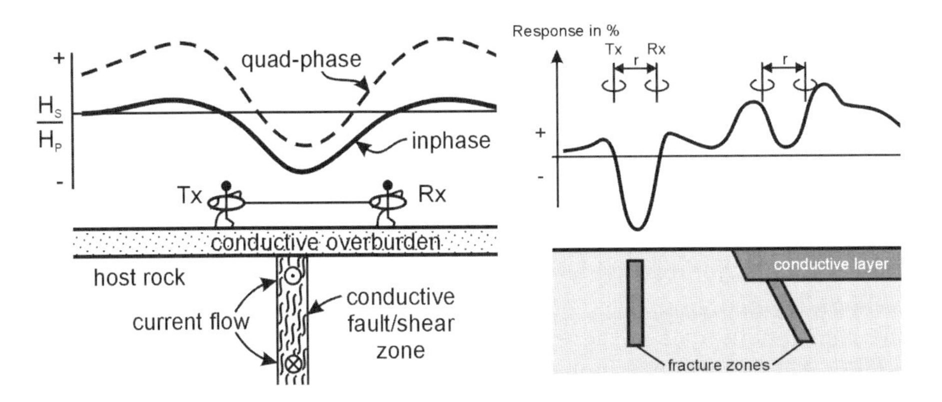

Fig. 5.8. left: Slingram response over a highly conductive fracture zone. The offset at the quadrature response is due to the conductivity of overburden (after McNeill 1990), right: influence of a good conductive layer on the Sligram response (after Grissemann and Ludwig 1986). Examples showing the influence of dipping layers and conducting sheets are shown in Chap. 13, Figs. 13.15 and 13.17

The depth of penetration is characterised by the skin-depth, where the amplitude of the primary electromagnetic wave is reduced by the factor e compared to the transmitted wave. Skin depth δ depends on signal frequency ω, ground conductivity σ in mS/m, and magnetic permeability μ:

$$\delta = \sqrt{\frac{2}{\sigma \cdot \omega \cdot \mu}} \tag{5.13}$$

which leads to a rule of thumb for the assessment of skin depth:

$$\delta = 503 \cdot \sqrt{\frac{\rho}{f}} \tag{5.14}$$

skin depth δ in m, specific resistivity ρ of the ground in Ωm, and frequency f in s^{-1}.

If the coil space is smaller than the skin depth, the term *operation at low induction numbers* is used (McNeill 1990). Then, the inphase component of the signal becomes very small and the quadrature component is directly related to the conductivity of the ground. This enables a direct reading of the mean ground conductivity. Examples for ground conductivity measurements for the mapping of a waste dump are shown in Chap. 17 Figs. 17.1 and 17.2.

However, for an interpretation of the so obtained conductivities it should be kept in mind that the induced current flow is mainly horizontal. If electrical anisotropy due to alternating layering of good and poor conductors (e.g. clay, sand) occur, then current flow is mainly in the good conducting layers resulting in high measured conductivities (Seidel 1997). Therefore, ground resistivity determination by electromagnetic methods can result in lower resistivities than obtained by VES measurements.

For a layered ground the measured conductivity is a weighted mean of the conductivities of the layers in which currents were induced. This is similar, but not identical to the apparent resistivity obtained by VES measurements. Data inversion to obtain the conductivity depth structure is possible, if measurements were done with different frequencies or coil orientations. For EM-34 and EM-31 instruments, a simple method to calculate the layered ground response is given by McNeill (1980).

From a sensitivity function R(Z) the contribution of the layers below the depth Z (normalised by the coil separation) can be calculated (Fig. 5.9). As an example, for a coil separation of 10 m the depth range below 10 m (Z = 1) contributes to the measured apparent conductivity by 42%, whereas the contribution of the depth range below 20 m (Z = 2) is only 25%.

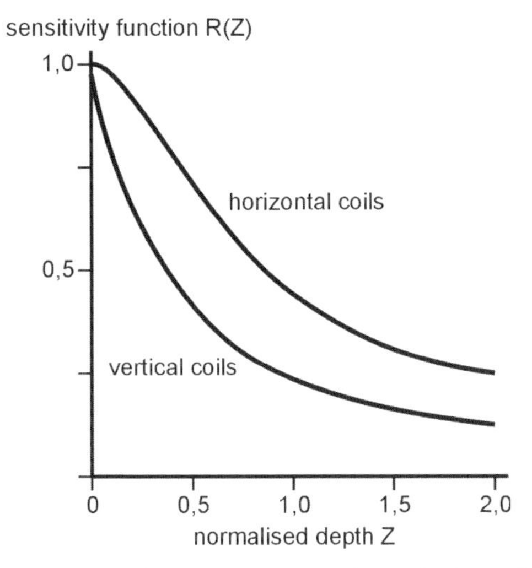

Fig. 5.9. Sensitivity function for horizontal and vertical coplanar loops after McNeill (1980), Z = depth normalized by the coil separation

The sensitivity function R(Z) can be calculated after McNeill (1980) for horizontal loops by:

$$R(Z) = \frac{1}{\sqrt{4 \cdot Z^2 + 1}} \tag{5.15}$$

and for vertical loops by:

$$R(Z) = \sqrt{4 \cdot Z^2 + 1} - 2 \cdot Z \tag{5.16}$$

For a layered ground consisting of n layers with conductivities σ_i and normalized depths Z_i to the bottom of layer i, the measured apparent conductivity σ_a is given by:

$$\rho_a = \sigma_1 \cdot [1 - R(Z_1)] + \sigma_2 \cdot [R(Z_2) - R(Z_1)] ..... + \sigma_n \cdot R(Z_{n-1}) \tag{5.17}$$

This can be used for the planning of field surveys, if the depth ranges of the target layers are known. Also a rough data interpretation is possible. However, it must be kept in mind that the resolution of these measurements for highly resistive layers is poor.

5.2.2 VLF, VLF-R, and RMT

VLF (very low frequency)

VLF techniques use the signal of very long wavelength radio transmitters. These transmitters are world-wide distributed, their signal is mainly used for marine navigation and communication. The transmitter frequencies are in the range of 15 – 25 kHz with transmitter powers in the range of 100 – 1000 kW.

The transmitted electromagnetic wave consists of electric field components E_x and E_z and a magnetic component H_y (Fig. 5.10). The electric component E_x leads to current flow in the ground, especially in elongated conductive underground structures like fracture zones stretching roughly in the direction to the transmitter. These currents induce an additional magnetic field component H_Z which results in a tilted magnetic field vector in the vicinity of the good conductor with a crossover at the center of the good conductor. The tilt angle is measured by a radio receiver with the antenna tilted to the direction of maximum received signal.

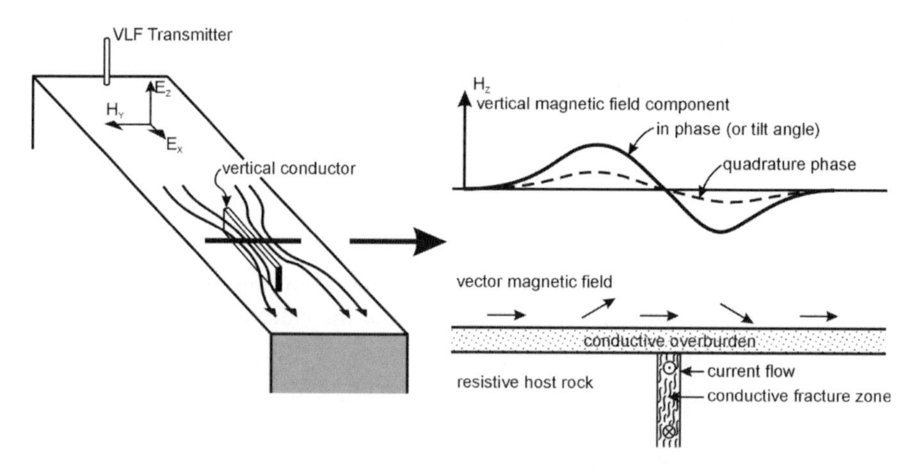

Fig. 5.10. Left: field components from a remote VLF transmitter and electric current flow in the ground, right: electric field component along the profile indicated in the left picture (after McNeill 1990)

Tilt angle data can be plotted as profiles where the crossovers indicate narrow conducting zones. Examples for the use of VLF-techniques for fracture zone detection are given in Chap. 13, Figs. 13.20, 13.21, 13.22, and 23. However, if the data density is sufficiently large to produce a contour map, the crossover should be converted to peak values to give a clear picture on the map. This can be done by applying an averaging filter to the data, e.g. the Fraser filter which converts a sequence of equidistant data $M_1....M_4$ into:

$$F_1 = (M_3 + M_4) - (M_1 - M_2) \tag{5.18}$$

F_1 is plotted at the center of this data sequence. The application of Fraser-filtering is demonstrated by Hutchinson and Barta (2002) and in Chap. 13 (Fig. 13.22).

VLF-resistivity and radio magnetotelluric (RMT)

From the ratio of the horizontal components of the electric and the magnetic field the apparent resistivity of the ground can be determined by (McNeill 1980):

$$\rho_a = \frac{1}{\mu\omega}\left(\frac{|E|}{|H|}\right)^2 \tag{5.19}$$

E = Amplitude of the horizontal electric field, V/m
H = Amplitude of the horizontal magnetic field, A/m
μ = magnetic permeability of free space, $\mu = 4\pi \times 10^{-7}$ H/m
$\omega = 2\pi f$, f = frequency, Hz

H is measured by an antenna and E is measured by an electrode pair orientated towards the radio transmitter (Fig. 5.11). The electrode spacing is in the range of 1 – 5 m. Radio transmitters are chosen in the VLF-frequency range (VLF-R) or in the radio-frequency range 10 – 300 kHz (RMT). As the penetration depth of the radio signal depends on the frequency, a VLF-R or RMT multi-frequency data set can be inverted to the resistivity depth structure. For inversion, the apparent resistivity $\rho_a(f)$ and the phase-shift $\Phi(f)$ (between E and H) is used. Examples for RMT-measurements on waste dumps are given by Hördt et al. (2000).

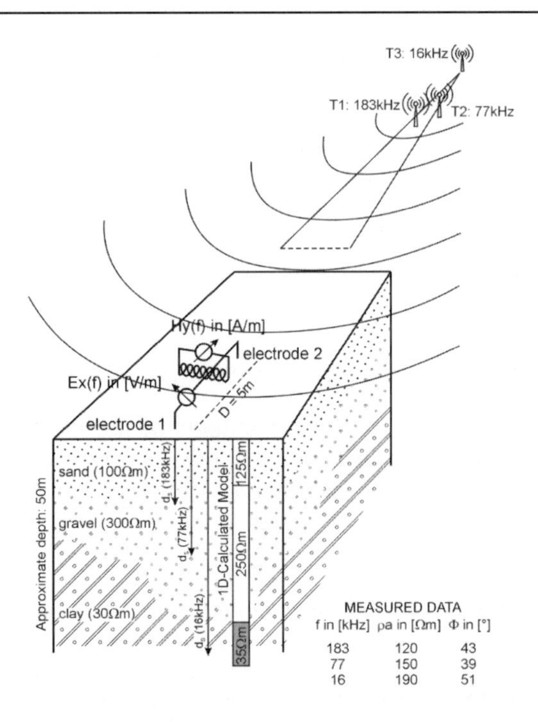

Fig. 5.11. Schematic principle of radiomagnetotelluric sounding (after Bosch and Gurk 2000)

5.3 References

Ahl A (2003) Automatic 1D inversion of multifrequency airborne electromagnetic data with artificial neural networks: discussion and case study. Geophysical Prospecting 51:89-97

Anderson WL (1989) A hybrid fast Hankel transform algorithm for electromagnetic modelling. Geophysics 54:263-266

Avdeev DB, Kuvshinov AV, Pankratov OV, Newman GA (1998) Three-dimensional frequency-domain modelling of airborne electromagnetic responses. Exploration Geophysics 29:111-119

Beamish D (2003) Airborne EM footprints. Geophysical Prospecting 51:49-60

Beard LP, Nyquist JE (1998) Simultaneous inversion of airborne electromagnetic data for resistivity and magnetic permeability. Geophysics 63:1556-1564

Bosch FP, Gurk M (2000) Comparison of RF-EM, RMT and SP measurements on a karstic terrain in the Jura Mountains (Switzerland). Proceed. of the Seminar *Electromagnetische Tiefenforschung* Deutsche Geophysikalische Gesellschaft:51-59

Fluche B, Sengpiel KP (1997) Grundlagen und Anwendungen der Hubschrauber-Geophysik. In: Beblo M (ed) Umweltgeophysik, Ernst und Sohn, Berlin pp 363-393

Fluche B, Siemon B, Grinat M (1998) Transfer of airborne electromagnetic interpretation methods to ground horizontal loop electromagnetic (HLEM) measurements. In: Proceedings of EAGE, 60th Conference & Technical Exhibition, Leipzig

Fraser DC (1978) Resistivity mapping with an airborne multicoil electromagnetic system. Geophysics 43:144-172

Frischknecht FC (1967) Fields about an oscillating magnetic dipole over a two-layer earth, and application to ground and airborne electromagnetic surveys. Q Col Sch Mines 62

Fountain D (1998) Airborne electromagnetic systems – 50 years of development. Geophysical Prospecting 29:1-11

Gabriel G, Kirsch R, Siemon B, Wiederhold H (2003) Geophysical investigation of Pleistocene valleys in Northern Germany. Journal of Applied Geophysics 53:159-180

Grissemann Ch, Ludwig R (1986) Recherche sur la fracturation profonde en zone de socle cristllin a partir de forage a gros debit et de lineaments landsat a l áide de methodes geophysiques avancees. Cooperation techniques projet no. 82.2060.0, Bundesanstalt für Geowissenschaften und Rohstoffe, Hannover

Huang H, Fraser DC (1999) Airborne resistivity data levelling. Geophysics 64:378-385

Huang H, Fraser DC (2003) Inversion of helicopter electromagnetic data to a magnetic conductive layered earth. Geophysics 68:1211-1223

Hutchinson P, Barta LS (2002) VLF surveying to delineate longwall mine-induced fractures. The Leading Edge 21:491-493

Hördt A, Greinwald S, Schaumann G, Tezkan B, Hoheisel A (2000) Joint 3D interpretation of radiomagnetotelluric (RMT) and transient electromagnetic (TEM) data from an industrial waste deposit in Mellendorf, Germany. European Journal of Environmental and Engineering Geophysics 4:151-170

Johanson HK, Sørenson K (1979) The fast Hankel transform. Geophysical Prospecting 27:876-901

Kuster H, Meyer KD (1979) Glaziäre Rinnen im mittleren und nordöstlichen Niedersachsen. Eiszeitalter und Gegenwart 29:135-156

McNeill JD (1980) Electromagnetic terrain conductivity measurements at low induction numbers. Geonics Technical Note TN-6

McNeill JD (1990) Use of Electromagnetic Methods for Groundwater Studies. In. Ward SH (ed) Geotechnical and Environmental Geophysics, vol 1. Soc Expl Geophys, Tulsa, pp 191 – 218

Mundry E (1984) On the interpretation of airborne electromagnetic data for the two-layer case. Geophysical Prospecting 32:336-346

Newmann GA, Alumbaugh DL (1995) Frequency-domain modelling of airborne electromagnetic responses using staggered finite differences. Geophysical Prospecting 43:1021-1041

Qian W, Gamey TJ, Holladay JS, Lewis R, Abernathy D (1997) Inversion of airborne electromagnetic data using an Occam technique to resolve a variable number of multiple layers. Proceedings from the High-Resolution Geophysics Workshop, Laboratory of Advanced Subsurface Imaging (LASI), Univ of Arizona, Tucson, CD-ROM

Sasaki Y (2001) Full 3D inversion of electromagnetic data on PC. Journal of Applied Geophysics 46:45-54

Seidel K (2005) Elektromagnetische Zweispulen-Systeme, Prinzip der Methode. In: Knödel K, Krummel K, Lange G (eds) Handbuch zur Erkundung des Untergrundes von Deponien und Altlasten, Bd 3 Geophysik. Springer, Berlin Heidelberg NewYork, pp 241-265

Sengpiel KP, Siemon B (1998) Examples of 1D inversion of multifrequency AEM data from 3D resistivity distributions. Exploration Geophysics 29:133-141

Sengpiel KP, Siemon B (2000) Advanced inversion methods for airborne electromagnetic exploration. Geophysics 65:1983-1992

Siemon B (2001) Improved and new resistivity-depth profiles for helicopter electromagnetic data. Journal of Applied Geophysics 46:65-76

Siemon B, Binot F (2002) Aerogeophysikalische Erkundung von Salzwasserintrusionen und Küstenaquiferen im Gebiet Bremerhaven-Cuxhaven – Verifizierung der AEM-Ergebnisse. In: Hördt A, Stoll JB (eds) Protokoll über das Kolloquium Elektromagnetische Tiefensondierung Burg Ludwigstein, pp 319-328

Siemon B, Eberle DG, Binot F (2004) Helicopter-borne electromagnetic investigation of coastal aquifers in North-West Germany. Zeitschrift für Geologische Wissenschaften (in press)

Stuntebeck C (2003) Three-dimensional electromagnetic modelling by free-decay mode superposition. PhD thesis, TU Braunschweig

Valleau N (2000) HEM data processing – a practival overview. Exploration Geophysics 31:584-594

Wait JR (1982) Geo-electromagnetism. Academic Press, New York

Ward SH, Hohmann GW (1988) Electromagnetic theory for geophysical applications. In: Nabighian MN (ed), Electromagnetic methods in applied geophysics, vol 1, Theory. Society of Exploration Geophysicists, IG no 3, Tulsa, pp 130-310

Xiong Z, Tripp AC (1995) Electromagnetic scattering of large structures in layered earth using integral equations. Radio Science 30:921-929

6 The transient electromagnetic method

Anders Vest Christiansen, Esben Auken, Kurt Sørensen

6.1 Introduction

6.1.1 Historic development

The transient electromagnetic (TEM) method has been developed and refined most intensively since the mid-1980s. This makes the method relatively "young" compared to the frequency domain method, the magnetotelluric method and the geoelectric method. The reason is twofold: firstly the TEM response covers a very large dynamic range, which makes it difficult to measure without sophisticated electronics. Secondly, the interpretation of TEM data is approximately 50 - 100 times more computer intensive compared to interpretation of frequency and geoelectric data. Though, with modern computers the interpretation of TEM data can be done interactively, which was not possible 15-20 years ago when large computers were only available to research institutions and a few large companies.

The inductive methods (TEM and frequency domain methods) were originally designed for mineral investigations. Back in the 60s and 70s frequency domain methods were dominating being very sensitive to low resistivity mineral deposits settled in a high-resistive host rock (>100000 Ωm). This is a typical mineralogical setting in North America. However, most of Australia is covered with a relatively thick layer of Rhyolite (up to 100 m) with low resistivity. Frequency domain methods have difficulties penetrating this layer because of the resistivity. This fact pushed the development of the TEM method in Australia in spite of the electronic difficulties. At that time only qualitative interpretations were possible, but they gave indicative information on mineralizations.

Over the last two decades the TEM method has become increasingly popular for hydrogeological purposes as well as general geological mapping. The frequency domain methods have found extensive use using a helicopter (see Chap. 5) whereas ground based frequency methods are used only very limited for hydrogeological purposes.

A key point in using the TEM method for hydrogeological purposes is the requirement for accurate data with high spatial density. The accuracy must be met by instrumentation, data processing and the interpretation algorithm, in order to have an optimum basis for the geological and hydrogeological interpretation. Most often it is not enough to map the groundwater reservoir, but also the internal geological structures, aquifer size, volumetric parameters and the geophysical character of cover materials must be mapped.

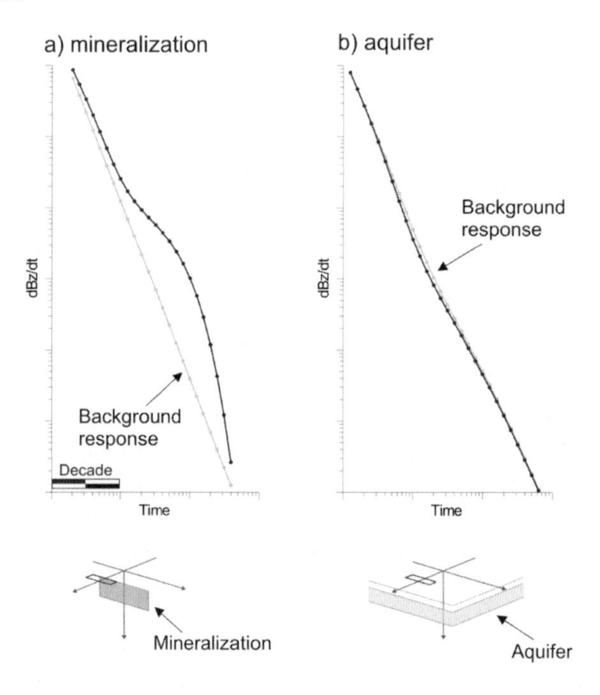

Fig. 6.1. Comparison of the responses of a base metal mineral exploration and a hydrological target as approximated by a vertical thin sheet and layered-earth model, respectively. The mineral exploration target is a vertical sheet measuring 90 m by 30 m at a depth of 20 m, with a conductance of 100 S, in a 100 Ωm half space. The parameters for a three-layer hydrological model with a layer representing a sandy aquifer are: $\rho_1 = 50$ Ωm, $\rho_2 = 100$ Ωm, $\rho_3 = 10$ Ωm, $t_1 = 30$ m, and $t_2 = 50$ m, where t is the layer thickness. The parameters for the background model (without an aquifer or a sheet) are: $\rho_1 = 50$ Ωm, $\rho_2 = 10$ Ωm and $t_1 = 80$ m

To illustrate the need for accurate data, Fig. 6.1 displays sounding curves from typical TEM soundings over a mineral deposit and over a groundwater reservoir. A thin-sheet model is used to compute the response of a mineral exploration target and compare it to the layered-earth response of a hydrological model. For both models the same central-loop TEM array is used. The response with the aquifer layer differs from that of the background response by a factor of approximately 1.2 or 20%, whereas the response from the mineralized sheet is roughly a factor of 100 above the background response. Hence, absolute data accuracy is crucial for hydrogeophysical investigations. From an instrument and interpretation point of view the difference between a factor of 1.2 and 100 is, of course, huge. An error of a few percent plays an important role for the hydrogeological response whereas it is almost negligible for the response over the mineral deposit.

6.1.2 Introduction

The electromagnetic phenomena are all controlled by the Maxwell equations. One of the most fundamental electromagnetic phenomena is that a varying magnetic field will result in a varying electric field which again will create another varying magnetic field which ...... etc., indefinitely. This phenomenon (amongst others) determines the propagation of electromagnetic fields which we call electromagnetic radiation. The light bulb, microwaves in the oven and long wave radio transmitters are all examples of electromagnetic radiation.

The electromagnetic geophysical methods are all based upon the fact that a magnetic field varies in time - the primary field - and thus, according to the Maxwell equations, induce an electrical current in the surroundings – e.g. the ground which is a conductor. This current and the associated electrical and magnetic fields are often called the secondary fields. When, for instance, we push a spear in the ground, we measure the electrical fields, or when we put a coil on the ground we measure the magnetic field. Naturally, the total sum of the fields is measured, i.e. the sum of the primary and secondary fields, without any possibility of separating them. The part of the total field which originates from the secondary field contains information about the conductivity structure of the ground, because the current induction is different for different earth materials. The assignment of electromagnetic geophysics is to extract this information from measured data and translate it to resistivity images of the subsoil.

Most electromagnetic methods can be classified as belonging either to the frequency-domain methods (FEM) or the time-domain methods (TEM). The FEM methods usually work by a transmitter transmitting a harmonic primary signal with a particular frequency and a receiver measuring the resulting secondary in-phase (real) and out-of-phase (quadrature) fields. The real part is the signal which is in-phase with the primary field, and the quadrature part is the part of the signal which is 90-degrees out of phase with the primary field. An advantage of this method is that by measuring one particular frequency, electromagnetic noise in the surroundings can be filtered off in a very efficient manner by synchronous detection. The main challenge is the presence of the primary field during the measuring procedure. As, in general, this field is much bigger than the secondary field which contains most of the information about the conductivity structure of the ground, you must either measure very accurately (which is very difficult), or compensate for the primary field before measuring. However, this results in great demands of accuracy to the coil-geometry of the equipment, and these are extremely difficult to meet in a field situation.

The TEM methods measure, as the name suggests, the amplitude of a signal as a function of time. The TEM methods work by a transmitter transmitting a pulse - typically a current switched-off very quickly - and the measurements are then made after the primary fields disappear, i.e. only on the secondary fields. This eliminates the difficulties with geometry of the setup related to the FEM method. On the other hand, it is necessary to measure the secondary field in a time interval which is so long that the amplitude of the signal varies greatly. It has a large *dynamic range*. Typically, the variation is a factor 1,000,000, and this field has to be measured from 0.00001 s to 0.001 s. In addition to this the characteristics of the signal do not allow for direct filtration of the signal to avoid surrounding electromagnetic noise. However, this problem is solved by measuring the transient signal in gates followed by averaging the gate values (stacking).

6.1.3 EMMA - ElectroMagnetic Model Analysis

Numerical examples presented in this chapter (Figs. 9, 12, 13 and 17) have been generated using the program EMMA and we suggest to use the program in connection to this chapter. EMMA is a geophysical electrical and electromagnetic modelling and analysis program. EMMA has a user-friendly interface allowing non-experts to calculate responses and perform model parameter analyses with a few clicks of the mouse. EMMA is freeware and is provided by the HydroGeophysics Group at University of Aarhus as a design and learning tool. EMMA can be downloaded from http://www.hgg.au.dk.

6.2 Basic theory

This section deals with the most basic electromagnetic theory forming the basis of an expression for the transient response. The aim is not a full derivation of the pertinent equations, but a presentation of the central parts relevant for a basic understanding of the TEM method. The derivation is general until the Schelkunoff potentials for all electromagnetic methods, but the application of the potentials requires the characteristics of the source to be defined, and the derivation becomes specialized for the TEM method. A detailed description of the theory is found in Ward and Hohmann (1988).

In this theoretical chapter, capital letters are used in the frequency-domain, normal letters in the time-domain.

6.2.1 Maxwell's equations

An electromagnetic field is defined by the five vector functions: **e** (electric field intensity), **b** (magnetic induction), **d** (dielectric displacement), **h** (magnetic field intensity) and **j** (electric current density). The interaction between the elements is governed by Maxwell's equations, describing any electromagnetic phenomenon. Maxwell's equations are uncoupled first-order linear differential equations and are in the time domain given by:

$$\nabla \times \mathbf{E} + \frac{\partial \mathbf{b}}{\partial t} = 0 \tag{6.1}$$

$$\nabla \times \mathbf{h} - \frac{\partial \mathbf{d}}{\partial t} = \mathbf{j} \tag{6.2}$$

$$\nabla \cdot \mathbf{b} = 0 \tag{6.3}$$

$$\nabla \cdot \mathbf{d} = \rho \tag{6.4}$$

where ρ is electric charge density [C/m^3]. Here Maxwell's equations are stated as differential equations, meaning that boundary conditions apply for a full description of the fields. We will not discuss the boundary conditions here.

These equations contain five field quantities, but can be simplified to only two using the frequency domain constitutive relations, which for applications with earth materials can be stated as:

$$\mathbf{D} = \left[\varepsilon'(\omega) - i\varepsilon''(\omega)\right]\mathbf{E} = \varepsilon\mathbf{E} \tag{6.5}$$

$$\mathbf{J} = \left[\sigma'(\omega) - i\sigma''(\omega)\right]\mathbf{E} = \sigma\mathbf{E} \tag{6.6}$$

$$\mathbf{B} = \mu_0 \mathbf{H} \tag{6.7}$$

The dielectric permittivity ε, and the electric conductivity σ, are complex functions of angular frequency ω, while the magnetic permeability μ_0 is frequency independent, real and assumed equal to the free-space value.

A Fourier transformation of Eqs. 6.1 and 6.2 applying the constitutive relations given in Eqs. 6.5 to 6.7, yields the Maxwell equations in the frequency domain

$$\nabla \times \mathbf{E} + i\omega\mu_0\mathbf{H} = \nabla \times \mathbf{E} + \hat{z}\mathbf{H} = 0 \tag{6.8}$$

$$\nabla \times \mathbf{H} - (\sigma + i\varepsilon\omega)\mathbf{E} = \nabla \times \mathbf{H} - \hat{y}\mathbf{E} = 0 \tag{6.9}$$

The impedivity $\hat{z} = i\omega\mu_0$ and the admittivity $\hat{y} = \sigma + i\varepsilon\omega$ will be used for convenience.

The homogeneous Maxwell equations in Eq. 6.8 and 6.9 apply only in source-free regions. In regions containing sources they are replaced by the inhomogeneous equations

$$\nabla \times \mathbf{E} + \hat{z}\mathbf{H} = -\mathbf{J}_m^S \tag{6.10}$$

$$\nabla \times \mathbf{H} - \hat{y}\mathbf{E} = \mathbf{J}_e^S \tag{6.11}$$

where $\mathbf{J}_m{}^S$ is magnetic source current, and $\mathbf{J}_e{}^S$ is electric source current.

6.2.2 Schelkunoff potentials

The inhomogeneous frequency domain Maxwell equations presented in Eq. 6.10 and Eq. 6.11 may be solved for homogeneous regions if $\mathbf{J}_m{}^S$ and $\mathbf{J}_e{}^S$ can be described. Expressing **E** and **H** in terms of the Schelkunoff potentials **A** and **F** facilitates the derivation of **E** and **H** by differentiation. Using potentials makes the differential equations easier to solve because the potentials are parallel to the source fields, unlike the fields themselves.

In general, the electric and the magnetic fields in each homogeneous region are described as a superposition of sources of either electric or magnetic type

$$\begin{aligned} \mathbf{E} &= \mathbf{E}_m + \mathbf{E}_e \\ \mathbf{H} &= \mathbf{H}_m + \mathbf{H}_e \end{aligned} \tag{6.12}$$

Thus, an electromagnetic field is described by the pairs of vector functions $\mathbf{E}_m$, $\mathbf{H}_m$ and $\mathbf{E}_e$, $\mathbf{H}_e$. For the first pair $\mathbf{J}_e{}^S$ is assumed to be zero, and for the latter $\mathbf{J}_m{}^S$ is assumed to be zero, meaning that the electric source current is zero for electric and magnetic fields due to a magnetic source, and vice versa.

All above mentioned equations are valid in general, but now we will be more specific on the TEM method. The TEM method considered uses a magnetic source ($\mathbf{J}_m^S$) transmitting a transverse electric field. This simplifies the situation because only the Schelkunoff potential **F** is necessary for calculations. The Schelkunoff potential **F** is defined as

$$\mathbf{E}_m \equiv -\nabla \times \mathbf{F} \tag{6.13}$$

Using this relation in Eq. 6.10 allows derivation of the inhomogeneous Helmholz equation

$$\nabla^2 \mathbf{F} + k^2 \mathbf{F} = -\mathbf{J}_m^S \tag{6.14}$$

where the wave number k is defined as:

$$\mathrm{k}^2 = \mu_0 \varepsilon \omega^2 - i\mu_0 \sigma \omega = -\hat{z}\hat{y} \tag{6.15}$$

For earth materials, and for frequencies less than 10^5 Hz, the displacement current is much smaller than the conduction current. Hence, $\mu_0 \varepsilon \omega^2 << \mu_0 \sigma \omega$ so that $k^2 \approx -i\mu_0 \sigma \omega$. This is called the quasi-static approximation.

The total electric and magnetic fields from a magnetic source can now be derived using the Schelkunoff potential **F**

$$\mathbf{E}_m = -\nabla \times \mathbf{F} \tag{6.16}$$

$$\mathbf{H}_m = -\hat{y}\mathbf{F} + \frac{1}{\hat{z}} \nabla(\nabla \cdot \mathbf{F}) \tag{6.17}$$

Under the assumption of a one-dimensional (1D) layered earth, **F** consists of one component only, the z-component

$$\mathbf{F} = F_z \mathbf{u}_z \quad ; \quad \mathrm{TE}_z \tag{6.18}$$

where F_z is a scalar function of x, y and z, while $\mathbf{u}_z$ is the unit vector in the z-direction. TE_z denotes the transverse electric field, which is the field propagating in the xy-plane. Substituting Eq. 6.18 into Eqs. 6.16 and 6.17 enables the expression of the field components

$$H_x = \frac{1}{\hat{z}} \frac{\partial^2 F_z}{\partial x \partial z} \qquad E_x = -\frac{\partial F_z}{\partial y}$$
$$H_y = \frac{1}{\hat{z}} \frac{\partial^2 F_z}{\partial x \partial z} \qquad E_y = \frac{\partial F_z}{\partial x} \tag{6.19}$$
$$H_z = \frac{1}{\hat{z}} \left(\frac{\partial^2}{\partial z^2} + k^2 \right) F_z \qquad E_z = 0$$

6.2.3 The transient response over a layered halfspace

The field transmitted in the TEM-method is a transverse electric field. Hence, only the *F*-potential is necessary for a derivation of the vertical magnetic field in the centre of a circular loop. A circular loop is a good approximation for a square loop of the same area. The latter is more often used in practice.

For a plane parallel 1D model consisting of source free regions as well as regions containing sources, a derivation of the appropriate Schelkunoff potential requires definition of source characteristics.

A circular or square transmitter loop can be calculated as an integration of vertical magnetic dipoles over the area of the loop. The Schelkunoff potential, in this frame of reference, is for a vertical magnetic dipole

$$F(\rho, z) = \frac{\hat{z}_0 m}{4\pi} \int_0^\infty \left[e^{-u_0|z+h|} + r_{TE} e^{u_0(z-h)} \right] \frac{\lambda}{u_0} J_0(\lambda\rho) d\lambda \tag{6.20}$$

with m being the magnetic moment of the dipole and J_0 being the Bessel function of order zero. $\lambda = \sqrt{k_x^2 + k_y^2}$ where k_x and k_y are the spatial frequencies in the x- and y-direction. $u_n = \sqrt{\lambda^2 - k_n^2}$ where k_n^2 is the wavenumber in layer n, which from Eq. 6.15 is, in the quasi-static approximation, $k^2 = -i\mu_0\sigma_n\omega$. ρ is the radial distance from source to receiver, $\rho = \sqrt{x^2 + y^2}$. r_{TE} is called the reflection coefficient and is a quantity expressing how the layered halfspace *modifies* the source field.

The terms of Eq. 6.20 are visualized in Fig. 6.2. The first term of the integral in Eq. 6.20 is the free-space response from the source dipole to the receiver. The second term has three parts as shown in the figure. The first part takes the field from the source dipole to the interface of the earth, the second term modifies it with the reflection coefficient, and the third part returns it to the receiver. The reflection coefficient, as visualized in Fig. 6.2, is calculated recursively starting from the bottom of the model.

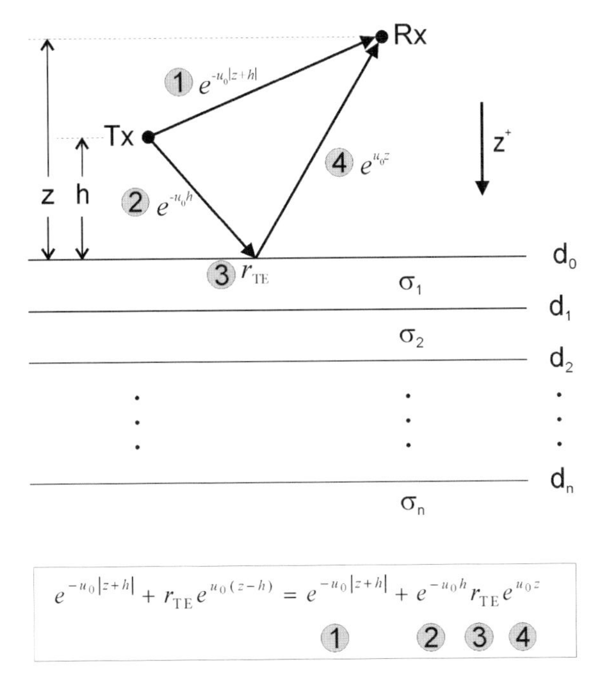

Fig. 6.2. Visualization of the terms in Eq. 6.20. z is positive downwards. Rx is in the height h above the ground. Tx at elevation z. The layers in the model are plane parallel and homogeneous. This model type is called a 1D model

Integrating Eq. 6.20 over a circular loop with radius a and current I yields after some manipulation

$$F(\rho, z) = \frac{\hat{z}_0 I a}{2} \int_0^\infty \frac{1}{u_0} \left[e^{-u_0|z+h|} + r_{TE} e^{u_0(z-h)} \right] J_1(\lambda a) J_0(\lambda \rho) d\lambda \qquad (6.21)$$

where J_1 is the first order Bessel function. Eq. 6.21 applies for a circular transmitter loop with the dipole receiver in distance ρ from the centre of the loop. Using Eq. 6.17 and simplifying further so the receiver is in the centre of the transmitter loop (central-loop configuration) we get for the vertical magnetic field

$$H_z = \frac{I a}{2} \int_0^\infty \left[e^{-u_0|z+h|} + r_{TE} e^{u_0(z-h)} \right] \frac{\lambda^2}{u_0} J_1(\lambda a) d\lambda \qquad (6.22)$$

Eq. 6.22 is expressed in the frequency domain (r_{TE} is a function of frequency). The transient response, the response in the time-domain, is obtained by inverse Laplace transform or inverse Fourier transform.

The integral in Eq. 6.22 is called a Hankel integral. This integral cannot be solved analytically, but must be evaluated using numerical methods.

6.2.4 The transient response for a halfspace

No analytic expression is derived in the general case, because of the complexity of Bessel functions and the integral in general. However, for the central loop configuration on the surface of a homogeneous halfspace r_{TE} becomes

$$r_{TE} = \frac{\lambda - u}{\lambda + u} \tag{6.23}$$

and Eq. 6.22 simplifies to

$$H_z = Ia \int_0^{\infty} \frac{\lambda^2}{\lambda - u} J_1(\lambda a) d\lambda \tag{6.24}$$

Using the simple relation $\mathbf{b}=\mu_0\mathbf{h}$ Eq. 6.24 can now be solved for b by evaluating the integral and applying an inverse Laplace transform

$$b_z = \frac{\mu_0 I}{2a}\left[\frac{3}{\pi^{1/2}\theta a} e^{-\theta^2 a^2} + \left(1 - \frac{3}{2\theta^2 a^2}\right)\mathrm{erf}(\theta a)\right] \tag{6.25}$$

where $\theta = \sqrt{\mu_0 \sigma / 4t}$ and erf is the error function. b_z may be evaluated for $t \to 0$ as $b_z = \mu_0 I/2a$. This is the size of the primary field in free space, i.e. the magnetic intensity before the current is turned off.

The time derivative, or the impulse response, db_z/dt is found through differentiation to be

$$\frac{\partial b_z}{\partial t} = -\frac{I}{\sigma a^3}\left[3\mathrm{erf}(\theta a) - \frac{2}{\pi^{1/2}}\theta a\left(3 + 2\theta^2 a^2\right)e^{-\theta^2 a^2}\right] \tag{6.26}$$

These are the central equations for the TEM method, and we will return to them later. The equations above apply only to the vertical field in the centre of the transmitter loop.

6.3 Basic principle and measuring technique

TEM methods use a direct current, usually passed through an ungrounded loop. The current is abruptly interrupted, and the rate of change of the secondary field due to induced eddy currents is measured using an induction coil. The primary field is absent while measuring. Fig. 6.3 summarizes the basic nomenclature and principles.

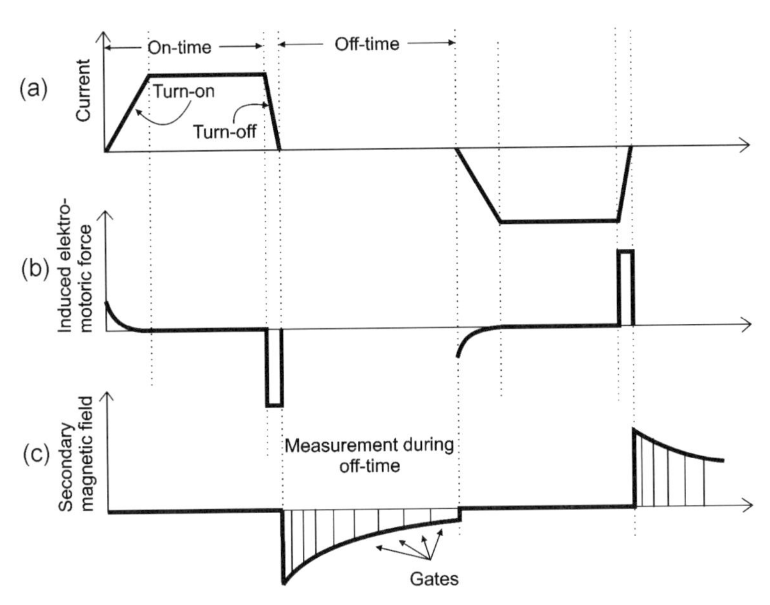

Fig. 6.3. Basic nomenclature and principles of the TEM method. (a) shows the current in the transmitter loop. (b) is the induced electromotoric force in the ground, and (c) is the secondary magnetic field measured in the receiver coil. For the graphs of the induced electromotoric force and the secondary magnetic field, it is assumed that the receiver coil is located in the centre of the transmitter loop

Typical values for a ground based system are a 50 - 200 µs long turn on ramp, 1- 40 ms on-time, 1 – 30 µs turn-off ramp and 1 - 40 ms off-time to measure. The depicted waveform is often referred to as a square waveform. Other waveforms with sine or triangular shapes are used, but mainly in airborne systems.

The data recording is done in time-windows, often called gates. The gates are arranged with a logarithmically increasing length in time to improve the signal/noise (S/N) ratio at late times. This principle is called log-gating and 8-10 gates per decade in time are often used.

As depicted in Fig. 6.3, the current direction shifts for each single pulse. A typical sounding consists of 1,000-10,000 single pulses (transients). The sign changes subdue 1) the unwanted signals from power lines if the repetition frequency is chosen as a subharmony of the power line frequency and 2) static shift in the amplifiers in the instrument. This is called synchronous detection. Typical repetition frequencies for a 50 Hz power line frequency is 25 Hz. When higher repetition frequencies are used, the frequency and the number of transients must be matched so that the 50 Hz power line frequency is exactly subdued.

In the field, a transient sounding can be conducted by placing a wire in a square loop on the ground. When investigating the top 150 m of the ground, a square with an area of 40 x 40 m^2 is commonly used. The wires are connected to the transmitter, and the receiving coil with a diameter of approximately 1 m is placed in the middle of the transmitter loop. The receiving coil is connected to the receiver and the receiver, in turn, is connected to the transmitter allowing for synchronization between the transmitter and the receiver (Fig. 6.4).

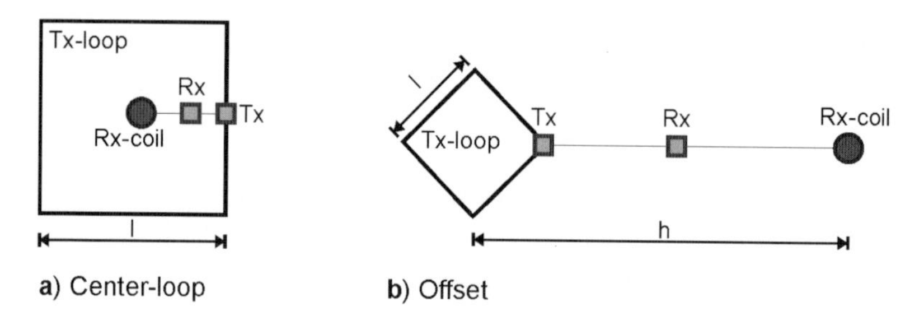

Fig. 6.4. Field setup of a TEM system. **a**) shows a central loop, **b**) an offset-loop configuration. Rx denotes receiver, Tx transmitter, l the side length of the loop and h the offset between Tx-loop and Rx-coil centres

The measurements are made by transmitting a direct current through the Tx-loop. This results in a static primary magnetic field. The current is shut off abruptly which due to Faradays Law induces an electrical field in the surroundings. In the ground, this electrical field will result in an electrical current which again will result in a magnetic field, the secondary field. The current will behave so that, just after the transmitter is switched off, the size of the secondary magnetic field from the current in the ground will be equivalent to the size of the primary magnetic field (which is no longer there). During this first phase the lapse of current is independent of the conductivity structure of the ground. Therefore the magnetic field does not *jump* when the current is switched off, but changes continuously. As time passes, the resistance in the ground will still weaken the current (which is converted to heat), and the current density maximum will eventually move outwards and downwards, leaving the current density still weaker. After the first phase, the diffusion is dependent on the conductivity of the ground. In a well conducting ground the current diffuses more slowly down into the ground compared to a poor conducting ground where the currents will diffuse and decay fast.

The decaying secondary magnetic field is vertical in the middle of the Tx-loop (at least if the ground consists in plane and parallel layers). Hereby an electromotoric power is induced in the Rx-coil - a voltage – and this is the signal which is measured as a function of time in the receiver.

Just after the current in the Tx-loop is shut off, the current in the ground will be close to the surface, and the measured signal reflects primarily the conductivity of the top layers. At later times the current will run deeper in the ground, and the measured signal contains information about the conductivity of the lower layers. Measuring the current in the receiving coil will therefore give information about the conductivity as a function of depth – this is often called a sounding.

The configuration shown in Fig. 6.4a has the receiver coil placed in the centre and is called a central loop or an in-loop configuration. The receiver coil can be placed outside the current loop which results in an offset loop configuration (Fig. 6.4b). This configuration is always used when the side length of the Tx-loop is shorter than approximately 40 m. The receiver cannot be placed inside the Tx-loop when it is small because the primary field becomes too big and interferes with the electronics in the receiver coil and the receiver.

Some transient equipment utilizes the current loop as a receiver coil, and apart from the electronically related problems, this causes no problems since the current in the current loop is shut off when the measurements are taken. This is called a coincident loop.

The advantages and drawbacks of the different configurations are discussed in larger detail in Sect. 6.8.2.

6.4 Current diffusion patterns

6.4.1 Current diffusion and sensitivity, homogeneous halfspace

To obtain a physical understanding of the current flow in the ground during a transient sounding, we will investigate the current density at different times after the current in the Tx-loop has been shut off. For a 1D ground the current flows cylindrically and symmetrically around the vertical axis through the centre of the Tx-loop. In the following only the right side of the current system is shown.

Fig. 6.5 shows the current density in a homogeneous halfspace at different times.

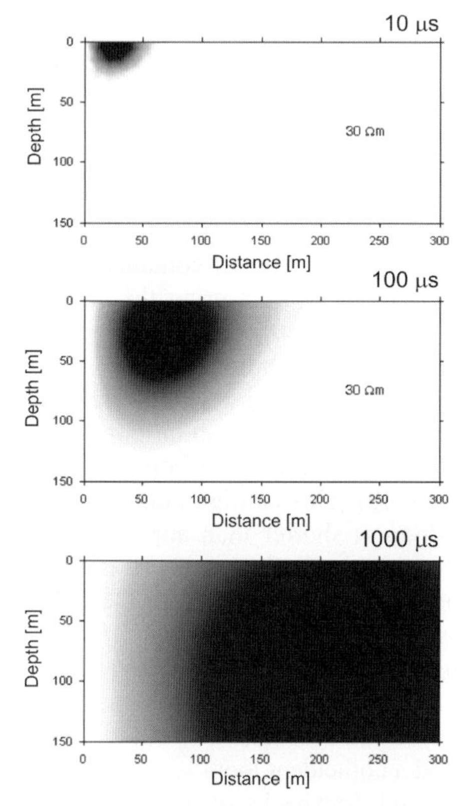

Fig. 6.5. Current density for an impulse response in an homogeneous halfspace at various times. The field is symmetrical around the centre-line of the transmitter. The current densities are normalized with the maximum value at that time giving the same maximum amplitude in all plots. In reality the maximum amplitude is approximately 1,000,000 times larger at 10 μs than at 1000 μs. The grey shaded scale indicates larger current densities for dark grey regions

You see that the maximum for the current moves downwards and outwards as time passes. Note that the maximum is fairly widespread in the halfspace. This maximum moves asymptotically along a cone which always creates an angle of 30° with the horizontal axis, meaning that the current has diffused about twice as much outwards as downwards. This diffusion pattern is often referred to as a *smoke ring*, for obvious reasons.

Every little part of the circular current system will contribute to the magnetic field on the surface which is dependent of the current density together with the radius and the depth of the current circle. Summarizing the contribution of all current density elements of Fig. 6.5 and integrating over thin horizontal layers results in a function which describes the relative contribution of different depths to the measured field at a given time. This function is shown in Fig. 6.6 at four different times. Often, this is referred to as the sensitivity function.

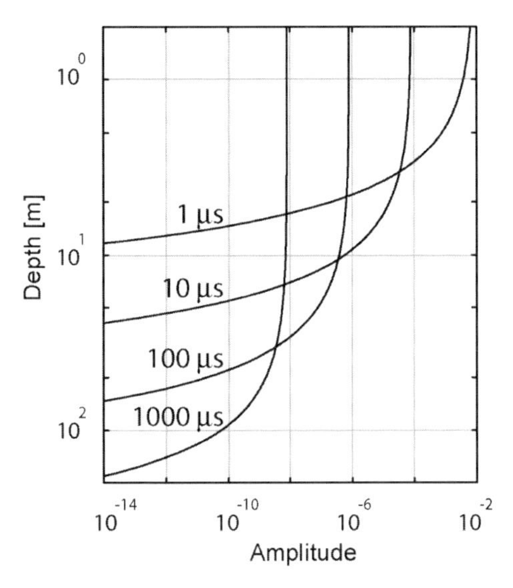

Fig. 6.6. 1D sensitivity function of the TEM system in the central loop configuration. The absolute values of the sensitivity function is plotted at four different times

It is seen that the function averages the conductivity down to a relatively well defined depth z', and that the sensitivity below this point decreases at a very steep rate (like $\exp(-z^{-2})$). Within a reasonable degree of approximation, one can say that a transient measurement to a given time is an average of conductivities from the surface down to a specific depth, and then no deeper.

These observations are only strictly valid for a vertical magnetic dipole source over a homogenous halfspace, but it can be proven that with reasonable approximations they retain their validity when the source is a current loop over a ground divided into layers.

6.4.2 Current densities, layered halfspaces

The earth can rarely be approximated by a homogeneous halfspace, so let us investigate the current density distributions for more complicated models.

The first model is a three-layer model with a low-resistivity layer embedded in high-resistivity layers.

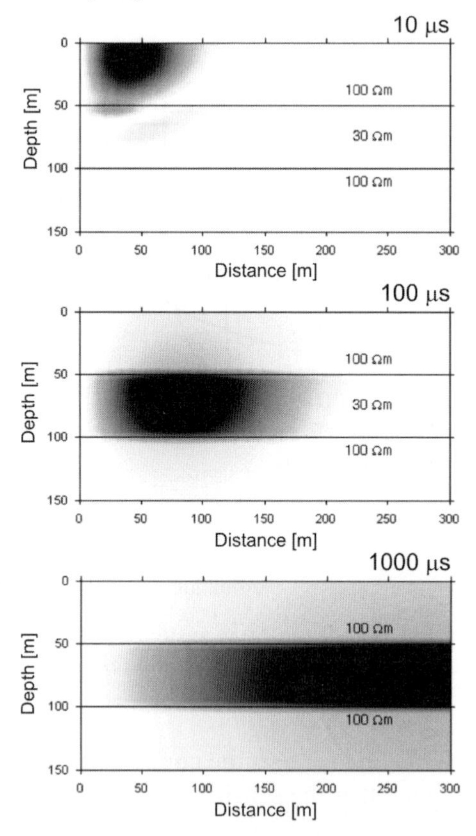

Fig. 6.7. Current density distribution in a three-layer model with a low-resistivity layer embedded in a high-resistive background. The current densities are normalized with the maximum value at that time giving the same maximum amplitude in all plots. The grey shaded scale indicates larger current densities for dark grey areas

Fig. 6.7 shows that already at 10 μs after the turn-off of the current pulse the maximum of the current density is located in the low-resistivity layer. At later times the maximum stays in the low-resistivity layer. Compared to the homogeneous halfspace (in Fig. 6.5), the low-resistivity layer interrupts the normal diffusion pattern and acts as a shield for the high-resistivity layer below. The faint grey structure in layer two at 10 μs is an evidence of currents running in the opposite direction.

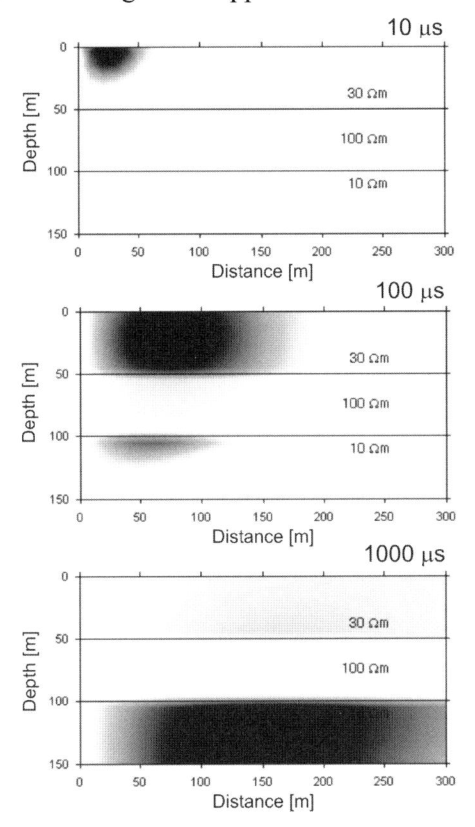

Fig. 6.8. Current density distribution in a three-layer model with a high-resistivity layer embedded in a low-resistivity background. The current densities are normalized with the maximum value at that time giving the same maximum amplitude in all plots. The grey shaded scale indicates larger current densities for dark grey areas

The model in Fig. 6.8 is the reverse of Fig. 6.7. Now we have a high-resistivity layer imbedded between two low-resistivity layers. At early times the current density maximum is divided between the top and the bottom layers. The current density in the high-resistivity layer is very low. At later times the current density distribution *jumps* entirely to the bottom layer skipping the high-resistivity layer in the middle. This implies that the resulting sounding curve retains very limited information on this layer.

Based on these examples we conclude that the TEM method is highly sensitive to low-resistivity layers (conductive layers) simply because a larger amount of the current flows in these layers. The diffusion speed depends on the resistivity of the layers, i.e. the diffusion speed is high for high-resistivity layers and low for low-resistivity layers.

6.5 Data curves

6.5.1 Late-time apparent resistivity

The decaying secondary magnetic field is referred to as *b* or the step response (Eq. 6.25). However, because an induction coil is used for measurements in the field, the actual measurement is that of *db/dt* (the induced electromotoric force is proportional to the time derivative of the magnetic flux passing the coil). The impulse response, *db/dt* (Eq. 6.26) of the magnetic induction is plotted in Fig. 6.9a for a variety of halfspace resistivities.

Fig. 6.9a indicates a power function dependence at late times. When θ approaches zero, i.e. at late times, Eq. 6.26 is approximated by

$$\frac{\partial b_z}{\partial t} \approx \frac{-I\sigma^{3/2}\mu_0^{5/2}a^2}{20\pi^{1/2}} t^{-5/2} \tag{6.27}$$

As seen the time derivative of *b* exhibits a decay proportional to $t^{-5/2}$. Observation of the curve of the decaying magnetic field in Fig. 6.9a is not very informative and the same applies for actually measured sounding curves. A plot of apparent resistivity, ρ_a, is more illustrative. It is derived from the late time approximation of the impulse response in Eq. 6.27 to be

$$\rho_a = \left(\frac{Ia^2}{20\,\partial b_z/dt}\right)^{1/2} \frac{\mu_0^{5/3}}{\pi^{1/3}} t^{-5/3} \tag{6.28}$$

The response curves plotted in Fig. 6.9a are shown as ρ_a-converted curves in Fig. 6.9b.

It is important to note that oscillations on an apparent resistivity curve are not necessarily reflections of variations in geology and cannot be interpreted as such. For instance, a ρ_a-curve always goes up before it goes down and vice versa (see the overshoot in Fig. 6.9b at $2\cdot10^{-5}$ s). Even though keeping in mind that the apparent resistivity is not equal to the true resistivity for a layered earth, it does provide a valuable normalization of data, with respect to source and the measuring configuration.

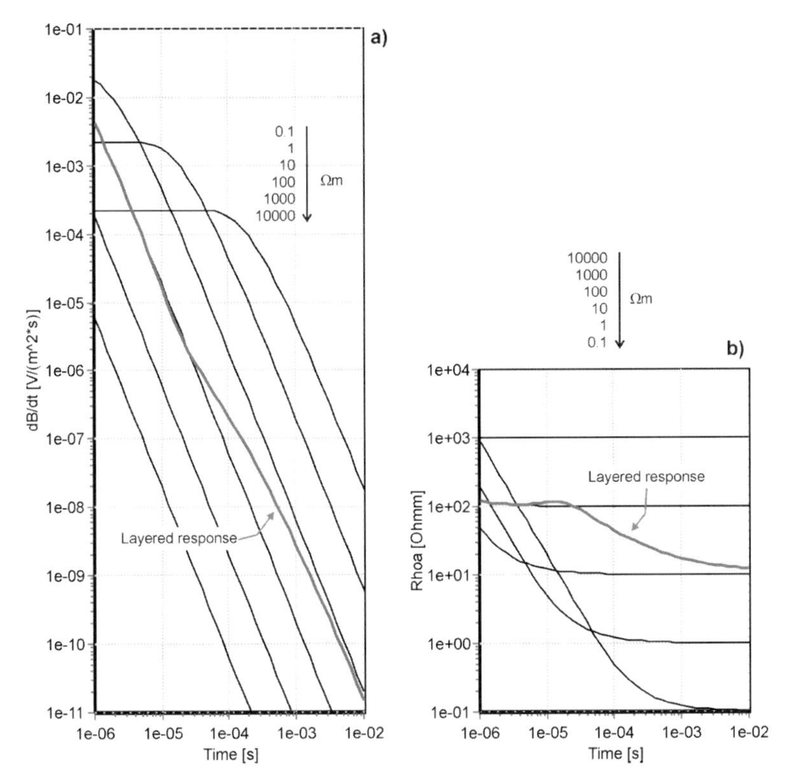

Fig. 6.9. In **a)** impulse responses (*db*/*dt*) for a homogeneous halfspace with varying resistivities are shown (black lines). The same curves converted to ρ_a are shown in **b)**. The grey line is the response of a two-layer earth with 100 Ωm in layer 1 and 10 Ωm in layer 2. Layer 1 is 40 m thick

6.6 Noise and Resolution

6.6.1 Natural background noise

A geophysical datum always consists of two numbers – the measurement itself and the uncertainty of the measurement. A measurement is never 100% certain (or 0% uncertain) because the measured data consist of both the earth response and the background noise.

One single transient in a TEM-sounding is most often affected significantly by noise. By repeating the measurement the noise is decreased and the signal enhanced. An ordinary TEM sounding consists of 1,000 to 10,000 single transients. Fig. 6.10 shows stacks of 50 single transients and a stack of 5,000 single transients. It is obvious that a sounding with 50 stacked transients would be very hard to interpret compared to the stack with 5,000 transients.

If using a log-gating technique, the S/N ratio is proportional to $\sqrt{N}$ where N is the number of measurements in the stack. We assume that the noise follows a Gaussian distribution. Thus, doubling the number of measurements in the stack improves the S/N ratio by a factor 1.41.

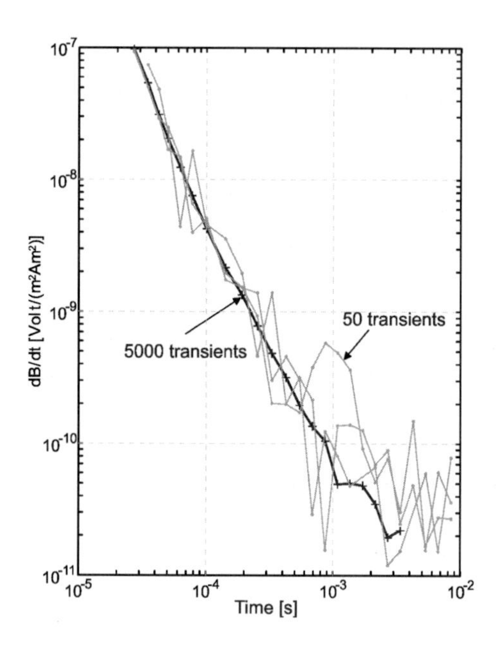

Fig. 6.10. TEM sounding curves stacked with 50 transients (grey) and 5,000 transients (black)

The surrounding electromagnetic noise originates from various sources. Some sources are near but most sources, such as lightening, are very distant. The fields from these sources travel around the globe in the cavity between the surface of the Earth and the ionosphere. This cavity is an efficient wave guide for electromagnetic radiation.

There is a natural level of electromagnetic noise which derives from fluctuations in the Earth's magnetic field originating from sun winds. Their frequencies are, however, so low that they do not influence the transient measurement. More important are the so-called spherics which originate from flashes of lightening from thunderstorms everywhere on the Earth, especially in tropical areas. Around 100 flashes of lightening hit the Earth per second. This noise has a random character and it is more powerful during the day than during the night and stronger during summer compared to winter.

Noise also originates from the supply of electricity and the related electrical installations. There is partly the harmonic 50 or 60 Hz signal, which is deterministic, partly the transient magnetic fields, which are of a random character and arise due to changes in the current power when electrical installations are turned on or off. In addition to this, the electromagnetic fields from communication equipment (radio, TV, telephone etc.) will act as sources of noise.

With well-working equipment the noise originating from the electronics in the instrument itself is negligible compared to the noise described above.

6.6.2 Noise and measurements

Using the log-gating technique, random noise will fall off proportional to $t^{-1/2}$. Fig. 6.11 summarizes some central principles of a TEM sounding.

The effective noise, i.e. the noise after stacking, is in the area of 1 nV/m2, varying between 0.1 and 10 nV/m^2. In Fig 6.11 the level is approximately 3 nV/m^2*s at 1 ms. A suite of noise measurements is shown grey on the figure. The trend of the curves is close to the predicted $t^{-1/2}$ dependence shown with the grey dashed line.

It is clear that the measurements at early times are many times bigger than the noise level. This means that the S/N ratio is high, and thereby the ability to repeat the measurements is good at early times. At the time when the measured signal passes the level of noise, it is normally proportional to $t^{-5/2}$ which means that the transition from a good S/N ratio to a very poor S/N ratio happens quite suddenly.

There are two ways to get data at later times, i.e. information from larger depths: 1) reduce the noise by stacking or 2) increase the moment of the transmitter. Stacking reduces the noise proportional to $\sqrt{N}$ where N is the number of measurements in the stack. The effect of increasing the moment is also shown in Fig. 6.11 with the black dotted line. The line indicates the level of a sounding at the same location with a 10 times larger moment and it is clear that the S/N ratio is much higher at later times.

6.6.3 Penetration depth

In relation to TEM soundings it is difficult, as for all other geophysical methods, to speak quantitatively and unambiguously about the penetration depth. In the following we will state some rules of thumb.

The depth down to which the current system diffuses is called the diffusion depth. This depth, z_d, is defined by

$$z_d = \sqrt{\frac{2t}{\mu\sigma}} \approx 1.26 \times \sqrt{\rho t}\ [\mathrm{m}] \quad , \quad \rho[\Omega\mathrm{m}], t[\mu\mathrm{s}] \tag{6.29}$$

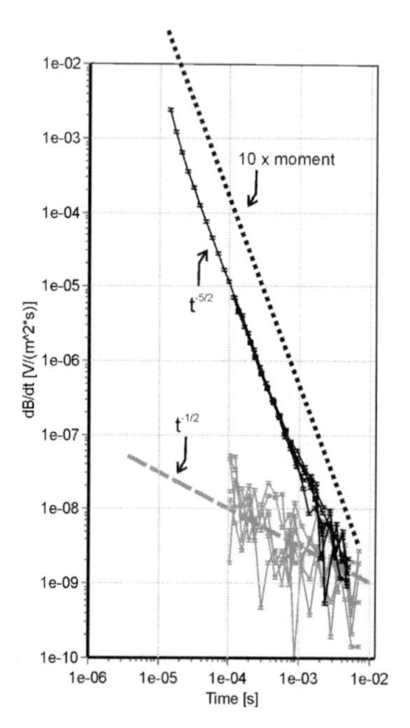

Fig. 6.11. TEM sounding and noise measurements. The grey curves are noise measurements with the $t^{-1/2}$ trend plotted with the thick dashed grey line. Error bars are 5%. The earth response is the black curves. The black dotted line indicates the approximate level of a sounding with a 10 times higher moment

This is an exact equation for plane fields only. For circular or quadratic loop sources the diffusion depth is about 1.8 times smaller than estimated by Eq. 6.29.

The diffusion time, i.e. the time at which the current has diffused to a certain depth, is expressed as

$$t_d = \frac{\mu\sigma z_d^2}{2} \approx 0.628 \times \frac{z_d^2}{\rho}\ [\mu s]\ ,\ \ \rho[\Omega m], z_d[m] \tag{6.30}$$

As seen in the Fig. 6.9a, the signal decreases in a homogenous halfspace by $t^{-5/2}$, and when the signal passes the level of the natural noise, we can no longer use the measurements. Thus, the level of the natural noise sets the limits for how late we can make our measurements, and thereby also how deep the current can diffuse into the ground. By using the expression given for db_z/dt for late times (Eq. 6.27) we find a relationship between the noise signal, V_{noise}, and the latest time, t_L at which we can make measurements:

$$\begin{aligned} V_{noise} &= \frac{dB}{dt} = \left(\frac{M}{20}\right)\left(\frac{\sigma}{\pi}\right)^{3/2}\left(\frac{\mu}{t_L}\right)^{5/2} \Rightarrow \\ t_L &= \mu\left(\frac{M}{20V_{noise}}\right)^{2/5}\left(\frac{\sigma}{\pi}\right)^{3/5} \end{aligned} \tag{6.31}$$

When t_L is equivalent with the diffusion time t_d

$$\begin{aligned} t_L &= \mu\left(\frac{M}{20V_{noise}}\right)^{2/5}\left(\frac{\sigma}{\pi}\right)^{3/5} = t_d = \frac{\mu\sigma z_d^2}{2} \Rightarrow \\ z_d &= \left(\frac{2}{25\pi^3}\right)^{1/10}\left(\frac{M}{\sigma V_{noise}}\right)^{1/5} = 0.551\left(\frac{M}{\sigma V_{noise}}\right)^{1/5} \end{aligned} \tag{6.32}$$

From these expressions it is seen that the maximal diffusion depth, which is a measure for the penetration depth, is proportional to the fifth root of the ratio between the moment of the current loop and the product of the conductivity and the noise level. The only way to increase the penetration depth is to increase the moment of the transmitter or decrease the effective noise level. The surrounding noise is a relatively unchangeable size, but the way in which we gather and process our data, by stacking many measurements, reduces the effective noise as discussed in Sect. 6.6.2. From Eq. 6.32 appears also that to double the penetration depth, the effective noise has to be reduced - or - the moment of the transmitter has to be increased by a factor 32.

6.6.4 Model errors, equivalence

Models that within the measuring error produce almost identical responses are called equivalent models. Sometimes equivalences can be very pronounced in the sense that very different models give rise to almost identical responses. In an ideal situation with no measuring error, equivalences do not exist, meaning that a given data set can be fitted with one and just one model. In real situations there is always noise on the data, and we assign uncertainty to the data according to the noise level. In that situation multiple models will fit the data set within the assigned uncertainty.

The TEM method shows only to limited degree equivalence for a low-resistivity layer between two high-resistivity layers (low-resistivity equivalence), a high-resistivity layer between two low-resistivity layers (high-resistivity equivalence) and layers where the resistivity is gradually decreasing (double descending equivalence).

In Sect. 6.4 we saw that the TEM method has only limited sensitivity to high-resistivity layers. This, of course, has an implication when we try to interpret data sets measured over models with such high-resistivity layers. In Fig. 6.12 a suite of responses is plotted.

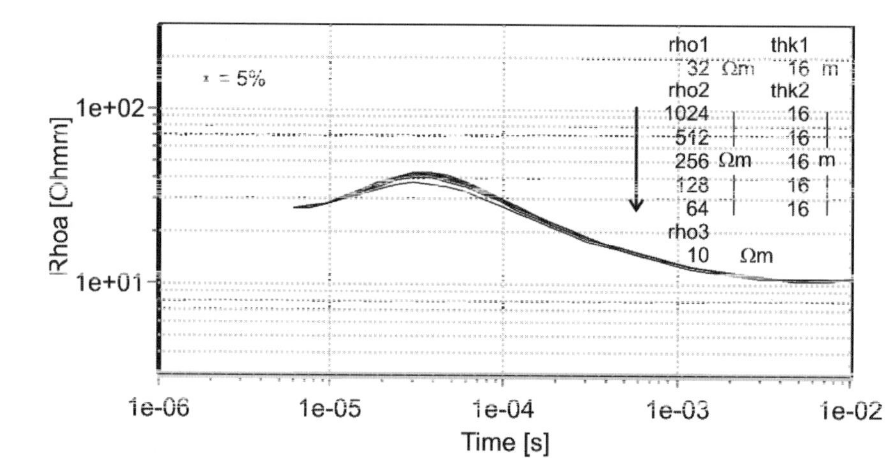

Fig. 6.12. Resistivity equivalence. The resistivity of the second layer is varied from 64 Ωm to 1024 Ωm. The base model is a maximum type model with 32 Ωm n the top layer, a high-resistivity middle layer (64 - 1024 Ωm) and 10 Ωm in the bottom layer. The thickness of layer one and two is 16 m

The response curves of the different models are only slightly separated. This is explained by the fact that the high-resistivity layer does not produce any significant amount of response. It is therefore not possible to say anything about the resistivity of that layer. However, the thickness will be fairly well determined because the thickness of the first layer and the depth to the third layer can be determined as having relatively low resistivity.

In Fig. 6.13 response curves for a series of double ascending models are shown. The ratio between thickness and resistivity (the conductance) for the second layer is constantly at 0.25 S. It is clearly very difficult to distinguish between these models, which is caused by the poor sensitivity to changes of the depth to the bad conductor at the bottom.

Other types of models including high-resistivity layers will be equally difficult to interpret with the TEM method. As a rule of thumb it is said that the TEM method is unable to distinguish between resistivities higher than 80 - 100 Ωm. They are just *high*.

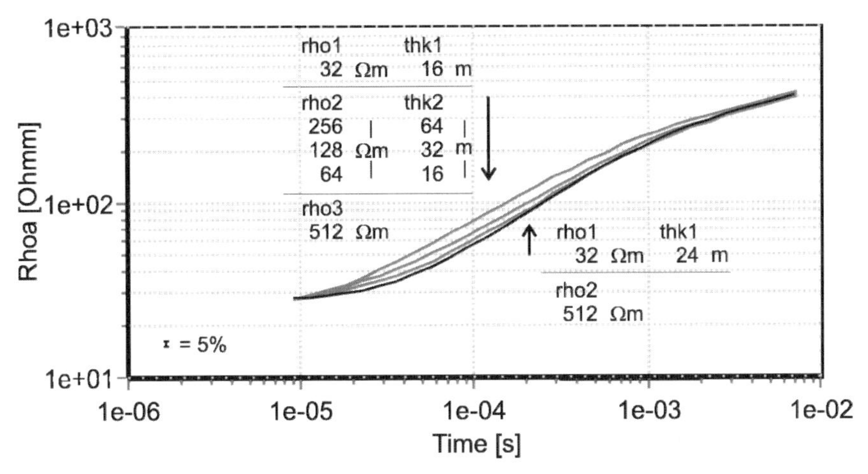

Fig. 6.13. Layer suppression in double ascending models. The resistivity of the second layer is varied from 64 Ωm to 256 Ωm and the thicknesses are varied from 16 to 64 m keeping the conductivity of the second layer constant. The resistivity of layer one is 32 Ωm and the thickness is 16 m. The resistivity of layer three is 512 Ωm . The black line is the two-layer model with the same total conductivity over the bad conductor at the bottom as the three-layer models

6.7 Coupling to man-made conductors

Coupling noise is not noise in the same sense as the noise described in Sect. 6.6.1. Coupling noise appears due to induced currents in all man-made electrical conductors within the volume in which the transmitted electromagnetic field propagates. The disturbance is deterministic, arising at the same delay time for all decays summed in the stacking process. Coupling effects in data cannot be accurately removed to provide a reliable interpretation; therefore soundings located close to pipelines, cables, power line, rails, auto guards and metal fences cannot be interpreted, and data should be culled.

The safe distance, defined as the minimum distance where good data can be measured, is counted as the distance between any point on the transmitter-receiver setup and the man-made conductor. The safe distance to any man-made conductor is at least 100 m over an earth with an overall resistivity of 40 – 60 Ωm. The safe distance increases with the resistivity.

Spatially dense sampling is the only way of adequately identifying and removing distorted data while still leaving enough data for a meaningful interpretation.

6.7.1 Coupling types

A general model for the disturbance from man-made structures is that of an oscillating circuit, and it is normally categorized into two types, galvanic and capacitive coupling, both shown in Fig. 6.14.

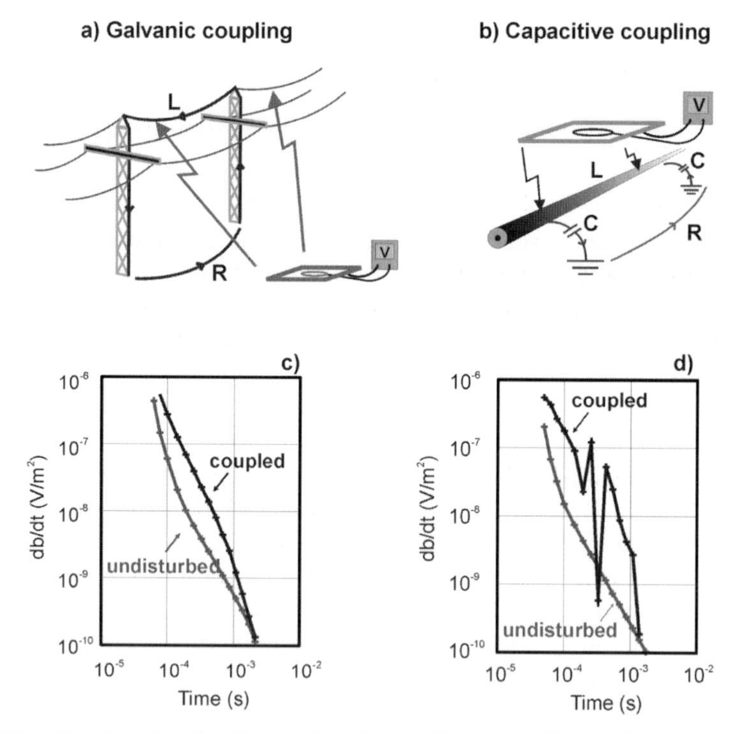

Fig. 6.14. Graphic sketch of the galvanic coupling **a)** and capacitive coupling **b)** with the typical imprints in the measured data in **c)** and **d)**

In the galvanic coupling the current has a galvanic return path to the ground as shown in Fig. 6.14a. It is characterized by an L-R circuit, with the decay decreasing exponentially. The disturbance depends on the time constant of the circuit, but can be very hard to identify on non-profile data. A galvanic-type coupling could arise from high-voltage power lines, grounded at each pylon. Animal fences and highway crash barriers are other examples. It can be very hard to recognize on single-site soundings because the whole sounding curve is shifted, as plotted in Fig. 6.14c. The shift can easily be confused by that of a low resistive layer at shallow depth.

The capacitive-type coupling is characterized by a L-C-R circuit having an inductive return path to the ground (Fig. 6.14b). A capacitive-type coupling could arise from buried polyurethane-isolated cables. This type of coupling is easily recognized because of its oscillating character, as seen in Fig. 6.14d.

Distance dependence of couplings and handling of coupled data sets are described under advanced topics.

6.7.2 Handling coupled data

Field work carried out in culturally developed areas will always give rise to coupled soundings, even if carried out with great care, keeping the safe distance at all times. This implies that *all* soundings must be evaluated individually in order to identify possible couplings. Data sets identified as influenced by a coupling must be excluded *before* the data interpretation. If coupled data sets are not identified, but assumed to be results of geophysical properties of the earth, an erroneous interpretation will be the result as shown in Fig. 6.15.

It is clear that the coupled response, S2, can be interpreted as if it was an undisturbed earth response (Fig. 6.15b and c). However, assuming that the data set is not coupled results in an erroneous interpretation (Fig. 6.15d). In this case the knowledge of the houses nearby and the close separation between the soundings is enough to identify the coupled data set, but often the only certain way to identify couplings is to achieve profile data.

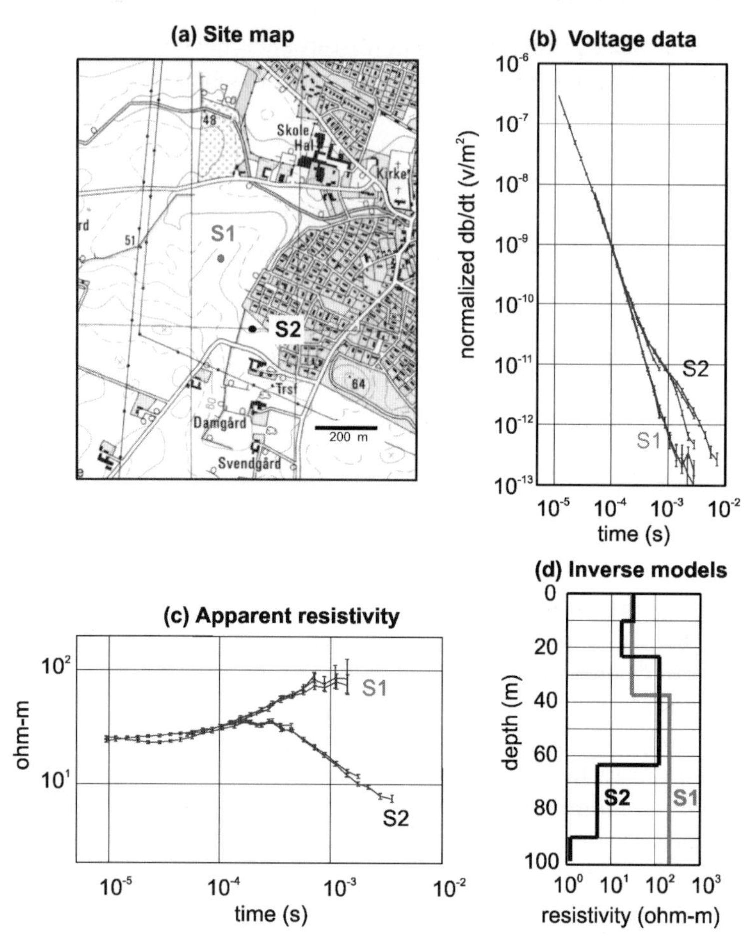

Fig. 6.15. Coupled and uncoupled soundings. Panel a) shows the position of two TEM soundings on a map of which S1 is not coupled and S2 is galvanically coupled due to the houses nearby. Plot b) is db/dt decay curves for the two soundings and plot c) is the same converted to apparent resistivity. Panel d) shows the inverted model from the two responses. The geology of the area is known to be very flat and slowly varying

6.8 Modelling and interpretation

6.8.1 Modelling

As mentioned in the introduction, a key point in using the TEM method for hydrogeological purposes is the requirement of accurate data with high spatial density. Insufficient data quality makes it impossible to obtain a reliable geophysical model for use in a hydrogeological interpretation.

To obtain data of sufficient quality, the following instrument parameters must be known and modelled in the data modelling algorithm:

- The transmitted waveform characteristics, including the exact appearance of the current turn-off and turn-on ramps and precise timing between the transmitter and the receiver. Timing parameters must be known to an accuracy of 100 ns, because of their severe impact on early-time data.
- The receiver transfer function, which is modelled by one or more low-pass filters, often has a strong influence on early-time data. Low-pass filters are included in the receiver system to stabilize the amplifiers and to suppress the noise from long-wave radio stations.
- The geometry of the transmitter-receiver configuration must be accurate, especially for the offset array configuration. Central-loop data are relatively insensitive to deviations in geometry as long as the transmitter area is unchanged.

Measuring data with a high spatial density serves two purposes: 1) the resolution of geological structures is improved and 2) erroneous data caused by instrument malfunction and transmitter-induced coupling to man-made conductors can be identified and eliminated. The latter is by far the most important.

6.8.2 The 1D model

It is at present time not possible to do inversions on TEM-data in more than one dimension on a routine basis. 3D inversion codes have been developed lately, but they are still computationally very intensive, and they require densely measured data sets. Therefore, it is inevitable that geological noise is present when describing a 3D world by a 1D model. The distribution of 2D and 3D structures decides the amount of geological noise.
Fig. 6.16 shows a TEM sounding curve and the best fitting 1D model. The model has 5 layers. More layers would also fit the data but, in order not to

make an erroneous interpretation the, *simplest model* fitting the data to an appropriate degree is chosen.

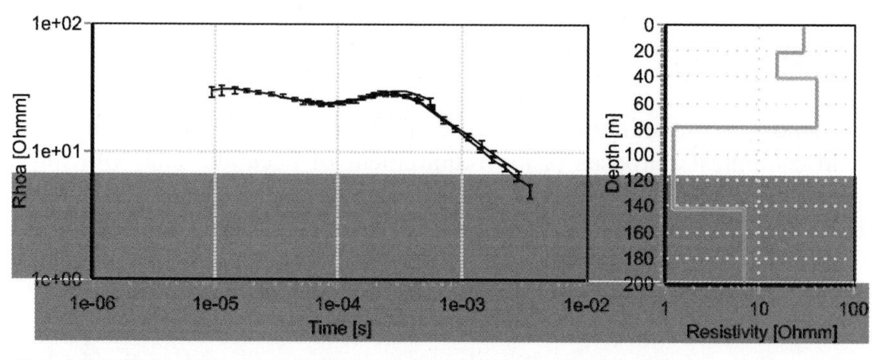

Fig. 6.16. Sounding curve and inverted model for a TEM sounding

6.8.3 Configurations, advantages and drawbacks

Ground-based TEM systems using a high transmitter moment normally utilize a transmitter loop of 40 - 100 m. The advantages of a large loop are that measurements can be made inside the loop, and the magnetic moment is large; the drawbacks are the low field efficiency and the pronounced coupling with cultural features. A small transmitter loop with a high current is very field efficient, but four issues must be tackled in the configuration design:

- Measuring in the central-loop configuration with a small transmitter loop and high output current saturates the receiver amplifiers due to high voltages arising from the turn-off of the primary field. After saturation, amplifiers produce distorted signals for hundreds of microseconds. Furthermore, currents on the order of nA leak from the transmitter coil after the current is turned off, adding to the earth's response and thereby distorting the data. Both effects become negligible because of geometry when using either a small output current with a large transmitter loop or a large offset between the transmitter and the receiver coils. Thus high-output current data must be measured in the offset configuration, while low current data can be measured in the central-loop configuration.
- The induced polarization (IP) effect is pronounced in some sedimentary environments. The IP effect, which is most pronounced in the central-loop configuration, moves to later times as the offset between the transmitter and the receiver coil increases.

- At early times, the offset configuration is extremely sensitive to small variations in the resistivity in the near surface. Extensive 3D modelling of such variations shows a pronounced influence on the measured fields before the current system passes beneath the receiver coil. In many cases these data are not interpretable with a 1D model, even if the section is predominantly 1D. At later times, after the current system has passed, the distorting influence has decayed. The central-loop configuration is much less affected by near-surface resistivity variations.
- The offset configuration is sensitive to small deviations in the array geometry. For a simple 60 m half-space model, a 30% error in the response is apparent near the sign change if the receiver coil is located 71 m instead of 70 m from the transmitter. In a routine field situation, it is next to impossible to work with such accuracy. After the sign change, the offset configuration is essentially equivalent to a central-loop configuration; the central-loop configuration is insensitive to the placement of the receiver inside the transmitter loop.
- A compromise is to use a high-power system where early times are measured in the central loop configuration with a small current of 1 – 3 A. Late times is, in turn, measured in the offset configuration with maximum output current. In this way the four issues are addressed, and the field production can still be kept high.

6.9 Airborne TEM

In this chapter we will give an overview of the airborne TEM system and discuss the specific topics where the airborne and the ground based techniques differ. We will focus on the relatively new helicopter systems as they have the sufficient accuracy necessary for groundwater investigations.

6.9.1 Historical background and present airborne TEM systems.

Airborne electromagnetic systems (AEM) have been used for more than 50 years. The development was driven by the exploration for minerals with its needs for surveying large areas within reasonable cost. The first attempts with airborne TEM systems in the 1950s were quite successful in base-metal exploration in Canada, and in that decade over 10 systems were in the air. The most successful system resulting from the 1950s was the INduced PUlse Transient (INPUT). Canada and the Nordic countries led the development and use of AEM systems, and by the 1970s the methodology was seen used worldwide.

With the decline in exploration for base metals, the use of AEM methods turned from anomaly detection to conductivity mapping, and frequency-domain helicopter EM (HEM) systems appeared. By the 1990s base-metal exploration was concerned with deep targets, and AEM systems began to follow two paths: fixed-wing time-domain systems designed for detection of deep conductive targets, and frequency-domain HEM systems intended for high-resolution, near-surface, conductivity mapping.

Of the more than 30 systems appeared since the inception of the AEM method, few are currently in routine use. The GEOTEM and the MEGATEM systems are digital enhancements of the INPUT system, which uses a half-sine transmitter waveform. The TEMPEST system uses a square transmitter waveform as is common for ground-based TEM systems.

Table 1. Key parameters of different airborne transient systems.

Name of equipment	Moment in kAm2	Transmitted waveform	Configuration and measured components.	Type of Calibration	Carrier type
GEOTEM	450	Half-sine	Offset-loop, Z and X	Relative	Fixed-wing
MEGATEM	1500	Half-sine	Offset-loop, Z and X	Relative	Fixed-wing
TEMPEST	55	Trapezoid	Offset-loop, Z and X	Relative	Fixed-wing
AeroTEM	40	Triangular	Central-loop, coplanar, Z and X	Absolute	Helicopter Sling-load
HoisTEM	120	Trapezoid	central-loop coplanar, Z	Relative	Helicopter Sling-load
VTEM	400	Trapezoid	central–loop coplanar, Z	Relative	Helicopter Sling-load
SkyTEM	120	Trapezoid	central-loop coplanar, Z and X	Absolute	Helicopter, Sling-load

The TEM systems mentioned above are fixed-wing systems, i.e. systems with the current-loop strung around an airplane from the nose, tail and wing tips. Only recently has the concept of a transient helicopter system come of age, and new systems are emerging making broadband measurements with a small footprint possible. Transient helicopter systems carry the transmitter loop as a sling load beneath the helicopter. Recently developed helicopter TEM systems are the AeroTEM, NEWTEM, Hoistem,

VTEM and SkyTEM systems. The AeroTEM, NEWTEM, Hoistem, and VTEM systems are designed primarily for mineral exploration. The SkyTEM system is designed for mapping of geological structures in the near surface for groundwater and environmental investigations and was developed as a rapid alternative to ground-based TEM surveying. Table 1 summarizes the key parameters of the airborne TEM systems currently in operation.

6.9.2 Special considerations for airborne measurements

In groundwater exploration, data with precision and quality are required as the decisive data changes can be as low as 10 – 15 %. When operating in the air a number of key issues need to be addressed to achieve the required data quality. The issues are all related to the calibration, the altitude and the flight speed of the system.

Calibration

In the context of high data quality, the calibration of the transmitter/receiver system plays a central role.

When airborne systems operate in the frequency domain, the strong primary field has to be compensated in order to measure the Earth response. Because of drift in the system the compensation changes in time, and its size has to be determined successively during the survey by high-altitude measurements. Furthermore, it is necessary to perform measurements along tie lines perpendicular to the flight lines and by post-processing to provide concordance between adjacent lines. This process is called levelling, and because of this a frequency system is said to be relatively calibrated.

When airborne systems are operating in the time domain, it is possible to reduce the interaction between the transmitter and the receiver system to a level, at which the distortion of the measured off-time signals is negligible. In this case, a calibration of the instruments can be performed in the laboratory and/or at a test site before the equipment is used in surveys. Neither high altitude measurements nor performing tie lines for levelling are then necessary during the survey. Such a system is said to be absolute calibrated.

The relatively calibrated systems have a lower S/N ratio and a lower data accuracy because of the levelling and the filtering of data compared to the absolutely calibrated systems.

Altitude

The Earth response decays with increasing altitude. This is illustrated in Fig. 6.17a). The model resembles a conducting clay cap above a resistive aquifer layer situated on a good conducting clay basement.

The random noise contribution from natural and man-made sources has no significant change within the operating range. Therefore, the decay in the Earth response solely causes a lower S/N ratio at late times resulting in a poorer resolution of the deeper part of the Earth.

The determination of the near surface layers also decreases with higher altitude because the fields have weakened. Fig. 6.17c) shows the standard deviation as a factor for the model parameters of the model in Fig. 6.17b). The determination of the resistivity of the first and the second layer and the thickness of the first layer decrease when the system moves from the ground to an altitude of 100 m. The thickness of the second layer remains well determined because it is very thick. In general, increasing altitude means a lower resolution of the upper layers. Related to groundwater investigations, the above figures show that high resolution of near-surface protecting clay layers requires operation at low altitudes.

Another implication of the decaying Earth response with altitudes is increased distortions of the Earth response due to coupling to man-made installations. As mentioned in chapter 6.7, a safety distance to installations of at least 100 m, depending on the model, has to be maintained in order to avoid distorted data sets. Airborne electromagnetic measurements introduce larger safety distances to installations compared to ground based equipment because of the lower Earth responses. The larger the flying altitude is the larger are the safety distances. If the signal at late times has decreased by a factor of X, the safety distance must be increased by a factor of $\sqrt{X}$ (assuming the coupling is caused by an infinite wire with field decay proportional to $1/r^2$, r being the distance to the wire). For the model in Fig. 6.1, the safety distance at an altitude of 50 m is approximately 1.4 times larger than at the surface. At an altitude of 100 m it has increased to approximately 1.7.

Flight speed

An important tool for increasing the S/N ratio in electromagnetic measurements is to perform stacking and filtering of measurements (see chapter 6.6).

In TEM measurements the noise is reduced by stacking the individual transient decays. To achieve a certain S/N ratio, a certain number of transient decay curves are necessary in the stack.

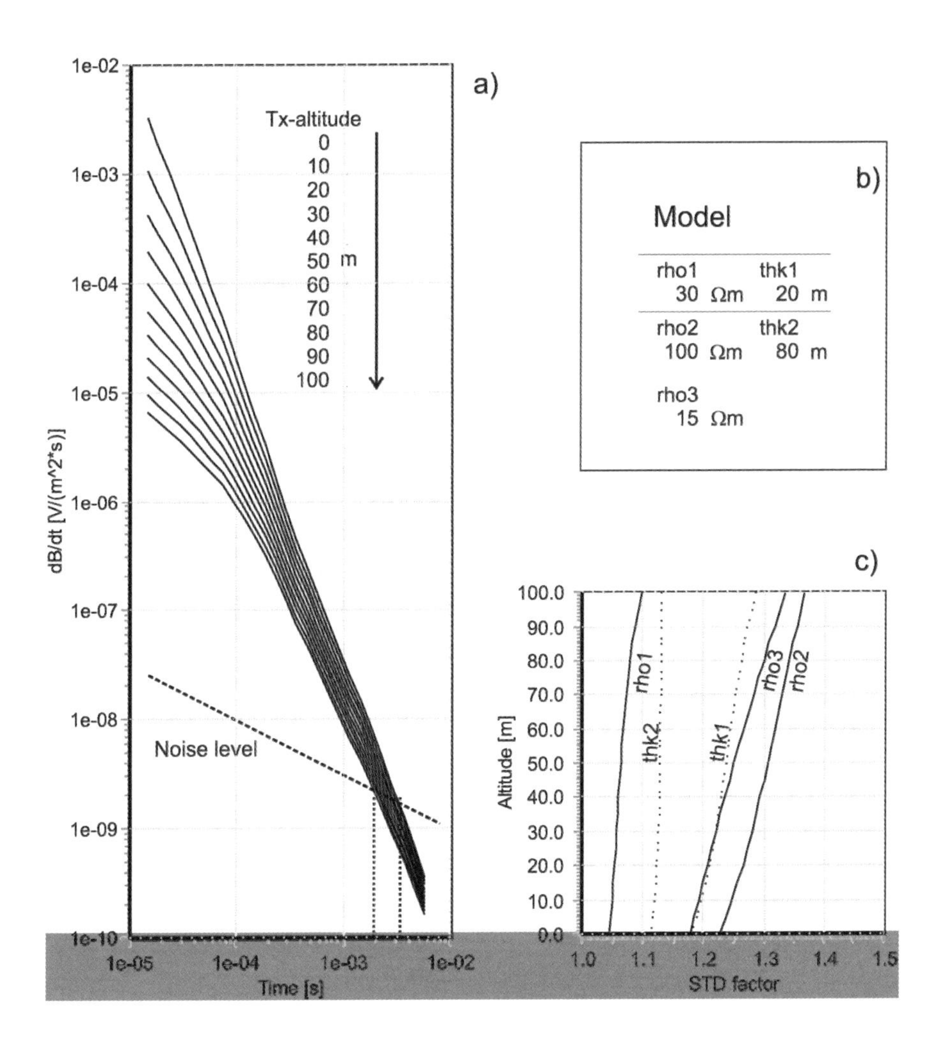

Fig. 6.17. Altitude and resolution. a) shows the Earth response as a function of altitude for the model in b). The transmitter moment is 22,500 Am^2 and responses are measured in the central-loop configuration. The transmitter height is varied from 0 m to 100 m in steps of 10 m. The response decreases more at early times than at late times. Data above the noise indicated by the dashed line are obtained until 1.8 ms at an altitude of 100 m and 3.2 ms at the surface (dotted lines). Plot c) shows the standard deviation as a factor (a factor of 1 means 0% uncertainty) for the parameters of the model in b) assuming the noise model indicated with the dashed line in a). Resistivities are solid lines, thicknesses dotted lines

If the noise affecting the data sets is uncorrelated Gaussian noise, we have for the standard deviation, *STD*, of the average of the stack that

$$STD \propto \frac{1}{\sqrt{N}} \tag{6.33}$$

where N is the number of transients in the stack.

When the system is moving while measuring, a trade-off between the lateral resolution and the vertical resolution of the Earth parameters exists because a certain time is needed to collect the individual transient decays for a sounding. A given vertical resolution is related to the time interval in which measurements are made, and the time interval is determined primarily by the noise level in the data sets (see ch. 6.6). In other words, the noise level is related to a certain stack size, which in turn is related to a certain acquisition time.

The lateral system movement while achieving the stack increases with velocity which decreases the lateral resolution. Hence, a higher vertical resolution inevitably means a decreased lateral resolution if the flight speed is unchanged. On the contrary, the same lateral resolution at a higher velocity results in an increase of the noise of the stacked transients because

$$Noise \propto \sqrt{\frac{V2}{V1}} \tag{6.34}$$

$V2/V1$ being the velocity increase factor.

Finally, the S/N ratio of the data sets is proportional to the size of the transmitted moment. This implies that increasing the transmitted moment by a factor of $\sqrt{V2/V1}$ maintains the S/N ratio at the same level.

The considerations above assumes uncorrelated gaussian noise, but when the receiver coil is moving in the static Earth magnetic field, non-gaussian noise is also induced in the coil. This is called microphonic buffeting. The microphonic buffeting noise increases with increased velocity and is reduced by many ingenious constructions. The total noise is the sum of the gaussian noise and the microphonic buffeting.

Data quality and post processing.

Airborne electromagnetic surveys are mostly cost effective. As the data acquisition is extremely fast, and large amounts of data are collected over a short period of time, the data quality control has to be automatic.

As discussed in Sect. 6.9.2, the use of an absolutely calibrated TEM system implies that no high-altitude measurements have to be carried out and subsequently used for compensating the data for the effects from transmitter-receiver interactions. Nor is it necessary to perform levelling of the data sets.

In order to maintain the high data quality demanded for groundwater surveys, the geometrical setup of the equipment has to be well determined at all times, as well as the transmitted current. The geometrical setup is determined by the altitude and the inclination of the transmitter and receiver coils. Furthermore, it is essentially that the calibration and the functionality of the instruments are well documented, and that all setup parameters are saved for the subsequent interpretation.

The post-processing of the measured data sets relates to two tasks. The first task is to process the altitude, inclination and position data in order to remove outliers and to provide continuity. Especially the altitude data need processing as they are, in many cases, affected by the vegetation on the surface. If the altimeter reflections from vegetation are not identified and corrected, errors will be introduced in the interpreted models. Fig. 6.18 shows a section of altimeter data from the SkyTEM system. The dots are reflections as picked up by one of the laser altimeters mounted on the transmitter frame. The solid line is the processed altimeter data. The effects of the erroneous reflections obtained over the forest are removed in the processed altitude curve. Data from the processed altimeter data are used in the interpretation of the data sets.

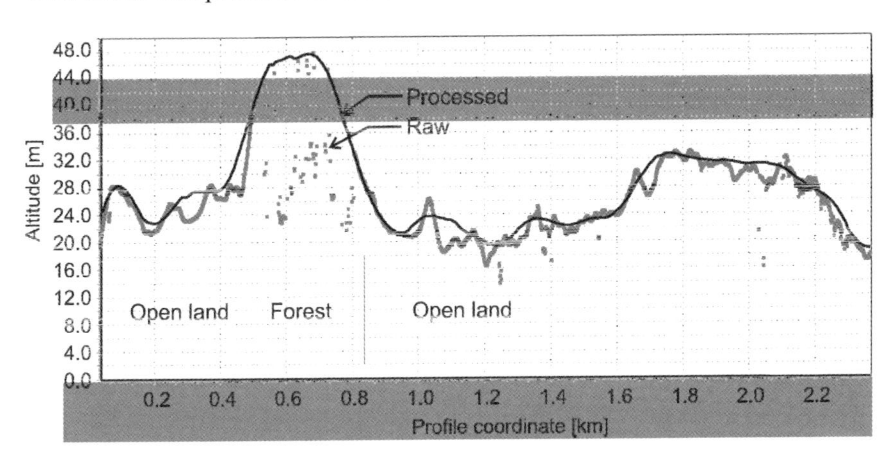

Fig. 6.18. Processing of altitude data. Dots are the actual reflections picked up by a laser altimeter. The solid line is the processed height data. Over the forest a large number of reflections come from the tree-tops

The inclination of the frame is used both to correct the altitudes and the data. Altitudes are measured assuming that the laser beam is normal to the ground surface. When the laser tilts, the normal altitude has to be calculated. The data compensation arises because it is assumed that the transmitter and the receiver are z-directed. This is not true when the frame is tilted, and a correction has to be applied as 2*cosine(inclination).

The second task is related to the distorting of the data sets by the coupling responses from man-made installations. This is a very time consuming process when operating in culturally developed countryside and involves a significant part of the post-processing time. However, the removal of coupling-distorted datasets is crucial for the quality of the interpreted datasets.

Fig. 6.19 is an example from a SkyTEM survey where the survey line crosses two couplings associated with roads. The data marked with grey in Fig. 6.19a) are coupled, and like the sounding curve in Fig. 6.19c) they can not be used for interpretation. The uncoupled data in Fig. 6.19b) have a smooth appearance in the whole time range until they reach the noise level for the last couple of gates.

6.10 Field example

In the following we will show results from a groundwater survey in Denmark. The survey was carried out using the airborne SkyTEM system.

6.10.1 The SkyTEM system

The SkyTEM system has been developed for groundwater investigations by the HGG group at the University of Aarhus, Denmark. During the last 3 years, the system has been intensively used for groundwater surveys. A key issue for the system is that it must produce data of the same quality as groundbased systems.

The transmitter and receiver coils, power supplies, laser altimeters, global positioning system (GPS), electronics, and data logger are carried as a sling load on the cargo hook of the helicopter. SkyTEM in operation is pictured in Fig. 6.20.

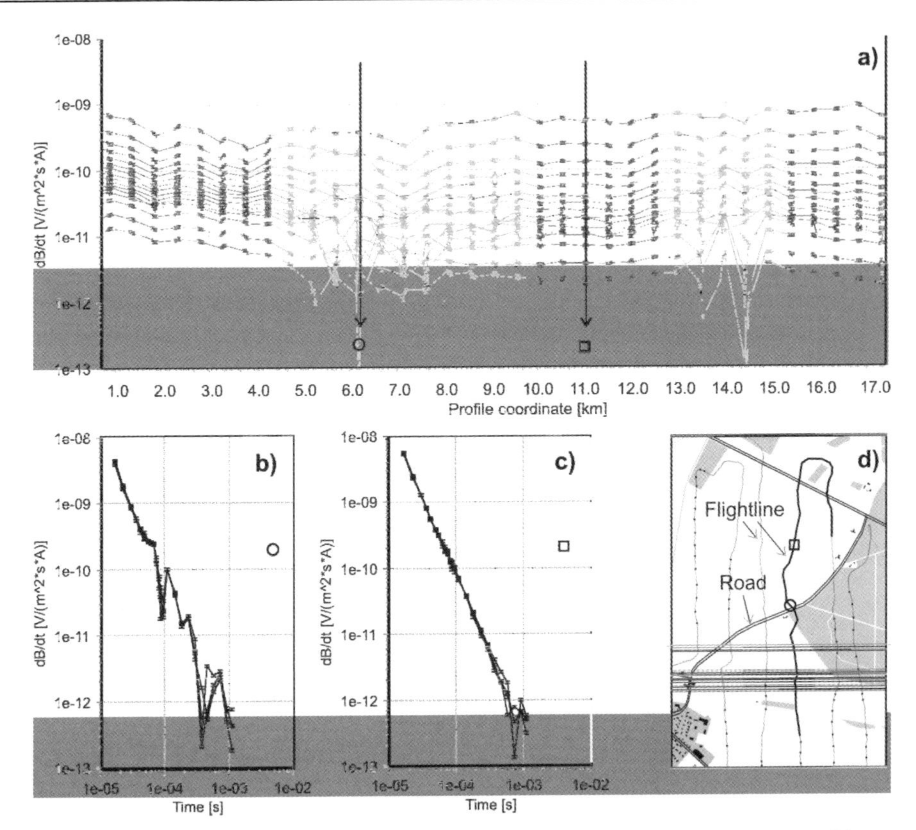

Fig. 6.19. Coupled data. Panel a) is a plot of selected time gates along a profile from a SkyTEM survey. Data are normalized with the transmitter moment. Data marked with grey are identified as coupled whereas black data are uncoupled. The coupled data are associated with the crossing of two roads. Plot b) shows a coupled data set, and for comparison an uncoupled data set is shown in c). Profile and position of selected soundings are shown on the inserted map in d). The coupled sounding is marked with a circle, the uncoupled sounding with a square. The thick solid line marks the profile section shown in a)

The array is located using two GPS position devices. Altitude is measured using two laser altimeters mounted on the carrier frame, as well as inclinometers measuring in both the x and the y directions. The measured data are averaged, reduced to data subsets (soundings) and stored together with GPS coordinates, altitude and inclination of the transmitter/receiver coils and transmitter waveform. Transmitter waveform information and other controlling parameters of the acquisition process are recorded for each data subset, thereby ensuring high data-quality control.

The transmitter loop is a four-turn 300 m^2 loop divided into segments to allowing transmittance of a low and a high moment. The transmitter loop is attached to a rigid wooden lattice frame construction. The receiver coil is located on the rudder, 1.5 m above the corner of the transmitter loop as shown in Fig. 6.20.

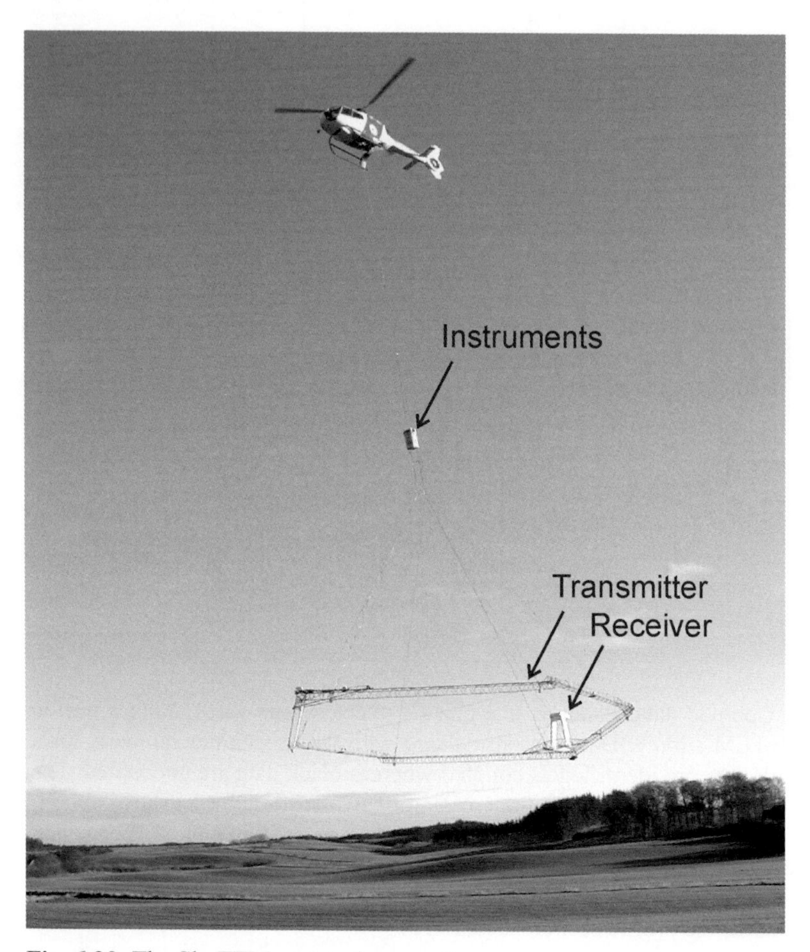

Fig. 6.20. The SkyTEM system in operation

The operational flying speed of the SkyTEM system in groundwater surveys is 15 - 20 km per hour (4.1 - 5.5 m/s) providing a high-moment stack size of approximately 1000 transients. This is sufficient to obtain data out to 2 - 4 ms. Consequently, high- and low-moment data segments yield an average lateral spacing of 35 - 45 m. A compromise between vertical resolution and safety concerns for the helicopter operation is to maintain an altitude of 15 - 20 m for the carrier frame and about 50 m for the helicopter. In forest areas, the flying altitude increases with the height of the trees.

The SkyTEM systems are absolutely calibrated at the national test site in Denmark before used in surveys. Occasionally the systems are brought to the site to ensure that the equipment is operating correctly.

As part of the standard field procedure for data quality check, repeated data are acquired every time the helicopter refuels and gets fresh batteries, at about 1.5 hour intervals. The repeated measurements when corrected for the geometrical parameters (altitude, inclination etc.) are expected to be within 5%.

6.10.2 Inversion of SkyTEM data

A sounding consists of a low and a high-moment segment. As the two segments are spatially separated, the data sequences are inverted with different altitudes. The flying altitude is included as an inversion parameter with a prior value and a standard deviation determined from the altimeters. All data sequences along the profile lines are inverted in one step using soft bands on the model parameters. This approach allows for smooth transitions along the profile line resembling the actual changes in geology.

6.10.3 Processing of SkyTEM data

Navigation and status data for the SkyTEM system make up a substantial amount of data. The basis for the processing is the following:

- GPS data are measured every second with two independent devices.
- The angle of the frame is measured every second by two independent devices.
- The altitude of the frame is measured 20 times per second with two laser devices.
- The transmitter current is stored for every 50 - 100 transients. The transmitter also monitors parameters like battery voltage and temperature.
- Every transient is stored and saved for further processing.

Processing of GPS and angle data is done by adaptive filtering of the data. The angle data are used to calculate *normal* reflection altimeter data and for calculation of exact transmitter and receiver altitudes and a field correction factor. The field correction factor accounts for the reduction in the z-directed magnetic moment caused by the movements of the transmitter/receiver-plane when flown in the wind. The angle from horizontal is normally between 0 and 15 degrees.

Processing of the altitude data is more critical and has to be evaluated by the user. A precise determination of the altitude is crucial to obtain the required resolution of the upper approximately 30 m of the subsurface. The main problem is that the lasers receive reflections from the treetops and not the forest floor when the system flows over both hard-wood and soft-wood forests. This is seen as abrupt drops in the altitude measurements. An adaptive filtering scheme has been designed to eliminate the unwanted reflections, but also this scheme fails when forest-floor reflections are absent in tens of seconds. In this case, the user has to either draw an altitude line on a profile plot of the altitude data or, if impossible, to mark the altitude as a free parameter in the inversion.

The processing of the transient data is done in two steps. The first step uses adaptive filters to eliminate noise. The stack size after step one is approximately 100 - 200 transients. At this stage, all data that are coupled due to man-made installations are removed. This process is quite time consuming and requires a close integration of gate profile plot, individual data set plots and a GIS-map.

In step two, the five - ten data sets from step one are averaged into sequences. The data sequences are the final soundings used in the inversion. A final sounding consists of about 600-800 SkyTEM transients yielding data from 17 μs to 2 to 4 ms. The soundings are on the average separated 30 - 50 m on the flight line.

6.10.4 The Hundslund Survey

In the following we will present the results from the Hundslund survey. The Hundslund survey covers about 40 km^2. The average line spacing is 250 m, which gives approximately 250 line kilometres of data. The data processing yields one sounding about every 40 m, totally about 1800 soundings.

Aquifers in this part of the country are often associated with buried valleys incised into the low-resistive Tertiary clays. The valleys, filled with sand and gravel deposits, are the primary aquifers. The primary purpose of the Hundslund survey is to find and delineate the buried valley structures, the secondary purpose to evaluate the vulnerability of the aquifers. The vulnerability is evaluated by mapping clay layers in the upper 30 m as clay layers prevent and delay contamination to reach water bearing sand and gravel layers below.

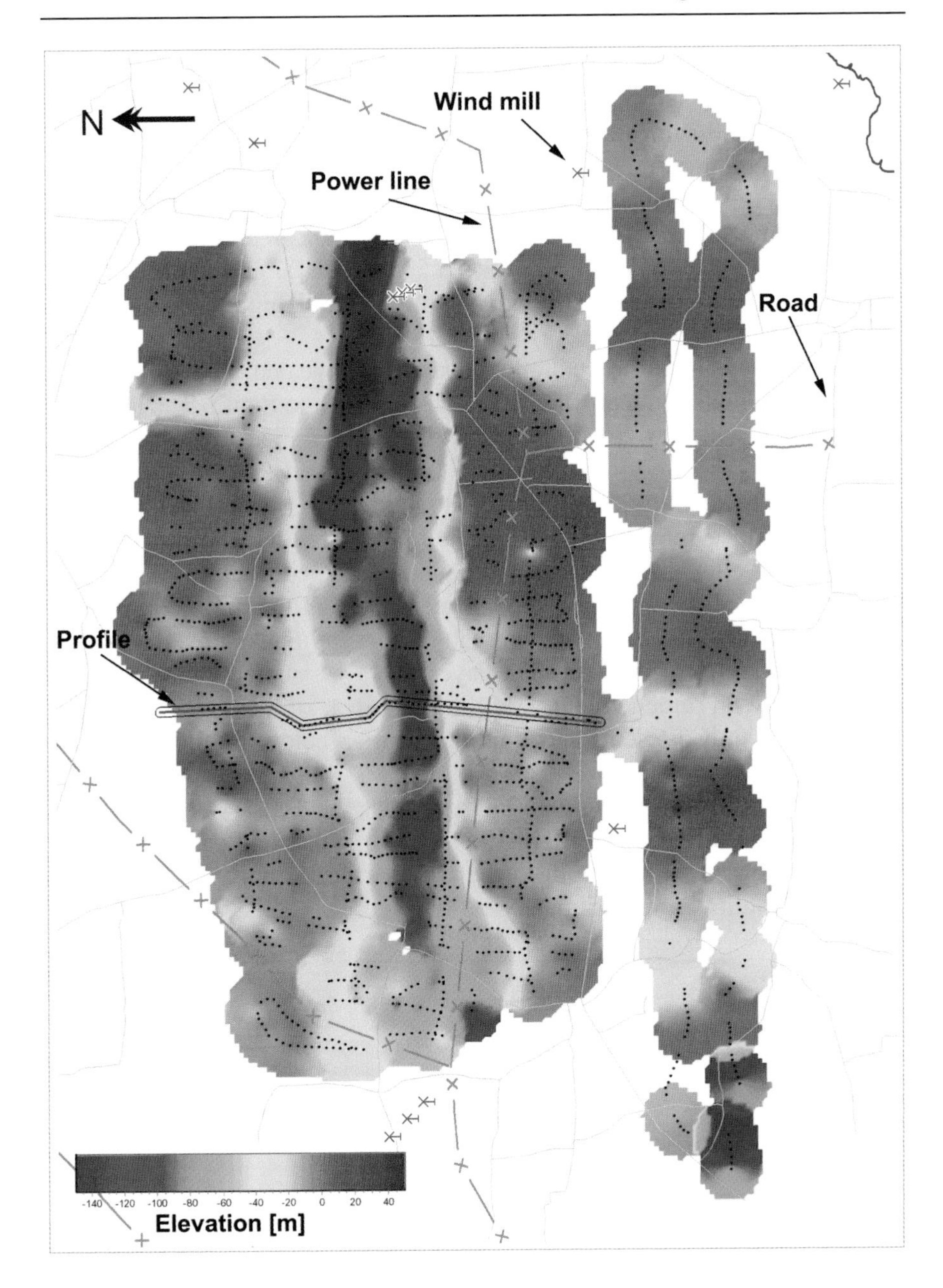

Fig. 6.21. The map shows the elevation of the low-resistive Tertiary clay. A distinct valley structure incised in the tertiary clay is striking E-W in the area. The area depicted is approximately 13 x 8 km

An impression of the flight paths is given in Fig. 6.21 by observing the black dots which show the SkyTEM sounding points. Data are absent along roads, power lines and windmills due to capacitive or galvanic couplings. The coloured theme map in Fig. 6.21 shows the elevation of the Tertiary clay. Experience shows that the Tertiary clay has a resistivity well below 12 Ωm. A distinct valley structure is striking E-W in the central part of the area. The valley is incised between 50 m and 100 m into the Tertiary clay plateau. Sub-channels striking N-S are seen at higher levels in the Tertiary clay surface.

To further evaluate the geological structures revealed by the SkyTEM survey, we have drawn a profile striking from North to South. The location of the profile (with a 50 m search radius) is shown in Fig. 6.21. The profile is shown in Fig. 6.22.

First, note the continuity from model to model and, second, that the Tertiary clay surface is well defined and is found all along the profile. The centre of the valley structure is seen around profile coordinate 3000 m. In this part, most of the models have 3 layers of which the middle layer is a sand and gravel layer. The valley structure is at the flanks covered with a 10 – 15 m clay-rich till layer. The layer has a resistivity in the interval from 30 – 40 Ωm (green colours). In the centre part of the valley, the layer is not consistently present, and thus the aquifer below might be unprotected.

The left part of the profile (0 – 2500 m) finds the Tertiary surface about 70 m below the surface (elevation -10 – -30 m). In the right part of the profile (3500 – 5000 m) the depth to the clay surface is only 10-30 m. The width of the valley is approximately 1000 m.

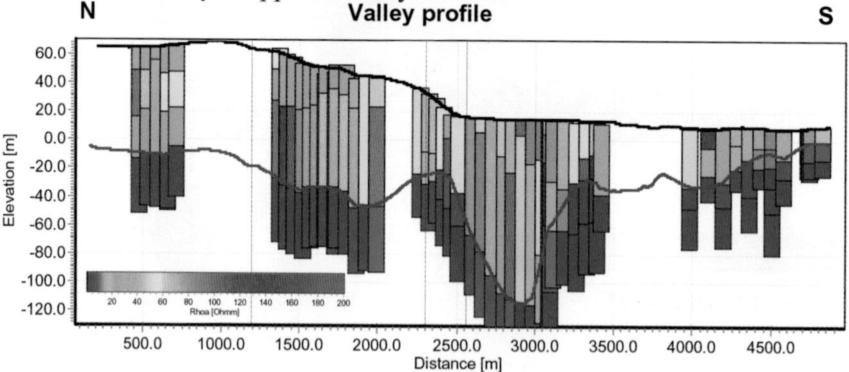

Fig. 6.22. North-South profile in the survey area. The coloured bars show the 1D models projected onto the profile line. The blue line indicates the elevation of the good conductor depicted in Fig. 6.21. The vertical lines indicate bend points on the profile line

We finish our presentation of this survey showing a processed data set. The final data sequences of the low and the high moment and the inverted model is shown in Fig. 6.23. The model to the left is a 4-layer model with a high-resistive second layer. The last layer in the model with a resistivity of about 5 Ωm is the Tertiary clay. The data sequences are shown to the right with the data uncertainty drawn as error bars. The solid line is the model forward response.

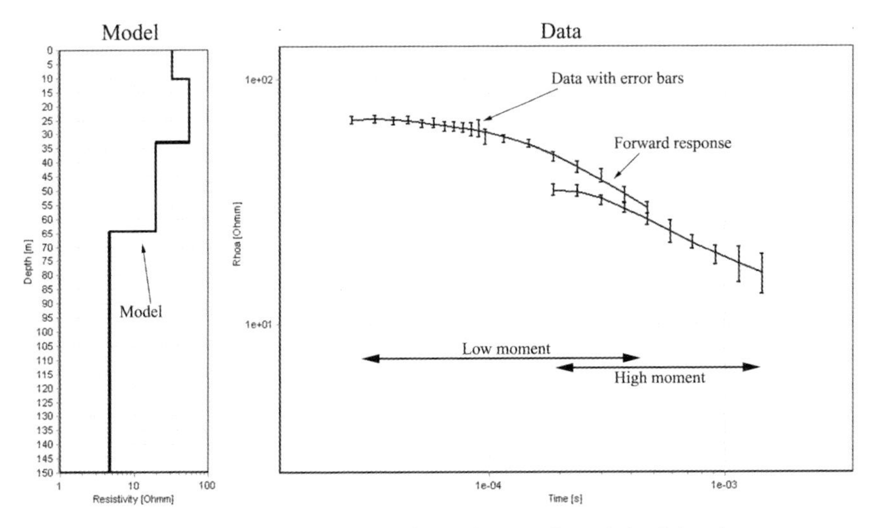

Fig. 6.23. The figure shows to the left the inverted model of the data sequences shown to the right. The two curves are the low- and high-moment data segment. The standard deviations are calculated from the data stack. The solid line is the forward response for the model shown to the left. The offset between the two segments is caused by different current levels

Acknowledgements

We would like to thank Niels Bøie Christensen who gave us inspiration to many sections of this manuscript with his lecture notes for a course in environmental geophysics. Tom Hagensen prepared the current density figures at short notice. Jens Danielsen helped with parts of the text and, Lone Davidsen improved the language. We would also like to thank the county of Aarhus for letting us use the SkyTEM field example from Hundslund. Max Halkjær supplied the field photo of the SkyTEM system.

6.11 References

The following references are recommended for further reading on the TEM method.

Auken E, Jørgensen F, Sørensen KI (2003) Large-scale TEM investigation for ground-water. Exploration Geophysics 33:188-194

Auken E, Nebel L, Sørensen KI, Breiner M, Pellerin L, Christensen NB (2002) EMMA - A Geophysical Training and Education Tool for Electromagnetic Modeling and Analysis. Journal of Environmental & Engineering Geophysics 7:57-68

Asten MW (1987) Full transmitter waveform transient electromagnetic modelling and inversion for soundings over coal measures. Geophysics 52: 279 – 288

Christensen NB (2002) A generic 1-D imaging method for transient electromagnetic data. Geophysics 67:438-447

Christensen NB, Sørensen KI (1998) Surface and borehole electric and electromagnetic methods for hydrogeophysical investigations. European Journal of Environmental and Engineering Geophysics 3 (1):75-90

Danielsen JE, Auken E, Jørgensen F, Søndergaard VH, Sørensen KI (2003) The application of the transient electromagnetic method in hydrogeophysical surveys. Journal of Applied Geophysics 53:181-198

Effersø F, Auken E, Sørensen KI (1999) Inversion of band-limited TEM responses. Geophysical Prospecting 47:551-564

Farquharson CG, Oldenburg DW (1993) Inversion of time-domain electromagnetic data for a horizontally layered Earth. Geophysical Journal International 114:433-442

Fitterman DV, Stewart MT (1986) Transient electromagnetic sounding for groundwater. Geophysics 51:995-1005

Fitterman DV, Anderson WL (1987) Effect of transmitter turn-off time on transient soundings. Geoexploration 24:131-146

Flis MF, Newman GA, Hohmann GW (1989) Induced-polarization effects in time-domain electromagnetic measurements. Geophysics 54:514-523

Fountain D (1998) Airborne electromagnetic systems - 50 years of development. Exploration Geophysics 29: 1–11

Frischknecht FC, Labson VF, Spies BR, Anderson WL (1991) Profiling methods using small sources *in* Electromagnetic methods in applied geophysics: Nabighian MN (Ed.) Society of exploration geophysicists

Goldman M, Tabarovsky L, Rabinovich M (1994) On the influence of 3-D structures in the interpretation of transient electromagnetic sounding data. Geophysics 59:889-901

Jørgensen F, Sandersen P, Auken E (2003) Imaging buried Quaternary valleys using the transient electromagnetic method. Journal of Applied Geophysics 53:199-213

McCracken KG, Hohmann GW, Oristaglio ML (1980) Why time domain. Bull Aust Soc Explor Geophys 11:176-179

McCracken KG, Pik JP, Harris RW (1984) Noise in EM exploration systems. Exploration Geophysics 15:169-174

McCracken KG, Orstaglio ML, Hohmann GW (1986) Mininization of noise in electromagnetic exploration systems. Geophysics 51:819-832

Macnae JC, Lamontagne Y, West GF (1984) Noise processing techniques for time-domain electromagnetic systems. Geophysics 49:934-948

McNeill JD (1990) Use of electromagnetic methods for groundwater studies. In: Ward SH (Ed.), Geotechnical and environmental geophysics 01, Society Of Exploration Geophysicists, pp 191-218

Munkholm MS, Auken E (1996) Electromagnetic noise contamination on transient electromagnetic soundings in culturally disturbed environments. Journal of Environmental & Engineering Geophysics 1:119-127

Nabighian MN, Macnae JC (1991) Time domain electromagnetic prospecting methods. In: Electromagnetic methods in applied geophysics, Nabighian MN (ed) Society of exploration geophysicists

Palacky GJ, West GF (1991) Airborne electromagnetic methods, in Nabighian MN (ed) Electromagnetic Methods in Applied Geophysics, vol. 2. Society of Exploration Geophysicists, pp 811–879

Spies BR (1989) Depth of investigation in electromagnetic sounding methods. Geophysics 54: 872-888

Sørensen KI, Auken E, Christensen NB, Pellerin L (2005) An Integrated Approach for Hydrogeophysical Investigations. New Technologies and a Case History. Near-surface Geophysics, vol 2. Application and case histories. SEG publication, in press

Sørensen KI, Auken E (2004) SkyTEM - A new high-resolution helicopter transient electromagnetic system. Exploration Geophysics 35: 191-199

Ward SH, Hohmann GW (1988) Electromagnetic theory for geophysical applications. Electromagnetic Methods in Applied Geophysics, vol. 1, (ed MN Nabighian), pp 131-311, SEG publication

West GF, Macnae JC (1991) Physics of the electromagnetic induction exploration method. In: Electromagnetic methods in applied geophysics: Nabighian MN and Corbett JD (eds) Society of exploration geophysicists

7 Ground Penetrating Radar

Norbert Blindow

Introduction

Ground penetrating radar (GPR) is a geophysical method for shallow investigations with high resolution which has undergone a rapid development during the last two decades (cf. e.g. GPR Conference Proceedings 1994 to 2004). Although its use in groundwater investigations is often limited by severe constraints, the benefits are great in areas with favourable conditions (e.g. Davis and Annan 1989, van Overmeeren 1994, Wyatt and Temples 1996, Tronicke et al. 1999).

GPR is an electromagnetic pulse reflection method based on physical principles similar to those of reflection seismic surveys. There are several synonyms and acronyms for this method like EMR (electromagnetic reflection), SIR (subsurface interface radar), georadar, subsurface penetrating radar and soil radar. GPR has been used since the 1960s with the term radio echo sounding (RES) for ice thickness measurements on polar ice sheets. The method has been increasingly applied for geological, engineering, environmental, and archaeological investigations since the 1980s.

Reflections and diffractions of electromagnetic waves occur at boundaries between rock strata and objects that have differences in electrical properties. The electric permittivity (dielectric constant) ε and the electric conductivity σ are petrophysical parameters which determine the reflectivity of layer boundaries and penetration depth.

In its simple time domain form GPR uses short pulses which are transmitted into the ground (Fig. 7.1). A part of this energy is reflected or scattered at layer boundaries or buried objects. The direct and reflected amplitudes of the electric field strength $\mathbf{E}$ are recorded as a function of traveltime. A high pulse rate enables quasi-continuous measurements by pulling the antennas along a profile with a progress of up to several kilometers per hour. In many cases, a preliminary interpretation of the profile data (radargram) is possible in the field.

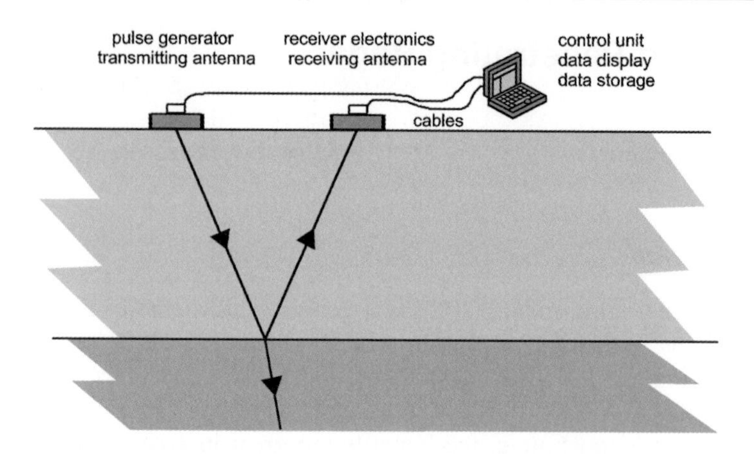

Fig. 7.1. GPR measurement setup

7.1 Electromagnetic wave propagation

7.1.1 Electric permittivity and conductivity

Electromagnetic wave propagation is governed by Maxwell's equations. These four coupled differential equations (which will not be discussed here) provide relations between the electric field **E**, the magnetic field **H**, time, space, and material related equations such as Ohm's law.

Material properties relevant for electromagnetic wave propagation are the electric permittivity ε and the specific electric conductivity σ. To be consistent with the usual notation in electrodynamics (von Hippel 1954a) we introduce a quantity called complex permittivity ε^*, which takes into account both the loss free displacement of electric charges in its real part and the dissipation (heating) caused by dielectric and conduction losses in its imaginary part:

$$\varepsilon^* = \varepsilon' - i\varepsilon'' \ (\mathrm{AsV^{-1}m^{-1}}) \text{ with } i = \sqrt{-1} \tag{7.1}$$

Normalization with the vacuum permittivity $\varepsilon_0 = 8.8542 \cdot 10^{-12}$ $\mathrm{AsV^{-1}m^{-1}}$ yields the dimensionless relative complex permittivity

$$\varepsilon_r^* = \varepsilon^*/\varepsilon_0 = \varepsilon_r' - i\varepsilon_r'' \tag{7.2}$$

Note that the real part ε_r' is equal to the commonly used permittivity (see also Sect. 1.3). Here, we will use ε as an abbreviation for ε_r'.

The conductivity σ is related to ε_r'' and the frequency f by

$$\varepsilon_r'' = \frac{\sigma}{\omega\varepsilon_0} \tag{7.3}$$

with $\omega = 2\pi f$ [s^{-1}].

The loss tangent is the ratio of conduction to displacement currents and defined as

$$\tan\delta = \frac{\varepsilon_r'}{\varepsilon_r''} = \frac{\sigma}{2\pi f\varepsilon_0\varepsilon_r} \tag{7.4}$$

For $\tan\delta = 1$ the conduction currents are equal to the displacement currents. Electromagnetic wave propagation utilized in GPR takes place for $\tan\delta < 1$, while for $\tan\delta > 1$ diffusive processes govern (the case of low frequency electromagnetics like RMT or AEM). Typical values for electric permittivity and conductivity of different aquifer materials are given in Table 1.1 of Sect. 1.4.

Examples for $\tan\delta$ of three materials may illustrate the use of Eq. 7.4. The properties of dry sand (σ =1 mS/m, ε = 4), saturated sand (σ =10 mS/m, ε = 25), and wet clay (σ =50 mS/m, ε = 25) are inserted at a frequency of 50 MHz. For dry sand we get $\tan\delta = 0.09$, which makes it well-suited for GPR. The saturated sand with $\tan\delta = 0.14$ is still fairly good, but the clay shows $\tan\delta = 0.72$ which is at the limit for wave propagation.

The magnetic permeability μ of geological materials has little effect with respect to GPR applications. It usually can be assumed to be the permeability of free space ($\mu = \mu_0 = 4\pi\cdot 10^{-7}$ $VsA^{-1}m^{-1}$).

Complex permittivity of water

Not included in the bulk conductivity of geological materials is a high frequency effect due to the dipole character of water molecules (which is used for heating in the microwave oven). This phenomenon is called Debye relaxation. While the real part of the permittivity ε_r' of water drops from about 80 at low frequencies down to 5.8 at 1000 GHz, the imaginary part goes through a maximum at about 10 GHz which corresponds to a high frequency conductivity of more than 20 S/m (von Hippel 1954b, Hoekstra and Delaney 1974). Although the frequencies involved in GPR for hydrogeological mapping do not exceed 400 MHz, Debye relaxation is present in moist or saturated materials already at 100 MHz. This is one of the reasons why low frequencies provide deep penetration (see below).

7.1.2 Electromagnetic wave propagation

Important and simple solutions for Maxwell's equations are harmonic plane waves in a rectangular coordinate system, the electric field $\mathbf{E} = (E_x, 0, 0)$ and the magnetic field $\mathbf{H} = (0, H_y, 0)$ with propagation in z-direction (e.g. von Hippel 1954a):

$$E_x(z,t) = E_{x0} e^{\gamma z - \omega t} \quad [\mathrm{Vm}^{-1}] \tag{7.5}$$

$$H_y(z,t) = H_{y0} e^{\gamma z - \omega t} \quad [\mathrm{Am}^{-1}] \tag{7.6}$$

with the complex propagation constant

$$\gamma = \alpha + i\beta \quad [\mathrm{m}^{-1}] \tag{7.7}$$

The **E**-field then is written as

$$\vec{E} = \vec{E}_0 e^{i(\beta z - \omega t)} e^{-\alpha z} \tag{7.8}$$

E and **H** are orthogonal to each other and orthogonal to the direction of propagation. The first exponential function contains the phase variations of the wave with respect to time and space. Hence, β is called the phase constant. The second exponential term describes an exponential attenuation of the wave with the attenuation constant α. Both parameters are frequency-dependent and are related to the complex permittivity via

$$\alpha = \frac{\omega}{c_0} \sqrt{\frac{\varepsilon_r'}{2} (\sqrt{1 + \tan^2 \delta} - 1)} \quad [\text{Neper m}^{-1}] \tag{7.9}$$

and

$$\beta = \frac{\omega}{c_0} \sqrt{\frac{\varepsilon_r'}{2} (\sqrt{1 + \tan^2 \delta} + 1)} \quad [\text{rad m}^{-1}] \tag{7.10}$$

with the speed of light in vacuum (or air) $c_0 = 2.998\ 10^8$ m/s ≈ 300 m/µs =0.3 m/ns. Instead of the attenuation factor α the absorption coefficient α' = 8.686α (in dB/m) is used. Note that α is small if $\tan\delta \ll 1$, which means that the conductivity σ has to be small for low absorption.

The unit dB (deciBel) is a logarithmic measure for ratios of power or voltage. In the case of voltage, 20 dB means a ratio of 10, 40 dB means a ratio of 100 and so on. In the case of power, 10 dB means a ratio of 10, 20 dB means a ratio of 100 and so on. Using dB has the advantage of avoiding huge numbers, if the ratios get large.

The propagation velocity v (phase velocity) of the radar waves is calculated from Eq. 7.10 by

$$v = \frac{2\pi f}{\beta} \quad [\text{ m ns}^{-1} \text{ for f in GHz}] \tag{7.11}$$

which collapses to:

$$v \approx \frac{c_0}{\sqrt{\varepsilon}} \tag{7.12}$$

for low-loss materials with $\tan\delta \ll 1$.

The wavelength λ is calculated from the well known relation $v = \lambda f$ for low-loss materials as:

$$\lambda \approx \frac{c_0}{f\sqrt{\varepsilon}} \tag{7.13}$$

For a low loss ground with $\varepsilon = 9$ we calculate from (7.12) that the velocity in the ground is one third of the speed of light: v = 0.1 m/ns. At a frequency of f = 100 MHz = 0.1 GHz the wavelength in the ground will be 1 m.

To give examples for the frequency dependence of the propagation parameters absorption α' and velocity v, we again look at the properties of dry sand, saturated sand and wet clay as at the end of Sect. 7.1.1. In Fig. 7.2, v and α' are plotted versus frequency in a double-logarithmic scale. The calculation of absorption was done with and without the effect of Debye relaxation to show the difference between reality and a too simplifying view taking into account only DC conductivity.

Velocity and absorption of the electromagnetic waves in the low frequency region rise with the square of frequency. For higher frequencies the velocities reach constant values and it can be shown that this takes place for $\tan\delta < 1$. To get dispersionless propagation one should take care that the GPR frequency is high enough to meet the condition $\tan\delta \leq 0.5$. If there was no water in the materials the absorption would show plateaus for higher frequencies. However, the Debye relaxation starts to raise absorption significantly above 100 MHz. This means that the high frequencies of pulses propagating in a wet environment are heavily attenuated. From Fig.7.2 it can be learned that frequencies around 50 MHz provide a good tradeoff with respect to absorption and dispersion (frequency dependence of propagation parameters).

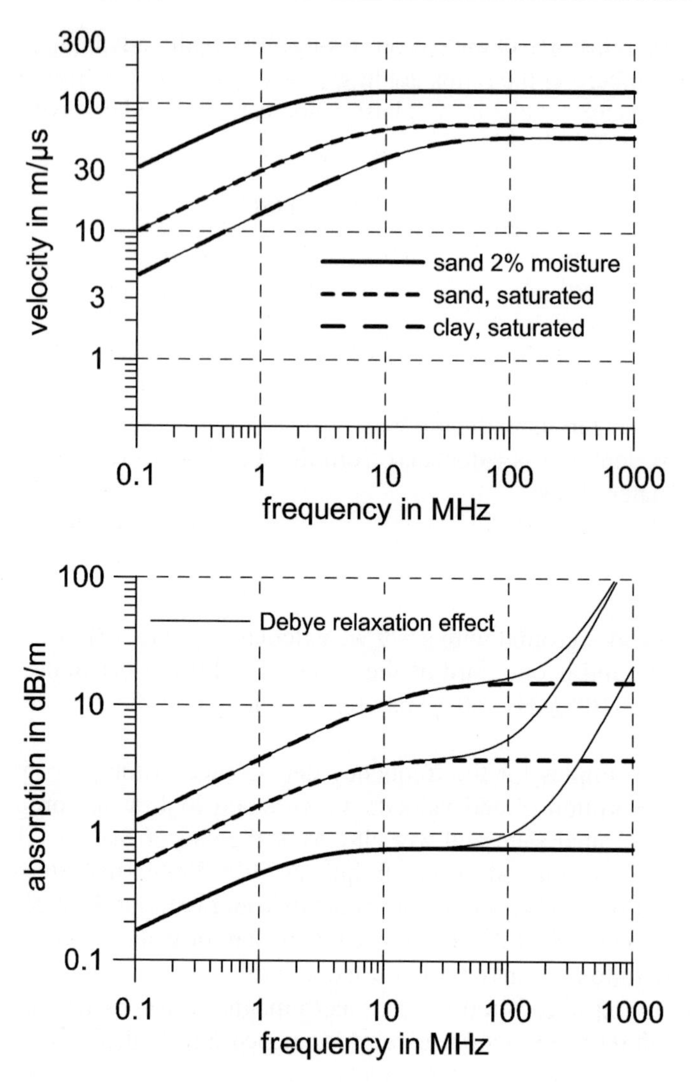

Fig. 7.2. Frequency dependence of propagation velocity and absorption coefficient

7.1.3 Reflection and refraction of plane waves

At a plane boundary between two media (1) and (2) with different properties of ε and σ an incident electromagnetic wave is both reflected in (1) and refracted into (2). The directions of incidence, reflection, and refraction are all in one plane. The angle of incidence θ_i is equal to the angle of reflection

θ_r'. The angle of refraction θ_r is related to the angle of incidence θ_i by Snell's law. The index of refraction n_{21} is related to the permittivity:

$$\frac{\sin\theta_i}{\sin\theta_r} = n_{21} = \sqrt{\frac{\varepsilon_2}{\varepsilon_1}} = \frac{v_1}{v_2} \tag{7.14}$$

Refraction from lower to higher velocity - especially from the subsurface with velocity v into air with velocity c_0 - means that there is a critical angle θ_c where θ_r reaches 90°. The refracted wave then propagates along the surface. For all $\theta_i > \theta_c$ there is total reflection.

In the far field of a transmitting antenna electromagnetic waves are approximated by plane waves. Here we will only regard differences of ε in the special case of normal incidence ($\theta_i = 0°$) which is contained in the more general Fresnel's equations (von Hippel 1954a). The coefficient of reflection or reflectivity (reflected part of incident unit amplitude) r is given by

$$r = \frac{v_2 - v_1}{v_2 + v_1} = \frac{\sqrt{\varepsilon_1} - \sqrt{\varepsilon_2}}{\sqrt{\varepsilon_1} + \sqrt{\varepsilon_2}} \tag{7.15}$$

and the refracted (transmitted) part t by

$$t = \frac{2v_2}{v_2 + v_1} = \frac{2\sqrt{\varepsilon_1}}{\sqrt{\varepsilon_1} + \sqrt{\varepsilon_2}} = 1 + r \tag{7.16}$$

It should be noted that these simple relations hold for ideal dielectrics only. If conductive (lossy) materials are involved, even the simple case of normal incidence the reflection coefficient may become complex leading to a deformation of the reflected wave.

We will apply Eq. 7.12 for the case of the boundary between moist sand ($\varepsilon_1 = 9$) and saturated sand ($\varepsilon_2 = 25$) which gives r = -0.25. The negative sign means a reversal of amplitude (180° phase shift) and 25% of the incident amplitude start to travel back to the surface. Such rather strong groundwater reflections are sometimes met in practice, but it is wise to calculate with reflectivities of 10% for estimates e.g. of penetration depth (see Sect. 7.1.7).

Multiple reflections bouncing between the surface and a strong reflector are not significant for GPR because there is usually considerable absorption in the soil and rock.

7.1.4 Scattering and diffraction

As in reflection seismic data, diffraction occurs at discontinuities of reflectors (interrupts, faults with a shift) and objects whose dimensions are small compared to the wavelength. Low frequencies reduce the disturbing influence of inhomogeneities in the rock (geological noise, clutter) if the wavelength resulting from Eq. 7.13 is considerably larger than the size of the inhomogeneities. A compromise has to be made with respect to resolution.

In order to localize linear diffractors (e.g. pipelines or cables) high frequencies have to be used. The polarization of the electric field should be parallel to the target objects for maximal sensitivity. If the orientation of the targets is not known, measurements must be made along orthogonal profiles.

7.1.5 Horizontal and vertical resolution

Resolution is a measure of the ability to distinguish between signals from closely spaced targets. In ground penetrating radar the resolution depends on the center frequency (or wavelength, which is proportional to the pulse period) and on the range of frequencies (bandwidth), as well as on the polarization of the electromagnetic wave, the contrast of electromagnetic parameters (mainly conductivity and relative permittivity), and the geometry of the target (size, shape, and orientation). Important are also the coupling to the ground, the radiation patterns of the antennas (especially the diameter of the Fresnel zone), and the noise conditions in the field. As a rule of thumb, the vertical resolution is theoretically one quarter of the wavelength $\lambda = v/f$, where v is the velocity of the electromagnetic wave in the medium and f is the frequency. Because the velocity is lowered by the presence of water (see Table 1.1 in Sect. 1.4), the resolution will be better in wet materials. The actual resolution is in the order of one half wavelength. The horizontal resolution is thoroughly discussed in the chapter on reflection seismic (Chap. 2). Objects can be separated laterally if their distance is larger than the diameter d_f of the first Fresnel zone. For example, a GPR system operating at f = 200 MHz in an environment with v = 0.1 m/ns has a wavelength of $\lambda = 0.5$ m. The horizontal resolution at a depth h = 4 m will be $d_f = (2\lambda h)^{1/2} = 2$ m, which is much poorer than the vertical resolution of better than 0.5 m.

Horizontal resolution can be improved by data processing with migration or SAR (synthetic aperture radar, the same process as migration) to a theoretical value of $\lambda/4$ in the case of real 3D-measurements.

7.1.6 Wave paths, traveltimes, and amplitudes

Ray geometry and traveltimes

The propagation of electromagnetic waves can be described by a ray representation as in optics and reflection seismics. A simple horizontal two-layer model with a reflector at depth h requires four wave paths and traveltime curves. The ray path scheme for GPR is shown in Fig. 7.3.

There are two direct waves with different phase velocities and amplitudes traveling along the ground surface: the air wave and the ground wave (Baños 1966). Since the air wave travels with the largest possible velocity for electromagnetic waves – the speed of light in vacuum c_0 – it can be used to determine time zero (like the time break in seismic surveys). The velocity v in the uppermost stratum is determined from the ground wave. Changes in the ground wave indicate changes in the uppermost stratum (e.g., moisture content, type of rock).

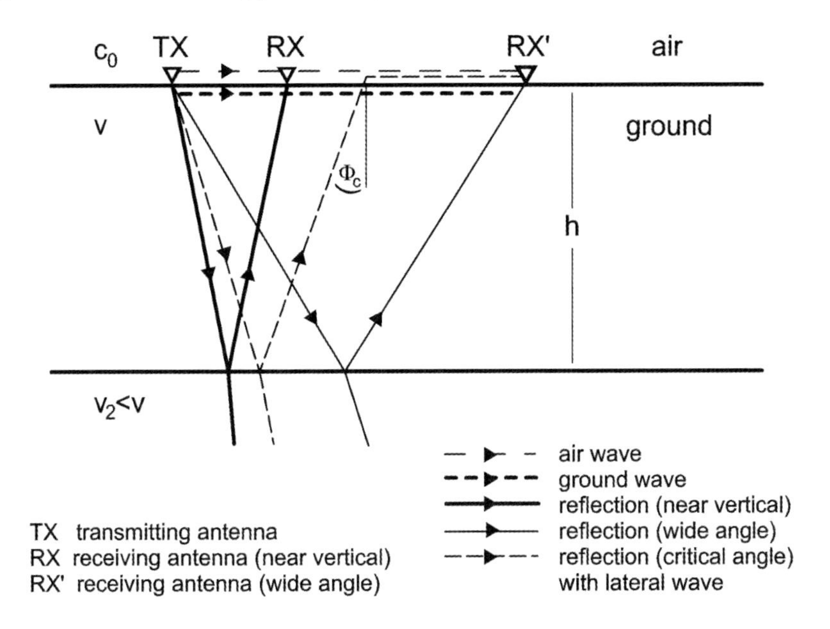

Fig. 7.3. GPR ray paths for the horizontal two-layer case

The traveltimes of the air wave t_a and the ground wave t_g as a function of distance x are given by:

$$t_a = \frac{x}{c_0} \text{ and } t_g = \frac{x}{v} \tag{7.17}$$

Because v is always less than c_0, a lateral wave is generated when the critical angle $\phi_c = \arcsin(v/c_0)$ is reached, analogous to the head wave in refraction seismic surveys. This wave propagates in air parallel to the ground surface. The critical angle ϕ_c is related to the critical distance x_c, which is given by:

$$x_c = \frac{2hv}{\sqrt{c_0^2 - v^2}} \tag{7.18}$$

For $x > x_c$ the lateral wave exists with the traveltime t_l:

$$t_l = \frac{x}{c_0} + \sqrt{\frac{1}{v^2} - \frac{1}{c_0^2}} \tag{7.19}$$

Refracted waves originating at subsurface boundaries are rarely observed, because in most cases the velocity decreases with depth.

The most important wave is the reflection at the layer boundary at depth h. Like in seismic surveys, the traveltime t_r is given by the hyperbola:

$$t_r = \frac{1}{v}\sqrt{x^2 + 4h^2} \tag{7.20}$$

For $x = 0$ we get the zero offset traveltime $t_0 = 2h/v$. The traveltimes are calculated in ns if the velocities are in m/ns and the distances are in m. A traveltime diagram of the different types of waves is shown in Fig. 7.4.

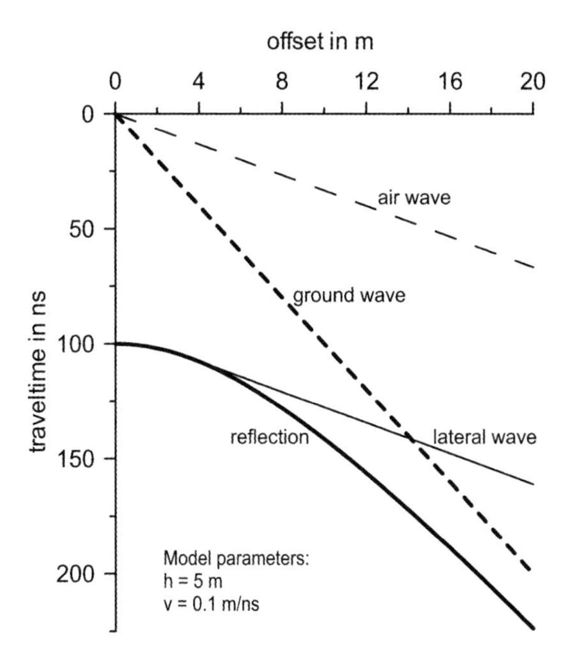

Fig. 7.4. GPR traveltime diagram for the horizontal two-layer case

Amplitudes

For the interpretation of GPR measurements and for the estimation of penetration depth it is useful to know the main features of amplitude decay with distance from the source.

The source emits a spherical wave into the lower halfspace. The amplitude decreases inversely proportional with distance due to geometric spreading and is subject to absorption due to the attenuation coefficient $\alpha = \alpha'/8.686$. Hence, a wave reflected at depth h with a reflectivity r has travelled a distance 2h and its amplitude E_r compared to the amplitude at unit depth E_1 will be

$$\frac{E_r}{E_1} = \frac{r}{2h} e^{-2\alpha h} \tag{7.21}$$

For small values of h the effect of geometric spreading is dominant while the exponential function is stronger at larger depths.

The field strength of the direct wave decrease in the far field with the square of the distance from the source (Baños 1966). The ground wave is diminished in addition by absorption and scattering. Also the lateral wave decreases in the far field region approximately with the square of the distance (Brekovskikh 1980).

In the case of diffractors, the so-called radar-equation (e.g., Daniels 1996) is a suitable alternative to Eq. 7.20 for waves diffracted back towards the source:

$$\frac{P_r}{P_t} = q_s \frac{G^2 \lambda^2}{(4\pi)^3 h^4} e^{-4\alpha h} \tag{7.22}$$

with P_t transmitted power [W], P_r received power [W], q_s scatter cross section [m²], G antenna gain (relative to that of a spherical dipole), λ wavelength at the center frequency [m], h distance between the antenna and the diffractor [m], and α attenuation coefficient [m^{-1}].

7.1.7 Estimation of exploration depth

To avoid dispersion, the operating frequency should be chosen according to $\tan\delta < 0.5$. Rearranging Eq. 7.4 for frequency and letting $\tan\delta = 0.5$, we get:

$$f_m \geq \frac{36000}{\rho \varepsilon} \tag{7.23}$$

where f_m is the central frequency in MHz, ρ the DC resistivity in Ωm (inverse of conductivity σ), and ε is the real part of the relative permittivity.

If no other information is available, the attenuation coefficient for frequencies that are low enough to avoid Debye relaxation can be estimated from the DC conductivity σ and the relative permittivity ε. Solving Eq. 7.9 with the approximation $\tan\delta \ll 1$ yields

$$\alpha' = 1640 \frac{\sigma}{\sqrt{\varepsilon}} \tag{7.24}$$

with α' in dB/m and σ in S/m.

For a successful application of GPR the two-way absorption along the raypath (from the transmitter down to the reflector and back to the receiver) should not exceed a maximum of 60 dB. The other losses due to geometric spreading, reflection and scattering sum up to occupy the remaining part of the dynamic range of existing GPR systems.

Hence, we can try to put up a rule of thumb for the maximum exploration depth h_{max} by letting $2h_{max}\alpha' < 60$dB or, after rearranging:

$$h_{max} \leq 0.018 \rho \sqrt{\varepsilon} \tag{7.25}$$

with h_{max} in m and ρ in Ωm.

Taking again the examples of dry sand, saturated sand and wet clay we get from Eq. 7.23 minimum frequencies of 9, 14, and 72 MHz, respectively. Using Eqs. 7.24 and 7.25 the absorption coefficient for dry sand is 0.82 dB/m and the exploration depth 36 m, for saturated sand the respective values are 3.3 dB/m and 9m, and finally for wet clay the absorption is 16.4 dB/m and the exploration depth 1.8 m.

7.2 Technical aspects of GPR

7.2.1 Overview of system components

GPR systems consist of a pulse generator, an antenna for transmission of electromagnetic pulses, a second antenna for reception of the direct and reflected signals, and receiving electronics to amplify and digitize the waveforms to be recorded and displayed (see Fig. 7.1). In some cases, both antennas are contained in one housing or there is one single antenna with a fast switch for changing between transmission and reception. Timing and triggering of the measurement process is done by a control unit which may also interface to external components like a survey wheel or GPS.

The cables connecting the different parts of a system are preferably nonmetallic fiber optic cables. If metallic cables are to be used some ringing or internal reflections may occur. Shielding and ferrite beads are used to suppress such artefacts.

Although these components may look different for systems of different GPR manufacturers the principles of operation and even electronic circuitry is generally the same for time domain (i.e. short pulse) measurements. A novel class of frequency domain equipment (stepped frequency radar) not described here uses a different concept to get similar results.

7.2.2 Antennas and antenna characteristics

Broadband antennas are needed to transmit and receive short electromagnetic pulses because of the broad frequency spectrum contained in short waveforms - for a monocycle pulse the bandwidth is typically in the order of the center frequency. Conventional broadband systems with directional antenna gain as used for radio and television in the VHF and UHF ranges are unsuitable for short pulses.

Center frequencies f_c between 20 and 1000 MHz are used for geological and engineering investigations. A broad bandwidth is generally achieved by resistive loading (damping) of electric dipoles or by using extended electric dipole elements. Several antenna designs have shown to be useful, e.g., dipoles with hyperbolic resistive loading (Wu and King 1965) and their modifications, unshielded and shielded bowtie-antennas and combinations of bowtie and loop dipoles with resistivity loading (Fig. 7.5).

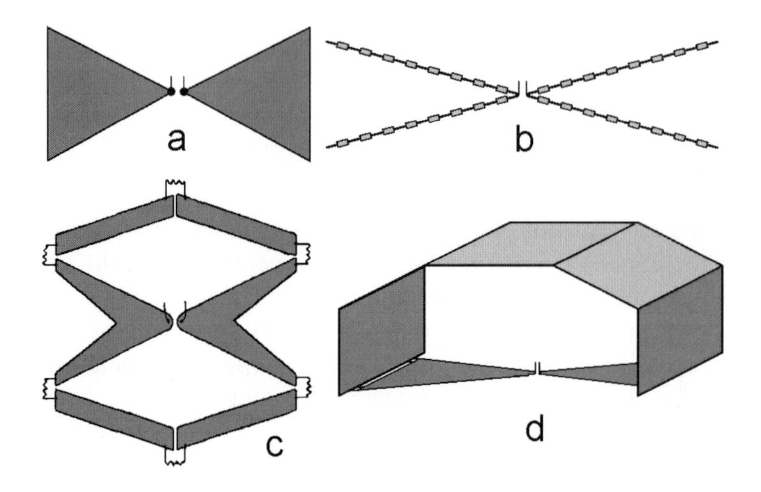

Fig. 7.5. Broadband antennas used for GPR: a) bowtie, b) Wu & King vee dipole with resistive loading c) folded loaded Dipole, d) shielded bowtie

Antennas with loading according to Wu & King have a signal loss of 20 dB per pair relative to a simple dipole. The radiated wavelet is the time derivative of the excitation function of the pulse generator, which is the shortest possible pulse. The frequency spectra of loaded loop dipoles and bowtie designs are narrower and the pulses are slightly longer. The waveforms are similar to the second derivative of a Gauss-function (Ricker wavelet) with some ringing and the amplitude losses per antenna pair are up to 10 dB.

Measurements in buildings, below power lines and trees are affected by "air reflections" from objects in the upper half-space. Shielded antennas are used to suppress these unwanted disturbances. Shielding is accomplished by placing absorbing materials and/or metal boxes above the dipoles.

GPR antennas couple to the ground by the concentration of electric field in the lower medium, best coupling is achieved by laying the antenna flat on the ground (Annan et al. 1975). Typical far field antenna patterns for short dipoles are shown in Fig. 7.6 for two orientations. The less pronounced near and intermediate field patterns (from one to several tens of wavelengths) which are important for typical GPR exploration depths are calculated using the FDTD (finite difference time domain) numerical method (Radzevicius et al. 2003, Lampe et al. 2003).

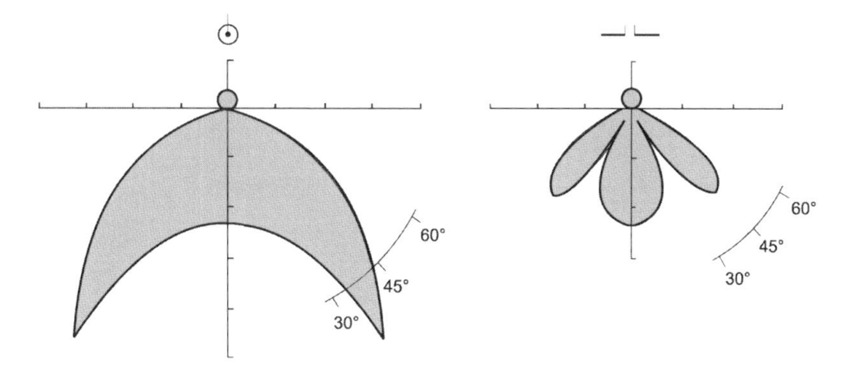

Fig.7.6. Far field radiation of short horizontal dipoles perpendicular and parallel to the dipole axis

7.2.3 Electronics

Pulse generators

Pulse generators for producing short, high-energy impulses for use in the field are often designed with transistor switches using the semiconductor avalanche effect. In principle, a number of condensers is charged in parallel and then discharged to the antenna in series by the switches. Pulses with nanosecond rise times and amplitudes of up to 2 kV can be produced for deep sounding (Fig. 7.7); amplitudes of 400 V with repetition frequencies of about 100 kHz are typical for most commercial systems. The pulse length of a transmitted electromagnetic wave is <20 ns, depending on the antenna frequency and type. It is important that the pulses are generated and repeated with high accuracy. The timing noise (jitter) within radar systems used for geophysics is usually better than 1 ns.

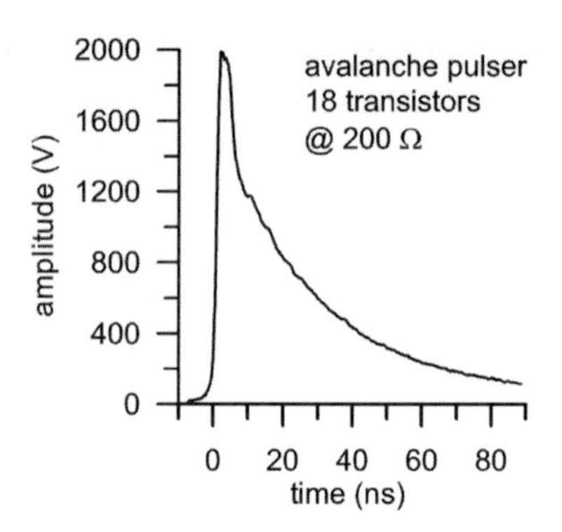

Fig. 7.7. GPR pulse for deep sounding (University of Muenster)

Receiver systems

Sampling of the waveform at the receiving antenna can be done by analog sample and hold converters or fully digital systems. For reasons of weight and power consumption, most systems utilize an analog sequential sampling method with digital control. Sequential sampling means that only one point of the total time series (trace) is recorded per transmitted pulse. This is not economic for the pulser but it is for the low power system as a whole. Continuous measurements are possible because the pulse rates are very high and the pulse generators offer a long lifetime.

The sequentially sampled signals are in the frequency range of audio waves so that they can be digitized by a 16-bit A/D-converter and stored in computer memory. The achievable dynamic range of good sampling devices is 80 –90 dB, which is low compared to seismic systems. The sensitivity limit is set by electronic noise due to the large bandwidth which can only be reduced by stacking (averaging of traces).

Digital real time receivers used for special purposes record one trace per shot, which has the slight advantage that high amplitude pulse generators can be used and the stacking rates may be higher.

The time window for hydrogeological investigations ranges from several 10 ns (for travel paths of several meters) to 1000 ns for a maximum depth of about 60 m in unsaturated or 30 m in saturated low loss material. The sampling intervals depend on the center frequency involved. One cycle should be sampled at least at 10 points. For a 100 MHz-pulse with 10 ns cycle time a sampling interval of 1 ns is sufficient.

Since most commercial GPR systems have only one to four channels, multiple coverage measurements with different source-receiver distances (offsets) like those in reflection seismic surveys are done sequentially.

Normally, georadar measurements are carried out as a profiling with constant offset (single coverage).

The dynamic range of GPR-systems is given by some manufacturers as the performance factor PF. It is the logarithmic ratio of the peak voltage at the pulse generator and the minimum recordable voltage at the receiver in dB. Performance factors of 130 to 140 dB are typical for existing systems. Because antenna efficiency, ground coupling, and spectral content of the broadband transmitter pulse are not specified, the PF is of little use to estimate penetration depth.

7.2.4 Survey practice

Standard procedures

There are two main procedures used for GPR surveys: constant offset (CO) profiling and common-midpoint measurements (CMP) which have replaced wide-angle reflection and refraction (WARR).

For constant offset profiling two antennas are moved over the ground at a fixed distance (bistatic configuration). The necessary antenna separation (offset) from center to center is typically about one wavelength in free space for unshielded antennas and about half a wavelength for shielded designs. If only one antenna with transmit/receive-switch is used, a true zero offset section is recorded (monostatic configuration) at the cost of penetration depth and dead time at the beginning of the time window.

Measurements are taken at constant intervals along the profile triggered by a survey wheel or at constant time intervals at a rate of one to one hundred measured traces per second proceeding at several kilometers per hour. It is possible and convenient to make measurements by car or tractor. This ability combined with its high resolution makes GPR one of the fastest methods of near surface geophysics.

For CMP measurements transmitter and receiver are moved stepwise and simultaneously away from a fixed midpoint to obtain a sequential multifold coverage. The evaluation of traveltime curves (like Fig. 7.4) from CMPs gives velocity-depth functions which are necessary to calibrate the conversion of traveltime into depth (see Sect. 2.3 reflection seismic surveys for details). WARR soundings (not recommended!) keep the transmitter at a fixed location while the receiver is moved away from it. Other methods to get velocity information are the use of borehole data (geological information, depth to groundwater) or for very near surface applications the evaluation of diffraction hyperbolas originating at underground objects like pipes.

To avoid pitfalls in the application of GPR, the suitability of an area or target for GPR measurements should be checked by a test measurement or by resistivity data from geoelectric soundings, from which the absorption can be estimated using Eq. 7.24. Layers with high conductivity (low resistivity) prevent successful soundings. It is useful to avoid e.g. heavily fertilized fields, roads paved with slag, and soils with high clay content.

Some GPR systems provide bandpass filtering of the data during acquisition for signal enhancement. However, the filter settings should kept as broadband as possible to maintain the information content during the acquisition phase. Modern systems have gain-setting options and a stacking function which should be carefully set to optimize the data quality.

Special techniques

Structural investigations or detailed studies may require real 3D-surveys with GPR. The profile spacing has to be dense (a quarter wavelength) to avoid spatial aliasing. The amount of data, the necessary processing steps, and the possibilities for representation are comparable to 3D seismic surveys. Special programs are available for processing and presenting the data. Repeated 3D-measurements are able to detect subsurface changes with time and are often referred to as 4D-measurements. The monitoring of DNAPL spills shows the potential of this technique (Brewster and Annan 1994, Brewster et al. 1995).

Radar tomography operates from borehole to borehole or from borehole to surface with slim watertight borehole antennas. The principles of tomography require a very dense ray coverage of the illuminated area. The theoretical limit of resolution (Williamson 1991, Williamson and Worthington 1993) depends upon wavelength and distance from the borehole and is in the order of the first Fresnel zone (see Sect. 7.1.5). This means that resolution is high close to the boreholes and lowest in between. Some successful case studies (e.g. Hubbard et al. 1997, Tronicke et al. 2002) are reported.

7.3 Processing and interpretation of GPR data

7.3.1 General processing steps

Processing of GPR data has much in common with seismic processing (see Chapt. 2.3). Because GPR systems are different from seismic systems and GPR signals may behave different due to frequency dependent absorption and phase changes at reflections there are some special features which will be addressed in this section. GPR-specific processing is contained in data-processing packages offered by manufacturers of GPR equipment or as add-ons for seismic software.

Time zero

Due to the electronic generation of pulses and the signal delays on cables a time zero of GPR pulses is not generated a priori. However, the constant speed of light c_0 and the known antenna offset x makes it possible to calculate the time $t_a = x/c_0$ which the air wave (the first signal) takes from the transmitter to the receiver. Time zero therefore is t_a in advance of the first signal in a radar trace. This implies a time shift of the data which has to be determined and corrected for each set of measurements.

Filtering

Filtering is a main step in post-acquisition data processing. In many cases this is sufficient to prepare the data for presentation and interpretation. Special filters for GPR take care of eliminating effects of the receiver electronics (removal of DC offsets and low frequencies, called "dewow") and of suppressing unwanted stripes due to system ringing. These have to be applied before the well-known bandpass filters for optimizing the signal to noise ratio.

Amplitude equalization

The amplitude decay of GPR signals (Eq. 7.21) has to be compensated for a proper presentation of the measured data (radargram) on a screen or printout. This can be done by setting a gain function with time, by an automatic gain control (AGC), or by a combination of both. A gain function has the advantage of maintaining relative differences along a profile; the AGC helps to get equal levels for signals which shall be correlated. Hence, a gain function is preferably used for profile data while a proper AGC is suitable for CMP data.

Static and dynamic correction

Static correction of GPR data means the application of time shifts to correct for surface topography. In many cases a surface-undulating layer (e.g. the vadose zone) overlies horizontal or inclined strata (e.g. Quaternary deposits or the water table). Correction of these undulations is done with the velocity of the top layer to enable or ease a proper interpretation.

Dynamic correction of GPR data means a variable time shift within a trace to correct for the effects of antenna offset. By producing such a pseudo zero-offset section distortions of the first signals occur which usually are suppressed by muting. The advantage for CO data is a linear depth scale for the top layer.

Further processing

As GPR rarely practises real multifold coverage the technique of CMP stacking, one of the most important processing steps in seismic surveys, is applied only for single CMP measurements necessary for velocity determination (see below). Advanced processing like migration may sometimes be helpful. For GPR the same techniques as for reflection seismic are used. Deconvolution to sharpen the signals has not been very successful because GPR signals are subject to rapid phase changes and dispersion effects stronger than experienced with reflection seismic data.

Presentation

Due to the high data density from profile measurements, the data are often displayed compressed in grey-scale or raster plots instead of individual wiggle traces. For the presentation of GPR data it is helpful to give both vertical scales (traveltime and depth), as well as time-zero, antenna offset, and the velocity or velocity-depth function used for depth conversion. CMP-measurements are preferably displayed as wiggle traces with a short description of processing.

7.3.2 Examples for GPR profiling and CMP data

Desert areas are supposed to be hot, dry and non-conductive thus providing optimal conditions for GPR investigations. This was supposed at beginning of a project to investigate fossil groundwater reserves in the Nubian Desert at the Egyptian-Sudanese border in an area called Great Selima Sand Sheet. It soon turned out that the fossil soils overlying the mainly quartzitic Nubian sandstone contain a considerable amount of clay which was still conductive enough to prevent successful GPR measurements for 85% of the 1800 profile kilometers we investigated in total. On the other hand, there are 15% of data showing deep reflections and geological strata. They concentrate in areas where the fossil soil cover was removed by wind deflation. From these data (Blindow et al. 1987) a GPR profile and a CMP will be shown and explained as examples for deep sounding.

Constant offset profiling

The continuous profiles were measured with a center frequency of 35 MHz utilizing a 2kV pulser and resistively loaded vee-dipoles mounted on wooden runners at an offset of 10m. Data were collected by towing the antennas behind a four-wheel-drive car at a speed of 10 km/h while doing one measurement per meter triggered by a large surveying wheel. The sampling interval is 1 ns, the original time window 1000 ns truncated to 512 ns for display.

Data processing was done by a time-zero correction of 33 ns (corresponding to 10 m offset), bandpass filtering with a second order frequency domain Butterworth filter (20 to 80 MHz) and applying a square-law gain function.

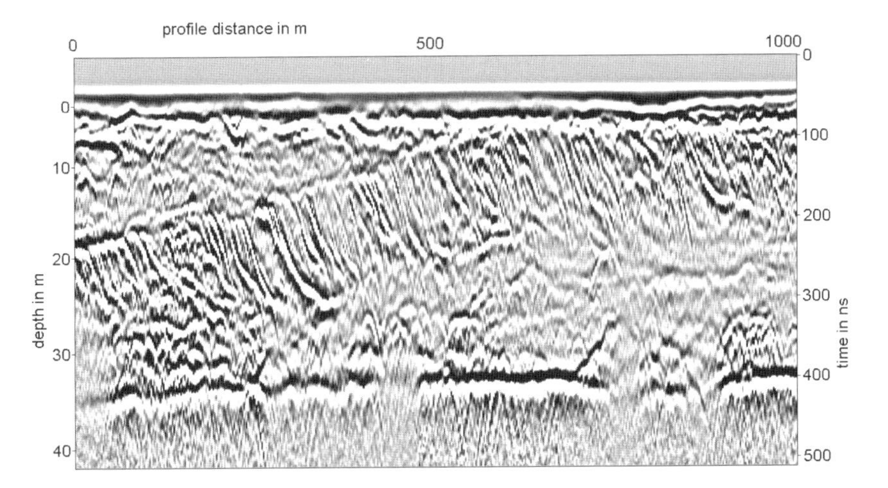

Fig.7.8. GPR profile showing a water table reflection in Nubian sandstone, Eastern Sahara

The depth scale was calculated from the mean velocity v = 0.167 m/ns measured at a nearby CMP. The depth scale is nonlinear especially in its upper part because the offset x is part of the traveltime-depth relation (Eq. 7.20). Depth zero is at $t_g = x/v$. The raster plot is vertically exaggerated by a factor of 13.

The velocity already gives a clue about the subsurface. Natural quartz has a permittivity of about $\varepsilon = 4$ which would correspond to v=0.15 m/ns. The velocity is raised by the air content due to porosity. Using Sen's mixture formula (see Sect. 1.3) it follows that the Nubian sandstone at this site has a porosity of about 20% which makes it a good reservoir rock. The velocity in the saturated rock is calculated to be 0.09 m/ns. With Eq. 7.15 we get a high negative reflectivity of 30% in amplitude.

Looking again at the GPR profile we see one prominent reflector at about 35 m depth which shows a slight upward tendency with growing profile distance. This reflection is strong - and it has reversed polarity compared with the direct air wave. To prove that this really is the fossil water table we did long distance levelling to the next well and found that the interpretation was right. We were able to check the accuracy of GPR with respect to locating the water table at or close to several other wells. It turned out that depth to groundwater measured by GPR compared to the direct measurement differed only by a few decimeters.

There are other features above the water table which can be interpreted as a band of cross-bedding strata and some other layering. Note that the frequency content of the cross-bedding signals is higher than that of the ground water reflection and the direct waves. This effect is most likely explained by reflections at thin strata, which return differentiated signals thus raising the spectral frequency maximum. Finally, we can postulate from diffraction hyperbolas originating at or close to the water table that there are porosity changes within the continuous reflection, perhaps due to fissures in the matrix.

CMP-measurements

CMP-measurements for the determination of propagation velocity have to be made at representative points in a survey area. This means in practice that first the profiling is done. After an inspection of profile data the CMP locations are selected. If no reflection is present on a profile there is little use to measure a CMP in this area. On the other hand, a multifold coverage always improves the signal to noise ratio by stacking of different ray geometries. The location of the following CMP is about 25 kilometers away from the profile Fig. 7.8. Continuous profiling there gave a very faint reflection at a depth of more than 50 m. The CMP measurement along with the according velocity stack is shown in Fig. 7.9.

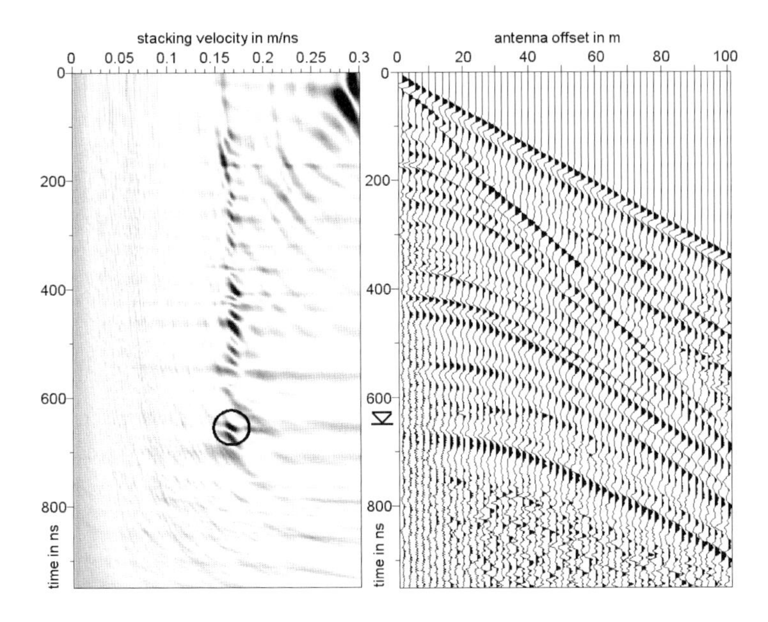

Fig. 7.9. CMP-measurement and velocity stack to evaluate a water table reflection in Nubian sandstone, Eastern Sahara

On the right hand side of the panel the processed data of the CMP measurement are plotted as wiggle traces as a function of offset and time. The strong linear traveltime curve is the air wave, followed by a number of reflection hyperbolas. The last and deepest one is indeed very faint for the profiling offset of 10 m, but increases in amplitude for larger distances. The reason is that the reflection coefficient increases with the angle of incidence.

The groundwave is superposed by wide angle reflections of numerous internal layers. Between the air wave and these reflections there is a number of lateral waves travelling also at the speed of light.

Velocity analysis of CMP-data can be done by matching a synthetic reflection hyperbola to reflection onsets visible in the CMP section. This method or the even more basic x^2-t^2-method has been replaced by velocity stacking. Either the envelopes or coherency measures between adjacent traces (semblance) or the filtered data after amplitude normalization are summed up (stacked) along theoretical hyperbolas whose zero offset time and velocity are changed systematically. If the parameters correspond to a reflection hyperbola of the CMP the stacking amplitude will be high and low amplitudes will occur elsewhere.

The result of this process is displayed at the left hand side of the panel in Fig. 7.9. Picking the maxima along with the according velocities and zero offset times yields a function of velocity vs. zero offset time. To be more precise, the velocities are so called stacking velocities. If they are very similar like in this example, there is little effect of refraction and we can regard them as RMS-velocities which have a physical meaning. The velocity taken from the encircled maximum in Fig. 7.9 is 0.165 m/ns which is slightly lower than for the profile of Fig. 7.8. The resulting depth to groundwater at this site is 54 m.

The examples show that GPR measurements sometimes require a strategy which is different from other geophysical methods. Because profiling is done fast and mobile, an area covered by a conductive layer can be searched for locations with sufficient penetration. At the same time, the existence of the conductive layer is mapped which also might provide valuable information.

As the physics of GPR can not be changed the success of the method will always be limited to areas with increased resistivity. Within these limits GPR can be used for measuring depth to groundwater, finding lenses of perched groundwater, mapping of stratigraphy and confining beds. It also has a great potential for the detection and monitoring of groundwater hazards as shown in Sect. 17.4.

7.4 References

Annan AP, Waller WM, Strangway DW, Rossiter JR, Redman JD, Watts RD (1975) The electromagnetic response of a low loss, 2-layer, dielectric earth for horizontal electric dipole excitation. Geophysics 40:285-29

Baños A (1966) Dipole radiation in the presence of a conducting halfspace. Pergamon Press, New York

Blindow N, Ergenzinger P, Pahls H, Scholz H, Thyssen F (1987) Continuous profiling of subsurface structures and groundwater surface by EMR methods in Southern Egypt. Berliner Geowiss Abh (A) 75.2:575-627

Brekovskikh LM (1980) Waves in layered media, 2nd edn. Academic Press, New York

Brewster ML, Annan AP (1994) Ground-penetrating radar monitoring of a controlled DNAPL release: 200 MHz radar. Geophysics 59:1211-1221

Brewster ML, Annan AP, Greenhouse JP, Kueper BH, Olhoeft GR, Redman JD, Sander KA (1995) Observed migration of a controlled DNAPL release by geophysical methods. Ground Water 33:977-987

Daniels JJ (1996) Surface Penetrating Radar. The Institution of Electrical Engineers, London
Hubbard SS, Petersen JE, Majer E, Zawislanski TP, Williams KH, Roberts J, Wobber F (1997) Estimation of permeable pathways and water content using tomographic radar data. The Leading Edge 16:1623-1628
Hoekstra P, Delaney A (1974) Dielectric properties of soils at UHF and microwave frequencies. J Geophys Res 79:1699-1708
Lampe B, Holliger K, Green AG (2003) A finite difference time-domain simulation tool for ground-penetrating radar antennas. Geophysics 68:971–987
Radzevicius SJ, Chen CC, Peters L, Daniels JJ (2003) Near-field dipole radiation dynamics through FDTD modelling. Journal of Applied Geophysics 52:75-91
Tronicke J, Blindow N, Gross R, Lange MA (1999) Joint application of surface electrical resistivity- and GPR-measurements for groundwater exploration on the island of Spiekeroog - northern Germany. Journal of Hydrology 223:44-53
Tronicke J, Dietrich P, Wahlig U, Appel E (2002) Integrating surface georadar and crosshole radar tomography: A validation experiment in braided stream deposits. Geophysics 78:1516–1523
Van Overmeeren RA (1994) Georadar for hydrogeology. First Break 12:401-408
von Hippel AR (1954a) Dielectrics and waves. MIT Press, Cambridge MA, USA
von Hippel AR (ed) (1954b) Dielectric materials and applications. MIT Press, Cambridge MA, USA
Wu TT, King RWP (1965) The cylindrical antenna with nonreflecting resistive loading, IEEE Trans Antennas and Propagation AP-13:369-373
Wyatt DE, Temples TJ (1996) Ground-penetrating radar detection of small-scale channels, joints and faults in the unconsolidated sediments of the atlantic coastal plain. Environmental Geology 27:219-225
Williamson PR (1991) A guide to the limits of resolution imposed by scattering in ray tomography. Geophysics 56:202–207
Williamson PR, Worthington MH (1993) Resolution limits in ray tomography due to wave behavior: Numerical experiments. Geophysics 58:727–735

General Reading

Blindow N (2005) Bodenradar. In Knödel K, Krummel H, Lange G (eds) Handbuch zur Erkundung des Untergrunds von Deponien und Altlasten, 2nd edn. Springer, Berlin Heidelberg
Daniels JJ (2004) (ed) Ground Penetrating Radar, 2nd edn. The Institution of Electrical Engineers, London
Davis JL, Annan AP (1989) Ground penetrating radar for high-resolution mapping of soil and rock stratigraphy. Geophys Prosp 37:531-551
GPR 1994, Proceedings of the Fifth International Conference on Ground Penetrating Radar; 12-16 June, 1994, Kitchener, Ontario Canada
GPR 1996, Proceedings of the Sixth International Conference on Ground Penetrating Radar; September 30- October 3, 1996, Tohoku Japan
GPR 1998, Proceedings of the Seventh International Conference on Ground Penetrating Radar; 27-30 May, 1998, Lawrence, Kansas USA

GPR 2000, Proceedings of the Eighth International Conference on Ground Penetrating Radar; 23-26 May, 2000, Gold Coast, Australia

GPR 2002, Proceedings of the Ninth International Conference on Ground Penetrating Radar; April 29 - May 2, 2002, Santa Barbara, California USA

GPR 2004, Proceedings of the Tenth International Conference on Ground Penetrating Radar; 21 - 24 June, 2004, Delft, The Netherlands

Olhoeft GR (2004) www.g-p-r.com Webpage on ground penetrating radar with a GPR Tutorial, a Bibliography and links to GPR Manufactures

8 Magnetic Resonance Sounding

Ugur Yaramanci and Marian Hertrich

8.1 Introduction

Magnetic Resonance Sounding (MRS) is a new technology which just passed the experimental stage to become a promising surface measurement tool to investigate directly the existence, amount and productiveness of ground water. The principle of Nuclear Magnetic Resonance (NMR), well known in physics, physical chemistry as well as in medicine, has successfully been adapted to assess the aquifer parameters.

The first high-precision observations of NMR signals from hydrogen nuclei were made in the forties of the last century. Meanwhile it is a standard investigation technology on rock cores and in boreholes (Kenyon 1992). The first ideas for making use of NMR in groundwater exploration from the ground surface were developed as early as the 1960s, but only in the 1980s was effective equipment designed and put to operation for surface geophysical exploration (Semenov et al. 1988, Legchenko et al. 1990). Extensive surveys and testing have been conducted in different geological conditions particularly in sandy aquifers but also in clayey formations as well as in fractured limestone and at special test sites (Schirov et al. 1991, Lieblich et al. 1994, Goldman et al. 1994, Legchenko et al. 1995, Beauce et al. 1996, Yaramanci et al. 1999, Meju et al. 2002, Plata and Rubio 2002, Supper et al. 2002, Vouillamoz et al. 2002, Yaramanci et al. 2002, Dippel and Golden 2003, Baltassat et al. 2005, Lange et al. 2005, Shushakov et al. 2005, Voillamoz et al. 2005).

8.2 NMR-Principles and MRS technique

The physical property used in NMR applications is the Spin of the nuclei under investigation, i.e. hydrogen protons of water molecules. This is to possess an angular momentum, without physically rotating, and an associated magnetic moment. The magnitude of this magnetic moment μ_0, its equilibrium orientation and the oscillation frequency ω_L are related to the static field B_0 by:

$$\vec{\mu}_0 = \frac{\gamma^2 \hbar \vec{B}_0}{3kT} \frac{N}{V} \tag{8.1}$$

$$\omega_L = -\gamma B_0$$

where $\gamma = 0.267518$ Hz/nT is the gyromagnetic ratio for hydrogen protons, N/V is the number of Spins per volume and h, k, and T are common physical constants. By the application of a secondary electromagnetic field B_1, perpendicular to B_0 and oscillating at the Larmor frequency ω_L, the spinning magnetic moment is deflected to an oscillating motion around the static field direction (Fig. 8.1a). During the excitation of the Spin it retains its magnitude and changes only the orientation describing a helix-shaped line on a unit sphere (Fig. 8.1b). The components of the magnetization vector perpendicular and parallel to the static field direction are harmonic functions of the magnitude of the secondary field B_1 and the pulse duration τ_p:

$$\mu_\perp(\tau_p) = \mu_0 \sin\left(\frac{1}{2}\tau_p B_1(\vec{r})\right) \quad \mu_\parallel(\tau_p) = \mu_0 \cos\left(\frac{1}{2}\tau_p B_1(\vec{r})\right) \tag{8.2}$$

Consequently the respective components of μ achieve arbitrary values between $+\mu$ and $-\mu$, and for long pulse durations or large magnitudes flip several times around. The perpendicular component continuously oscillates at the Larmor frequency in the plane perpendicular to B_0, whereas the parallel component shows no such oscillating component.

According to the semi-classical solutions of the Bloch equations, describing the quantum mechanical phenomenon of NMR, both components of the magnetization decay individually and independently from their achieved magnitudes at pulse duration τ_p. The perpendicular component decays to zero with time constant T_2, still precessing in the x-y plane, the parallel component builds up exponentially to its equilibrium magnetization with time constant T_1. Consequently the magnitude of the total magnetization is not constant during the relaxation process. Its initial value at pulse cutoff is μ_0 and the final value after full relaxation is μ_0 again, but during the transient relaxation process its magnitudes varies with decreasing $\mu_\perp$ and increasing $\mu_\parallel$. The corresponding trajectory in a unit sphere for an excitation angle of 90° is shown in Fig. 8.1c. In general T_1 is $\geq T_2$, but at low static field strength, as given in Earth's field applications, they can be assumed to be equal.

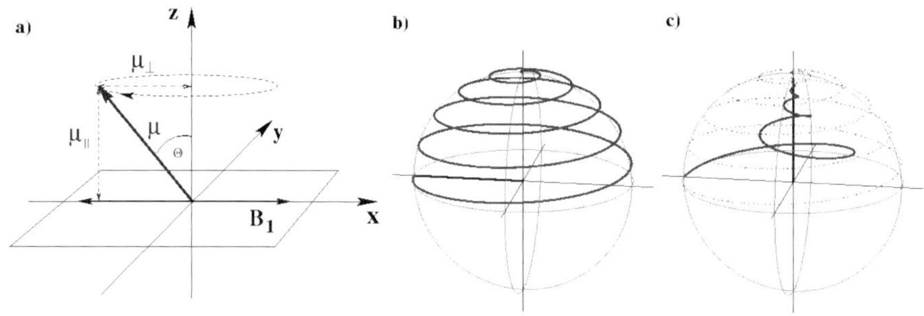

Fig. 8.1. a) Components of the precessing magnetic moment of the Spin. Unit trajectory of the Spin magnetic moment during the excitation process b) and during the relaxation process c) (after Levitt 2002). Direction of the static field B_0 is along the z-axis

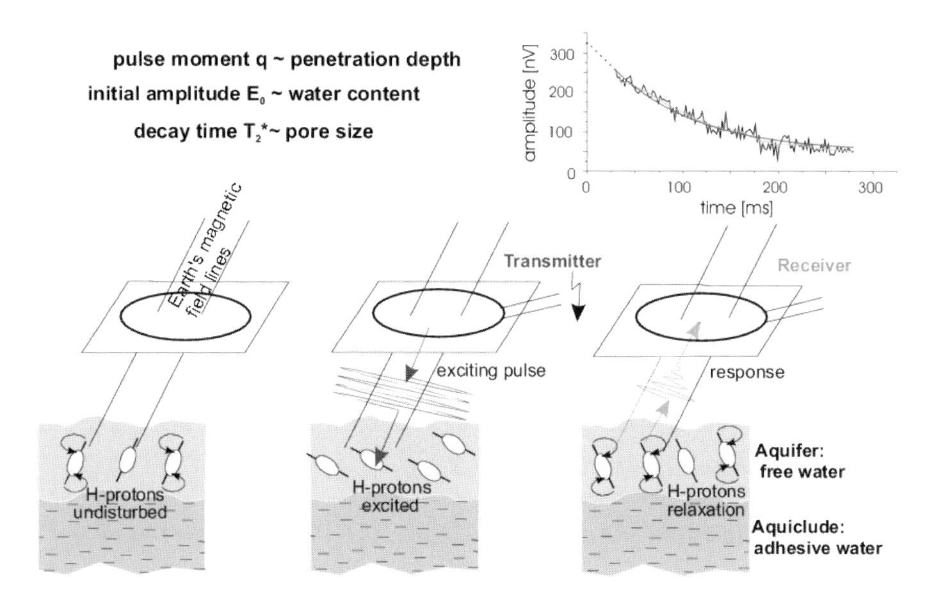

Fig. 8.2. Principle sketch of a Magnetic Resonance measurement with stages of protons undisturbed, excited and relaxing and relaxation signal

The relaxation follows the functions:

$$\mu_\perp(t) = \mu_\perp(\tau_p)e^{-\frac{t}{T_2}} \qquad \mu_\parallel(\tau_p) = \mu_\parallel(\tau_p)\left(1 - e^{-\frac{t}{T_1}}\right) \tag{8.3}$$

The technique of Magnetic Resonance Sounding in geophysical applications makes use of the NMR phenomenon of the hydrogen protons of the groundwater (Fig. 8.2). The Spin system is perturbed by a pulse in the

transmitter coil and emits after pulse cutoff the response signal from the subsurface. Here, the Earth's magnetic field is the static field, leading to a Larmor frequency of some 1.2-2.5 kHz. The secondary field perturbing the Spin system is given by the electromagnetic fields of large surface loops. The response signal is also recorded in a surface loop, usually of the same dimension. In conventional depth soundings measurements are realized with transmitter and receiver being the same loop, the coincident configuration, but for 2D investigations both, transmitter and receiver are individual loops at arbitrary separation.
From a single recording the curve characteristics initial amplitude and decay constant of the transient curve are derived as well as the phase shift. By increasing the excitation intensity the depth sensitivity can be controlled.

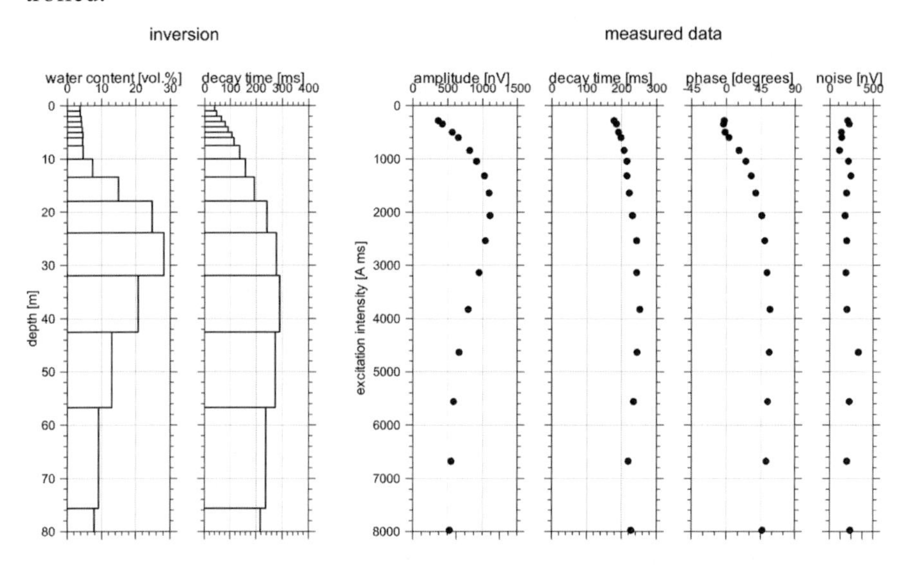

Fig. 8.3. Sketch of the pulse and signal registration sequence of a MRS recording (left). Data obtained after processing of the MRS recordings in dependency of the excitation intensity (right) and inverted subsurface parameters water content and decay time in depth (middle)

According to the previous explanation the tipping angle depends on the magnitude of the secondary field. Due to the nature of the transmitter loop field to decay rapidly in distance from the loop, Spins at shallow depths perform multiple rotations and by loss of coherence do interfere such that no resulting signal is emitted. In deeper layers the Spins achieve their maximum tipping angle and contribute the major part of the entire signal. A set of increasing pulses thus does not only provide an increasing penetration depth but also a depth selective focus. The thereby achieved sounding curve can now be interpreted by means of inversion to a water content and decay constant distribution with depth.

The basic equation describing the voltage response in the receiver loop in dependency of the excitation intensity, i.e. the pulse moment q, and the recoding time t, is given according to Weichman et al. (2000) by:

$$\begin{aligned} V_1(q,t) = \omega\,\mu_0 \int_V d^3r f(r) \sin\left(-\gamma \frac{q}{I_0}\left|B_1(r)\right|\right) e^{-\frac{t}{T_2^*(r)}} \\ \times \frac{2}{I_0}\left|B_1(r)\right| e^{i[\zeta_T(r)+\zeta_R(r)]} \\ \times [\vec{b}_R(r)\cdot\vec{b}_T(r) + i\vec{b}_0\cdot\vec{b}_R(r)\times\vec{b}_T(r)] \end{aligned} \tag{8.4}$$

Here, the pulse moment q is defined as the product of pulse duration τ_p and current through the loop I_0, $q = \tau_p I_0$.

In this form the accounting terms of the equation can be physically interpreted as follows:

- The first line of the equation gives the signal amplitude of a Spin system, emitting the NMR response. It consists of the magnitude of the magnetization vector, i.e. the number of hydrogen nuclei, and the excitation angle they have achieved by the energizing pulse. The excitation angles of the resulting magnetization is a harmonic function, determined by the normalized amplitude of the co-rotating part of the transmitter field and the pulse moment q.
- The second line describes the sensitivity of the receiver to a signal in the subsurface. This spatial sensitivity is independent from signal generation, i.e. the pulse moment. It is simply given by the magnetic field distribution associated to the receiver loop. Additionally the signal undergoes a phase lag from the transmitter to the sample and from the sample to the receiver due to electromagnetic attenuation.
- The most important part in the treatment of the mathematical foundation of separated loop configuration is shown in the third line. Whereas the first two lines only cover the scalar values of excited signals and their associated induced voltage in the loop, this parts accounts the vectorial nature of the evolution of the MRS signal. The expression in the bracket describes the dependency of the evolving signal on the directions of transmitter and receiver field and their orientation in respect to the E-arth's field. Note, that only the unit direction vectors of the magnetic fields enter this part of the function. The first part of the sum, the real part of the whole bracket, is simply a scalar vector multiplication, whereas the second, the vector product of both fields and their scalar product with the Earth's field, scales the imaginary part of the expression. In

case of coincident loop soundings, the latter part vanishes whereas the former one approaches unity; the complete expression becomes a real unity value. For all other loop layouts the value of this last term, the geometric term, scales the NMR effect and the signal response by a value in the range [-1,1 ; -i, i]. Consequently, the evolving signal for separated loops is in general complex valued. The resulting phase shift of the recorded signals is the geometrical phase of the MRS signal. It acts in addition to the quantum mechanical phase due to frequency deviations (Hertrich and Yaramanci, 2003) and frequency spectra of the pulse (Legchenko, 2004) and the electromagnetic phase lag caused by signal propagation in conductive media.

The water content entering this equation is only part of the total water content in the pore space. Water that is bound to the internal surface or stored in very fine pores does not contribute to the MRS response signal since its relaxation constants are too small to be detected by the current MRS technique. The water content contributing to the response signal is thus closely related to the mobile water content. The remaining parameters in the integral of Eq. 8.4 then describe the constant settings of the measurement as loop shape, Earth's electrical conductivity and choice of pulse moments and are commonly combined to the MRS kernel function:

$$
\begin{aligned}
V_R(q,t) &= \omega_L \mu_0 \int_V d^3 r f(r) K(q,t)\, e^{\frac{t}{T_2^*(r)}} \\
K(q,t) &= \sin\left(-\gamma \frac{q}{I_0} \left| B_1(r) \right| \right) \\
&\times \frac{2}{I_0} \left| B_1(r) \right| e^{i[\zeta_T(r)+\zeta_R(r)]} \\
&\times [\vec{b}_R(r) \cdot \vec{b}_T(r) + i\vec{b}_0 \cdot \vec{b}_R(r) \times \vec{b}_T(r)]
\end{aligned}
\tag{8.5}
$$

In case of 2D or 1D conditions, the formulation is commonly simplified by reducing the kernel function to the necessary dimension by integrating the general kernel in direction of the respective Cartesian dimension. The expression for 3D is then

$$
V_R(q,t) = \int_0^\infty \int_{-\infty}^\infty \int_{-\infty}^\infty f(x,y,z) K_{3D}(x,y,z) e^{-\frac{t}{T_2^*(x,y,z)}} dxdydz \tag{8.6}
$$

and can be reduced to 2D as a section in depth and profile direction like

$$V_R(q,t) = \int_0^{\infty}\int_{-\infty}^{\infty} f(x,z)K_{2D}(q;x,z)e^{-\frac{t}{T_2^*(x,z)}}dxdz \tag{8.7}$$

$$K_{2D}(q;x,z) = \int_{-\infty}^{\infty} K_{3D}(q;x,y,z)dy; \quad \partial f(y)/\partial y = 0$$

Reducing the kernel to 1D allows only a water content variation in depth

$$V_R(q,t) = \int_0^{\infty} f(z)K_{1D}(q;z)e^{-\frac{t}{T_2^*(z)}}dz \tag{8.8}$$

$$K_{1D}(q;z) = \int_{-\infty}^{\infty}\int_{-\infty}^{\infty} K_{3D}(q;x,y,z)dxdy; \quad \partial f(x)/\partial x = \partial f(y)/\partial y = 0$$

The principle how the multiplication of a water bearing layer and the kernel function leads to a characteristic sounding curve is shown in Fig. 8.4. Here the 1D kernel function according Eq. 8.8 for selected pulse moments is represented by line plots. The shaded area below the curves of the kernel function represents the resulting signal amplitude.

The characteristics of the kernel function by means of depth sensitivity is significantly determined by the loop size. Fig. 8.5 shows the kernels for loop sizes of 100 m, 50 m and 20 m above an insulating halfspace as contour plots. The characteristic depth focus changes to shallower depth for smaller loops. However, it is visible that the penetration depth can not be arbitrarily increased by only increasing the pulse moment since the focus depth converges to a maximum depth. Due to the reduced investigated volume for smaller loops the magnitudes of the kernels and consequently the maximum possible response signals decrease with loop size and currently limit the method for very shallow applications in the range of the usually observed natural noise.

A high potential of the technique of separated loop MRS measurements lies in the investigation of 2D structures. Similar to the previous formulation of the 1D kernel, the general formula for the MRS response is reduced by summation according Eq. 8.7 along one dimension. The resulting 2D kernel then represents the sensitivity of a sounding to a 2D water content distribution, assumed to be infinitely extended in the summation direction. In contrary to the depth focus of the coincident soundings in Fig. 8.6, separated loops soundings provide a sensitivity to shallow structures in the range of the loops. This pattern is more prominent for increasing loop separation. The combination of several soundings along a profile with different loop separations consequently provides a high coverage of 2D sensitivity and allows the derivation of a 2D water content distribution from a profile like survey.

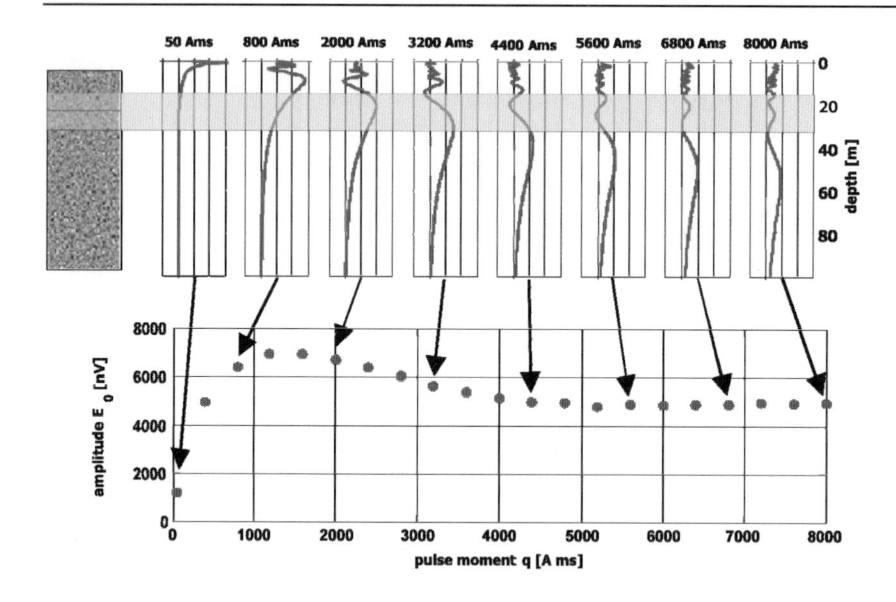

Fig. 8.4. Kernel functions (i.e. sensitivities) for different excitation intensities (pulse moments) and constitution of a sounding curve for amplitudes

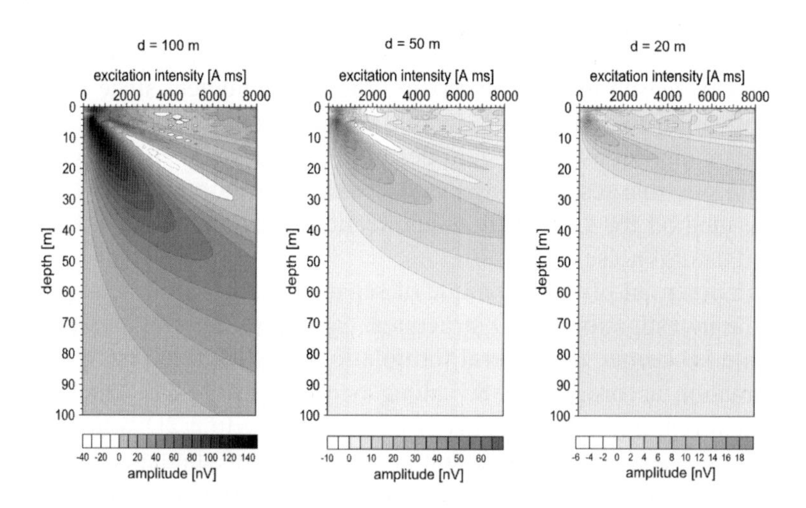

Fig. 8.5. Kernel functions (for 1D soundings) for different loop sizes

Such a reconstruction of the physical parameters from a set of measurements with overlapping and variable coverage of a subsurface section makes it, according the generally accepted convention in geophysics, a tomography application, e.g. geoelectrics, and therefore we call it Magnetic Resonance Tomography (MRT).

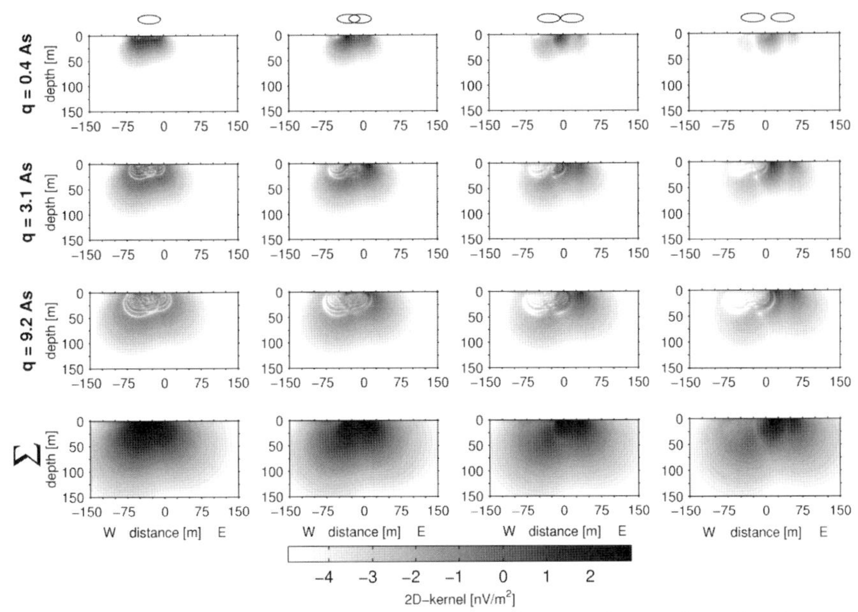

Fig. 8.6. 2D kernels for increasing loop separation for three selected pulse moments (lines 1-3) and as the cumulative kernels for an entire sounding. Measurement configuration is 48m loop diameter, 2 turns, Earth's field at 48000 nT and 60°inclination, profile direction 45°N. Kernel magnitude is in logarithmic scale; dark represent higher sensitivities

8.3 Survey at Waalwijk / The Netherlands

The geology of the area mainly consists of well-sorted sands. At this site, a borehole nearby confirms a shallow aquifer with a water table at about 6 m. The lower boundary of the aquifer is at 52 m depth to clay and silt units (Lubczynski and Roy 2003). All units are heterogeneous and consist of thin layers.

The electrical subsurface model at the site, as derived from TDEM measurements shows moderate to low ground resistivities, thus reducing the effective penetration depth of the measurements. Hereby, the low resistivity values correspond to clayey material between 52 and 80 m depth. The local geomagnetic field intensity during the survey was 48360 nT (corresponding to a Larmor frequency f_L=2059 Hz) with an inclination of approximately 60 degrees.

Table 8.1. Electrical subsurface model as derived from TDEM and lithology as derived from a nearby borehole for the site Waalwijk.

depth [m]	resistivity [Ωm]	lithology
0 – 6	120	fine sands
6 – 52	60	fine – coarse sands
52 - 85	20	clayey sands
> 85	60	medium sands

The measurements were conducted using a square loop with a side length of a = 113.5 m. Excitation intensities (24 pulse moments) ranged from 57 to 5350 A ms. Considering the resulting MRS sensitivity for this configuration a reliable depth of investigation of approximately 60 m can be expected. The total recording length of the decaying signals is 450 ms with a 'dead time' between the exciting pulse and start of recording of 40 ms. The data generally show a high S/N ratio (example for pulse moment q = 500 A ms plotted in Fig. 8.7). Therefore, the relaxation data is of good quality even for lower amplitudes, i.e. later recording times.

MRS relaxation signals can exhibit a multi-exponential relaxation behaviour (Fig. 8.7) due to a signal superposition of layers having different decay times (~ grain sizes) and / or due to a possibly multi-modal decay time distribution within the layers. Whereas the signal amplitudes of MRS are proportional to the amount of water in the subsurface, the decay times are a function of grain sizes (e.g. Shirov et al. 1991, Yaramanci et al. 1999) and pore sizes (e.g. Kenyon 1992). While small grain sizes, i.e. pores, correspond to small decay times large grain sizes, i.e. pores, will exhibit long decay times.

The so far available inversion schemes make use of an approximated mono-exponential fitting approach of the NMR relaxation. However, this practice does not take into account that, analogous to borehole and laboratory NMR, every signal contribution of a single sample or volume element is a superposition of signal contributions due to a specific grain size i.e. pore size distribution within the investigated sample / volume element in the subsurface. Therefore, a newly developed inversion scheme with regard to the multi-exponential nature of decay in the layers and that of the recorded signal too (Mohnke and Yaramanci 2005) has been used by the inversion.

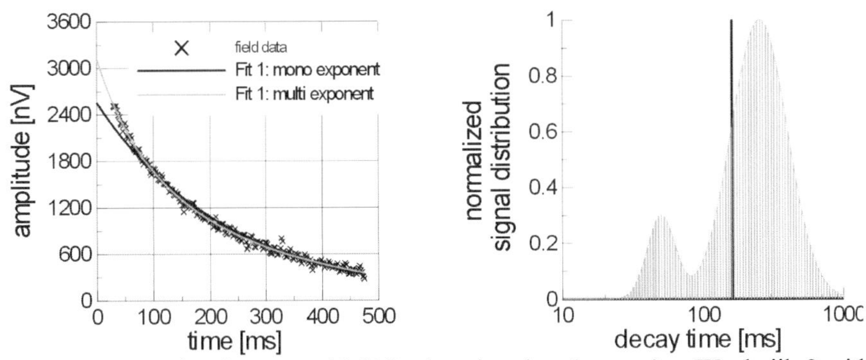

Fig. 8.7. Example of measured MRS relaxation data (crosses) at Waalwijk-2 with mono-exponential (black line) and multi-exponential (grey line) fitting of the data and corresponding decay time distributions (right)

Inversions have been carried out using both conventional mono-exponential inversion (Fig. 8.8a) and multi-exponential inversion approaches using free (Fig. 8.8b) and fixed decay time distributions (Fig. 8.8c). Mono-exponential fitting shows a shallow aquifer between 6 and 50 m with a mean integrated water content of about 28 %. The inverted (mono-exponential) decay times range from 200-300 ms for the aquifer layers and is about 100-150 ms for neighbouring layers. For greater depths (> 50 m) the decay times increase to values of about 400 ms coinciding with an increase of data misfit and do not comply with the fine grained material expected from borehole data.

Multi-exponential inversion results (Fig. 8.8 b,c) show an improved data fit, especially at the beginning of the signal (crosses) as the consequence of a better fit of the short decay times. The integrated water content, being about 34 % within the aquifer, is generally higher. The multi-exponential decay time distribution within the aquifer layers show that the main fraction of the water content can be associated to prominent decay times around 400 ms, (medium to coarse material). In addition a small fraction of the water content is associated to rather small decay times between 40 – 60 ms and consequently resulting in higher integrated water content when extrapolating to initial amplitude (t = 0 ms) values (see also Fig. 8.8 left). This may indicate small intrusions of clayey / fine grained material or more likely be of an artificial origin occurring due to low S/N for small q values that focus on these more shallow depths. For depths below 50 m the decay time distribution indicate a different type of lithology. Here, decay times of about 90-100 ms (while slightly overstated for pure clay and silt with 50 ms) reflect the clayey / fine-grained material as expected from borehole measurements.

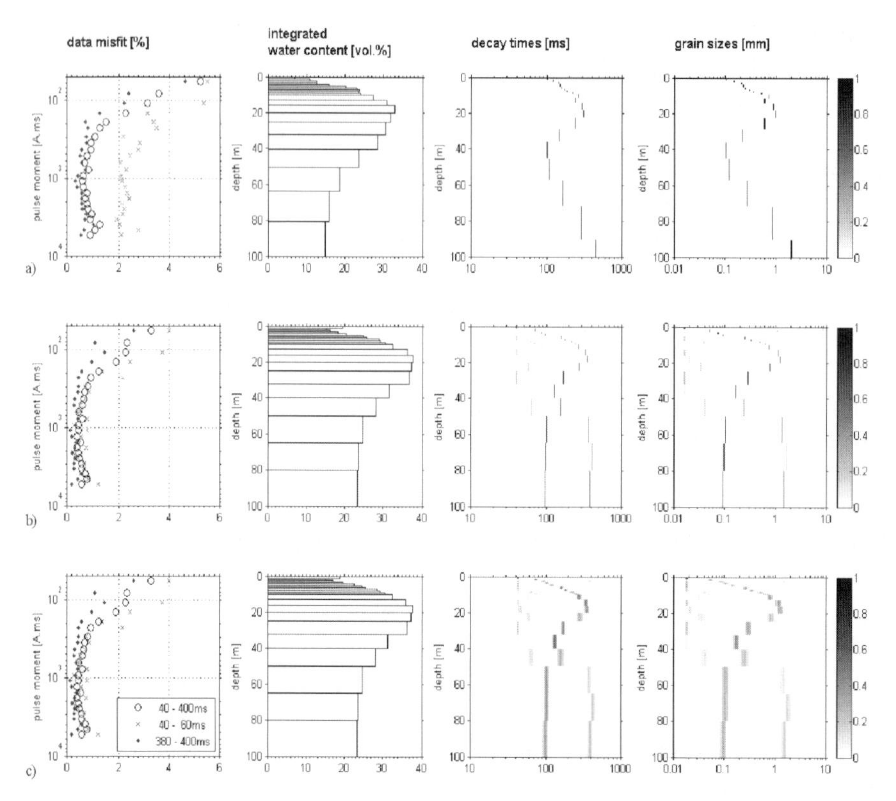

Fig. 8.8. Inversion results for the site Waalwijk-2. a) with conventional mono-exponential inversion, b) with multi-exponential inversion using two free decay times, c) with fixed decay times. Left: Percentage data misfit for recorded relaxation signals (windows: total [o] start [x] and end [.]); Center left: Integrated mobile water content; Center right: Decay time distributions; Right: Derived grain size distributions. The grey scale gives the fractional water volume in the material with corresponding decay times and grain sizes

However, besides these prominent maxima at low decay times a distinct, albeit smaller, maximum appears at decay times of about 450 ms. This may be an indication of interbedded layers of coarse material within the layers, that cannot be discriminated in detail due to a decreasing resolution of the inversion for these depth. This would correlate to the heterogeneous layering derived from the borehole.

Another possibility for this bi-modal distribution may be just a 'shadow'-effect from the high decay times of the aquifer. Fig 8.8 at right shows grain size distributions as derived from MRS decay times according to an empirical relation introduced by Shirov et al. (1991) and Yaramanci et al. (1999). While the aquifer shows grain sizes in the range of 1 mm (coarse sands, gravel) the grain sizes < 0.1 mm below 50 m represent the fine material of the aquiclude.

While the conventional approximated inversion approaches a priori assume mono-exponential MRS relaxation data, a comprehensive inversion scheme allows to take into account the multi-exponential features.

Whereas conventional inversion will reduce the resolution of MRS using more or less averaged decay times, comprehensive inversion can yield a more realistic decay time distribution with depth. Moreover, since the water content is proportional to the initial signal amplitude (t=0 ms) the significant improvement of data fitting by multi-exponential inversion of MRS signals is basic for an improved quantitative determination of the integrated mobile water content in the subsurface, that so far is usually underestimated by the conventional approach.

8.4 Survey at Nauen / Germany with 2D assessment

The survey in Nauen was performed on a well investigated test site, where a variety of additional data are available (Yaramanci et al. 2002). From previous surveys a shallow aquifer from about 2 m down to some 20 m is known, that slowly crops out towards the surface. From geoelectrical and radar sections the upper boundary of the clayey confining bed is known. However, the deeper structures are neither visible in radar-section nor can they be interpreted by a resistivity contrast. The MRS survey was designed to cover the outcrop of the aquifer and spread towards the as aquiclude assumed direction. There, previous individual MRS measurements did already point to larger water contents than they were presumed from geoelectrics and geological conditions.

The survey was performed (Hertrich 2005) with 4 individual positions of loops with 48 m diameter, adjacent half overlapping. All sixteen possible permutations of transmitter and receiver positions have been realized. The result of various inversions is shown in Fig. 8.9 and the corresponding measured data set is shown in Fig. 8.10.

In the sounding data matrix on the right hand side the modelled data are plotted (Fig. 8.10) in comparison to the measured data. The solid lines show the data adaptation with all 16 soundings. The dashed lines are calculated from the model that has been reconstructed by using only the 4 coincident soundings. The deviation between both therefore shows the gain in model resolution by fitting all available data. The found model (Fig. 8.9 3. from bottom) represents the presumed outcropping aquifer on the lefthand side around P1. Its boundary obviously crops out at around P2 with an immediate transition into a second undulating water bearing layer towards right side. Both structures have their lower boundary at a layer of very low water content at some 20 m. Below this zone a second aquifer appears at some 40-50 m depth.

The lower boundary of the low water content zone shows a significant depression at around +20 m on the profile. The trisection of the section into a first aquifer, confining bed and second aquifer can be roughly estimated from the 2D inversion using only the coincident soundings as shown in the second model in the left hand side of the figure. A resolution of a depth variation of layer boundaries or a discrimination of structures within the layers is, however, not given. The single 1D inversions and the pseudo-2D plot (Fig. 8.9 bottom and above) do also just roughly show the structure of a low water content layer between two aquifers but does not reflect the structure of this body as it is rendered by full 2D inversion of the complete dataset.

Comparing the uppermost 30 m of the complete inversion result of the MRS survey including all available data to the inversion result of a resistivity section in Wenner configuration of 2 m basic electrode spacing allows a combined interpretation. The top layer, showing water contents close to zero is represented by resistivities of some 4000 Ωm from the left to P3 and less than 100 Ωm from P3 to the right. This corresponds to a dry sand layer at the left changing to clay material of the outcropping till layer to the right. The first aquifer, intersected at around P2 has water contents of about 35% in the left area and slightly less in the right part. However, only the left part is represented by increased resistivities from the resistivity of the surrounding clayey material. From the vanishing difference in resistivities of the right part of the aquifer a significant clay content and a corresponding reduced hydraulic permeability can be assumed here.

From the conventional resistivity investigation, the aquifer system at this location would not have been characterized correctly due to the ambiguous correlation of resistivities and lithology. Here, the definition of aquifer structures from MRS and the combined interpretation of the subsurface properties by resistivity and water content leads to a reliable estimation of aquifer characteristics and delineation of clayey materials.

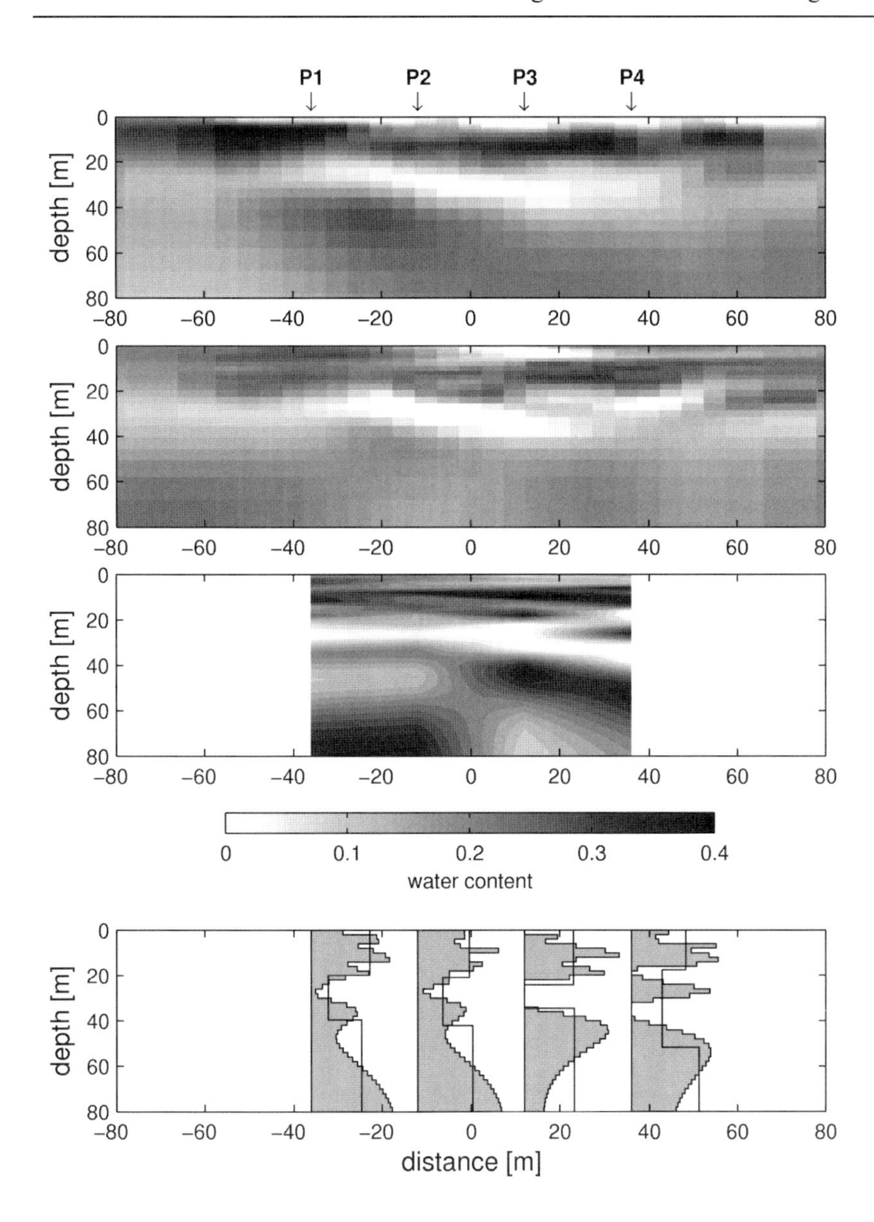

Fig. 8.9. Results of the 2D survey in Nauen by different inversion approaches. From bottom to top: 1) Inversion of coincident loop measurements (smooth and 3 layer block inversion), 2) contours of smooth inversion in 1), 3) Joint inversion of coincident loop measurements 4) Joint inversion of coincident loop and separated loop measurements

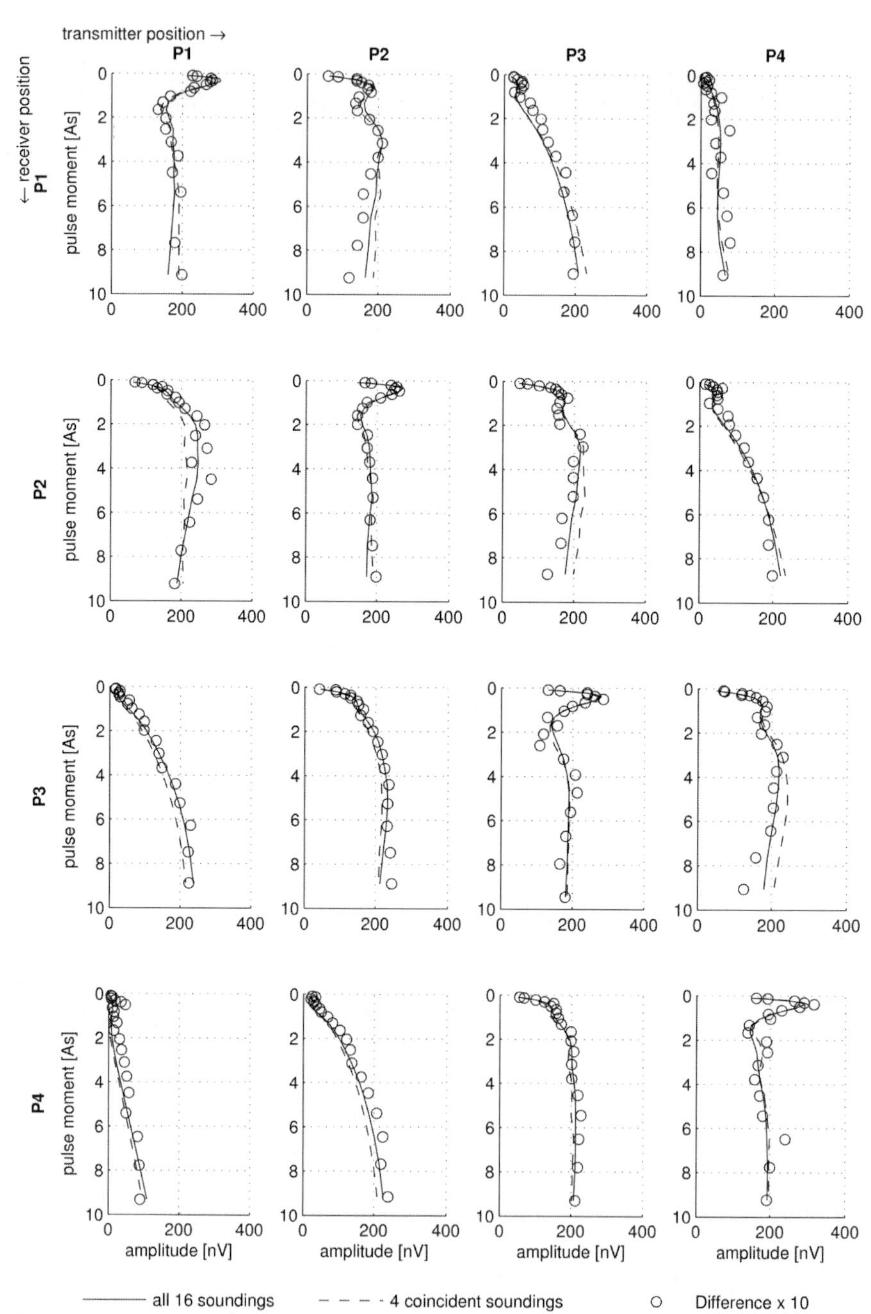

Fig. 8.10. Measured amplitude data and modelled curves for the inversion results (in Fig. 8.9) for joint inversion of coincident loops and for joint inversion of coincident and separated loops (full data)

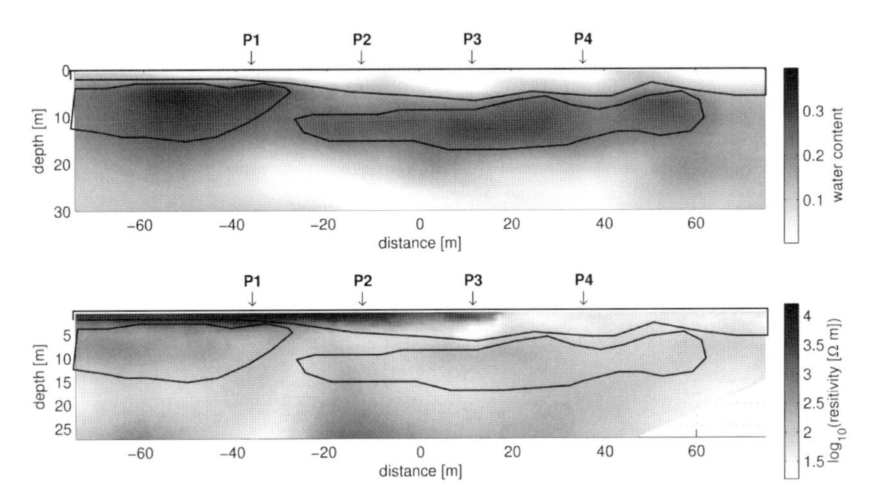

Fig. 8.11. Comparison of MRS and geoelectrics at the test site Nauen. Distinguishable structures in water content in combination with their electrical pattern allow a more profound interpretation of hydraulic properties

8.5 Current developments in MRS

The method of SNMR has just passed the experimental stage to become a powerful tool for groundwater exploration and aquifer characterization. Some further improvements are still necessary and just in work.

The most concern currently is the effect of resistivities and their appropriate inclusion into the analysis and inversion. In MRS the exciting field will be distorted and polarized considerably due to induction in the presence of conductive structures. Earlier considerations of this (Shushakov and Legchenko 1992, Shushakov 1996) have led to new improved theoretical description and numerical handling of this problem (Weichmann et al. 2002, Valla and Legchenko 2002). For an appropriate analysis of MRS the resistivity structure should always be available by which geoelectrical method ever and incorporated into the analysis. The experience shows that even moderate conductive structures may have large distortion effects.

In fact the incorporation of resistivities allows an improved modelling of the complex MRS signal, i.e. the phases, (Braun et al. 2002, 2005) which is not only useful for understanding the measured phases but is also the basis for a successful inversion of complex signals to yield the resistivity information directly from SNMR measurements.

In the analysis of SNMR generally the relaxation is assumed to be mono-exponential. Even if individual layers show a mono-exponential decay, the integration results in a multi-exponential decay in the measured signal. The most comprehensive way taking account for this is to consider a decay time spectra in the data as well as in the inversion as shown in Chap. 8.3. This leads to the pore size distribution which is a new information and allows improved estimation of hydraulic conductivities. The determination of water contents will also be significantly improved as the initial amplitudes are much better fitted using a decay spectra approach.

Currently SNMR is carried out mainly with a 1D working scheme using coinciding loops. The 2D surveying as shown in Chap. 8.4 is not standard yet. The errors might be very large by neglecting the 2D or even 3D geometry of the structures (Warsa et al. 2002, Hertrich et al. 2005) which have to be considered in the analysis and inversion in the future. Measurement layouts are to be modified to meet the multi-D conditions, which is easier to accomplish for nonconductive structures. In multi-D structures the actual difficulty is the numerical incorporation of the electromagnetic modelling for the exciting field even when the electrical data is available. The full 2D inversion for decay times are not available yet.

As in any geophysical measurement also with SNMR the inversion plays a key role by interpreting the data. The limits of inversion and also the imposed conditions in terms of geometrical boundary conditions as well as differences in the basic physical model used may lead to considerable differences. The inversion of SNMR data may be ambiguous, since not only different regularizations in the inversion impose a certain degree of smoothness upon the distribution of water content (Legchenko and Shushakov 1998, Yaramanci et al. 1998, Mohnke and Yaramanci 1999) but also the number of layers and the size of layers forced in the inversion may considerably influence the results. The rms-error is not necessarily a sufficient measure for assessing the quality of the fit of a model to the observed data. The most recent research suggests that a layer modelling with free boundaries avoids the problems associated with regularization and takes into account the blocky character of the structure where appropriate (Mohnke and Yaramanci 2002). Similar consideration are needed for 2D inversions considering the imposed cell sizes and shapes as well as smoothing constraints.

Further improvement in the inversion can be achieved in case of geoelectrical measurements are available and they can be incorporated into a joint inversion with SNMR. Examples of joint inversion of SNMR with Vertical Electrical Sounding show considerable improvement in the detectability and geometry of the aquifers and allows also by utilizing of appropriate petrophysical models the separation of mobile and adhesive water (Hertrich and Yaramanci 2002).

Finally the importance of the SNMR method lies in its ability to detect water directly and allowing reliable estimation of mobile water content and hydraulic conductivity. In this respect it is unique, since all other geophysical methods yield estimates, if ever, indirectly via resistivity, induced polarization, dielectric permittivity or seismic velocity. Using SNMR in combination with other geophysical methods not only allows direct assessment but is complementary to the information yielded by other geophysical methods.

8.6 References

Baltassat JM, Legchenko A, Ambroise B, Mathieu F, Lachassagne P, Wyns, R (2005) Magnetic resonance sounding (MRS) and resistivity characterisation of a mountain hard rock aquifer: the Ringelbach Catchment, Vosges Massif, France. Near Surface Geophysics 3 (in print)

Beauce A, Bernard J, Legchenko A, Valla P (1996) Une nouvelle méthode géophysique pour les études hydrogéologiques: l'application de la résonance magnétique nucléaire. Hydrogéologie 1:71-77

Braun M, Yaramanci U (2003) Inversion of Surface-NMR signals using complex kernels. Proceedings of the 9th European Meeting on Environmental and Engineering Geophysics, Prague

Braun M, Hertrich M, Yaramanci U (2002) Modeling of phases in Surface NMR. Proceedings of 8th European Meeting on Environmental and Engineering Geophysics, Aveiro

Braun M, Hertrich M, Yaramanci U (2005) Complex inversion of MRS data. Near Surface Geophysics 3 (in print)

Dippel S, Golden H (2003) MRS and TEM for shallow aquifer definition at Phosphate Hill, NW Queensland, Australia. Proceedings of the 2nd international workshop on MRS, Orleans/France

Goldman M, Rabinovich B, Rabinovich M, Gilad D, Gev I, Schirov M. (1994) Application of the integrated NMR-TDEM method in groundwater exploration in Israel. Journal of Applied Geophysics 31:27-52

Hertrich M (2005) Magnetic Resonance Sounding with separated transmitter and receiver loops for the investigation of 2D water content distributions. PhD Thesis, Technical University Berlin

Hertrich M, Yaramanci U (2002) Joint inversion of Surface Nuclear Magnetic Resonance and Vertical Electrical Sounding. Journal of Applied Geophysics 50:179-191

Hertrich M, Yaramanci U (2003) Complex transient Spin dynamic in MRS appli-

Hertrich M, Yaramanci U (2005) Magnetic Resonance soundings with separated transmitter and receiver antennas. Near Surface Geophysics 3 (in print)

Kenyon WE (1992) Nuclear magnetic resonance as a petrophysical measurement. Int. Journal of Radiat. Appl. Instrum, Part E, Nuclear Geophysics 6:153-171

Lange G, Mohnke O, Grisseman C (2005) SNMR measurements in Thailand to investigate low porosity aquifers. Near Surface Geophysics 3 (in print)

Legchenko A (2004) Magnetic Resonance Sounding: Enhanced Modeling of a Phase Shift. Applied Magnetic Resonance 25:621–636

Legchenko A, Shushakov O (1998) Inversion of surface NMR data. Geophysics 63:75–84

Legchenko AV, Semenov AG, Schirov MD (1990) A device for measurement of subsurface water saturated layers parameters (in Russian). USSR Patent 1540515

Legchenko AV, Shushakov OA, Perrin JA, Portselan AA (1995) Noninvasive NMR study of subsurface aquifers in France. Proceedings of 65th Annual Meeting of Society of Exploration Geophysicists:365-367

Levitt MH (2002) Spin dynamics - Basics of Nuclear Magnetic Resonance. John Wiley & Sons, LTD

Lieblich DA, Legchenko A, Haeni FP, Portselan AA (1994) Surface nuclear magnetic resonance experiments to detect subsurface water at Haddam Meadows, Connecticut. Proceedings of the Symposium on the Application of Geophysics to Engineering and Environmental Problems, Boston, 2:717-736

Lubczynski M, Roy J (2003) Hydrogeological interpretation and potential of the new magnetic resonance sounding (MRS) method. Journal of Hydrology, 283:19-40

Meju MA, Denton P, Fenning P (2002) Surface NMR sounding and inversion to detect groundwater in key aquifers in England: comparisons with VES–TEM methods. Journal of Applied Geophysics 50:95-111

Mohnke O, Yaramanci U (2005) Interpretation of relaxation signals using multi-exponential inversion. Near Surface Geophysics 3 (in print)

Mohnke O, Yaramanci U (1999) A new inversion scheme for surface NMR amplitudes using simulated annealing. Proceedings of 60th Conference of European Association of Geoscientitists &Engineers:2-27

Plata J, Rubio F (2002) MRS experiments in a noisy area of a detrital aquifer in the south of Spain. Journal of Applied Geophysics 50:83-94

Semenov AG, Burshtein AI, Pusep AY, Schirov MD (1988) A device for measurement of underground mineral parameters (in Russian). USSR Patent 1079063

Shirov M, Legchenko AV, Creer G (1991) A new direct non-invasive groundwater detection technology for Australia. Exploration Geophysics 22:333-338

Shushakov OA, Fomenko VM, Yashchuk VI, Krivosheev AS, Fukushima E, Altobelli SA, Kuskovsky VS (2005) Hydrocarbon contamination of aquifers by SNMR detection. Near Surface Geophysics 3 (in print)

Shushakov OA (1996) Groundwater NMR in conductive water. Geophysics 61:998-1006

Shushakov OA, Legchenko AV (1992) Calculation of the proton magnetic resonance signal from groundwater considering the electroconductivity of the medium (in Russian). Russian Acad. of Sci., Inst. of Chem. Kinetics and Combustion, Novosibirsk, issue 36:1-26

Supper R, Jochum B, Hübl G, Römer A, Arndt R (2002) SNMR test measurements in Austria, Journal of Applied Geophysics 50:113-121

Valla P, Legchenko A (2002) One-dimensional modelling for proton magnetic resonance sounding measurements over an electrically conductive medium. Journal of Applied Geophysics 50: 217-229

Vouillamoz JM, .Descloitres M, Toe G, Legchenko A (2005) Characterization of crystalline basement aquifers with MRS: comparison with boreholes and pumping tests data in Burkina Faso. Near Surface Geophysics 3 (in print)

Vouillamoz JM, Descloitres M, Bernard J, Fourcassier P, Romagny L (2002) Application of integrated magnetic resonance sounding and resistivity methods for borehole implementation - A case study in Cambodia. Journal of Applied Geophysics 50:67-81

Warsa W, Mohnke O, Yaramanci U (2002) 3-D modelling of Surface-NMR amplitudes and decay times. In: Water Resources and Environment Resaerch, ICWRER 2002, Eigenverlag des Forums für Abfallwirtschaft und Altlasten e.V., Dresden:209-212

Weichman PB, Lavely EM, Ritzwoller MH (2000) Theory of surface nuclear magnetic resonance with applications to geophysical imaging problems. Physical Review E 62 (1, Part B):1290–1312

Weichman PB, Lun DR, Ritzwoller MH, Lavely EM (2002) Study of surface nuclear magnetic resonance inverse problems. Journal of Applied Geophysics 50:129-147

Yaramanci U, Lange G, Knödel K (1998) Effects of regularisation in the inversion of Surface NMR measurements. Proceedings of 60th Conference of European Association of Geoscientists&Engineers:10–18

Yaramanci U, Lange G, Knödel K (1999) Surface NMR within a geophysical study of the aquifer at Haldensleben (Germany). Geophysical Prospecting 47:923-943

Yaramanci U, Kemna A, Vereecken H (2005) Emerging Technologies in Hydrogeophysics. In: Rubin Y, Hubbard S (ed) Hydrogeophysics. Springer Verlag

Yaramanci U, Lange G, Hertrich M (2002) Aquifer characterisation using Surface NMR jointly with other geophysical techniques at the Nauen/Berlin test site. Journal of Applied Geophysics 50:47-65

9 Magnetic, geothermal, and radioactivity methods

Kord Ernstson

9.1 Magnetic method

The magnetic method is a versatile, easy-to-operate geophysical tool applicable to quite different subsurface exploration problems. It has much in common with the gravity method both theoretically and with regard to field work.

9.1.1 Basic principles

The earth's large-scale magnetic field is superimposed by small-scale *magnetic anomalies* related with magnetized rocks (Fig. 9.1). *Magnetization* is the parameter corresponding to density in the gravity method. Different from density, however, magnetization is a vector quantity which, simply speaking, is related with the concept of north and south pole of a magnet. The vector of magnetization may have arbitrary orientation in a rock, and that is why geometrically identical causative bodies can show quite different magnetic anomalies. There are two situations to be taken into account: the location and orientation of a causative body in various latitudes, and the *remanent magnetization*.

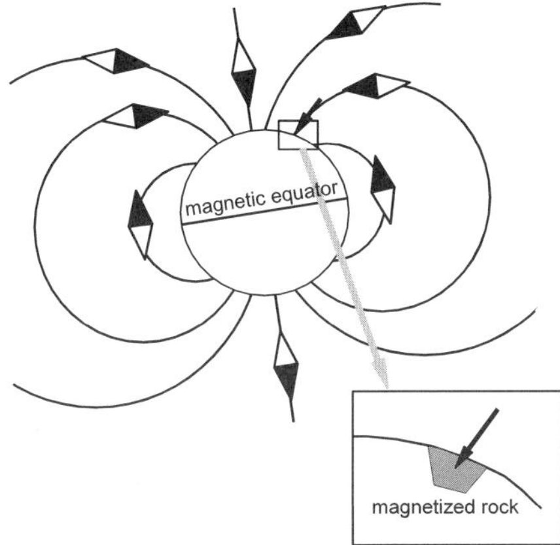

Fig. 9.1. The earth magnetic field and the magnetization of rocks

The first case is related with induced magnetization M_i that is characteristic of all rocks, because they are located in the inducing earth magnetic field F (Fig. 9.1). The vector $\mathbf{M_i}$ is in general parallel to the vector of the inducing field **F**, and it is

$$M_i = k \cdot F \quad (9.1)$$

k is the dimension-less *magnetic susceptibility* that can be positive or negative, and the unit of both magnetization and magnetizing field is the ampere/meter (A/m). Susceptibility as a magnetic petrophysical parameter is more common than induced magnetization because k is independent of the position of the measurement. M_i changes corresponding to the configuration and strength of the earth magnetic field which approximately is a dipole field (Fig. 9.1).

The relations of the various parameters and the consequences of varying induced magnetization are explained in Figs. 9.2 and 9.3. In Fig. 9.2, a body striking W - E is shown exposed in Central European latitudes to an inducing field strength of F = 38.2 A/m. By Eq. 9.1, the assumed susceptibility k = 0.013 is related with an induced magnetization of M_i = 0.5 A/m.

An instrument for measuring F (a magnetometer, see below) will read the related *magnetic induction* B.

$$B = \mu_0 \cdot F \quad (9.2)$$

μ_0 is the *magnetic permeability*, $\mu_0 = 4\pi * 10^{-7}$ T/Am^{-1}, and T(esla) the unit of the magnetic induction. Eq. 9.2 relates the magnetizing total field H = 38.2 A/m to the instrument reading of the total intensity of the undisturbed field, B = 48,000 nT, 1 nT (nanotesla) = 10^{-9} T (tesla). In practice, the nanotesla (nT) is the working unit for the magnetic field, its vector components (vertical intensity, horizontal intensity) and the measured magnetic anomalies (Fig. 9.2). Frequently, T denotes the total intensity not to be confused with the tesla (T).

In Fig. 9.3, the strong influence of the geometry of the magnetizing field becomes evident. Apart from the anomaly curve for the situation in Fig. 9.2, model curves have been calculated for the body exposed to the same field, however striking N - S (instead of W - E), and for the identical body striking W - E but located at equatorial latitudes. Significantly varying anomalies are the result. Different from gravity anomalies of causative bodies, magnetic anomalies are in general composed of combined positive and negative parts, which can again be attributed to the simple north pole - south pole concept of magnetization. Also different from gravity anomalies, the largest magnetic anomalies are concentrated on the lateral boundaries of a magnetized body (Fig. 9.4).

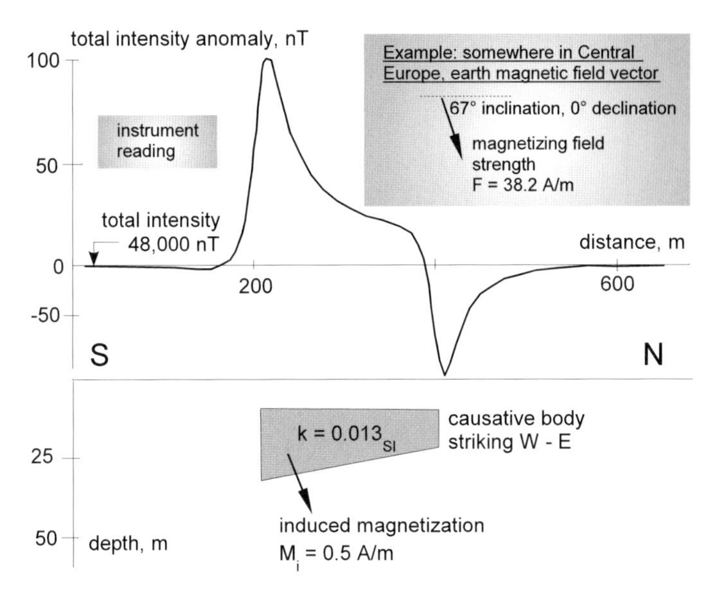

Fig. 9.2. Calculated total-intensity anomaly of a magnetized causative body located in Central European latitudes, and relations between magnetizing field and instrument readings. For explanation see text

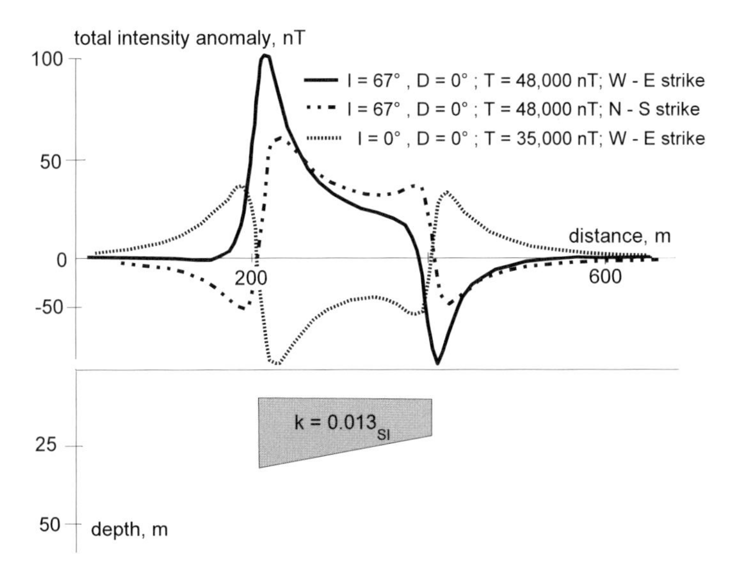

Fig. 9.3. Calculated total-intensity anomalies for the same but variously magnetized causative body. Inclination I = 67° and T = 48,000 nT total intensity of the undisturbed field in Central European latitudes, and I = 0° and T = 35,000 nT in equatorial latitudes. Declination in each case D = 0°

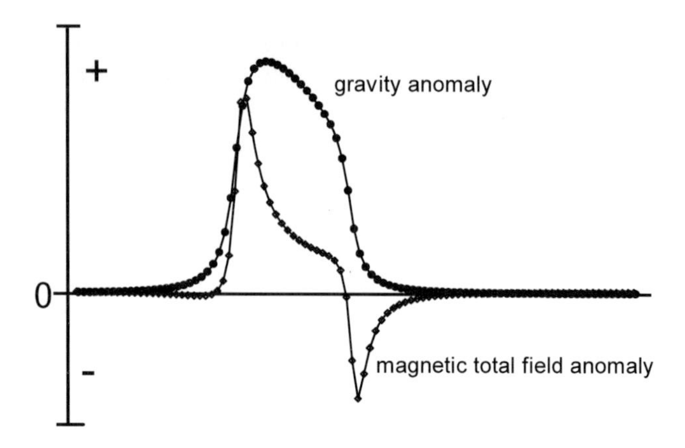

Fig. 9.4. Comparison of the gravity and magnetic anomalies over the causative body from Fig. 9.2

Remanent magnetization. Frequently, susceptibility describes the magnetic behavior of a rock only incompletely. Rocks containing minerals like magnetite may acquire a permanent or remanent magnetization by cooling from high temperatures, by chemical and several other processes, if they are exposed to a magnetizing field, normally the earth magnetic field. On the deposition of fine-grained sediments, magnetic particles may experience an orientation with regard to the earth magnetic field leading to a detrital remanent magnetization. The vector of the remanent magnetization fixed to the rock can be very stable and survive long geological times. On a later change of the earth magnetic field or a change of the spatial position of the rock by, e.g., tectonics, induced and remanent magnetizations may exhibit quite different orientations. They vectorially contribute to an effective magnetization responsible for the measured magnetic anomaly (Fig. 9.5). The remanent magnetization may be considerably higher than the induced magnetization, and ignoring it with the interpretation of magnetic anomalies basic errors can be the result.

9.1.2 Magnetic properties of rocks.

Magnetism of rocks is related with the magnetism of the rock-forming minerals. Diamagnetic minerals like quartz and calcite have negative susceptibilities in the order of 10^{-5}. Minerals like feldspars and micas are paramagnetic and have higher (positive) susceptibilities (10^{-4} - 10^{-2}). The positive susceptibility of ferrimagnetic minerals like magnetite (k ~ 1 - 10) and

Very roughly, the susceptibility of rocks in an ascending order may be addressed as follows: (k ~ 0.0001) sedimentary rocks - metamorphic rocks (0.001) acid volcanic and plutonic rocks - basic plutonic rocks (0.01) basic volcanic rocks (0.1).

Strong remanent magnetization is abundantly observed with young volcanic rocks, while in sedimentary and metamorphic rocks the remanent magnetization is in general much lower than the induced magnetization.

Water is diamagnetic ($k = -9.05 * 10^{-6}$) and has no importance for the actual magnetization of a rock. However, playing an important role in the alteration of rocks and being a carrier of iron-bearing and generally chemically active solutions, water can change the magnetization of rocks during geological times. Both an intensity increase and an intensity decrease are possible. Magnetization changes of rocks by contaminated groundwater may be addressed in environmental geophysical surveys.

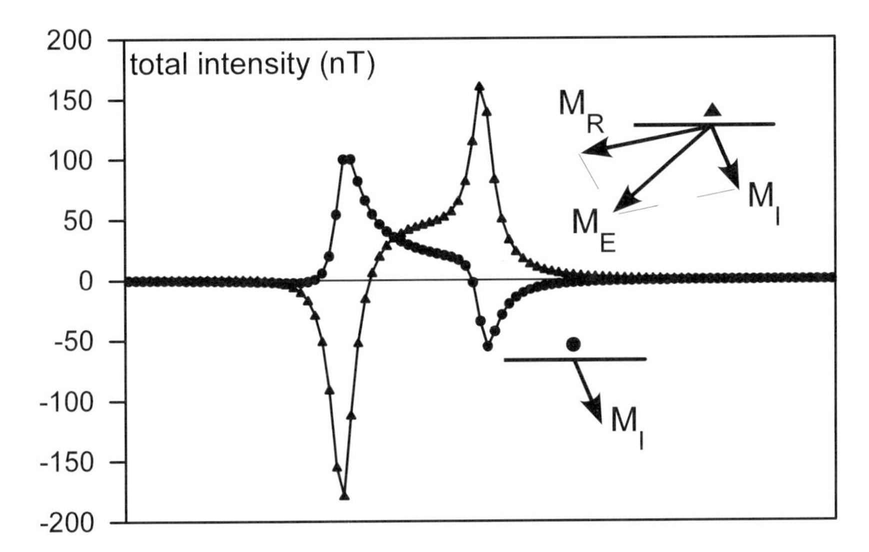

Fig. 9.5. Calculated total field anomaly curves over the causative body from Fig. 9.2 for both pure induced magnetization M_I and an effective magnetization M_E resulting from vector addition of M_I and remanent magnetization M_R. For reasons of simplicity, the vector of the remanent magnetization is orientated perpendicular to the strike of the causative body

9.1.3 Field equipments and procedures

Magnetic field measurements are carried out with magnetometers. Earlier purely mechanical magnetometers have completely been replaced by electronic instruments based on various physical processes. Most common are *proton precession magnetometers* (proton magnetometers, in short) composed of a portable sensor and an electronic unit for display and storage of the data. Proton magnetometers measure the total intensity of the earth magnetic field vector at a resolution of 0.1 - 1 nT. For *fluxgate magnetometers*, a sensor axis is defined for measuring the intensity of the field vector parallel to the sensor axis. Thus the vertical or various horizontal components can be addressed.

Cesium and rubidium vapor magnetometers define the group of so-called *optically pumped magnetometers* that measure the total intensity at a resolution of 0.1 - 0.001 nT and at an absolute precision of 1-0.2 nT. Because of geological and man-made noise and the need for exact spatial positioning of the sensor, geological surveys can rarely take advantage of this extreme precision.

For engineering and environmental geophysical surveys of near-surface layers and causative bodies, instruments for the measurement of the vertical gradient of the total intensity or of the magnetic vertical component have been developed. Because there is no possibility to directly measure the field gradient, the gradiometers for proton, fluxgate and optically pumped magnetometer measurements are composed of two sensors mounted vertically one above the other, and the measurement of the gradient is approximately replaced by the measurement of the field difference. Gradiometers can advantageously be used for a continuous digital data acquisition, they allow a better resolution of magnetic structures, and because of the registration of the field difference, there is no need to consider the time variations of the earth magnetic field.

Small hand-held instruments enable the measurement of the susceptibilities of exposed rocks and rock samples, and induced and remanent magnetization can be estimated from magnetometer readings if a roughly spherical sample is rotated near the sensor.

Similar to gravity surveys, magnetic measurements are performed in grids or on profiles perpendicular to elongated, so-called two-dimensional structures (see, e.g., the examples in Figs. 9.2 –9.5). Station spacing depends on the depth of investigation but is frequently distinctly smaller than in gravimetry, when near-surface and small-scale structures are considered. Station spacing may be especially important with technical and environmental magnetic surveys in areas where noise level from superficial magnetic material is considerable, or when short-wavelength magnetic anomalies are the target of the survey (e.g., anomalies related with waste deposits). In these cases, a station spacing being too large will produce spatial aliasing implying the measurement of purely fictitious anomalies and completely misleading modeling and interpretation (Fig. 9.6). The spatial-aliasing effect is a general problem with geophysical measurements especially when station spacing is a factor of cost.

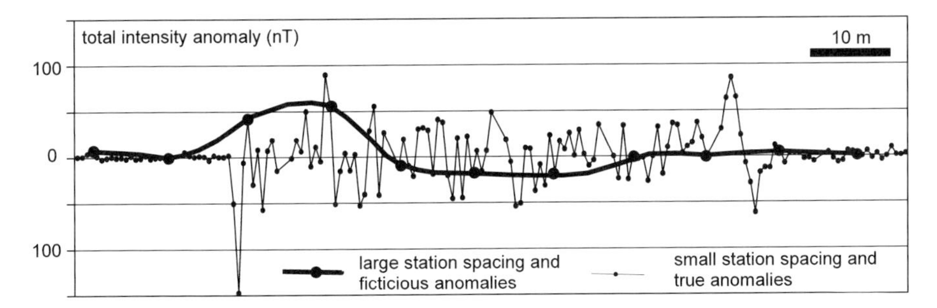

Fig. 9.6. Spatial aliasing from large station spacing, and the production of fictitious anomalies

As with other time-varying geophysical fields, repeated magnetometer readings at a base station have to be performed to record periodic and non-periodic magnetic fluctuations related with ionospheric processes. Since they are quite irregular and cannot be calculated, a second, permanently recording instrument at the base station is the best solution especially when only faint anomalies are to be measured.

Because of the high precision of magnetometers, a sufficient distance to all iron objects (fences, pylons, rails, cars, supply grids, reinforced concrete, personal utensils like coins, penknife, ballpoint, etc) is advisable. Bricks and generally pottery may be strongly magnetic, and roads and paths may be graveled with magnetic material.

Different from gravimetry, corrections of altitude, latitude, and topography can in most cases be omitted in small-sized areas of investigation typical of environmental and hydrogeological surveys. All in all, magnetic measurements are easily and rapidly done, and in a terrain being not too difficult, an experienced observer can take readings at several hundred stations per day. With a continuously reading gradiometer, several kilometers profile length per day are possible.

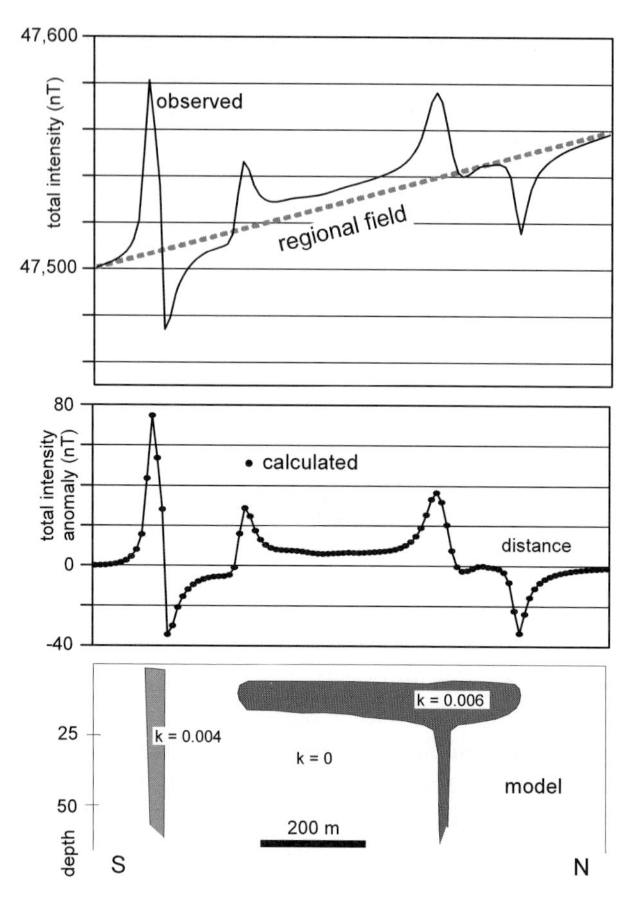

Fig. 9.7. Principle of magnetic data evaluation and interpretation. Inducing field data are 65° inclination and 0° declination, and the modeled 2-D bodies are striking W - E. k = susceptibility of the model bodies

9.1.4 Data evaluation and interpretation

Magnetic data are displayed as isomagnetic charts or on profiles. For reasons of simplicity, the basic scheme of magnetic data evaluation and interpretation is shown in Fig. 9.7 for a profile plot. In general, the observed data result from the superposition of the earth's normal field and the fields of one or more causative bodies of different shapes and depths of burial. In the simplest case, the trend of a regional field is roughly linear, and the subtraction centers positive and negative anomalies to a zero axis (Fig. 9.7, upper and middle). The computer modeling of magnetic anomalies may consider bodies of arbitrary shape and arbitrary magnetization parameters, but frequently a two-dimensional approach is sufficient (Fig. 9.7, middle and lower). Various programs are commercially available and may even be downloaded from the internet. Computer modeling is the first step in the interpretation of a magnetic survey first providing physical models as a base for geological modeling. Geological modeling may consider abrupt magnetization changes related with a fault, or magnetization distributions related with facies. Correspondingly, the magnetic anomalies from Fig. 9.7 may be interpreted by intrusive bodies in a host rock of assumed zero susceptibility.

In practice, modeling and interpretation may be more complex compared with the example from Fig. 9.7. The separation of regional and local anomalies my be difficult and in some cases basically ambiguous. In the case of a host rock being also magnetic, susceptibility differences and, if necessary, deviating vectors of remanence must be introduced to the modeling. Frequently, a filtering of measured data can help in the interpretation of magnetic anomalies. Filtering is based on the wavy character of profile plots of magnetic data (see, e.g., Fig. 9.6) and on the concept of anomaly wavelength. From Fig. 9.8 it is evident that with increasing depth of a magnetized body the related anomaly becomes increasingly broader, or, in other words, the wavelength of the anomaly becomes increasingly larger. A common definition of the wavelength of geophysical anomalies is the half-width taken from the anomaly as shown in Fig. 9.8. Half-width is a useful quantity to estimate the depth of a causative body, and here filtering and the computation of filters is concerned. Low-pass filters are applied to suppress short-wavelength anomalies while, to the contrary, they are accentuated by high-pass filters. As wavelengths are related with the depths of causative bodies, wavelength filtering may enhance or suppress the magnetic signature of deep-seated or near-surface bodies and structures in an isomagnetic map or on a profile. A large variety of mathematical filter operators exist from simple smoothing of data up to potential field calculations like analytical upward and downward continuation and the computation of vertical and horizontal gradient fields.

In Fig. 9.9, the isomagnetic map of a fluxgate gradiometer ground survey is shown displaying lots of small-scale anomalies related with scattered waste. A low-pass filtering is achieved by computing the analytical upward continuation simulating the gradiometer measurement at four meter height above ground. The contours of the horizontal gradient denote the strongest lateral field variations and thus reflect a kind of high-pass filtering. As can be seen from Fig. 9.8 (to the right), the anomaly is steepest for the uppermost model body.

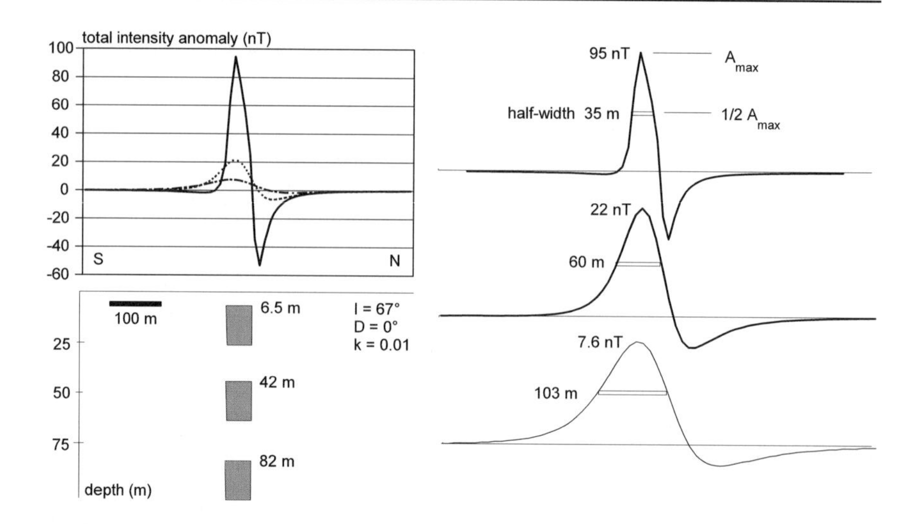

Fig. 9.8. Relation between anomaly width and depth of body, and the definition of anomaly wavelength by half-width

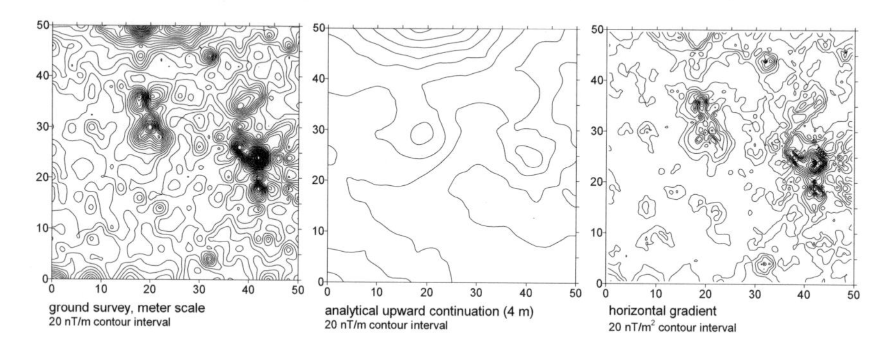

Fig. 9.9. Fluxgate gradiometer ground measurements and derived fields. Environmental-geophysical survey of waste scattered over karstified dolomites

Like gravity, geoelectrical and other geophysical methods, also magnetic measurements and their interpretation suffer from the principle of equivalence stating that a measured set of data may in general be modeled by quite different magnetized bodies. A simple example for a two-dimensional dike-like body striking W - E is shown in Fig. 9.10. The calculated total intensity anomaly is practically identical, if the width of the dike is doubled while halving the susceptibility. Similarly, the dip of a dike remains undetermined as long as the orientation of the vector of magnetization is unknown.

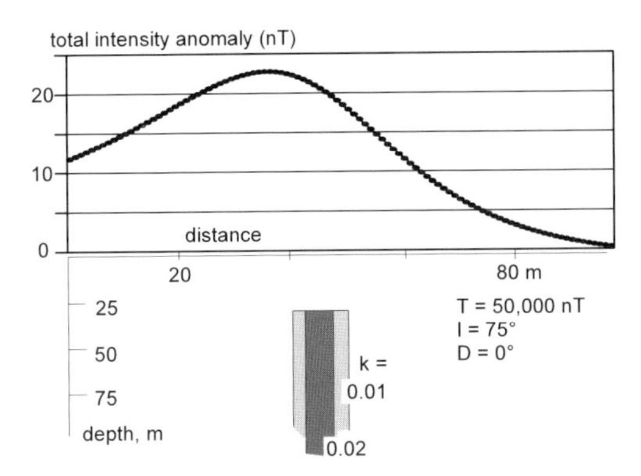

Fig. 9.10. Equivalence for the product of width times susceptibility of magnetized dike-like bodies

The modeled causative bodies were so far assumed to have a homogeneous magnetization. Practically, such bodies in general do not exist. Inhomogeneous magnetization means that the smooth magnetic anomaly for a homogeneously magnetized body is superimposed by small-scale fluctuations related with a magnetic texture of the rock. The magnetic texture may reflect rock textures originating from, e.g., flow, depositional and diagenetic processes. Hence, magnetic texture measured in the form of preferential wavelengths, amplitudes, and strike directions may also magnetically characterize a rock and can advantageously be used in the interpretation of field measurements as exemplified in Fig. 9.11 for a model of synthetic, distinctly different magnetic textures.

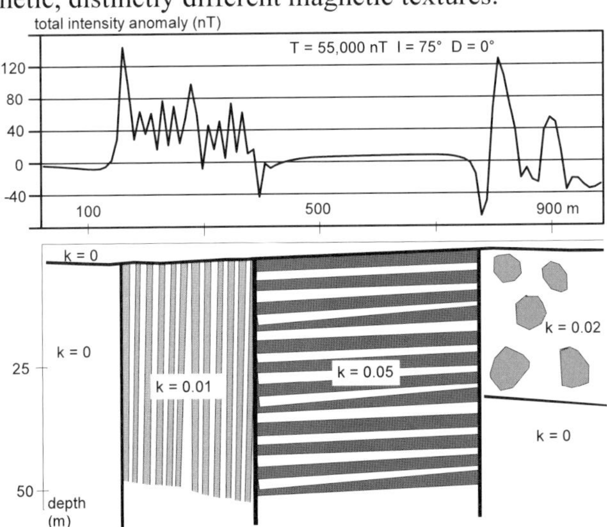

Fig. 9.11. Synthetic magnetic textures of larger rock units and calculated total intensity anomalies

9.2 Geothermal method

In applied geophysics, geothermics has never played a major role. This may be explained by the difficulty to obtain lots of data as a base for model calculations and quantitative interpretations. With the increasing interest in the exploitation of geothermal energy and in underground heat and cold storage, however, geothermics has more and more attracted attention. Since groundwater and its migration play an important role for the underground temperature field, the hydrogeological aspects of geothermics are obvious.

The four basic subjects of geothermics are temperature, heat, heat storage, and heat transfer. *Temperature* is the most important measuring quantity with unit Kelvin [K] related with the unit Celsius [°C] by

$$x\,[K] = (x-273)\,[°C] \quad (9.3)$$

The relations between the units Celsius and Fahrenheit are given by

$$x\,[°F] = ((x - 32) \cdot 5/9)\,[°C] \qquad x\,[°C] = (5/9\,x + 32)\,[°F] \quad (9.4)$$

In a medium, a temperature field is defined by a temperature distribution that may be temporally constant (stationary geothermal processes) or time-varying (instationary geothermal processes). Temperature fields are plotted as isotherm maps (Fig. 9.12).

Heat is a form of energy with unit joule (J) or watt seconds (Ws). Heat (energy) can be *stored* by a body, hence the body can take heat and emit it corresponding with a temperature change. The amount of taken and emitted heat Q is proportional to the temperature difference ΔT:

$$Q = m \cdot c \cdot \Delta T \quad (9.5)$$

where m is the mass of the body and c the specific heat. The specific heat c is a material constant with unit J/(K * kg). The heat capacity is defined by W = m * c.

Water has a large specific heat (c_W = 4186 J/[K * kg]) and, therefore, heat capacity, while dry rocks have specific heats of roughly 800 J/(K * kg) only. The specific heat c_R of water-containing rocks is calculated from

$$c_R = (W_d + W_w)/m \quad (9.6)$$

W_d heat capacity of dry rock
W_w heat capacity of the water content
m total mass.

The specific heat of rocks is important when heat and cold storage in the underground is considered.

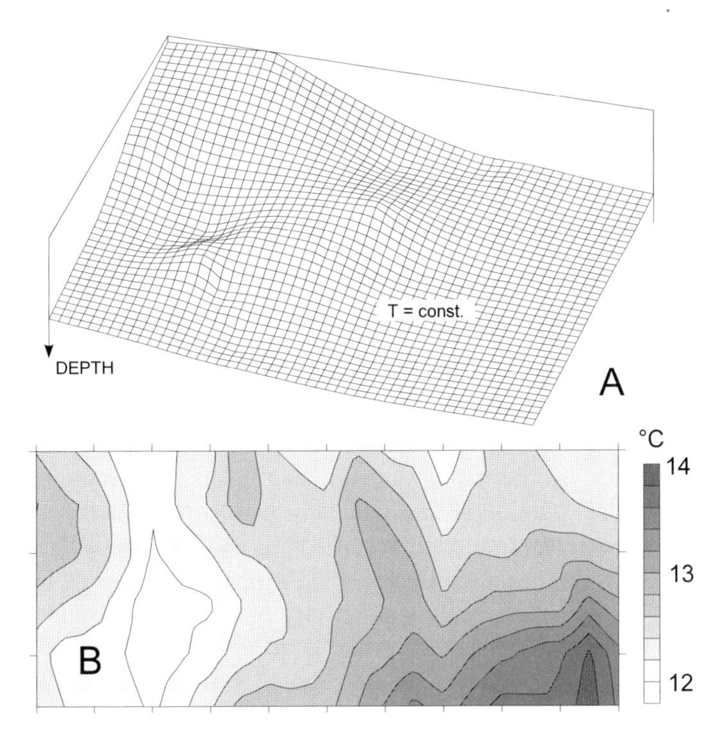

Fig. 9.12. Temperature fields as shown by isotherms. A: Isotherm of spatially constant temperature. B: Isotherms at a selected depth

Heat transfer takes place by heat conduction, heat convection and heat radiation. *Heat conduction* occurs between two points of different temperature and is defined by

$$Q = \lambda \cdot t \cdot \Delta T \cdot F / L \tag{9.7}$$

if the temperature difference ΔT is kept constant between the two points of separation L (Fig. 9.13A). Then, Q is the constant amount of heat flowing during the time t through a cross-section F. The proportionality factorλ is called *thermal conductivity* (unit W/K*m). Converting (9.7) into

$$Q / t \cdot F = \lambda \cdot \Delta T / L = q \tag{9.8}$$

q is the heat flow density (heat flow in short) with unit watt/m² (W/m^2).

In an arbitrary isotropic medium, the heat flow is a vector with orientation perpendicular to the isotherms (Fig. 9.13), and instead of the temperature drop $\Delta T/L$ the temperature gradient grad T has to be considered.

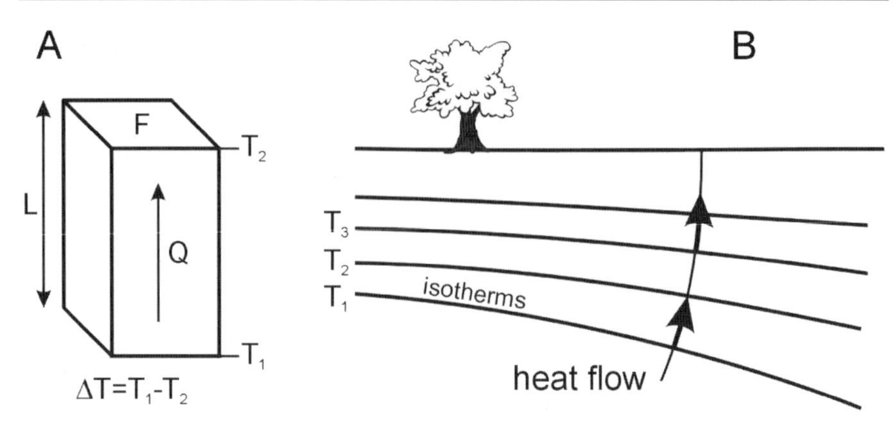

Fig. 9.13. Heat transfer and heat flow

For stationary geothermal processes, thermal conductivity is the decisive petrophysical parameter that relates heat flow to the temperature drop (9.8) or temperature gradient, respectively. Values between 1 and 5 W/K∗m are most common among rocks. Compared with, e.g., metals ($\lambda \sim$ 500 ... 5000 W/K∗m) , thermal conductivity of rocks is low, and it strongly depends on porosity, water content and texture. Water saturated sand, e.g., has a mean thermal conductivity of about 2.4 W/K∗m (VDI 1998). The thermal conductivities of water and air are $\lambda_W \sim 0.6$ W/K∗m and $\lambda_A \sim 0.03$ W/K∗m (at 293 K).

In geothermics, instationary thermal processes implying time-varying temperature fields frequently attract far more attention. Often, periodical temperature variations like heating and cooling of the ground by diurnal and annual solar radiation are considered.

Thermal diffusivity is the significant petrophysical parameter that describes instationary geothermal processes as for example the propagation of temperature changes or the balance of temperatures. Thermal diffusivity a (unit m^2/s) depends on thermal conductivity λ, specific heat c and density ρ, and it is

$$a = \frac{\lambda}{c \cdot \rho} \tag{9.9}$$

In many geothermal processes heat transfer by *heat convection* is far more important than by heat conduction. Heat convection may result from temperature-related density variations or from pressure gradients in both cases coupled to a mass transfer. Hence, temperature may be an important quantity with regard to all aspects of groundwater propagation.

Heat radiation is emitted from every body in the form of electromagnetic waves preferentially in the infrared domain. For practical geophysical purposes, heat transfer by heat radiation is unimportant. But depending on the temperature of a radiating body, heat radiation enables remote temperature measurements.

9.2.1 The underground temperature field

As is well known from boreholes and mining, temperature increases with increasing depth, and the gradient is roughly 0.03 K/m in many regions of the world. According to Eq. 9.8, the temperature gradient is related with a heat transfer towards the earth's surface, which in geothermics is called terrestrial heat flow as the vertical component of the heat flow density. The origin of the heat flow is disputed, but heat production related with radioactive decay in rocks probably plays an important role.

On average, the terrestrial heat flow is $q_E = 0.063$ W/m², which is small compared with the solar radiation energy J_E brought to the earth's surface: $J_E = 625$ W/m². Consequently, the roughly periodically acting annual and diurnal solar radiation has a considerable influence on all studies of the near-surface heat flow and temperature fields.

Different from the stationary heat flow, the solar radiation causes an instationary process of heat conduction related with thermal diffusivity as the petrophysical parameter. Similar to the propagation of electromagnetic waves, the period of the temperature variation basically determines the propagation in the ground, and a phase shift and an exponential amplitude decrease can be observed. Short-period temperature variations have a small depth penetration.

Details of the temperature-wave propagation can be computed. Thus, for normal soil conditions and climates a midday temperature maximum at the surface has propagated down to a depth of about 0.5 m at midnight, where the amplitude has decreased to only 5% of the surface amplitude. Likewise, the summer temperature maximum of the annual "wave" propagates down to a depth of roughly 10 m in winter time, where an amplitude decrease down to about 5% of the surface amplitude is observed. Obviously, these conditions are of basic importance for near-surface temperature measurements.

9.2.2 Field procedures

Within the frame of hydrogeology, heat flow, thermal conductivity and diffusivity measurements are presently of low importance, and thus temperature measurements and thermometers are addressed here only. Resistance thermometers based on temperature-sensitive metallic and semiconductor resistors (thermistors) with a resolution of 0.01 K are most common in temperature probes that may be connected to a borehole cable.

For infrared surface measurements, receivers for thermal radiation have been constructed using thermocouples or bolometric detectors. High-sensitive photo-detectors are used in temperature scanners and thermal imaging cameras that must be operated at very low temperatures of some 10 K.

While geothermal infrared measurements are generally performed at the earth's surface, thermometer measurements are carried out in shallow boreholes (1 - 4 m deep) to escape the diurnal temperature wave at least. Long-lasting periods of heat and cold may cause temperature waves of increased depth penetration, and data sampled over a longer time period in an area may not immediately be compared. Temperature recordings at a base station can serve for corrections. There are many more factors influencing temperature measurements in near-surface layers such as different vegetation, morphology, shadows and underground water content. Drilling of even shallow boreholes for thermometer emplacement may disturb the natural temperature field for half an hour or more.

Stations for temperature measurements can be arranged in a grid or on profiles perpendicular to the strike of two-dimensional geological structures. Station spacing depends on the depth of the expected anomalous geothermal zone and may be of the order of meters with hydrogeological campaigns.

Reasonably, infrared measurements are preferentially done during the night. Since, however, the temperature variations continue, the survey should be performed within a short time. Obviously, the influence of the surface properties, vegetation and moisture is much larger compared with shallow borehole measurements. Helicopter surveys enable infrared thermography of inaccessible areas. In favorable situations, thermal imaging cameras my be operated from raised locations such as bridges.

9.2.3 Interpretation of temperature data

Measured and corrected temperature data are plotted in the form of geo-isotherms (Fig. 9.12) or on profiles. Mathematical filtering as applied, e.g., in gravimetry or in the magnetic method, may be useful.

Frequently, the data evaluation is reduced to the identification of areas or sections of anomalously high or low temperatures. Anomalies having amplitudes as low as 1° or less may be significant. If they cannot be explained by superficial effects, models of anomalous heat conduction or heat convection or a superposition of both may be considered. The latter case may apply to an underground karst cave as a thermal causative body in the terrestrial heat flow (Fig. 9.14A) or by thermal heat convection (Fig. 9.14B). The convection model shows that depending on the near-surface temperatures in summer and winter a negative or a positive temperature anomaly may be measured and that there may be no anomaly at all during spring and fall.

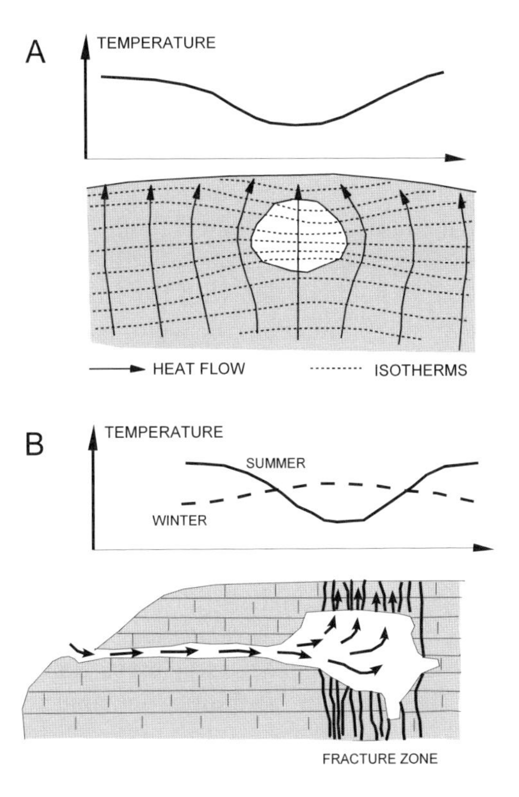

Fig. 9.14. Karst cave as a thermal causative body. A: Heat conduction model. B: Heat convection model

In general, the near-surface temperature field is much more influenced by heat convection than by heat conduction, and groundwater propagation possibly related with faults and fracture zones plays a major role. Especially in geothermal areas of mobile geological zones significant temperature anomalies may be measured that are related with the convection of thermal water. Bodies of anomalous thermal diffusivity that are located within the zone of the depth penetration of the long-period thermal waves may also cause temperature anomalies. Near-surface aquifers with low thermal diffusivity may behave as such a thermal causative body.

9.3 Radioactivity method

Natural radioactivity of rocks results from the content of uranium, thorium, and the potassium K^{40} isotope. By nuclear disintegration, both the uranium isotopes U^{238} and U^{235}, and the thorium Th^{232} isotope are related to various radioactive daughter elements in three decay series that end at stable lead isotopes. Within the series, the mobile radium (Ra) isotopes and the gaseous radon (Rn) isotopes are the radioactive daughter elements most significant in geology and hydrogeology. Different from the U and Th decay series, K^{40} immediately disintegrates to stable argon and calcium. Immediate disintegration to a stable daughter isotope is shown also by the radioactive carbon C^{14} and the radioactive tritium H^3 both serving in hydrogeology for dating and tracing purposes.

Nuclear disintegration is connected with the emission of helium nuclei (α radiation) or electrons (β radiation), or with the so-called K-capture. In all three processes, a high-energy electromagnetic radiation (γ-ray) is emitted. Gamma-ray energy peaks may be characteristic of the decay of a definite isotope.

In rocks the contents of uranium and thorium are of the order of ppm, and of the order of 100 ppm for K^{40} which makes up 0.012 % of the non-radioactive natural potassium. Apart from enrichments of radioactive minerals (e.g., zircon, monazite), acid rocks are in general more radioactive than basic rocks, and among sediments an increasing clay content correlates with increasing radioactivity. Groundwater radioactivity depends on the content of dissolved and suspended radioactive isotopes and their compounds. Generally, Th shows low mobility compared with U, Ra, K and the gaseous Rn.

For practical purposes the absorption of α-, β- and γ-rays is nearly as important as is the radiation itself. Various physical processes are responsible for the complete absorption of α- and β-rays by thin, micrometer- to millimeter-sized plates. Only γ-rays are capable to penetrate rocks of some decimeter thickness although the weakening is enormous, too.

The interaction of matter and radioactive radiation is also the base for the construction of portable measuring devices used in field surveys. Geiger-Müller counters and more sensitive scintillation meters are the most common instruments. Gamma-ray spectrometers are able to separate the characteristic energy peaks related with the nuclear disintegration and thus to identify the U, Th and K sources. The α-radiation of the gaseous Ra is measured in special ionization or scintillation chambers or by buried solid trace detectors. An important factor with all radioactivity measurements are the statistical nature of the nucleus disintegration and the related fluctuating count rates.

Radioactivity measurements are predominantly done in borehole geophysics, while radioactivity field surveys are relatively unimportant compared with other geophysical methods. They are normally carried out in a grid, and carborne and airborne surveys are common. The data (mostly recorded as counts per second, cps) are plotted on iso-radioactivity maps, and the varying parameters may be the total count or spectral counts in separate U, Th or K channels of a gamma-ray spectrometer. Apart from uranium prospecting, radioactivity measurements may serve as an aid in geological mapping for clarifying lithostratigraphical and structural conditions. Deep-reaching fracture zones may be special targets, especially for radon gas measurements (Fig. 9.15).

Radon gas emanometry has been performed since more than sixty years and has in recent times gained increased interest in environmental geology and geophysics. Radon originates from the uranium and thorium decay in rocks in the earth's crust and migrates through joints and fissures of fracture zones. Groundwater migration and carrier gases are assumed to facilitate the transport, which is important because of the rapid radioactive decay of the radon (3.8 day half-life). Finally, the migrated radon intermixes with the air in the soil where the concentration is measure by the alpha radiation. The radon distribution in the soil layers is complex and not completely understood. Important factors are the lithology of the soil-forming rocks, tectonic features, the geochemistry of the radioactive elements, soil physics and soil chemistry, changing groundwater table and the influence of topography and weather (see Gregg and Holmes 1990, Nielson et al. 1990).

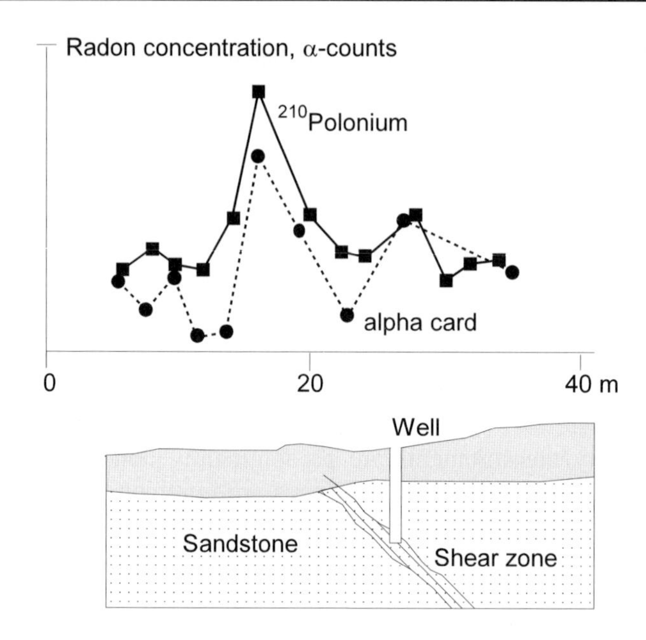

Fig. 9.15. Exploration of groundwater associated with a fracture zone by radon measurements. Both solid-trace detector (alpha-card) measurements of radon in the soil gas and α counts of the long-lived daughter element ^{210}Polonium exhibit a distinct concentration maximum over the shear zone. Simplified from Liang, 1986; in Nielson et al., 1990)

9.4 References

Gregg LT, Holmes JJ (1990) Radon detection and measurement in soil and groundwater. In: Ward SH (ed) Geotechnical and Environmental Geophysics, vol 1. Soc Expl Geophys, Tulsa, pp 251-262

Liang JH (1986) Nuclear techniques in prospecting for aquifers and its effectiveness: in Contributions to the exploration geophysics and geochemistry, 4, Geol. Publishing House, Beijing (in Chinese).

Nielson DL, Cui Linpei, Ward SH (1990) Gamma-ray spectrometry and radon emanometry in environmental geophysics. In: Ward SH (ed) Geotechnical and Environmental Geophysics, vol 1. Soc Expl Geophys, Tulsa, pp 219-250

VDI (1998) Thermische Nutzung des Untergrundes – Erdgekoppelte Wärmepumpenanlagen (VDI 4640-2). Verein Deutscher Ingenieure

10 Microgravimetry

Gerald Gabriel

Gravimetry is one of the classical and therefore well established methods in applied geophysics. It deals with the density distribution of the earth's crust. Advances in theory, technology and application were not only pushed by the need for geophysical exploration, but also by progresses in the field of geodesy. Many books and technical papers dealing with theory and application of gravimetry exist, covering all aspects in detail. This chapter focuses on topics being relevant for the application of this method to groundwater geophysics. The term "microgravity" points to the small magnitudes of gravity anomalies that often have to be expected in the context of groundwater geophysics making great demands on instruments, on the layout of field surveys, and on data processing (e.g. Debeglia and Dupont 2002). The successful application of the gravity method in groundwater geophysics is documented in many papers, for regional studies (e.g. Ali 1986, Oartfield and Czarnecki 1991, Rosselli et al. 1998) as well as for local studies (e.g. Valli and Mattson 1998, Murty and Raghavan 2002, Wiederhold et al. 2002, Gabriel et al. 2003), and time-dependent surveys (e.g. Strange and Carroll 1974, Hagiwara 1978, Talwani et al. 2001).

10.1 Physical Basics

Gravitation is a basic attribute of matter. Newton's law of gravitation (Isaac Newton, 1643–1727) defines the gravitational force F between two point masses m_1 and m_2. The force on m_2 is given by

$$F = -f \cdot \frac{m_1 m_2}{r^2} \tag{10.1}$$

with *r*: the distance between m_1 and m_2,

f: gravitational constant ($6.67 \cdot 10^{-11} \frac{m^3}{kg \cdot s^2}$).

For n masses m_i as product of density ρ_i and volume ϑ_i the gravitational force acting on a unit mass m_0 can be rewritten as

$$F = -f \cdot m_0 \cdot \sum_{i=1}^{n} \frac{\rho_i \cdot \vartheta_i}{r_i^2}. \qquad (10.2)$$

Therefore, for a continuum the gravitational acceleration g as the gravitational force per unit mass is given by

$$g = \frac{F}{m_0} = -f \cdot \iiint \frac{1}{r^2} \rho(r) \cdot \delta\vartheta. \qquad (10.3)$$

Due to the earth rotation the gravitational acceleration acting on a unit mass is not only caused by gravitation, but it is superimposed by the centrifugal acceleration a_c. The latter is the product of the square of the angular velocity ω of the earth and the perpendicular distance d of the unit mass from the rotation axis of the earth:

$$a_c = w^2 \cdot d. \qquad (10.4)$$

Therefore the centrifugal acceleration becomes maximal at the equator and zero at the poles. It acts perpendicular to the rotation axis and against the gravity acceleration that is directed to the centre of the earth. Consequently the resulting gravitational acceleration on the earth surface increases from the equator to the poles. Additionally, the earth's flattening strengthens this increase.

In geodesy or geophysics gravimetry deals with the measurement of the magnitude of the gravity acceleration, the "gravity". In SI-units gravity values are given in m/s^2. In honour of Galilei (Galileo Galilei, 1564–1642) in applied geophysics still the unit mGal (=1E-05 m/s^2) is used, in geodesy gravity values are often given in $\mu m/s^2$ (=1E-06 m/s^2).

10.2 Gravimeters

Gravimeters are instruments for measuring gravity. The past century was characterized by technological developments increasing the accuracy of the instruments by four decades. A comprehensive overview about gravimeters is given by Chapin (1998).

Absolute gravity measurements allow the establishment of high precision gravity networks, but are also used for high precision time-dependent observations, for example in the context of postglacial rebound. Modern instruments are based on the free-fall method or the rise and fall method. During the movement of a sensor in a vacuum chamber the fundamental properties distance and time are monitored. The measurement of the distance is realised by He-Ne-gaslasers and Michelson interferometers whereas the measurement of time is based on rubidium normals. By these methods accuracies better 10^{-8}g can be achieved (e.g. Faller et al. 1980, Torge 1989, Niebauer et al. 1995, Timmen 1996) by statistical analysis of some hundred or thousand single measurements; one station can be observed within two or three days. A new generation of miniaturized transportable absolute gravimeters reduce the observation time to some 10 minutes, at an undisturbed site the precision is supposed to be comparable to that of the well established absolute gravimeters (Brown et al. 1999).

Due to economic reasons and effectiveness in exploration surveys relative gravimeters are in use. With these instruments differences in gravity are observed. Because the gravity differences $\Delta\mathbf{g}$ caused by local density inhomogeneities are small compared to the undisturbed gravitational acceleration $\mathbf{g}$, the observed change in magnitude is approximately the projection of $\Delta\mathbf{g}$ to $\mathbf{g}$. The construction of relative gravimeters only enables the observation of the vertical component of $\Delta\mathbf{g}$.

Relative gravimeters can either be constructed to work in a dynamic mode or a static mode. Gravimeters based on the dynamic method observe the oscillation time of the sensor, whereas for gravimeters based on the static method the sensor is hold in a null-position and the change of equilibrium between two observations is a quantity depending on the gravity change. The majority of modern gravimeters used for field surveys are based on the static method.

Commonly in use are gravimeters from LaCoste & Romberg (Model G and D; metal spring), Scintrex, and Worden (both quartz spring). The measurement range of LaCoste & Romberg gravimeters Model G is about 7000 mGal, for Model D gravimeters it is restricted to 200 -300 mGal. For surveys of larger gravity differences Model D gravimeters must be reset to the requested measurement range. The LaCoste & Romberg gravimeter Model D has a precision better ± 10μGal (Torge 1989).

The measurement range of Worden gravimeters is also restricted to about 200 mGal (depending on the model) but can be extended to about 6600 mGal by resets. The precision is given to be about 10 μGal (Torge 1989).

Scintrex gravimeters, e.g. the CG-5 Autograv, are fully automatically including the correction of e.g. earth tides, tilt, and temperature. Their precision is about 5 μGal and they can be operated world wide without resetting.

The measuring principle of LaCoste & Romberg gravimeters is based on a small mass mounted to the end of a horizontal beam. The beam is held by an inclined *zero length spring* (e.g. Torge 1989, Chapin 1998). Gravity changes acting on the mass cause a displacement of the equilibrium position and therefore a change of the length of the spring. Using the dial on top of the gravimeter fixed on the measurement screw the mass can be readjusted to the null-position. After Torge (1989) a measurement precision of 10 μGal requires a positioning precision of 0.25 nm. This is achieved with the help of a special transmission system. The whole gravimeter system is air-tight sealed and – because a metal spring underlies thermal expansion – a heater provides a constant temperature.

LaCoste & Romberg gravimeters are astatized systems (e.g. Torge 1989). Astatized (rotational) systems are characterized by an increase of sensitivity due to construction layout. The force of gravity is maintained in an unstable equilibrium with the restoring force. Generally, the instability is provided by the introduction of a third force which intensifies the effect of any change in gravity from the value in equilibrium (e.g. Sheriff 1984).

For LaCoste & Romberg gravimeters electronic feed-back systems are available in addition to the integrated capacitive position indictor (CPI). The CPI consists of a plate mounted to the gravity sensor as the movable part of a three-plate capacitor. The other two plates are fixed and generate an electronic output depending on the position of the movable plate. An electronic feed-back system readjusts the gravity sensor to the null-position by the generation of an electrostatic force. The required voltage is a direct measure of the gravity difference between two observations. Depending on the type of electronic feed-back systems the measurement range is limited to some 10 mGal. Because these systems are not linear, in practise the feed-back voltage should be kept small (below 20 μGal). Feed-back systems can be used in combination with the mechanical system of the gravimeter; the coarse adjusting of the sensor to the null position is done over the measurement screw by the dial, the fine adjusting with the feed-back system. The resulting gravity difference is the sum of both readings, converting them into gravity changes by applying instrument specific calibration functions.

10.3 Gravity surveys and data processing

The significance of gravity data with respect to a hydrogeological situation depends mainly on the accuracy achieved in the field survey. Basic requirements for an accurate data set are gravimeters with well known physical characteristics, a well designed survey, high precision measurements, and state of the art data processing.

10.3.1 Preparation and performance of field surveys

Gravimeters have to be calibrated at least one or two times a year. Because the calibration function can change significantly with time, a confirmation of the known values is worthwhile. For local field surveys with small relative gravity differences the knowledge of the linear calibration factor might be sufficient. For LaCoste & Romberg gravimeters Torge (1989) states an error of 0.05 mGal for the residual nonlinear calibration terms (from the nonlinearity of the lever system) for gravity differences of a few 100 mGal. Nevertheless, also periodic calibration parameters, resulting from graduation errors and eccentricities in the screw and the gears, should be estimated by calibrations. Nowadays for the calibration of gravimeters established calibration lines (e.g. Kanngieser et al. 1983, Wenzel 1996) are used in combination with laboratory methods (calibration by changing the inclination of the gravimeter, calibration of the screw against feed-back values, calibration against absolute gravity meters). Calibration lines are required because the linear factor of the calibration function cannot be determined to a sufficient precision by laboratory methods alone (Torge 1989).

Furthermore, before each field survey, the adjustment of the levels must be checked. Knowledge of the inclination-insensitive point of each level is of great importance, because the gravity reading is very sensitive to inclination changes using astatized gravimeters. In practise this point is found by tilting the gravimeter systematically along both sides and plotting the readings against the tilt. The results are two parables with their angular points defining the inclination-insensitive points (Fig. 10.1). In order to increase the accuracy of the field survey the levels have to be re-adjusted if their inclination-insensitivity points are not close to the "zero-position". Another opportunity is to level the gravimeter during the survey according to the inclination-insensitivity points.

Prior to the field survey the acquisition parameters have to be defined. Therefore an initial idea about geology and expected anomalies is indispensable. For 2-dimensional (2-D) geological structures gravity measurements along profiles might be sufficient, for more complex structures 3-D surveys have to be performed. After Jung (1961) a source body can be considered 2-D if its ratio length/width exceeds 4:1. The spacing of adjacent observation points is significant not only for the success of a survey, but also for its effort. It depends on the geometry, depth, and density of the source body and the aimed resolution of the survey. The point spacing for engineering and environmental microgravity investigations can vary between some decimetres and some hundred meters. In order to plan the survey the magnitude and wavelength of the expected anomaly can be calculated applying formulas for simple source bodies (e.g. Telford et al. 1990). For most investigations in the frame of groundwater geophysics point spacing between 50 m and 100 m is sufficient. The surveyed area should exceed the dimension of the geological target in order to map the corresponding gravity anomaly completely and to distinguish between regional and residual gravity anomalies. In areas where less information is available a survey should start with a coarse net of points, in areas of interest additional measurements can be performed later.

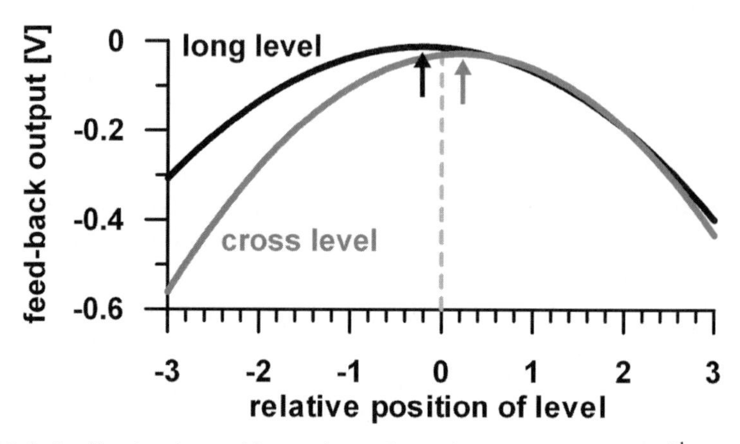

Fig. 10.1. Inclination-insensitive points of gravimeter G-662 on 02nd July 2002 (1 Volt ≅ 1 mGal)

During the survey gravity stations have to be selected properly. Readings can be affected by microseismics (natural and artificial) and wind (causing strong disturbances by trees). Gravity stations near steep slopes, buildings, frequented roads, and ditches should be avoided. Ideally each gravity station within the surveyed area is recorded twice in order to avoid outliers.

The reading procedure itself should be the same at each point of the survey. The dial has to be approached from the same side each time, otherwise its backlash will cause errors that can amount to some 10 μGal. The reading can be taken as soon as the gravimeter output is stable, best controlled by a feed-back system. The total time required for the observation of one single station is between 5 and 10 minutes. Therefore about 30 to 50 stations can be surveyed during a day, provided that the time for transport between the stations is negligible. Unfavourable environmental conditions like bad weather or microseismics can cause delays.

Local gravity investigations in the frame of groundwater geophysics will often be performed by densifying an existing data base by complementing measurements. In order to assure the compatibility of the gravity values resulting from different campaigns, relative gravity measurements should be tied to existing absolute gravity points, in general available from the ordnance surveys. In practise a local base point not influenced by temporal gravity changes should be established within the investigated area. The absolute gravity value for this local base point results from repeatedly measured ties to at least two absolute gravity points of a regional network. If possible a local gravity network should be established including the calculation and correction of loop misclosures in order to achieve a higher accuracy of the local base point.

A high accuracy of the observed gravity differences requires a precise recording of the gravimeter drift. The drift of a gravimeter is caused by external effects (e.g. temperature and air pressure changes, mechanical shocks) as well as internal effects (fading of spring tensions). Practical experiences show, that the drift should be observed at least every two hours by measurements on a control point (Fig. 10.2). Generally, the drift should not exceed 100 μGal/day.

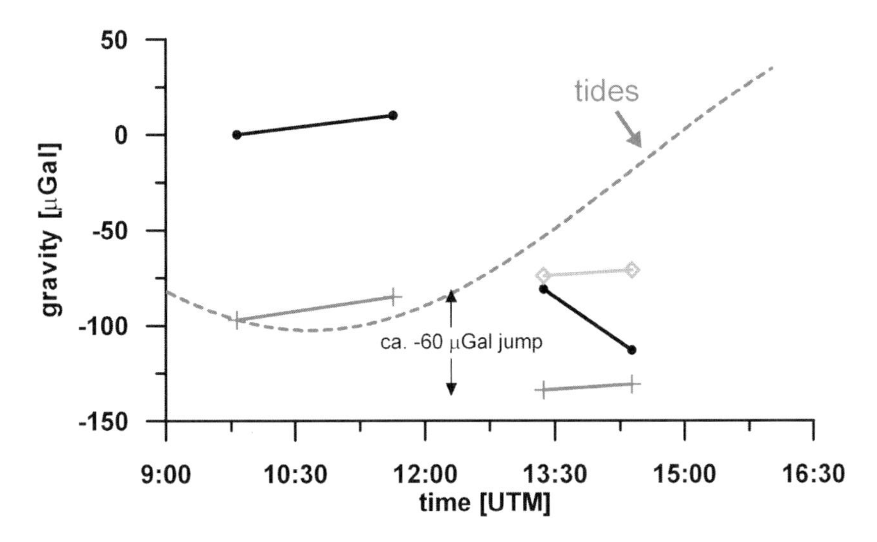

Fig. 10.2. Artificial jump observed with gravimeter G-662 on 16th June 2004. Black lines with bullets represent the observed gravity and grey lines with crosses represent the tide corrected gravity values. Knocking the gravimeter against the transport box caused a jump of about 60 μGal. Correcting the jump (grey line with diamonds) yielded a smooth, nearly linear drift

If the gravimeter is knocked or jolted, either during transport or during the measurement procedure, the drift has to be checked immediately. Furthermore these incidents should be noted in the protocol. Gravity observations made in a period where great drift rates indicate larger artificial jumps in gravity caused by external influences on the gravimeter (Fig. 10.2) must be cancelled and replaced by new observations. For recording the gravimeter drift different measurement schedules like the difference method, the star method, the step method, or the profile method are established (Watermann 1957). Parallel observation by operating more than one gravimeter also increases the accuracy and reliability.

10.3.2 Data processing

Absolute gravity values do not only reflect the density contrasts within the earth's crust, but also the shape of the earth itself. They are influenced by the geodetic latitude, the elevation, and temporal gravity changes. In order to get anomaly values that can be interpreted in geological terms, known temporal effects have to be corrected and known disturbing mass distributions such as topography have to be reduced.

Therefore, during a gravimetric survey following information must be noted in order to enable a reliable data processing: (a) station number (including repetition measurements for drift estimation), (b) date and time of gravity reading in UT (needed for earth tide correction and drift correction), (c) height of gravity sensor above the earth's surface, (d) scale units, (e) feed-back output (if available), and (f) air-pressure. At least for spring gravimeters it is recommended also to note the temperature of the instrument. Temperature changes caused by e.g low power supply affect the reading. The operating temperature is different for each individual gravimeter. Furthermore calibration factors for the used gravimeters (incl. feed-back systems) and earth tides must be known as well as the coordinates and elevations of the gravity stations and values for tied absolute gravity points.

In solid earth physics interpretations are commonly based on the Bouguer anomaly Δg (after the French scientist Pierre Bouguer, 1698–1758). The Bouguer anomaly is defined as the difference of the observed gravity value g_{obs} at a given station and the theoretical gravity value g_{th} due to a homogenous earth for this station:

$$\Delta g = g_{obs} - g_{th} \quad . \tag{10.5}$$

The theoretical gravity value g_{th} due to a homogenous earth is given by (Fig. 10.3)

$$g_{th} = \gamma + \delta g_h + \delta g_{bpl} - \delta g_{ter} \tag{10.6}$$

γ: normal gravity,
δg_h: free air reduction,
δg_{bpl}: Bouguer plate reduction,
δg_{ter}: terrain reduction.

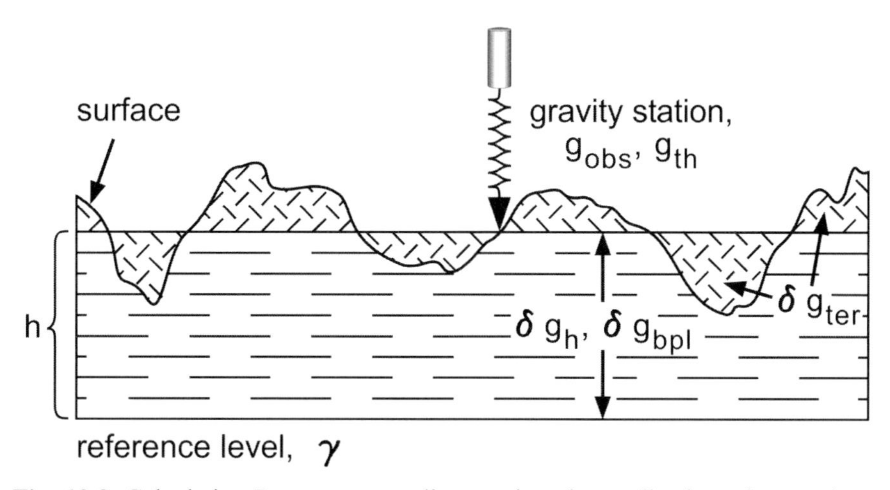

Fig. 10.3. Calculating Bouguer anomalies requires the application of corrections and reductions (further explanations are given in the text)

With Eq. 10.5 the Bouguer anomaly at the station height (Ervin 1977, Hinze 1990, Li and Götze 2001) is

$$\Delta g = g_{obs} - \gamma - \delta g_h - \delta g_{bpl} + \delta g_{ter}\,. \tag{10.7}$$

The resulting value reflects gravity anomalies due to inhomogeneous densities below the gravity station. They can either be plotted as profiles or as contour maps.

Correction of temporal gravity changes in the observed gravity

In Eqs. 10.5 and 10.7 g_{obs} represents the observed absolute gravity value derived from relative gravity measurements including ties to an absolute gravity point. Temporal effects like Earth tides and the gravimeter drift are corrected.

Earth tides are caused by the varying gravitation of the earth and celestial bodies – primarily moon and sun – at the different points of the earth and the centrifugal acceleration due to the rotation of all bodies around common centres of gravity (e.g. Melchior 1966). The maximum variation of the Earth tides for an elastic earth is 0.29 mGal, about 1.16 times the tides for a rigid earth. In addition to the earth tides also ocean tides occur. Besides their direct gravitation (movement of water masses), ocean tides also cause gravity changes due to periodic deformation (loading) of the earth crust and corresponding height changes – the ocean loading tides. Compared to the magnitude of Earth tides, correction for ocean tides and ocean loading tides are small. They can be considered by a corresponding choice of earth tide parameters.

The correction of Earth tides can either be done as part of the drift correction, or more accurately by using Earth tide models. The computation of earth tides is based on the expansion of the tidal potential (e.g. Cartwright and Taylor 1971, Cartwright and Edden 1973, Wahr 1980, Tamura 1993, Hartmann and Wenzel 1995) taking into account the elastic behaviour of the earth. Depending on the model used, earth tide corrections with accuracy better than 1 μGal can be provided applying correct amplitude factors (differing from 1.16) and phases for the different tidal constituents. Because earth tides strongly depend on the latitude, the ellipsoidal coordinates should be considered at least with a precision of 50 km.

The accuracy of gravity measurements can be increased by considering air pressure variations. Although gravimeters are air-tight sealed gravity measurements are affected by air pressure variations directly by the gravitation of the air masses as well as by the deformation of the surface due to loading effects. From high-resolution time dependent gravity recordings regression coefficients from -0.2 to -0.4 μGal/hPa are known. In practise a regression coefficient of -0.3 μGal/hPa can be used (e.g. Warburton and Goodkind 1977, Torge 1989). Generally, the necessity of air pressure corrections depends on the desired accuracy of each gravity survey. The correction should be considered at least for calibration or exploration surveys in mountainous areas. For local gravity surveys, where air pressure differences mostly remain smaller 10 hPa, air pressure correction is not mandatory.

Normal gravity reduction

The theoretical gravity value g_{th} strongly depends on the geodetic latitude φ, the elevation of the gravity station, and topographic masses. Gravity as a function of geodetic latitude can be calculated by the normal gravity formula:

$$\gamma = \gamma_e \cdot (1 + \beta_1 \sin^2 \varphi - \beta_2 \sin^2 2\varphi) . \tag{10.8}$$

Values for the gravity at the equator γ_e, the gravity flattening β_1 (a function of the gravity at the equator and the gravity at the poles γ_p), and the factor β_2 (a function of the earth's flattening f, the semi-major axis of the reference ellipsoid, the centrifugal acceleration at the equator, and gravity at the equator) depend on the used reference ellipsoid. For the normal gravity formula in the Geodetic Reference System 1980 (Moritz 1984)

$$\gamma_e = 978\,032\,677 \text{ mGal}$$

$$\beta_1 = 0.005\,302\,4 \qquad \text{with} \qquad \beta_1 = \frac{\gamma_p - \gamma_e}{\gamma_e}$$

$$\beta_2 = 0.000\,005\,8 \qquad \text{with} \qquad f = 1/298.257 \quad .$$

Normal gravity increases from the equator to the poles by about 5.186 Gal. For mean longitudes a north-south gradient of about 0.7 mGal/km can be estimated. Therefore the precision requirement for the position of the gravity station is about 15 m based on the precision of modern spring gravimeters.

Free air reduction and Bouguer plate reduction

The elevation h of the gravity station has to be considered within two terms: the free-air reduction (height reduction) and the Bouguer plate reduction. In general, gravity decreases with increasing distance from the earth surface by 0.3086 mGal/m. This mean gravity gradient is commonly used in data processing, although its true value can vary in a great range mainly depending on the near surface geology. The free-air reduction is given by

$$\delta g_h = h \cdot (-0.3086 \text{ mGal/m}) , \tag{10.9}$$

with h the height difference between a reference level (often mean sea level) and the station height.

The Bouguer plate reduction considers the masses between the reference level and the elevation of the gravity station. The masses depend on the density ρ assumed for the rocks; ρ is called Bouguer density or reduction density. The gravity effect of such a plate of constant thickness h and density ρ (in kg/m^3) is given by

$$\delta g_{bpl} = 2 \cdot \pi \cdot f \cdot h \cdot \rho$$
$$= 0.0419 \cdot E-03 \cdot h \cdot \rho \quad [mGal] \qquad (10.10)$$

f: gravitational constant,

or with ρ=2670 kg/m^3, the mean density of the earth's crust (Hinze 2003),

$$\delta g_{bpl} = 0.1119 \cdot h \quad [mGal] \quad .$$

The above formula only considers the effect of a plane Bouguer plate. In the case of regional surveys a spherical plate reduction is more accurate (Bullard 1936, Swick 1942, La Fehr 1998, Talwani 1998).

From the formulas for the free-air reduction and for the Bouguer plate reduction follows, that the elevation of the measurement point should be known with a precision better than 5 cm in order to be compatible with the precision of modern gravimeters. This can be achieved by geometric levelling, tachymetry, or by differential GPS measurements (e.g. Fairhead et al. 2003, Seeber 2003) including existing control points.

Especially in microgravity surveys the Bouguer density value is crucial, because incorrect values can lead to so-called reduction anomalies that exceed the anomalies caused by the source bodies. Nettleton (1939) developed a method that can be applied in order to derive the Bouguer density value directly from the observed gravity anomaly provided the gravity survey is conducted in an area with strong elevation changes and the topography is not correlated with isolated geological structures. Calculating Bouguer anomalies with different Bouguer densities and comparing them with topography, the Bouguer anomaly showing no correlation with the topography is assumed to be calculated with the correct Bouguer density value. If the density of the rocks considered in the reductions varies significantly, the use of different Bouguer densities should be contemplated.

Terrain reduction

The gravity at station height is additionally affected by the topography around the station. Valleys cause a mass deficit reducing gravity. Hills cause a mass excess, but the resulting force acts in opposite direction to the attraction of the masses below the station and therefore hills also reduce gravity. Therefore a terrain reduction δg_{ter} taking into account gravity effects due to topography is necessary. Nowadays digital elevation models are available. Computer programs decompose the topography in elementary bodies like cuboids and consider the $1/r^2$ dependence of gravity by increasing the bases of the cuboids with increasing distance to the gravity station (e.g. Bott 1959, Ehrismann et al. 1966, Ehrismann and Lettau 1971). Forsberg (1985) and Sideris (1985) presented computations in the spectral domain based on the Fast Fourier Transformation. For the terrain reduction the Bouguer density has to be used.

If only topographic maps are available methods after Hammer (1939) or Schleusener (1940) can be used. By concentric circles and radial lines the area around the gravity station is divided into sectors. The area of these sectors increases with increasing distance from the gravity station. For each sector a mean elevation has to be estimated from the contours of the topographic map. The difference to the elevation of the gravity station is considered in the calculation of the terrain reduction as well as the radii of the circles and the density of the rocks.

In microgravity special emphasis has to be put on the consideration of the near field topography around the gravity station (e.g. Schöler 1976). Debeglia and Dupont (2002) also consider the effect of buildings.

10.4 Interpretation

The following section deals with the interpretation of gravity anomalies. The different methods discussed are divided into direct and indirect methods. Direct methods derive information about the depth, geometry, and physical parameter of the source body from the observed gravity anomaly. Are these parameters well known, the corresponding gravity anomaly can be calculated by indirect methods. This definition is not unique throughout the literature.

10.4.1 Direct methods

Characteristic attributes of a gravity anomaly caused by a single source body are its amplitude, its wavelength, its inflection point, and its half-width. The apex of the anomaly is located above the centre of the source body. But nature is more complex. Generally the observed gravity anomaly is the integral effect of several source bodies in the subsurface, with different positions, depths, geometries, and density contrasts.

In applied geophysics different tools are available for an analysis of the observed anomaly. This includes empirical formulas as well as filter algorithms. The results of these direct methods are ambiguous. Whereas small-wavelength anomalies can only be caused by shallow structures, long-wavelength anomalies can either be caused by deep sources or extended sources near the surface. Spectral analyses and filtering procedures are common mathematical tools applied in geophysics and other sciences (e.g. Grant and West 1965, Jenkins and Watts 1968, Bath 1974, Buttkus 2000).

The maximum depth of a source can be estimated from the half-width $b_{1/2}$ of the anomaly defined as the width of the anomaly at its half-maximum (minimum). For a sphere the maximum depth of its centre is about $0.65 \cdot b_{1/2}$, for a horizontal cylinder it is $0.5 \cdot b_{1/2}$ (e.g. Sheriff 1984).

In order to improve the result of such estimations the separation of the regional field (characterised by "long-wavelength" anomalies) and the residual field (characterised by "short-wavelength" anomalies) is necessary (Fig. 10.4, 10.5). The separation between both can be achieved by wavelength filtering. Theoretical requirements are significantly differing depths of the different source bodies. Information referring to this can be derived from seismic surveys, drillings and other investigations. The separation of the regional and residual fields by wavelength filtering is achieved by eliminating fractions below or above a cut-off wavelength λ_c (e.g. Zurflüh 1966, Fuller 1967). Applying the Fast Fourier Transformation FFT (Brigham 1974) the spectrum $F(k_x, k_y)$ of the input data $f(x, y)$ – in this context the Bouguer anomaly – can be calculated (k_x, k_y: wavenumbers). The spectrum is multiplied by a function $\Psi(k_x, k_y)$ defining the filter process $\Phi(k_x,k_y)$ (high-, low-, or band-pass). The inverse transformation in the space domain gives the filtered anomaly $\varphi'(x,y)$:

$$\begin{aligned}\varphi'(x,y) &= \Im^{-1}\{F(k_x,k_y)\cdot\Psi(k_x,k_y)\}(x,y) \\ &= \frac{1}{2\pi}\iint e^{i(k_x\cdot x+k_y\cdot y)}\Phi(k_x,k_y)\,dk_x dk_y\ .\end{aligned} \tag{10.11}$$

The filter function $\Psi(k_x, k_y)$ of an ideal filter is characterised by points of discontinuity at its edges (e.g. low-pass filter: $\Psi(k_x, k_y)=0$ for $\lambda < \lambda_c$ and $\Psi(k_x, k_y)=1$ for $\lambda > \lambda_c$), producing high frequent overshoots (Jenkins and Watts 1968). Therefore, in practise smooth filter functions without points of discontinuity have to be applied, without changing the filter characteristic significantly (Bracewell 1965).

Field continuation also is a filter process. Generally, field continuation is a mathematical operation for the calculation of a field $f(x,y)$ on a surface $z_2(x,y)$, if this field is known on a surface $z_1(x,y)$. Assuming that the Bouguer anomaly Δg represents the gravity on a constant surface z_1 gravity on the surface z_2 can be calculated by the FFT using the function

$$\Psi(k_x,k_y) = e^{\sqrt{k_x{}^2+k_y{}^2}\cdot(z_1-z_2)}\ . \tag{10.12}$$

This function has a low-pass characteristic for $(z_1-z_2)<0$ (upward continuation), and a high-pass characteristic for $(z_1-z_2)>0$ (downward continuation). The application of field continuation is based on the assumption that no sources are located between the two surfaces. Therefore in practise downward continuation becomes problematic. In some applications it is used to estimate the depth of the source with the calculated gravity field becoming instable when exceeding the depth of the source.

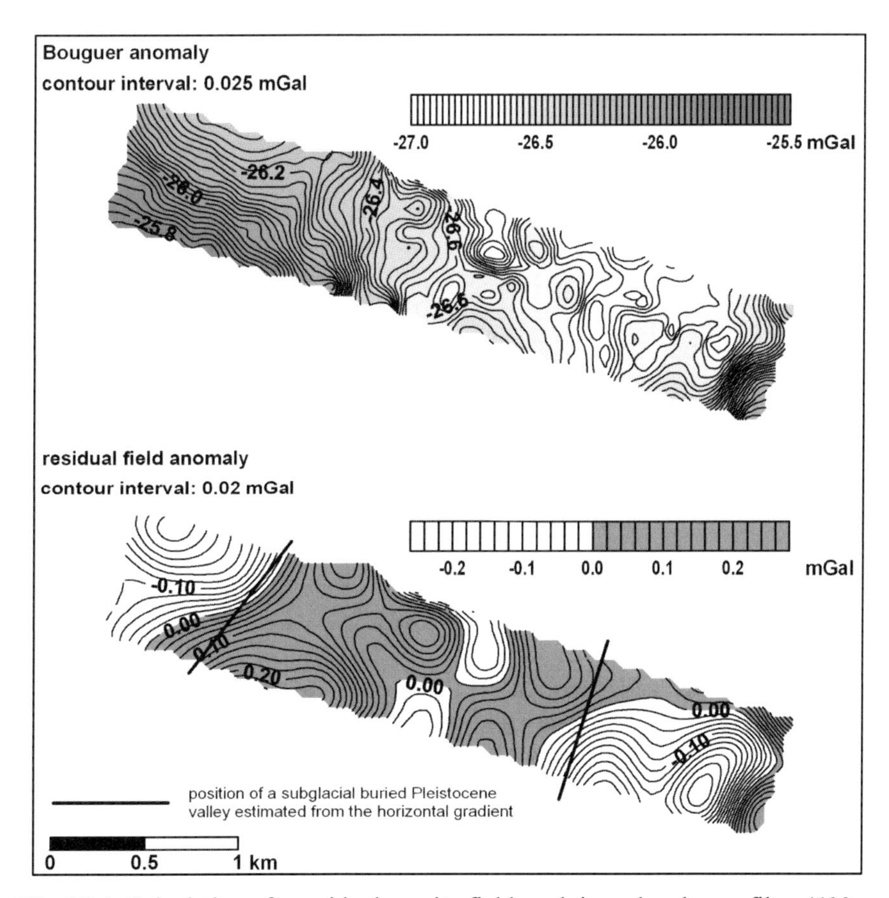

Fig. 10.4. Calculation of a residual gravity field applying a band-pass filter (400 m to 3000 m) to observed Bouguer anomalies. Gravity anomalies from a regional data base (not shown) complemented the local gravity data set. Positive anomalies in the residual field are supposed to be caused by a buried Pleistocene subglacial valley; local negative anomalies are caused by peat

Applying FFT techniques requires the Bouguer anomaly values given in a grid. In order to minimize the influence of overshoots – the so-called Gibb's phenomenon – the dimension of the grid must exceed the area under investigation by at least 20%.

From the calculation of the horizontal gravity gradient – approximated by the difference quotient $(g_{i+1}-g_i)/(x_{i+1}-x_i)$ – for a simple source body information about its horizontal position and change of thickness can be derived (Fig. 10.4, Fig. 10.5). In microgravimetry surveys the horizontal gradient often is affected by local density inhomogeneities not associated with the geological target. Therefore in practise the observed Bouguer anomalies must be also considered when interpreting the horizontal gradient. Constraints from other geophysical methods help to interpret the horizontal gradient.

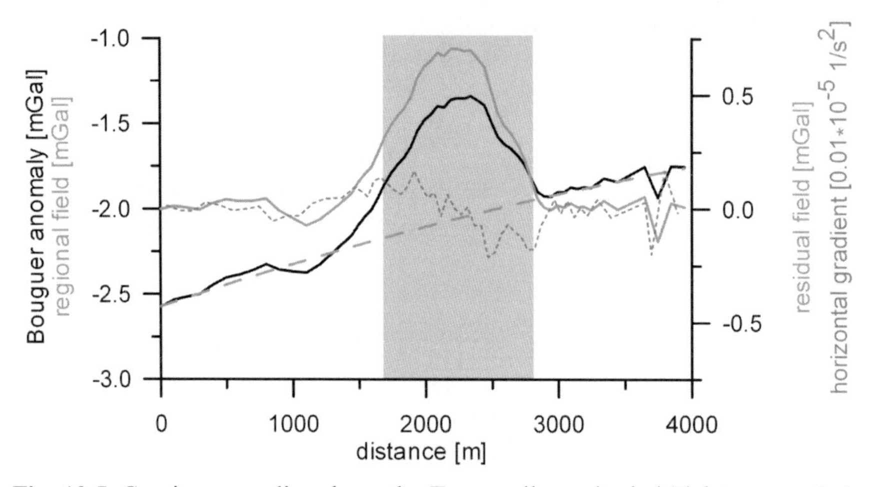

Fig. 10.5. Gravity anomalies above the Trave valley, a buried Pleistocene subglacial valley in North Germany: Bouguer anomaly (solid black line), regional field (dashed grey line), residual field (solid grey line), and horizontal gradient (fine dashed grey line). Field separation was done by fitting a 2nd order polygon to the regional field. The grey area indicates the location of the buried valley interpreted from the horizontal gravity gradient

A method closely related to the horizontal gradient is the maximum curvature method. Curvature is a 2-D property of a surface characterising the bending in a certain point (e.g. Rektorys 1969). Curvature is defined to be positive over anticlines and negative over synclines. The Bouguer anomaly can be analysed in terms of maximum curvature. One way to calculate curvature of a gridded surface is a quadratic approximation of the particular point of the surface by a least square technique (e.g. Roberts 2001). The six coefficients of the approximation can be calculated from the eight surrounding grid values and the size of the grid cells. In the theory of curvature the maximum curvature is lastly a function of five of these coefficients. The first time the maximum curvature method was applied to gravity data is reported by Stadtler et al. (2003). They analysed synthetic data as well as gravity anomalies caused by buried Pleistocene subglacial valleys. The buried valleys with an infill of higher density than the surrounding emerge as parallel elongated areas of negative curvature at the valley flanks and one area of positive values along its centre. The result is sensitive to the grid spacing and the data coverage.

10.4.2 Indirect methods

Indirect methods are either based on the application of simple analytic formulas for elementary source bodies like vertical and dipping faults, spheres, cuboids, horizontal and vertical cylinders (e.g. Militzer and Weber 1984, Telford et al. 1990), or more complex expressions for 2-D and 3-D forward modelling of arbitrary shaped bodies.

For 2-D calculations the algorithm after Talwani et al. (1959) is implemented in many computer programs. The gravity effect Δg_{2D} of the source body (density ρ) defined by a polygon of n vertices (Fig. 10.6) in P is given by

$$\Delta g_{2D} = \sum_{i=1}^{n} \Delta g_i \tag{10.13}$$

$$\Delta g_i = 2 \cdot f \cdot \rho \cdot a_i \cdot \sin \Phi_i \cdot \cos \Phi_i \cdot \left[\Theta_i - \Theta_{i+i} + \tan \Phi_i \cdot \ln \left\{ \frac{\cos \Theta_i \cdot (\tan \Theta_i - \tan \Phi_i)}{\cos \Theta_{i+1} \cdot (\tan \Theta_{i+1} - \tan \Phi_i)} \right\} \right] \tag{10.14}$$

with:

$$\Phi_i = \arctan \frac{z_{i+1} - z_i}{x_{i+1} - x_i}$$

$$\Theta_i = \arctan \frac{z_i}{x_i}$$

$$a_i = x_{i+1} + z_{i+1} \frac{(x_{i+1} - x_i)}{(z_i - z_{i+1})} \quad .$$

For many geological situations the assumption of two-dimensionality is not valid. The problematic nature of the 2-D modelling is well studied (e.g. Jung 1961, Götze 1984). Although the resolution of the geological structure is nearly always very high and the fit to the measured data convincing (Fig. 10.7), due to the integral effect of all masses around the observation point the results are often unrealistic. The gravity effect of the geological bodies beside the profile is present in the measured data but falsely explained by the resulting model.

Therefore 3-D modelling is often more suitable. Many algorithms used in 3-D software packages, subdivide the subsurface into elementary bodies (cuboids, cylinders) and calculate the gravity anomaly at a given station as the sum of the effects of all source bodies.

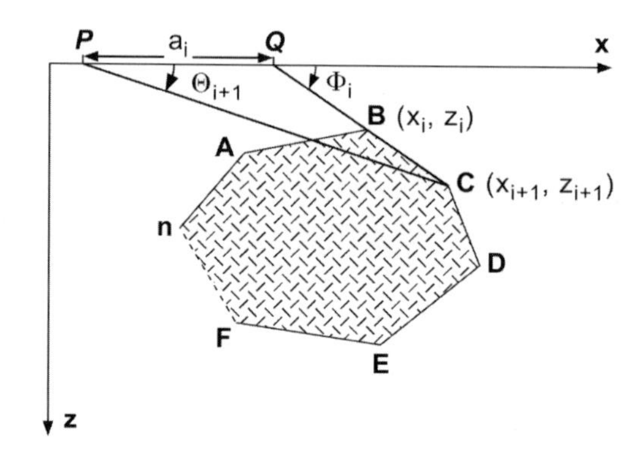

Fig. 10.6. Geometrical elements involved in the calculation of the gravitational attraction of a polygon (after Talwani et al. 1959)

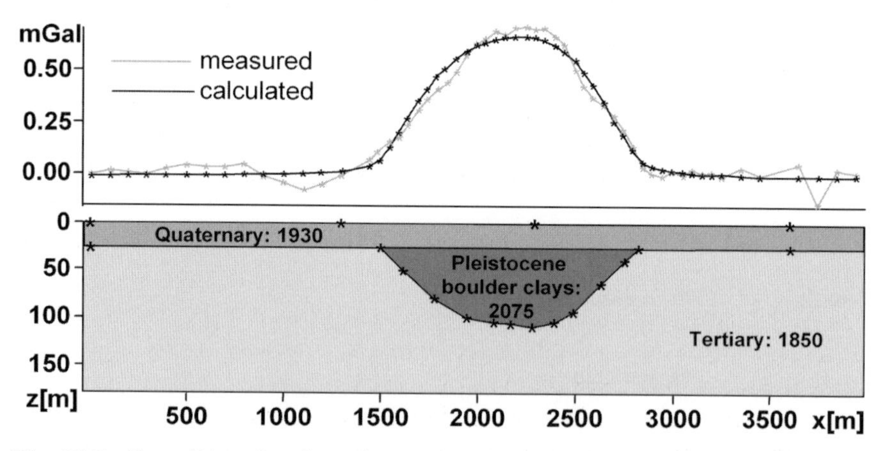

Fig. 10.7. Two-dimensional gravity model of the Trave valley in North Germany (densities in kg/m^3, vertical exaggeration: 5:1). The modeling is based on the residual gravity field (cf. Fig. 10.4)

An algorithm giving the exact solution for arbitrarily shaped bodies was presented by Götze and Lahmeyer (1988). The area under consideration is defined by several parallel vertical planes. The spacing between two adjacent planes is chosen according to the geological situation as well as the location of gravity anomalies. In each vertical plane the borders of bodies with constant density are described by polygons defined by vertices. The 3-D structure is realised by an automatic triangulation of the polygons describing same bodies in adjacent planes. Using the Gauss and Green theorem the calculation of the gravity effect is reduced to the solution of line-integrals.

Since gravity modelling is ambiguous, consideration of constraining geoscientific information is basic requirement for obtaining an appropriate interpretation. On the one hand the geology has to be taken into account as far as it is known; on the other hand the modelling based on gravity data should not contradict the results of other geophysical measurements. Within its resolution an accurately developed gravity model should be a synthesis of the available geoscientific information. From this point of view ambiguity of gravity needs not be a disadvantage inevitably.

10.4.3 Density estimation

For gravity modelling density information must be available. Whereas in literature generally density ranges and mean densities are summarised (e.g. Wohlenberg 1982, Telford et al. 1990), densities for local lithologies can be derived from laboratory experiments, logging, core scanners, or velocity-density relationships (e.g. Johnson and Olhoeft 1984).

The bulk density ρ can easily be determined in the laboratory by the buoyancy method. The sample to be analysed is weighed in dry condition (mass m) and in water (mass m_w). The bulk density can be calculated by

$$\rho = \rho_w \cdot \frac{m}{m - m_w} \tag{10.15}$$

with ρ_w the density of water. The obtained densities are very precise for the particular sample, but in order to get representative densities for a geological formation a sufficient number of samples needs to be analysed. For sediments this method is less practicable due to suspension of the sample.

The dry density of sediments can be estimated by dehumidifying the samples in a drying furnace. Afterwards they get paraffined. The density ρ_d of the sample is given by

$$\rho_d = \frac{m}{(m_p - m_{wp}) \cdot (m_p - m) \cdot \frac{1}{\rho_w \rho_p}} \tag{10.16}$$

- m: mass of the dry not paraffined sample,
- m_p: mass of the dry paraffined sample,
- m_{wp}: mass of the dived paraffined sample,
- ρ_w: density of water,
- ρ_p: density of paraffin .

Logging techniques (Serra 1984, Rider 1996, Fricke and Schön 1999) require the existence of boreholes in the investigated area that penetrate the relevant lithologies but are not cased. Density logging is very sensitive to small borehole irregularities. Therefore the calliper log should also be considered when interpreting density logs. The horizontal penetration depth for density logging is less than 15 cm depending on the distance between the gamma source and the detector, and the density.

If no samples or downhole density data are available this lack of (in situ) density information can partly be compensated by converting seismic velocities into densities. Velocity-density relationships are published by several authors (e.g. Nafe and Drake 1957, Ludwig et al. 1970, Gardner et al. 1974, Hamilton 1978). But the validity of the formulas is limited to single lithologies or certain depth intervals. Especially for near surface structures, consisting of high-porous, non-saturated sediments, it must be considered that the seismic velocity is more sensitive to porosity, pore pressure and water saturation than density is (Sheriff and Geldart 1996).

10.5 Time dependent surveys

Whereas (micro)gravimetric surveys in the frame of hydrogeophysics are mostly conducted in order to investigate geological structures by interpreting areal gravity changes, with time dependent gravity surveys gravity changes caused by mass variability can be monitored. Nowadays high-resolution continuous gravity field satellite missions provide insight into continental hydrology. Especially from the gravity data of the GRACE (Gravity Recovery and Climate Experiment) mission – a joint US-German partnership mission (Tapley et al. 2004b) – seasonal global scale continental water storage variations were successfully recovered (Tapley et al. 2004a, Wahr et al. 2004).

From high resolution gravity time series, e.g. superconducting gravimeters, temporal gravity changes caused by changes of the groundwater table or changes in soil moisture are known since many decades (Bonatz 1967, Elstner and Kautzleben 1982, Peter et al. 1995, Bower and Courtier 1998, Harnisch and Harnisch 1999, Virtanen 2000, Zerbini et al. 2001). These effects are well studied because they are often interpreted as noise and have to be corrected in order to study long-period phenomena like the polar motion or oscillations of the earth core. Hydrologically induced gravity signals have periods between some hours and more than one year (Kroner 2001) and amplitudes between some 0.1 μGal and 10 μGal. Takemoto et al. (2002) reported an increase in gravity of about 4.3 μGal caused by a rise of the groundwater table of 1 m for stations in Bandung/Indonesia and Kyoto/Japan. Strange and Carroll (1974) and Hagiwara (1978) correlate gravity changes between 10 μGal and 35 μGal with land subsidence caused by groundwater withdrawal. Gravity changes can also be correlated with precipitation (Imanishi 2000, Kroner 2001).

These magnitudes are resolvable with transportable relative gravimeters also. Therefore, repeatedly observed gravity networks might contribute to the understanding of local hydrological regimes. Prerequisites would be, among others, well calibrated gravimeters (including non-linear and periodic terms), a representative distribution of gravity stations, stable platforms for gravity registration, height control of the gravity station, correction of earth tides, and correction of air pressure induced gravity changes. The network must be tied to a reference point that is not affected by mass changes. The use of several gravimeters in such surveys is indispensable in order to derive significant results. If vertical ground movements are expected, they have to be monitored by additional levelling surveys in order to separate gravity changes due to height changes and gravity changes due to mass variability.

Recently time lapse gravity gradiometry was successfully performed by Talwani et al. (2001), e.g. monitoring the fluid/gas interface in the Prudhoe oil field during water injection. Gravity gradients are the spatial variation of the three components of the gravity vector. With gravity gradiometers all nine components of the gravity gradient tensor can be measured, in doing so only five components are independent. Talwani et al. (2001) deployed a gradiometer that measures a combination of only three components, all lying in the horizontal plane. Gradiometry is superior to gravimetry when shallow density contrasts cause short wavelength anomalies between some meters and a few kilometres, although the interpretation of the observed gradients is ambiguous, too. Compared to gravity, the demands concerning levelling precession and height control for gradiometry are less; the effects of tides and air pressure changes are of greater wavelength than those that can be resolved by gradiometry. Actually, the application of gradiometry is restricted by the cost price of the instruments.

Acknowledgement

This chapter benefits from the proof-reading by Corinna Kroner; Juliane Herrmann produced some of the figures. Both is gratefully acknowledged.

10.6 References

Ali OA (1986) Gravity and seismic refraction measurements for deep groundwater search in southern Darfur region, Sudan. The Journal of the University of Kuwait (Science) 13(2):245-257

Bath M (1974) Spectral analysis in geophysics. Elsevier, Amsterdam

Bonatz M (1967) Der Gravitationseinfluß der Bodenfeuchtigkeit. Zeitschrift für Vermessungswesen 92:135-139

Bott MPH (1959) The use of electronic digital computers for the evaluation of gravimetric terrain corrections. Geophysical Prospecting 74(3):565-572

Bower DR, Courtier N (1998) Precipitation effects on gravity measurements at the Canadian absolute gravity site. Phys. Earth Plant. Int. 106:353-369

Bracewell R (1965) The Fourier Transform and its Application. McGraw-Hill Book Comp., New York

Brigham EO (1974) The Fast Fourier Transform. Prentice Hall

Brown JM, Niebauer TM, Richter B, Klopping FJ, Valentine JG, Buxton WK (1999) Miniaturized Gravimeter May Greatly Improve Measurements. EOS Online supplement, http://www.agu.org/eos_elec/99144e.html

Bullard EC (1936) Gravity Measurements in East Africa. Phil. Trans. Roy. Soc. London 235:486-497

Buttkus B (2000) Spectral Analysis and Filter Theory in Applied Geophysics. Springer, Berlin Heidelberg New York

Cartwright DE, Edden AC (1973) Corrected tables of tidal harmonics. Geophys. J. R. Astr. Soc. 33:253-264

Cartwright DE, Taylor RJ (1971) New computations of the tide-generating potential. Geophys. J. R. Astr. Soc. 23:45-74

Chapin D (1998) Gravity instruments: Past, present, future. The leading edge January 1998:100-112

Debeglia N, Dupont F (2002) Some critical factors for engineering and environmental microgravity investigations. Journal of Applied Geophysics 50:435-454

Ehrismann W, Lettau O (1971) Topographische Reduktion von Schweremessungen in der näheren und weiteren Stationsumgebung mit Digitalrechnern. Archiv für Meteorologie, Geophysik und Bioklimatologie 20:383-396

Ehrismann W, Müller G, Rosenbach O, Sperlich N (1966) Topographic reduction of gravity measurements by the aid of digital computers. Boll. di Geof. Teor. ed Appl. 8:1-20

Elstner C, Kautzleben H (1982) Results of annual gravity measurements along a W-E profile inside the GDR for the period 1970-1980. In: Proc. Gen. Meeting of the I.A.G.:341-348, Tokyo

Ervin CP (1977) Theory of the Bouguer anomaly. Geophysics 42(7):1468

Fairhead JD, Gree CM, Blitzkow D (2003) The use of GPS in gravity surveys. The leading edge October 2003:954-959

Faller JE, Rinker RL, Zumberge MA (1980) Absolute gravity as a reconnaissance tool for vertical height changes and for studying density changes. In: Proc. Second Int. Symp. on Problems related to the Redefinition of North American Geodetic Networks, pp 919-932, Ottawa

Forsberg R (1985) Gravity field terrain effect computations by FFT. Bull. Géod. 59:342-360

Fricke S, Schön J (1999) Praktische Bohrlochgeophysik. ENKE im Georg Thieme Verlag, Stuttgart

Fuller B (1967) Two-dimensional frequency analysis and design of grid operators. Mining Geophysics 2:658-708

Gabriel G, Kirsch R, Siemon B, Wiederhold H (2003) Geophysical investigation of buried Pleistocene subglacial valleys in Northern Germany. In: Huuse M, Lykke-Andersen H, Piotrowski JA (eds) Geophysical Investigations of Buried Quaternary Valleys in the Formerly Glaciated NW European Lowland: Significance for Groundwater Exploration (Special Issue). Journal of Applied Geophysics 53:159-180, doi:10.1016/j.jappgeo.2003.08.005

Gardner GHF, Gardner LW, Gregory AR (1974) Formation velocity and density – the diagnostic basics for stratigraphic traps. Geophysics 39(6):770-780

Götze HJ (1984) Über den Einsatz interaktiver Computergraphik im Rahmen 3-dimensionaler Interpretationstechniken in der Gravimetrie und der Magnetik. Habilitationsschrift, Institut für Geophysik der TU Clausthal, Clausthal-Zellerfeld

Götze HJ, Lahmeyer B (1988) Application of three-dimensional interactive modelling in gravity and magnetics. Geophysics 53:1096-1108

Grant FS, West GF (1965) Interpretation theory in applied geophysics. McGraw Hill, New York

Hagiwara Y (1978) Recent non-tidal gravity changes during earthquake activities in Japan. Boll. Geof. teor. appl. XX:390-400

Hamilton EL (1978) Sound velocity-density relations in sea-floor sediments and rocks. Journal of the Acoustical Society of America 63(2):366-377

Hammer S (1939) Terrain corrections for gravimeter stations. Geophysics 4:184-194

Harnisch M, Harnisch G (1999) Hydrological influences in the registrations of superconducting gravimeters. Bull. Inf. Marees Terr. 131:10161-10170

Hartmann T, Wenzel G (1995) The HW95 tidal potential catalogue. Geophysical Research Letters 22(24):3553-3556

Hinze WJ (1990) The role of gravity and magnetic methods in engineering and environmental studies. In: Ward SH (ed) Investigation of geophysics no. 5: Geotechnical and environmental geophysics (Vol. 1). Society of Exploration Geophysicists, pp 75-126

Hinze WJ (2003) Bouguer reduction density, why 2.67? Geophysics 68:1559-1560

Imanishi Y (2000) Present status of SG T011 at Matsushiro, Japan. Cahier du Centre Européen de Géodynamique et de Séismologie 17:97-102

Jenkins GM, Watts DG (1968) Spectral Analysis and its Application. Holden-Day, San Francisco

Johnson GR, Olhoeft GR (1984) Density of rocks and minerals. In: Carmichael RS (ed) Handbook of physical properties of rocks. CRC Press 3:1-38

Jung K (1961) Schwerkraftverfahren in der angewandten Geophysik. Geest & Portig, Leipzig

Kanngieser E, Kummer K, Torge W, Wenzel HG (1983) Das Gravimeter-Eichsystem Hannover. Wiss. Arb. Univ. Hannover 120

Kroner C (2001) Hydrological effects on gravity data of the Geodynamic Observatory Moxa. J. Geod. Soc. Japan 47(1):353-358

La Fehr TR (1998) On Talwani's "Errors in the total Bouguer reduction". Geophysics 63(4):1131-1136

Li X, Götze HJ (2001) Ellipsoid, geoid, gravity, geodesy, and geophysics. Geophysics 66(6):1660-1668

Ludwig WJ, Nafe JE, Drake CL (1970) Seismic Refraction. In: Maxwell AE (ed) The Sea, vol. 4. Wiley-Interscience, New York, pp 53-84

Melchior P (1966) The Earth Tides. Pergamon Press Ltd., Oxford

Militzer H, Weber F (1984) Angewandte Geophysik, Band 1: Gravimetrie und Magnetik. Springer, Wien New York, and Akademie, Berlin

Moritz H (1984) Geodetic Reference System 1980. Bulletin Géodésique 58:388-398

Murty BVS, Raghavan VK (2002) The gravity method in groundwater exploration in crystalline rocks: a study in the peninsular granitic region of Hyderabad, India. Hydrogeology Journal 10(2):307-321

Nafe JE, Drake CL (1957) Variation with Depth in Shallow and Deep Water Marine Sediments of Porosity, Density and the Velocities of Compressional and Shear Waves. Geophysics 22:523-552

Nettleton LL (1939) Determination of density for the reduction of gravimeter observations. Geophysics 4:176-183

Niebauer TM, Sasagawa GS, Faller JE, Hilt R, Klopping F (1995) A New Generation of Absolute Gravimeters. Metrologia 32(3):159-180

Oartfield WJ, Czarnecki JB (1991) Hydrogeologic inference from drillers logs and from gravity and resistivity surveys in the Amargosa Desert, southern Nevada. Journal of Hydrology 124(1-2):131-158

Peter G, Klopping FJ, Berstis KA (1995) Observing and modeling gravity changes caused by soil moisture and groundwater table variations with superconducting gravimeters in Richmond, Florida, U.S.A. Cahier du Centre Européen de Géodynamique et de Séismologie 11:147-159

Rektorys K (1969) Differential geometry. In: Survey of Applicable Mathematics. Iliffe Books Ltd, MIT Press, Cambridge, pp 298-373

Rider M (1996) The Geological Interpretation of Well Logs (2nd edn). Whittles Publishing, Caithness

Roberts A (2001) Curvature attributes and their application to 3D interpreted horizons. First Break 19(2):85-100

Rosselli A, Olivier R, Veronese L (1998) Gravimetry applied to the Hydrogeological Research in the Large Alpine Valleys in Trentino Region. Studi Trentini di Scienze Naturali – Acta Geologica 75:53-64

Schleusener A (1940) Nomogramme für Geländeverbesserung von Gravimetermessungen in der angewandten Geophysik. Beitr. z. angew. Geophysik 8:415-430

Schöler W (1976) Die Beeinflussung der Schwere und des Vertikalgradienten durch das Gelände in Stationsnähe. Archiv für Meteorologie, Geophysik und Bioklimatologie 25:79-88

Seeber G (2003) Satellite Geodesy (2nd edn). de Gruyter, Berlin New York

Serra O (1984) Fundamentals of well log interpretation, Vol 1: the acquisition of logging data. Dev. Pet. Sci., 15A, Elsevier, Amsterdam

Sheriff RE (1984) Encyclopedic dictionary of exploration geophysics, 2nd edn. SEG, Tulsa

Sheriff RE, Geldart LP (1996) Exploration seismology Vol. 2: Data-processing and interpretation. Cambridge University Press, New York

Sideris MG (1985) A fast Fourier transform method for computing terrain corrections. Manuscripta Geodaetica 10:66-73

Stadtler C, Casten U, Thomsen S (2003) Anwendung der "maximum curvature" Methode auf Schweredaten zur Lokalisierung und Kartierung quartärer Rinnen in Südjütland (Dänemark). Extended abstract GGP02, 63th Annual Meeting of the German Geophysical Society (DGG)

Strange WE, Carroll DG (1974) The relation of gravity change and elevation change in sedimentary basins. EOS 56:1105

Swick CH (1942) Pendulum gravity measurements and isostatic reductions. U.S. Coast and Geodetic Survey Special Publication 232

Takemoto S, Fukuda Y, Higashi T, Abe M, Ogasawara S, Dwipa S, Kusuma DS, Andan A (2002) Effect of groundwater changes on SG observations in Kyoto and Bandung. Bull. Inf. Marées Terr. 136:10839-10848

Talwani M (1998) Errors in the total Bouguer reduction. Geophysics 63(4):1125-1130

Talwani M, Worzel JL, Landisman M (1959) Rapid gravity computations for twodimensional bodies with application to the Mendocino submarine fracture zone. J. Geophys. Res. 64:49-59

Talwani M, DiFrancesco D, Feldmann W (2001) System enables time-lapse gradiometry. American Oil & Gas reporter 44:101-108

Tamura Y (1993) Additional terms to the tidal harmonic tables. In: Hsu HT (ed) Proc. 12th Int. Symp. Earth Tides. Science Press, Bejing, pp 345-350

Tapley BD, Bettadpur SV, Ries JC, Thompson PF, Watkins MM (2004b) GRACE measurements of mass variability in the Earth system. Science 305:503-505

Tapley BD, Bettadpur S, Watkins M, Reigber Ch (2004a) The gravity recovery and climate experiment: Mission overview and early results. Geopys. Res. Lett. 31(5):L09607, doi:10.1029/2004GL019920

Telford WM, Geldart LP, Sheriff RE (1990) Applied Geophysics, 2nd edn. Cambridge University Press, Cambridge

Timmen L (1996) Absolutgravimetrie – Aufgaben und Anwendungen –. Zeitschrift für Vermessungswesen 121,6:286-159

Torge W (1989) Gravimetry. De Gruyter, Berlin New York

Valli T, Mattson A (1998) Gravity Method – an effective way to prospect groundwater areas in Finland. In: Casas A (ed) Proceedings of the IV meeting of the Environmental and Engineering Geophysical Society (European Section), Instituto Geográfico National, Madrid, pp 185-188

Virtanen H (2000) On the observed hydrological environmental effects on gravity at the Metsähovi station, Finland. Cahier du Centre Européen de Géodynamique et de Séismologie 17:169-176

Wahr JM (1980) Body tide of an elliptical, rotating, elastic and oceanless earth. Geophys. J. R. Astr. Soc. 64:677-704

Wahr J, Swenson S, Zlotnicki V, Velicogna I (2004) Time-variable gravity from GRACE: First results. Geophys. Res. Lett. 31(11):L11501, doi:10.1029/2004GL019779

Warburton RJ, Goodkind JM (1977) The influence of barometric-pressure variations on gravity. Geophys. J. R. astr. Soc (48): 281-292

Watermann H (1957) Über systematische Fehler bei Gravimetermessungen. Deutsche Geodätische Kommission, DGK C21, München

Wenzel HG (1996) The vertical gravimeter calibration line at Karlsruhe. Bulletin d'Informations Bureau Gravimetrique International 78:47-56

Wiederhold H, Agster G, Gabriel G, Kirsch R, Schenck PF, Scheer W, Voss W (2002) Geophysikalische Erkundung quartärer Grundwasserleiter im südlichen Schleswig-Holstein. Z. angew. Geol. 48:13-26

Wohlenberg J (1982) Density of rocks – Dichte der Gesteine. In Landolt-Börnstein – Zahlenwerte und Funktionen aus Naturwissenschaft und Technik, Neue Serie Va:113 - 119

Zerbini S, Richter B, Negusini M, Romagnoli C, Simon D, Domenichina F, Schwahn W (2001) Height and gravity variations by continuous GPS, gravity and environmental parameter observations in the southern Po Plain, near Bologna, Italy. Earth Planet. Sci. Lett. 192:267-279

Zurflüh E (1966) Application of two-dimensional linear wavelength filtering. Geophysics 32:1015-1050

11 Direct Push-Technologies

Peter Dietrich, Carsten Leven

It is widely accepted that a detailed knowledge of preferential flow and transport paths is essential for site characterization and for a reliable planning of site remediation. Conventional field investigations, however, are invariably hampered by insufficient information about these variations in the subsurface.

Commonly, boreholes and geophysical surface measurements are used for such investigation purposes. In case of boreholes, information can be obtained from cores and geophysical logging. Typically, borehole data have high vertical resolutions, but suffer from a lack of information in lateral directions, i.e. between the boreholes. This gap can be filled by the application of geophysical surface measurements which can provide horizontally continuous information. However, due to physical reasons the vertical resolution of surface methods decreases with depth.

An alternative approach for the site investigation is the use of Direct Push (DP) technology (also known as "cone penetration testing" or "direct drive technology"). This technology refers to a growing family of tools used to obtain subsurface investigations by pushing and/or hammering small-diameter hollow steel rods into the ground. By attaching gadgets to the end of the steel rods, it is possible to conduct high resolution logging of rock parameters as well as to collect soil, soil gas, and ground water samples. Besides the broad applicability of DP technology, it also allows for a reduction of costs for the installation of monitoring equipment and tomographic surveys.

Due to the development of new powerful machines and tools, the application of DP technology increased strongly during the last years and became a viable alternative to conventional methods for site investigation. With the new generation of DP machines several sounding locations can be completed per day. Furthermore, under ideal conditions (e.g. soft, unconsolidated sediments) depths of more than 50 m can be reached.

11.1 Logging tools

The most common application of DP technology is the recording of vertical profiles. In contrast to conventional borehole logging, DP methods are mainly set up to record data directly while driving the DP tools into the subsurface. As a consequence, the speed of data acquisition can be significantly increased and simplified by the use of DP technology which - in the end - also leads to a reduction of costs compared to the installation of drilled wells. In addition to the time aspect, the application of the DP technology also provides further advantages – compared to conventional drilling – as it allows the employment of much smaller and more flexible systems that require much less physical labor due to their simpler operation. As the DP tools are driven into the subsurface without the use of drilling to remove soil to make a path for the tool, the disturbance of the subsurface is less as well as no drill cuttings are generated (Thornton et al. 1997).

The tremendous variety of available equipment for Direct Push allows the logging of geophysical, geotechnical, hydrological and hydrogeochemical data. In this context it is worthwhile to mention also tools that are developed for measurements during auger drilling (e.g. Sørensen 1989, Sørensen et al. 2003). The combined interpretation of such logging data can therefore enable a reliable characterizing of subsurface structures (Schulmeister et al. 2003, Schulmeister et al. 2004, Sellwood 2005), e.g. as shown in Fig. 11.1. In the following sections an overview of Direct Push tools is given.

11.1.1 Geotechnical tools

One of the oldest applications of Direct Push technologies is the geotechnical investigation of the subsurface. Probably the best known technique is the Cone Penetrometer Testing (CPT) that is used for determining the subsurface stratigraphy in situ and to estimate geotechnical parameters of the subsurface material. In a CPT test, a cone-shaped probe (penetrometer) at the end of a string of steel rods is driven into the ground at a constant rate. This method was already developed in the 1920s in Holland and was originally used for measuring the sleeve friction and mechanical tip resistance, which occur when the cone is driven into the subsurface. Measurements of sleeve friction and mechanical tip resistance allow for distinguishing between different kinds of unconsolidated sediments (e.g. Robertson 1990). Sand, e.g. in general has low sleeve friction and high tip resistance (Robertson and Campanella 1983a), while till can be identified by high sleeve friction and low tip resistance (Robertson and Campanella 1983b).

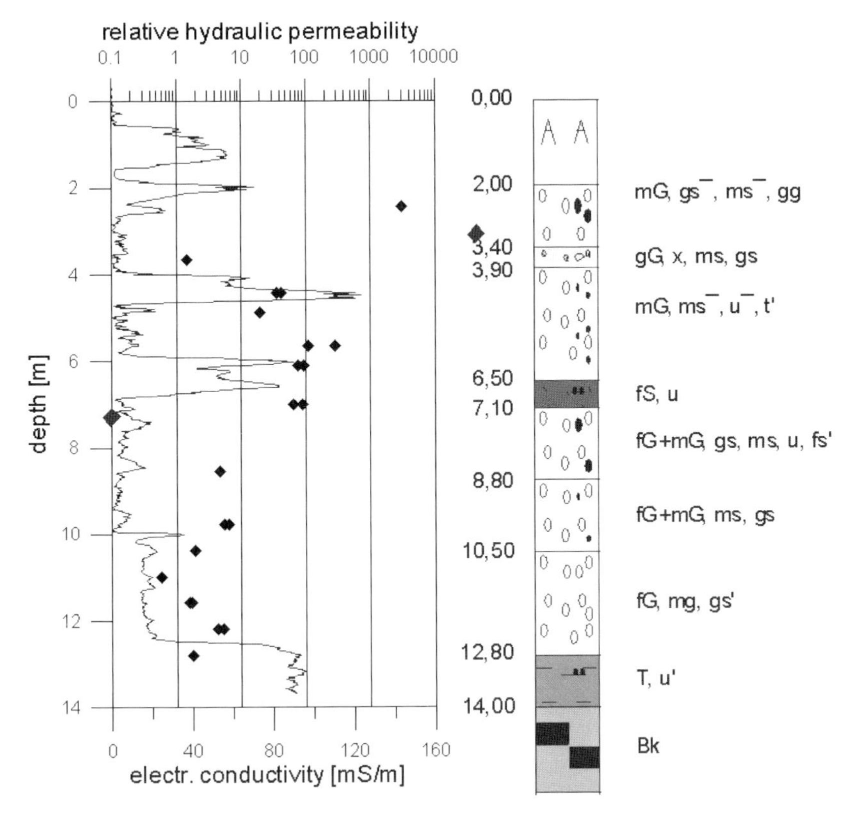

Fig. 11.1. Comparison of an electrical conductivity profile (solid line) and a profile of relative hydraulic permeability (squares) with the corresponding borehole profile

Modern CPT probes are additionally equipped with inclinometers that allow the determination of deviations from the vertical during the penetration, thus an exact determination of the three-dimensional position of each measuring point is possible. Other CPT tools include pore pressure sensors or are combined with other geophysical tools such as the electrical conductivity probe (section 11.1.2). Application of CPT measurements in groundwater investigations are described e.g. by Smolley and Kappmeyer (1991), Zemo et al. (1994), and Tillman and Leonard (1993).

Another geotechnical DP tool is the beat counter. With this tools the beats with a defined force are counted, which are necessary for a certain penetration distance. The interpretation of beat count profiles in terms of subsurface material is based on empirical relations.

11.1.2 Geophysical tools

Probably, the best known geophysical DP tool is the conductivity probe that measures the electrical conductivity while the DP rods are pushed into the subsurface (e.g. Beck et al. 2000, Schulmeister et al. 2003, Sellwood 2005).

Because clay and silty units have often a characteristically increased electrical conductivity (Chap. 1), the conductivity probe is mainly used for the delineation of low permeable zones in the subsurface. Usually, the results of the vertical profiles originating from several different locations are correlated and interpreted in terms of the extent of hydraulically low permeable zones (Fig. 11.2).

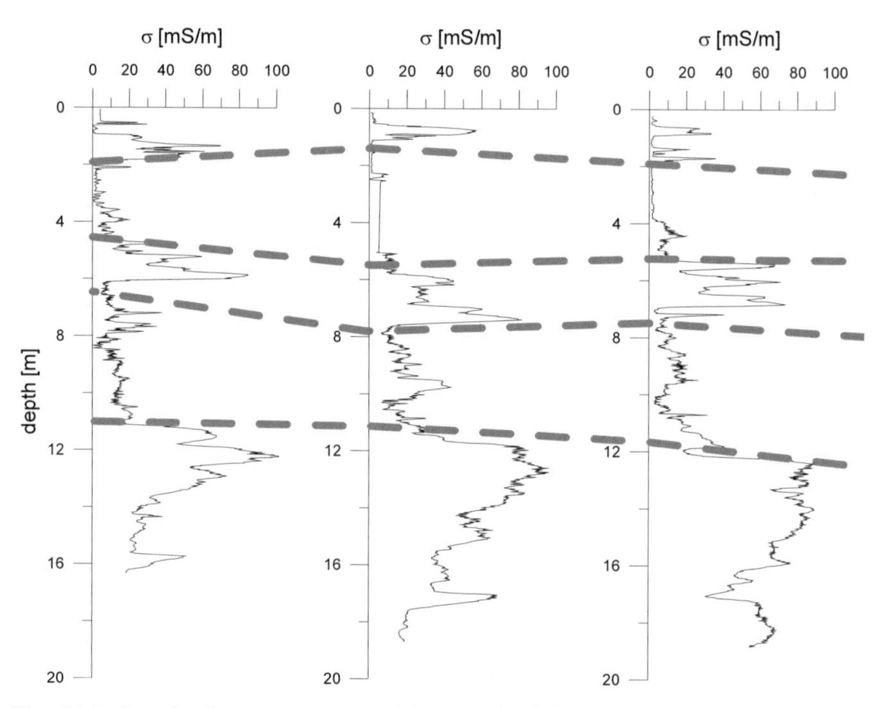

Fig. 11.2. Result of measurements with a conductivity probe with a correlation of main units indicated by the dashed lines. The distance between the sounding locations is 10 m

A further application of geoelectrical probes is the measurement of induced polarization or spectral induced polarization. For these measurements, the frequency behavior is recorded in addition. For background information we refer to Chap. 4.

Special tools equipped with geophones allow for a profiling of seismic velocities (Robertson et al. 1986b, Terry et al. 1996). While advancing the DP string with geophones positioned near the tip, seismic signals are generated at the surface and the signals are recorded with the geophones in the DP string. From this seismic record, the seismic velocities can be derived. If three-component geophone systems are used, the velocity of p- and s-wave can be determined (Fig. 11.3 and Fig. 11.4). With both seismic velocities, dynamic elastic modules can be calculated (Eqs. 13.1 – 13.4).

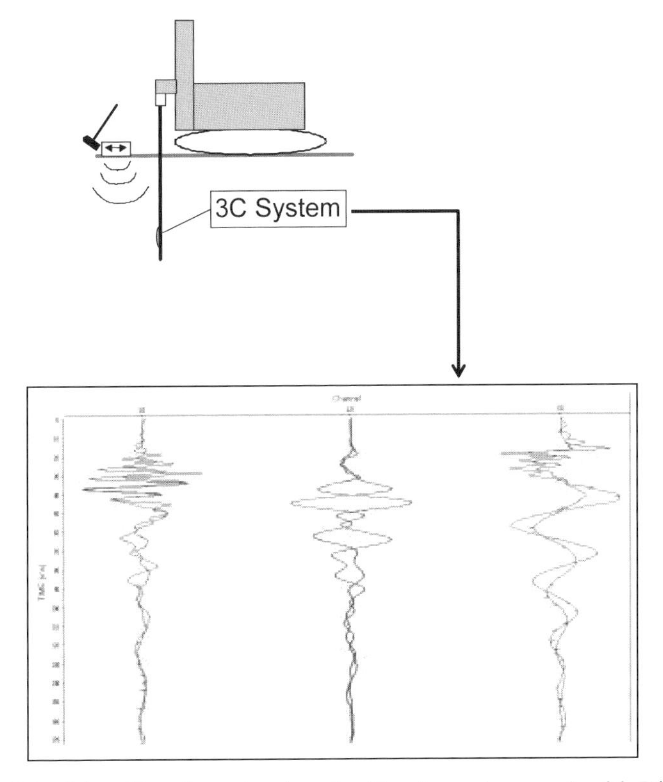

Fig. 11.3. Principle of logging seismic velocities (p- and s-wave) with Direct Push

Another possibility for the determination of velocity distributions using the DP technology is the use of a seismic source in the tip. Particularly, p-wave sources are very robust and can be applied for this purpose in case of rough and hard subsurface conditions. Another advantage of the source in the tip is that the arrival of the induced seismic signal can simultaneously be recorded with geophones at the surface using different geophone offsets. As it will be shown later, with this arrangement, the velocity distributions can be determined along vertical profiles and along two-dimensional vertical sections (Fig. 11.7).

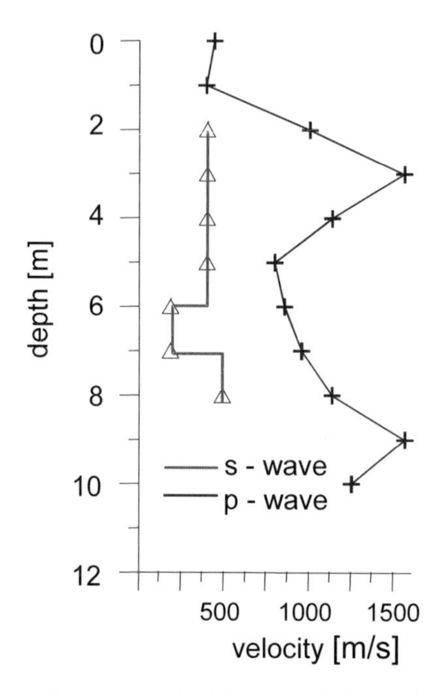

Fig. 11.4. Profiles of s- and p-wave velocities recorded with DP tools

Other geophysical DP tools comprise nuclear logging tools that either detect natural γ-radiation or emit radiation (γ or neutron) and measure the response of the surrounding material. They are applied in hollow steel rods which are pushed in the subsurface. The data of nuclear logging are interpreted in terms of clay content (natural γ-radiation), hydrogen content (neutron-neutron-measurements) or rock density (γ–γ-measurements).

11.1.3 Hydroprobes

One of the most important rock parameters for hydrogeological site characterization is the hydraulic conductivity k (see Chap. 15). A variety of approaches have been developed to determine this parameter with numerous technologies. Among these, the Direct Push technology is a very promising approaches to obtain k estimates at a resolution and accuracy that is rarely been possible by conventional field investigation techniques.

One group of DP tools for estimating hydraulic conductivity comprises CPT surveys. The simplest CPT approach is the use of empirical relationships which are based on sediment classification information (Robertson et al. 1986a, Farrar 1996). Generally, this approach delivers (at best) only order-of-magnitude estimates of formation conductivity. Further CPT methods for the determination of k values use pore-pressure dissipation tests data, which are obtained by arresting the CPT probe (e.g., Baligh and Levadoux 1980, Robertson et al. 1992, Abu-Farsakh et al. 1998, Sully et al. 1999). Estimates of hydraulic conductivity from these analyses involve great uncertainty (Lunne et al. 1997) in determining volume compressibility from one of many empirical equations. Alternatively, the use of maximum pore pressure magnitudes developed during advancing the CPT tool provides a one-step method to determine hydraulic conductivity magnitudes (Elsworth 1991, Elsworth 1993, Voyiadjis and Song 2003). The applicability of such methods depends on the validity of the underlying soil or stress distribution models.

A further group of DP tools for the characterization of hydraulic conductivity distributions includes injection tests. Fejes et al. (1997) suggest a method, which uses the infiltration rate for the estimation of k values. Other injection approaches use the ratio of injection rate to injection pressure. The required measurements (injection pressure and rate) are done during the advancement of the rods (Pitkin 1998, Pitkin and Rossi 2000) or at selected depth (Butler and Dietrich 2004). However, the effects of screen clogging and the zone of compaction can introduce considerable uncertainty into the results of injection methods.

The most reliable k estimates can be gathered from Direct Push slug tests (e.g., Hinsby et al. 1992, Henebry and Robbins 2000, Butler et al. 2002, McCall et al. 2002) and the Direct Push permeameter (e.g., Lowry et al. 1999, Mason and Lowry 1999, Butler and Dietrich 2004). A Direct Push slug test consists of measuring the recovery of head in the DP rod with an open screen after a near-instantaneous change in head in the DP rod. The recorded head data are analyzed using conventional methods for the interpretation of slug tests. For a detailed description of different methods see Butler (1998). In the case of a high hydraulic conductivity, corrections for slug tests must be applied due to the small-diameter of Direct Push pipes (Butler 2002).

The Direct Push permeameter is a tool with an injection port consisting of an unshielded screen at its lower end and pressure transducer ports placed along the body of the tool at different distances above the injection port. Constant-rate injection tests are performed at each test interval by injecting water through the screen at a constant rate, while monitoring pressure changes at the transducer locations. Once the induced head gradient between the transducers has stabilized, k estimates are obtained from the injection rate and the steady-state pressure drop between the pressure transducers. The DP permeameter is relatively rapid and is virtually insensitive to screen clogging and the zone of compaction (Lowry et al. 1999).

In Fig. 11.5 a comparison of DP permeameter data with DP slug tests and injection logger data is shown. The injection logging data are calibribtated against the DP slug test data (Dietrich et al. 2003).

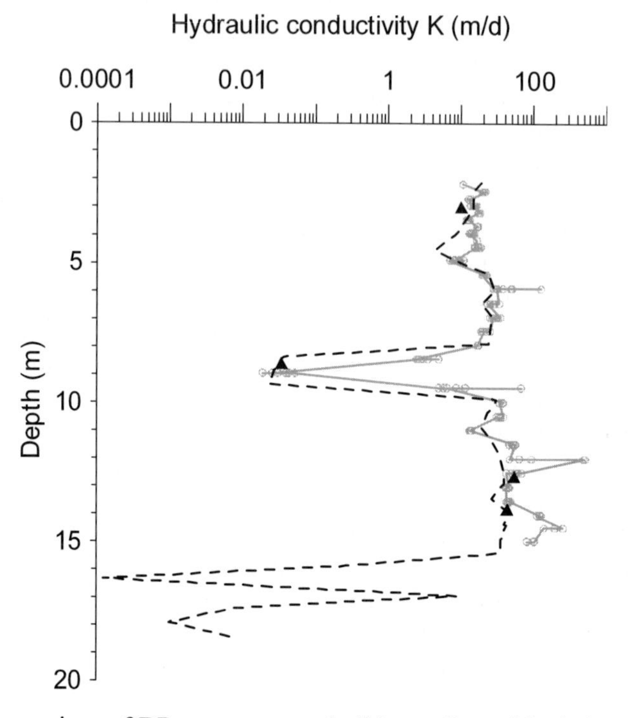

Fig. 11.5. Comparison of DP permeameter (solid grey line with circles) K profile with DP slug tests (black triangles) and calibrated DP injection logger results (dashed line) from nearby locations (after Dietrich et al. 2003)

11.1.4 Hydrogeochemical tools

Besides the determination of the rock parameters, DP technologies can be used for the exploration of the spatial extension of contaminations with volatile and semi-volatile organic compounds. In this context, the possibility to resolve very thin layers, which are critical in terms of the migration of contaminants, is an important advantage of DP versus well based investigations. In the following two different classes of hydrogeochemical DP tools are introduced:

The first class comprises in-situ spectroscopy tools, e.g. for the detection of hydrocarbons by fluorescence (e.g. Kram 1998, Kram et al. 2001). Since the fluorescence intensity is proportional to petroleum hydrocarbon concentration, this method allows for delineating the extent of affected subsurface material. The induced light is produced by a pulsed laser. Therefore, the applied method is also called Laser-Induced Fluorescence (LIF). One example for a LIF tools is the Rapid Optical Screening Tool (ROST™, Neuhaus 2001) that measures relative concentrations and a spectral product fingerprint continuously in real time.

The second class of chemical tools consists of systems, which have been designed to capture volatile contaminations. Examples are the Membrane Interface Probe (MIP), Hydrosparge™, and the Thermal Desorption Sampler (TDS).

For the detection of chlorinated hydrocarbons and BTEX in the subsurface, the Membrane Interface Probe system (MIP) is used. The MIP tool mobilizes volatile organic compounds (VOC) like PCE, TCE and their degradation products as well as BTEX - present as gaseous, dissolved, solid or free product phases – so that they move by diffusion through a heated membrane on the cone's sleeve. These compounds are then transported through a carrier gas stream up through the DP rod to a gas chromatograph where the VOC compounds are detected by special detectors, such as flame ionization detectors (FID), photo-ionization detectors (PID), and direct sampling ion trap mass spectrometers (DELCD). The major advantage of the MIP tool is that the VOCs are removed in situ from the soil matrix allowing the use in the saturated and unsaturated zone as well as in coarse-grained and fine-grained formations.

The Hydrosparge™ system is designed to collect VOCs from groundwater for on-site analysis. VOCs are extracted from groundwater out of a sampling interval by physically sparging and transporting them to the surface for analysis. As the Hydrosparge™ system is lowered through a DP groundwater sampler that has been advanced into the water column, VOCs can be analyzed during advancement without retrieving the DP rods or handling samples, thus leading to increased efficiency and precision as well as to a cost reduction over traditional sampling methods.

The Thermal Desorption Sampler (TDS) is specifically used for the in-situ characterization of vadose zone soils. The TDS system consists of a special probe that collects a soil plug into a chamber where it is heated while purging and transporting VOCs to the surface for analysis. The system can also be used to collect VOCs onto analytical traps for later analysis. The TDS is designed for screening on-site soils only.

Besides these analytical systems for detecting organic contaminants, other DP systems have been developed for detecting inorganic contaminants in the subsurface. One example for such DP systems is based on the x-ray fluorescence technology (XRF). XRF is a well-established, non-destructive laboratory and field screening method for the detection, identification, and delineation of heavy metal contaminants in the subsurface, and can be applied in both the unsaturated and saturated zones.

In XRF analysis the photoelectric effect is used. Fluorescent x-rays are produced by radiating the soils with an x-ray source with a definite excitation energy that will partly be absorbed by the target elements resulting in an energy emission in the form of x-rays. Hereby, each element emits a unique x-ray at a characteristic energy level or wavelength. The elements present in a sample can be identified qualitatively by analyzing the energy of the characteristic x-rays, while a quantitative analysis can be performed by measuring the intensity of the x-ray as the intensity of the x-rays is proportional to the concentration.

More information about chemical tools can be found in Nielson (1994) Lambson & Jaobs (1995), U.S. EPA (1995), Bujewsk & Rutherford (1997), Kram (1998), Neuhaus (2001), Kram et al. (2001a), and Rogge et al. (2001).

11.1.5 Miscellaneous other tools

For in situ analysis of lithological properties, video imaging systems have also been developed for Direct Push (Van den Boogaart et al. 2001). Two approaches for the systems can be distinguished. The first approach is characterized by mounting the camera directly into the probe. The images from the camera are sent to the surface and can be viewed at a video monitor. The second approach consists of an optical system with lenses and light fibers inside the rods. The camera and the video monitor are on site.

Temperature profiles can also be measured in combination with Direct Push. As for nuclear logging tools, the measurement is carried out by inserting a set of temperature sensors in hollow steel rods which are driven in the subsurface. However, the temperature sensors need a certain time for adaptation to the local temperature. Examples for application of temperature measurements in combination with Direct Push include locating leaks in embankments and containment as well as determination of plume extensions from waste disposals.

11.2 Sampling tools

Direct Push technology is already accepted and widely used for sampling (ASTM D6001, Chiang et al. 1992, U.S. A.C.E. 1996, Scaturo and Widdowson 1997, U.S. EPA 1998, Pitkin et al. 1999, Jacobs et al. 2000). In the following paragraphs, an overview of tools for sampling of soil, soil gas, and groundwater is given.

11.2.1 Soil sampling tools

For the collection of soil samples from discrete depths, DP sampling tools are designed in the way of retrieving the samples without the necessity of removing the overlying soil. Samplers with different diameters and lengths enable the sampling with different volumes.

The simplest sort of soil sampler consists of a hollow sampling tube with a special driving cone. The tool is driven to the desired depth, the driving cone of the sampling tube is loosened and drawn back, so that the hollow sampling tube with its sharpened end can be advanced to the required depth to be filled with unconsolidated material. The entire assemblage is then brought back to the surface.

Other systems of soil sampling tools use dual-tube samplers existing of hollow rods as an outer casing acting as a support for the borehole. With a smaller set of rods, a sample liner is inserted to collect the soil samples inside the outer casing.

11.2.2 Soil gas sampling tools

Soil gas samples are typically used to detect elevated concentrations of VOCs in soil gases which provides information about vadose-zone contaminants and the distribution and concentration of VOC in soils and groundwater.

DP soil gas samplers can be divided into two groups comprising continuous and discrete sampling tools. The first group of tools is continuously sampling the soil gas as the tool is driven into the subsurface. The gas samples, which are transported to the surface by pumps or inertial displacement, can be analyzed as they are collected using PID or FID detectors or they are collected for on- or off-site analysis.

The second group of soil gas sampling tools is employed for discrete sampling. The tool is pushed to the desired depth and the soil gas will be sampled from the exposed soil at the tip of the tool into a sample chamber.

The tool is then removed from the DP hole to collect the sample. Because the sampling occurs at exactly known depths with this kind of tool, contamination plumes or sources can be located with high precision.

11.2.3 Groundwater sampling tools

In the same way as soil gas sampling tools, groundwater sampling tools can basically be divided into two groups: continuous or discrete groundwater samplers. The first group of continuous tools consists of a screened sampler with a collection port to sample groundwater from the DP hole at different depths. The sample is transported to the surface using a pump, bailer, or any other suitable devices. After sampling, the tool is driven to the next sampling interval.
Tools for discrete sampling - the second group of groundwater sampling tools –consist of a sampling screen that is pushed out from the bottom of the DP rods. The tool is pushed to the desired depth for sampling and is then removed to collect the sample. Other tools for discrete sampling incorporate a sampling chamber, which is evacuated in advance, to preserve gases and volatile compounds dissolved in groundwater to interact with the atmosphere, and to prevent any alterations caused by pumping or bailing.

11.3 Tomographic applications

Despite the fact that DP technology leads to vertical information with a high resolution along the measured profiles, it suffers - as other borehole derived data – from the lack of information in lateral directions, i.e. between the individual vertical profiles. As already mentioned in the beginning, the application of geophysical surface measurements can provide horizontally continuous information but suffer from a lack in vertical resolution. An alternative approach to compensate the poor lateral resolution of single borehole methods and the poor vertical resolution of surface geophysical methods is the application of tomographic methods. Tomographic measurements are usually carried out between two boreholes. Therefore, tomographic measurements greatly depend on the availability and the positioning of boreholes on a site. Moreover, if no boreholes are available or their positioning is inappropriate with respect to the goal of investigation, the installation of one or more boreholes would be necessary for employing tomographic surveys. As a consequence, the costs of a survey would increase dramatically with the installation of new wells. In this context, the selection of the best location for drilling boreholes for a tomographic survey can be viewed as a challenging task, since the optimal position and extent of the tomographic plane strongly depends on the subsurface situation.

However, the use of DP technology offers new possibilities for tomographic measurements as tools that provide the source for a tomographic array can be driven in the same manner as logging tools. Therefore, only one borehole – as location for the receivers – is sufficient (Fig. 11.6).

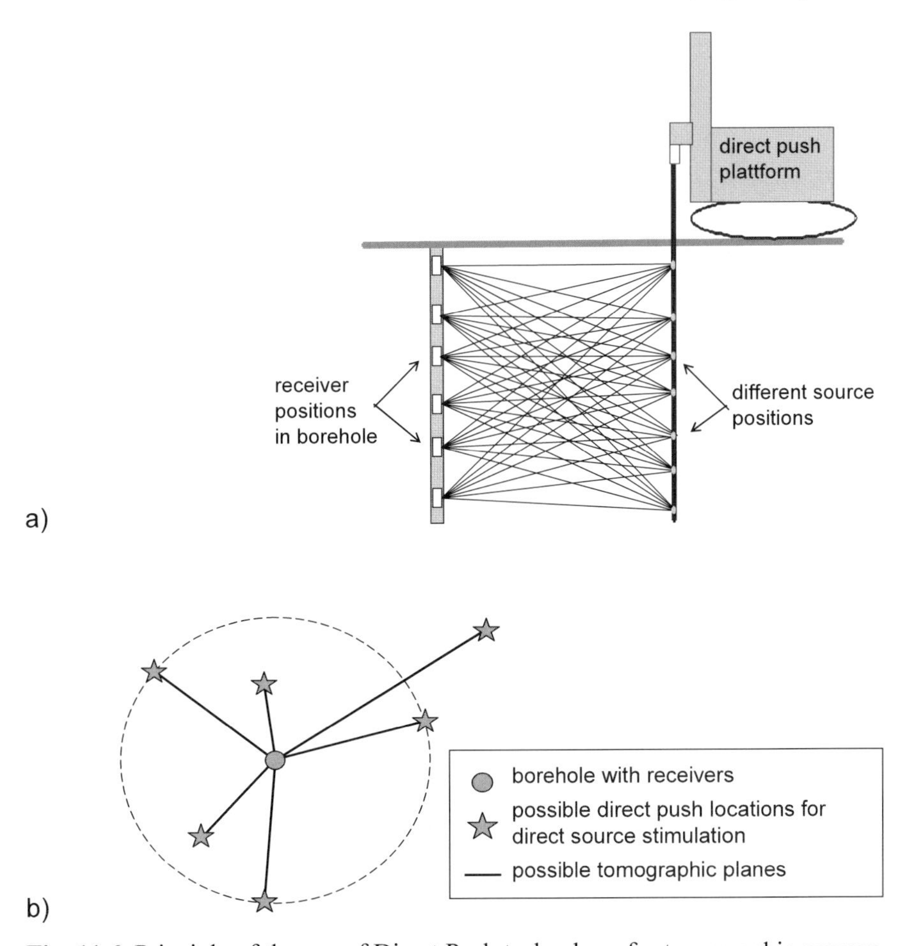

Fig. 11.6. Principle of the use of Direct Push technology for tomographic surveys. a) side view (each line represents a connection between a source and a receiver position), b) top view

The tomographic measurements are carried out by stimulating the source at different depths during the advancement of the DP rod (Fig. 11.6a), while the induced source signal is recorded by the receivers at fixed depth in the conventional borehole. A major advantage of the DP technology over conventional methods is the flexibility of the DP equipment, which allows performing tomographic surveys in different directions and over different distances in a flexible and efficient way (Fig. 11.6b) as well as the combination with surface surveys (Fig. 11.7).

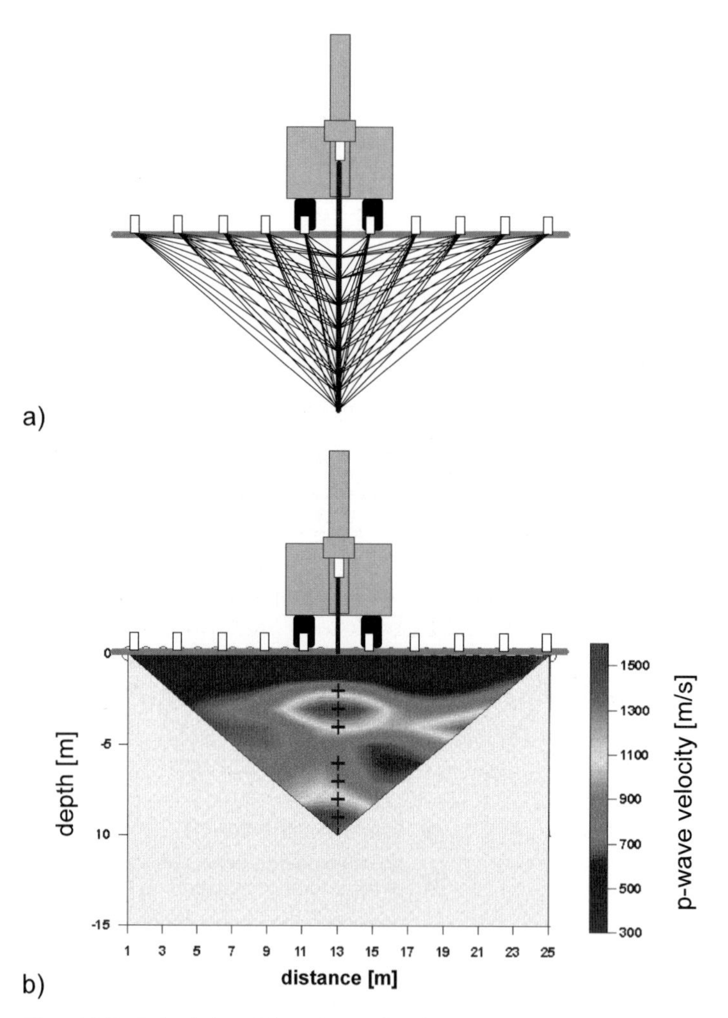

Fig. 11.7. Principle of logging seismic velocities with several geophones at the surface and a source in the tip of DP-rod. a) Scheme of measurements (each line represents a connection between a source position and a geophone). b) Result of the interpretation of measured travel times

11.4 Permanent installations

Direct Push technology also allows the installation of temporary and permanent wells and piezometers that are used for measuring and monitoring purposes (ASTM D6724, ASTM D6725, Kram 2001, Kram et al. 2001b, Kram et al. 2003). Such DP wells incorporate screened intervals with prepackaged filters and are driven into the subsurface with special DP rods that are redrawn. A gravel pack and grouting can be installed in the same manner as for a conventional well.

Another application of DP technology is linked to multi-port systems that are realized with inflatable membranes holding and pressing sampling or measuring devices such as geoelectrical probes against the wall of the DP hole. For water sampling, small pumps can be installed very inexpensive in different depths. The installation of packers at various depths isolating individual sampling intervals is another possibility.

Interesting applications for permanent DP installations include site characterization with pump or slug tests (Butler and Dietrich 2004, Butler et al. 2002), the observation of remediation processes (e.g. Ramirez et al. 1993, Spies and Ellis 1995), and detecting tank leaks (e.g. Binley et al. 1997).

Another example is the installation of geoelectrical electrodes in the subsurface for monitoring purposes in order to investigate ground water flow conditions (e.g. White 1988, White 1994, Dietrich 1999). For this purpose, electrode chains can be installed in an efficient way using DP technology downstream from a salt tracer injection (Fig. 11.8). The electrode chains are arranged in a control plane which allows the measurement of electrical breakthrough curves (Fig. 11.9).

11.5 Conclusions

The Direct Push technology allows a cost-effective, rapid sampling and data collection along vertical profiles in unconsolidated soils and sediments. By using different kinds of logging tools at the same location, it is possible to determine site specific relationships between geophysical, geotechnical and hydrogeological parameters. Furthermore, DP technology can be used for the installation of monitoring equipment and tomographic surveys. Based on its possibilities, Direct Push is a very useful supplement for borehole and surface measurements. Particularly, in combination with geophysical surface measurement, Direct Push investigations help to develop detailed three-dimensional subsurface models.

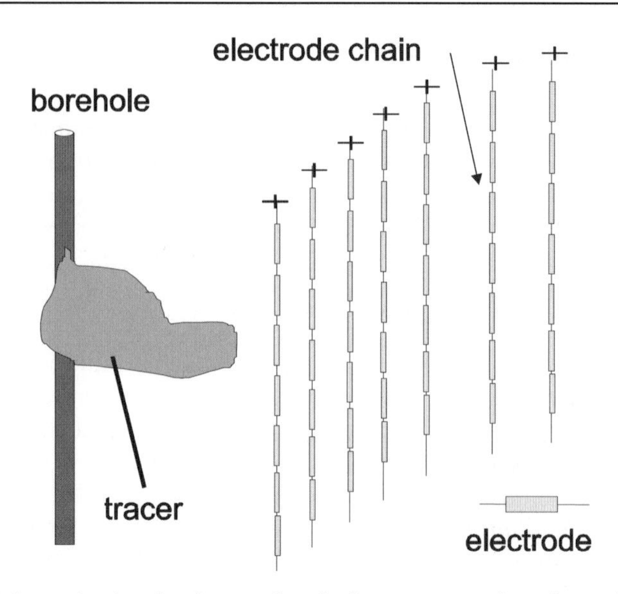

Fig. 11.8. Schematic sketch of a geoelectrical tracer test using electrode chains arranged which are installed by Direct Push along a control plain

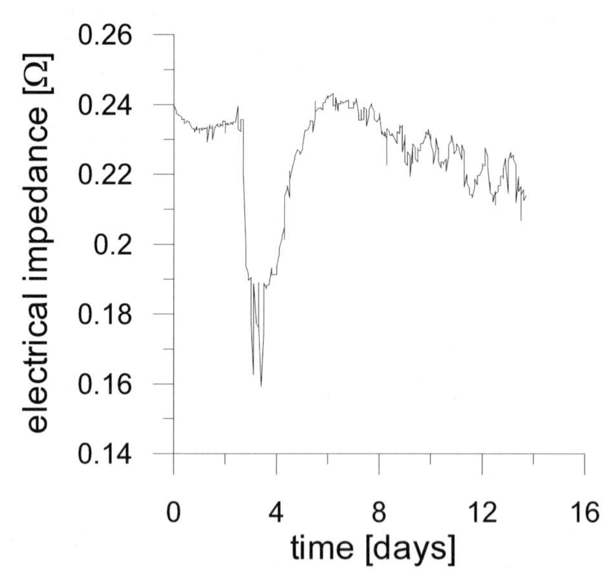

Fig. 11.9. Example of a "geoelectrical breakthrough curve". The breakthrough of a salt tracer can be detected by the decrease in electrical impedance

11.6 References

Abu-Farsakh MY, Voyiadjis GZ, Tumay MT (1998) Numerical analysis on the miniature piezocone penetration tests (PCPT) in clays. Int J Numer Analyt Meth Geomech 22:791–818

ASTM D6001, Standard guide for direct-push water sampling for geoenvironmental investigations. Annual Book of ASTM Standards, v.04.08

ASTM D6724, Standard guide for installation of direct push ground water monitoring wells. Annual Book of ASTM Standards, v.04.08

ASTM D6725, Standard practice for direct-push installation of prepacked screen monitoring wells in unconsolidated aquifers. Annual Book of ASTM Standards, v.04.08

Baligh MM, Levadoux JN (1980) Pore pressure dissipation after cone penetration. Report R80-11 Dept of Civil Engineering, Massachusetts Institute of Technology

Binley A, Daily W, Ramirez A (1997) Detecting leaks from environmental barriers using electrical current imaging. J of Environmental & Engineering Geophysics 2:11-21

Beck FP, Clark PJ, Puls RW (2000) Location and characterization of subsurface anomalies using a soil conductivity probe. Ground Water Monitoring and Remediation 20:55-59

Bujewski G, Rutherford B (1997) The Rapid Optical Screening Tool (ROST TM) Laser-Induced Fluorescence (LIF) System for screening of petroleum hydrocarbons in subsurface soils. — US EPA Innovative Technology Verification Report EPA/600/R-97/020

Butler JJ Jr (1998) The design, performance, and analysis of slug tests. Boca Raton, Lewis Publishers

Butler JJ Jr (2002) A simple correction for slug tests in small-diameter wells. Ground Water 40:303-307

Butler JJ Jr, Dietrich P (2004) New methods for high-resolution characterization of spatial variations in hydraulic conductivity. In: Proc. of International Symposium on Hydrogeological Investigation and Remedial Technology, National Central University, Jhongli, Taiwan, pp. 42-55

Butler JJ Jr, Healey JM, McCall GW, Garnett EJ, Loheide SP (2002) Hydraulic tests with direct-push equipment. Ground Water 40:25-36

Chiang CY, Loos KR, Klopp RA (1992) Field determination of geological / chemical properties of an aquifer by cone penetrometry and headspace analysis. Ground Water 30:428-436

Dietrich P (1999) Konzeption und Auswertung gleichstromgeoelektrischer Traceversuche unter Verwendung von Sensitivitätskoeffizienten.- Tübinger Geowissenschaftliche Arbeiten (TGA), Reihe C, ISSN 0935-4948, Nr. 50:130

Dietrich P, Butler JJ Jr, Yaramanci U, Wittig V, Tiggelmann T, Schoofs S (2003) Field comparison of direct-push approaches for determination of K-profiles. Eos 84: F661

Elsworth D (1991) Dislocation analysis of penetration in saturated porous media. Journal of Engineering Mechanics 117: 391-408

Elsworth D (1993) Analysis of piezocone dissipation data using dislocation methods. Journal of Geotechnical Engineering 119: 1601-1623

Farrar JA (1996) Research and standardization needs for direct push technology applied to environmental site characterization. In: Sampling Environmental Media, ed. by Morgan, J. H., ASTM Special Technical Publication 1282, American Society for Testing and Materials, Philadelphia:, pp 93-107

Fejes I, Szabadvary, Vero L (1997) Geophysikalische Penetrationssondierungen, In: Knödel K, Krummel H, Lange G, Handbuch zur Erkundung des Untergrundes von Deponien und Altlasten, Bd. 3 - Geophysik, pp 897-922

Henebry BJ, Robbins GA (2000) Reducing the influence of skin effects on hydraulic conductivity determinations in multilevel samplers installed with direct push methods. Ground Water 38: 882-886

Hinsby K, Bjerg PL, Andersen LJ, Skov B, Clausen EV (1992) A mini slug test method for determination of a local hydraulic conductivity of an unconfined sandy aquifer. Journal of Hydrology 136: 87-106

Jacobs JA, Kram M, Lieberman S (2000) Direct push technology sampling methods, In: Standard Handbook of Environmental Science, Health, and Technology, ed. by Lehr, J., McGraw Hill, pp. 11.151-11.163

Kram ML, Lorenzana D; Michaelsen J; Major W, Parker L; Antwort C; McHale T (2003) Direct-push wells prove effective for long-term ground water monitoring. Water Well Journal 57:16 - 19

Kram M (2001). Direct-push versus HSA drilled monitoring wells. RPM News, Spring 2001:6-7

Kram ML, Lieberman SH, Fee J, Keller AA (2001a) Use of LIF for real-time insitu mixed NAPL source zone detection. Ground Water Monitoring and Remediation 21:67-76

Kram ML., Lorenzana D, Michaelsen J, Lory E (2001b) Performance comparison: direct-push wells versus drilled wells. Naval Facilities Engineering Service Center Technical Report, TR-2120-ENV, January 2001

Kram ML (1998) Use of SCAPS Petroleum Hydrocarbon Sensor Technology for real-time indirect DNAPL detection. Journal of Soil Contamination 7:73-86

Lambson M, Jaobs PA (1995) The use of the Laser Induced Fluorescence Cone for environmental investigations. Proceedings of the International Conference on Cone Penetration Testing, CPT '95, Linköping, Sweden

Lowry W, Mason N, Chipman V, Kisiel K, Stockton J (1999) In-situ permeability measurements with direct push techniques: Phase II topical report, SEASF-TR-98-207 Rept to DOE Federal Energy Tech Center

Lunne T, Robertson PK, Powell JJM (1997) Cone penetration testing in geotechnical practice. London, Blackie Academic and Professional

Mason N, Lowry W (1999) In-situ permeability measurements with direct push techniques: Phase III topical report, SEASF-TR-99-223 DOE Fed Energy Tech Center

McCall W, Butler JJ Jr, Healey JM, Lanier AA, Sellwood SM, Garnett EJ (2002) A dual-tube direct-push method for vertical profiling of hydraulic conductivity in unconsolidated formations. Environmental and Engineering Geoscience: 8:75-84

Neuhaus M (2001) Spatial In Situ Delineation of Soil and Groundwater Contamination with Environmental CPT. In: Proceedings of the conference "Field Screening Europe 2001", Karlsruhe, Germany, pp 71-78

Nielsen BJ (1994) New tools to locate and characterize oil spills in aquifers. Symposium on Natural Attenuation of Ground Water, EPA/600/R-94/162

Pitkin SE (1998) Detailed subsurface characterization using the Waterloo Profiler. Proc of the 1998 Symp on the Application of Geophysics to Environmental and Engineering Problems, EEGS, pp 53-64

Pitkin SE, Cherry JA, Ingleton RA, Broholm M (1999) Field demonstrations using the Waterloo Ground Water Profiler. Ground Water Monitoring and Remediation 19:122-131

Pitkin SE, Rossi MD (2000) A real time indicator of hydraulic conductivity distribution used to select groundwater sampling depths (abstract). Eos 81:239

Ramirez A, Daily W, LaBrecque D, Owen E, Chesnut D (1993) Monitoring of an underground steam injection process using electrical resistance tomography. Water Resources Research: 29:73-87

Robertson PK (1990) Soil classification using the cone penetration test. Can Geotech. J 27: 151-158

Robertson PK, Campanella RG (1983a) Interpretation of cone penetration tests: Part I: Sand. Canadian Geotechnical Journal 20:719-733

Robertson PK, Campanella RG (1983b) Interpretation of cone penetration tests: Part II: Clay. Canadian Geotechnical Journal 20:734-745

Robertson PK, Campanella RG, Gillespie D, Greig J (1986a) Use of piezometer cone data. Proc. ASCE Spec. Conf. In Situ '86. Use of In Situ Tests in Geotechnical Engineering. Blacksburg, pp 1263-1280

Robertson PK, Campanella RG, Gillespie D, Rice A (1986b) Seismic CPT to measure in situ shear wave velocity. Journal of Geotechnical Engineering 112:791-803

Robertson PK, Sully JP, Woeller DJ, Lunne T, Powell JJM, Gillespie DG (1992) Estimating coefficient of consolidation from piezocone tests. Canadian Geotechnical Journal 29:539-550

Rogge M, Christy TM, De Weirdt F (2001) Site Contamination Fast De-lineation and Screening Using the Membrane Interface Probe. In: Proceedings "Field Screening Europe 2001", Karlsruhe, Germany, pp 91-98

Scaturo DM, Widdowson MA (1997) Experimental evaluation of a drive-point ground-water sampler for hydraulic conductivity measurement. Ground Water 35:713-720

Schulmeister MK, Butler JJ Jr, Healey JM, Zheng L, Wysocki DA, McCall GW (2003) Direct-push electrical conductivity logging for high-resolution hydrostratigraphic characterization. Ground Water Monitoring and Remediation 23:52-62

Schulmeister MK, Healey JM, Butler JJ Jr, McCall GW (2004) Direct-push geochemical profiling for assessment of inorganic chemical heterogeneity in aquifers. Journal of Contaminant Hydrology 69:215-232

Sellwood SM, Healey JM, Birk S, Butler JJ Jr (2005) Direct-push hydrostratigraphic profiling: Coupling electrical logging and slug tests. Ground Water 43:19-29

Smolley M, Kappmeyer JC (1991) Cone penetrometer tests and Hydro-Punch sampling: A screening technique for plume definition. Ground Water Monitoring Review 11:101-106

Sørensen K (1989) A method for measurement of the electrical formation resistivity while auger drilling. First break 7:403-407

Sørensen K, Pellerin L, Auken E (2003) An auger tool to estimate hydraulic conductivity using a resistivity analogy. ASEG 16th Geophysical Conference and Exhibition, February 2003, Adelaide

Spies BR, Ellis RG (1995) Cross-borehole resistivity tomography of a pilot-scale, in-situ vitrification test. Geophysics 60:886-898

Sully JP, Robertson PK, Campanella R, Woeller DJ. (1999) An approach to evaluation of field CPTU dissipation data in overconsolidated fine-grained soils. Can. Geotech. J. 36:369-381

Terry TA, Woeller DJ, Robertson PK (1996) Engineering soil parameters from seismic cone penetrometer tests; an overview. In: Bell RS; Cramer MH (Eds.) Proceedings of the symposium on the Application of geophysics to engineering and environmental problems pp 1279-1287

Thornton D, Ita K, Larson K (1997) Broader use of innovative ground water access technologies. In: Superfund XVIII Conference Proceedings, Vol. 2

Tillman N, Leonard L (1993) Vehicle mounted direct push systems, sampling tools and case histories: An overview of an emerging technology, In: Proc. of the 1993 meeting on Petroleum Hydrocarbons and Organic Chemicals in Ground Water, Ground Water Management 17, pp 177-188

U.S. A.C.E. - U.S. Army Corps of Engineers (1996) Tri-service site characterization and analysis penetrometer system. Eng. Tech. Letter 1110-1-171

U.S. EPA (1995) Rapid Optical Screen Tool (ROST). — Innovative Technology Verification Report EPA/540/R-95/519

U.S. EPA (1998) Environmental Technology Verification Report – Soil Sampling Technology – Art's Manufacturing and Supply – AMS Dual Tube Liner Sampler –EPA 600R-98/093

Van den Boogaart J, van Deen JK, Kinneging NA, Meyer JG, van Ree CC (2001) The camera cone as an effective site screening tool. In: Proceedings “Field Screening Europe 2001”, Karlsruhe, Germany, pp 107-111

Voyiadjis GZ, Song CR (2003) Determination of hydraulic conductivity using piezocone penetration test. International Journal of Geomechanics 3:217-224

White PA (1994) Electrode arrays for measuring groundwater flow direction and velocity. Geophysics 59:192-201

White PA (1988) Measurement of ground-water parameters using salt-water injection and surface resistivity. Groundwater 6:179-186

Zemo DA, Pierce YG, Gallinatti JD (1994) Cone penetrometer testing and discrete-depth ground water sampling techniques: A cost-effective method of site characterization in a multiple-aquifer setting. Ground Water Monitoring and Remediation 14:176-182

12 Aquifer structures – pore aquifers

For the water supply, pore aquifers are the most important kind of aquifers. After some general remarks on geophysical exploration of pore aquifers, case studies and field examples on pore aquifer surveys under different hydrogeological condition are presented. This includes buried Quaternary valleys in Northern Germany and Denmark (Helga Wiederhold), large TEM surveys for Quaternary aquifers on a Danish island (Flemming Jørgensen et al.), and a VES survey to map fractured limestone aquifers in Egypt (Mohamed Mabrouk et al.).

12.1 Pore aquifers – general

Reinhard Kirsch

12.1.1 Definition

Groundwater can be found in pore spaces of unconsolidated and consolidated sedimentary rocks and weathering layers, in joints and fissures of hard rock, in fault zones, and in karst caves. Aquifers with water reservoir stored in pore spaces are called pore aquifers or porous aquifers. Similar conditions for geophysical exploration are in aquifers connected to joints and fissures of rocks, e.g., originated by cooling of igneous rocks. So, also this type of aquifer will be treated as pore aquifer, while fault zones and karst caves, where the water reservoir is embedded in nearly impermeable material, are treated in Chap. 13 "fracture zones and caves".

12.1.2 Porosity – a key parameter for hydrogeology

The volume of open space (pore space) in rocks in relation to the total rock volume is called porosity Φ:

$$\Phi = \frac{V_{porespace}}{V_{total}} \tag{12.1}$$

Porosity due to pore space between mineral grains or clastic rock fragments is called primary porosity; an additional porosity due to tectonic fractures or dissolution caves is called secondary porosity. Both porosities are expressed in percent (0..100%) or, mostly in equations, as fraction (0..1).

Porosity values range from 0.1% for a dense limestone to more than 80% for some kinds of pyroclastic rocks. For unconsolidated sediments, porosity in general decreases with increasing grain size from more than 50% for clay to less than 10% for coarse gravel. For nonuniform material, the sorting is important for the porosity. The porosity of well sorted material is higher than the porosity of poorly sorted material, because small particles can fill the pore space between the larger particles and so reduce the porosity.

The waterfilled pore space cannot be drained totally by pumping or evaporation, a fraction of water is bound to the mineral grain surface by adhesion. The drainable pore space defines the effective porosity:

$$\Phi_{eff} = \Phi - V_{boundwater} \tag{12.2}$$

The amount of bound water is strictly related to the inner surface of the pore space. Material with small grain size has a high inner surface, e.g. clay up to 100 m^2/g compared with silt (about 1 m^2/g) and sand (about 0.1 m^2/g). A high amount of bound water reduces the effective porosity. Therefore, in spite of the high porosity of clay, the effective porosity of clay is very small (see Chap. 15, Fig. 15.3). Typical values for effective porosity are: clay < 5%, fine sand 10-20%, coarse sand 15-30%. For well sorted aquifers composed of coarse material, the effective porosity is only slightly smaller than the total porosity.

Due to the adhesive forces, a certain amount of pore water is also found in the unsaturated zone. The amount of pore water related to the pore volume is given by the saturation degree S_W.

An important parameter for water extraction from an aquifer is the storage coefficient S, which is defined by the water volume V_{water} which can be extracted by lowering the watertable by Δh:

$$S = \frac{V_{water}}{\Delta h} \tag{12.3}$$

A normalisation of S by the aquifer volume leads to the specific storage coefficient S_{sp}, which depends on the effective porosity Φ_{eff}, on the modulus of elasticity (Young´s modulus) of the rock matrix E_{matrix}, and on the pore water properties compressibility χ and density ρ:

$$S_{sp} = \rho \cdot \left[(1 - \Phi_{eff}) / E_{matrix} + \Phi_{eff} \cdot \chi \right] \tag{12.4}$$

12.1.3 Physical properties of pore aquifers

In general, physical properties of pore aquifers are clearly different to less permeable layers as solid rock, clay, or till. As discussed in Chap. 1, increasing porosity reduces seismic velocity. Density is also reduced, so a porosity change at a layer boundary results in significant impedance contrast and high seismic reflection amplitudes (Fig. 12.1b). Following Archie´s law, increasing porosity leads to decreasing electrical resistivity. However, it must be kept in mind that porosity of coarse material like gravel is, in general, lower than porosity of finer material like silt (although effective porosity and permeability is higher for coarse material than for fine material). Porosity of clayey material depends on clay content (Fig. 12.1e), but is in general higher than porosity of clay-free material.

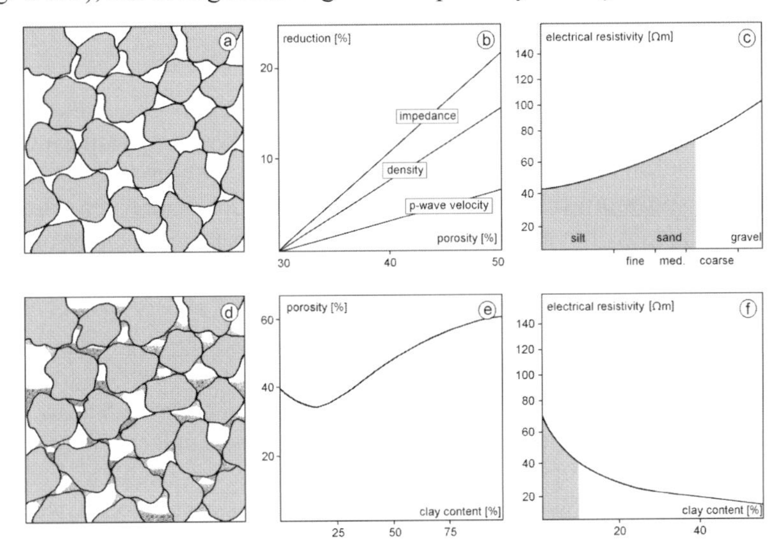

Fig. 12.1. Physical properties of pore aquifer material (Gabriel et al 2003, with permission from Elsevier): influence of porosity and clay content on density, seismic velocity, and electrical resistivity: (**a**) well sorted, clay free sediment, (**b**) reduction of p-wave velocity (after Morgan 1969), density, and impedance as a function of porosity, (**c**) electrical resistivity as a function of grain size for fresh water saturated material (after TNO 1976), (**d**) clayey sediments, pore space partly filled with clay minerals, (**e**) porosity related to clay content (artificial sand – clay mixture, Marion et al. 1992), (**f**) electrical resistivity related to clay content after Sen et al. (1988)

The grey shaded area in Figs. 12.1c and f denotes similar resistivities for clayey and clay-free material, this can lead to interpretation errors.

Although pore aquifers are treated here as homogenous bodies, physical anisotropy can be observed due to alternating sedimentation of coarser and finer material. For electrical resistivity, this can lead to different results obtained by VES measurements (mainly horizontal current flow) and borehole measurements (mainly vertical current flow).

12.1.4 Geophysical survey of pore aquifers

Pore aquifers are well suited as targets for geophysical surveys. Objectives of these surveys are, e.g., depth and thickness of aquifers and impermeable layers, depth to water table, or special underground structures like saltdomes influencing the aquifer. Seismic, resistivity and electromagnetic, gravity, and, for shallower underground structures, GPR methods are applied.

Seismic methods

Typical seismic velocities for p-waves are 300 – 600 m/s for unsaturated sand, 1500 – 2000 m/s for saturated sand, 1500 – 2500 m/s for saturated till or clay, and more than 3000 m/s for hard rock. Seismic velocities for fractured hardrock, however, can be similar to seismic velocities of saturated sand. Generally, saturated and unsaturated material can be clearly separated by p-wave velocities, while velocities for clayey and clay-free deposits are overlapping.

For the surveying of aquifer structures, refraction as well as reflection seismic measurements can be applied. The use of refraction seismic measurements is restricted to relatively simple underground structures, e.g., near surface aquifers underlain by basement rocks. An example from the Wadi Khor Baraka, Sudan, is shown in Fig. 12.2. Seismograms were recorded by a 24 channel Geometrics seismograph with 10 m geophone spacing. First arrivals of shot and reversed shot were interpreted by the GRM-method (Palmer 1981). The steeply dipping basement (5935 m/s) in the depth range 20 – 60 m is overlain by an aquifer (1748 m/s) and unsaturated sands (405 m/s). The aim of this survey was to find drill locations; a well drilled at the right side of this profile could use only the upper part of the aquifer leading to unsufficient yield.

Reflection seismic surveys can image the sequence of aquifers and impermeable layers down to depths of several hundred meters. However, depending on the field layout, a minimum depth is required for a layer to be visible in the seismic section. Drilling results are essential for a geological interpretation of a seismic section as shown in Fig. 12.3. Target of this

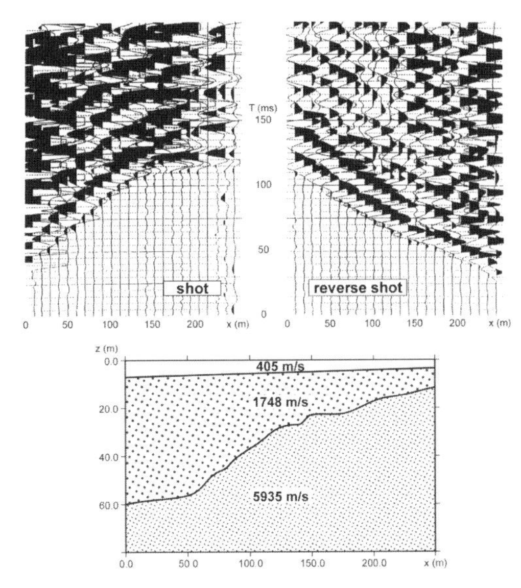

Fig. 12.2. Field example from a refraction seismic survey for aquifer and basement structure in Khor Baraka, Sudan (after THOR 1989)

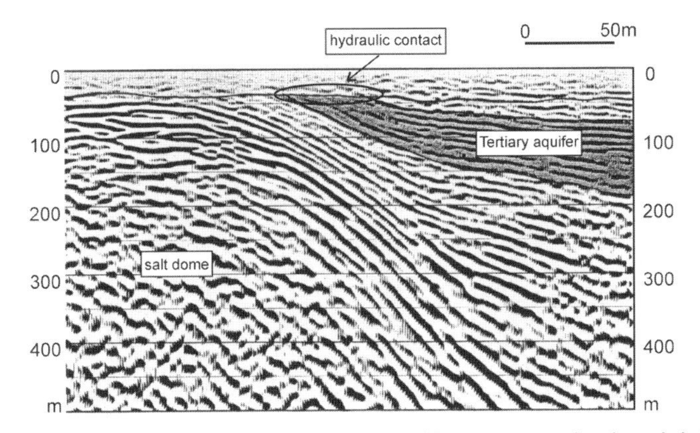

Fig. 12.3. Reflection seismic survey on the aquifer structure in the vicinity of a salt dome, grey shading indicates a Tertiary aquifer identified by drilling results (after Schultz-Rincke et al. 1997, with permission from BGR)

survey was the aquifer structure (Tertiary lignite sands embedded in mica clay layers) in the vicinity of a salt dome. The aquifer was identified by results of a drilling some hundred meters apart from the seismic profile. Due to uplift of the salt dome the Tertiary layers were uplifted too, resulting in a hydraulic contact of Tertiary and Quaternary aquifers.

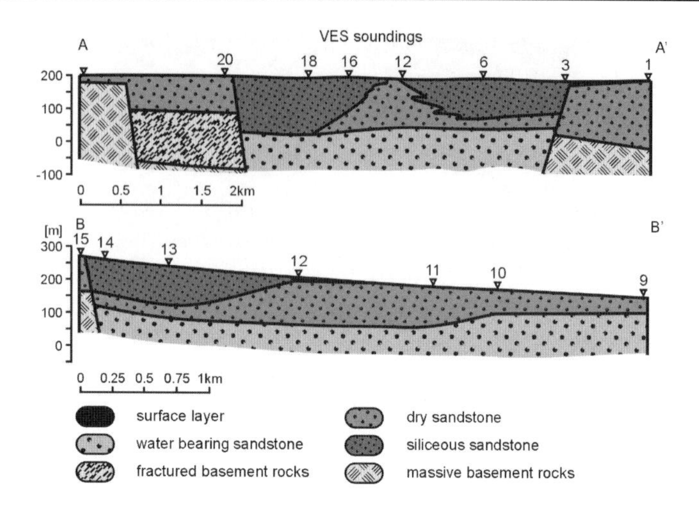

Fig. 12.4. VES results showing a sandstone aquifer, electrode spacing (AB/2) was 700 m (Mabrouk 1991). Typical resistivities are: *surface layer*: > 3000 Ωm; *dry sandstone*: 200 - 700 Ωm, *partly exceeding* 3000 Ωm; *water bearing sandstone (aquifer)*: 100 - 200 Ωm; *fractured basement rocks* 700 Ωm, *massive basement rocks*: 1200 Ωm

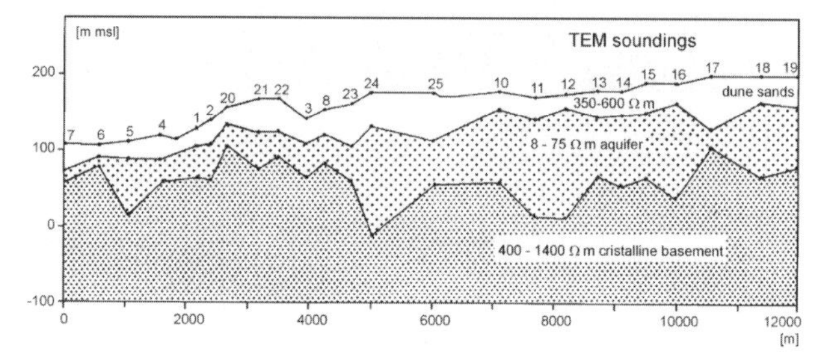

Fig. 12.5. TEM survey in Namibia, aquifer partly saturated with saline water (Schaumann 1997, with permission from BGR)

Resistivity methods

Resistivity of water saturated sediments is lower than resistivity of solid basement rock and unsaturated sediments, but higher than resistivity of clayey material. Two examples illustrating the use of resistivity methods are given in Figs. 12.4 and 12.5. Fig. 12.4 shows VES results to locate a sandstone aquifer in a tectonically structured region in Egypt (Mabrouk 1991). Fig 12.5 shows TEM results for aquifer exploration (sand and gravel) in Namibia, here partly saltwater occurred in the aquifer leading to resistivities as low as 8 Ωm (Schaumann 1997).

Ground penetrating radar

GPR is characterized by high resolution, but limited depth of penetration. Aquifer studies based on GPR are therefore restricted to shallow underground structures. Reflection coefficient for radar signals mainly reflects permittivity contrast, while permittivity depends on water content. Typical reflection horizons are water table and, in the saturated zone, layer boundaries with porosity contrast. An example for GPR surveys in areas with complex aquifer structures is shown in Fig 12.6, where 2 near surface aquifer divided by silty clay occur.

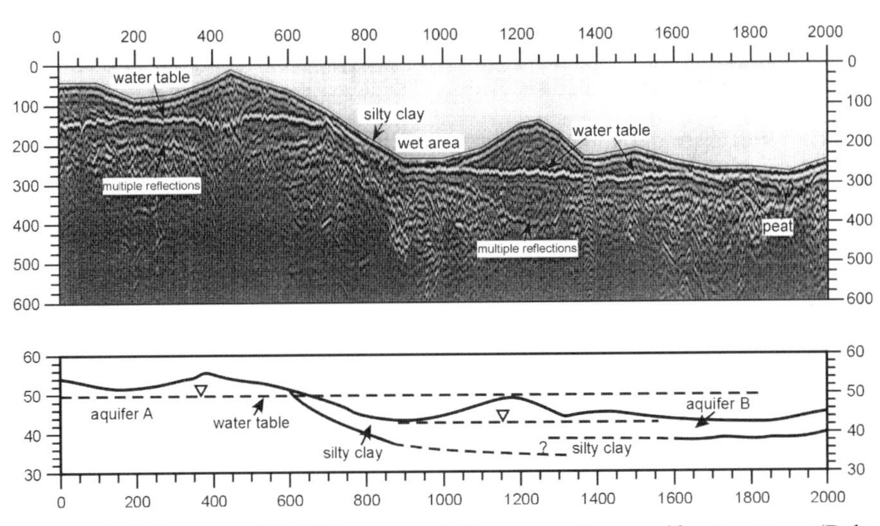

Fig. 12.6. GPR survey (60 MHz antenna) for near surface aquifer structures (Bahloul et al. 1999)

12.2 Buried valley aquifer systems

Helga Wiederhold

12.2.1 Introduction

Buried valleys are underground structures typical for areas affected by glaciations. Especially when the glaciers met poorly consolidated ground, i.e. in sedimentary basins, subglacial erosion could lead to deeply incised channels, which were refilled with glacial sands and tills when the glaciers retreated. These now buried valleys reach depths from only 30 m up to 600 m, and the length of the deeper valleys can be more than 100 km. Buried valleys are of great hydraulical importance. If they are filled with sandy material they are groundwater reservoirs of great yield, but often the valleys also incise deep groundwater reservoirs or cut through their covering layers.

The map of maximum glacial extent of the Quaternary glaciations shows that North America and Northern Europe are the predestined areas for these structures in the northern hemisphere. These areas with high population density have a high demand on good quality water. And as groundwater of near surface aquifers is increasingly endangered by pollution by industry or intense farming, deeper groundwater reservoirs like buried valleys, which are better protected to contaminations, become more and more of interest for the water supplies.

But looking deeper also needs new technologies as these aquifers are not that easy accessible by boreholes as near surface aquifers and the more expensive access by boreholes needs better pre-site studies. This is a chance and a challenge for geophysical methods. Especially as the valleys are anomalies compared to normal geological layering the conditions to visualize them by geophysical methods are good.

Knowledge of the nature and distribution of the glacially-deposited materials and its surrounding is important to spatial development and planning. Thus the aims of the geophysical exploration are the mapping of the lateral extent, form and internal structure of the valley as well as the determination of the hydrogeological parameters of the sedimentary fill.

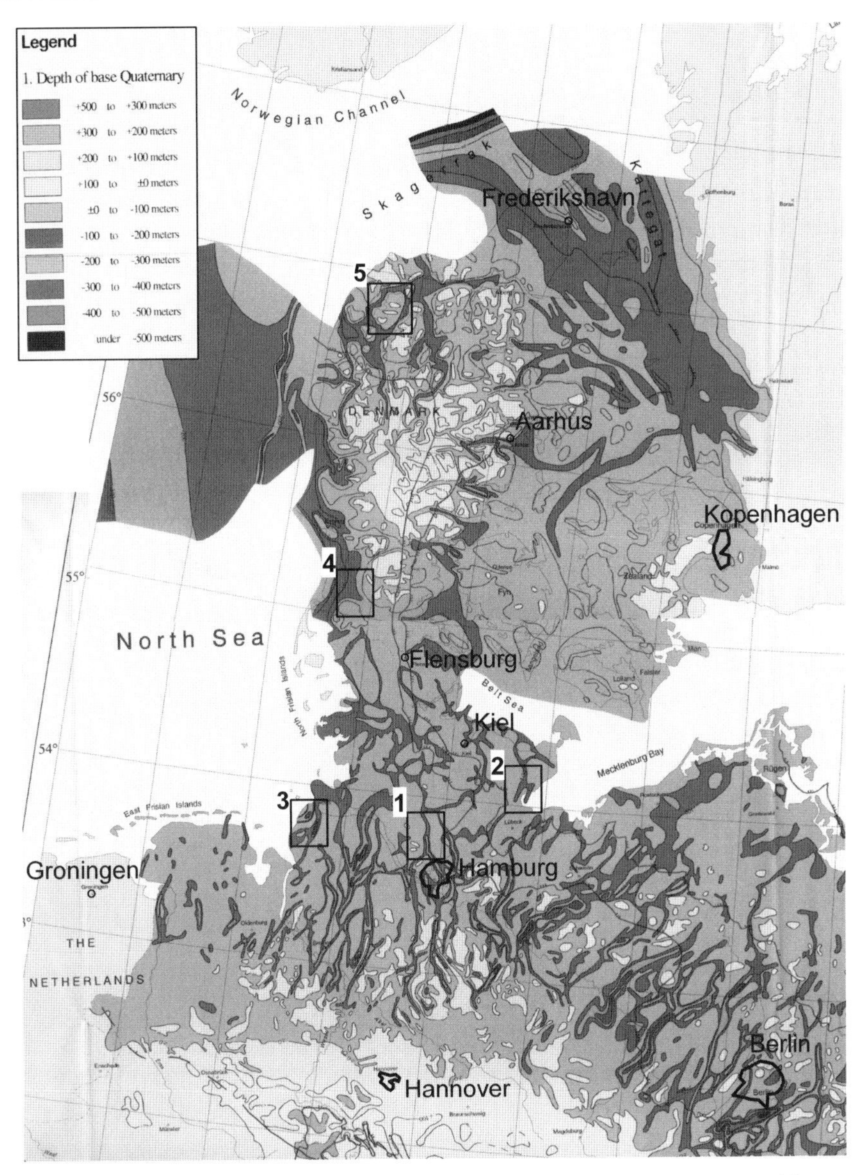

Fig. 12.7. Base of Quaternary deposits after Stackebrandt et al. (2001, with permission from LBGR) Study areas of examples shown in this book are outlined: 1 Ellerbek Valley, 2 Trave Valley, 3 Cuxhaven Valley (see example in Sect. 5.1.5), 4 Bredebro (Poulsen and Christensen 1999), 5 Mors (see Sect. 12.3)

12.2.2 Geological and hydrological background

The distribution of buried Pleistocene subglacial valleys in Northern Germany and Denmark is shown in the map of the base of Quaternary deposits after Stackebrandt et al. (2001) in Fig. 12.7. The deeper valleys (> 100 m) were formed during the oldest of the three major glaciations, the Elsterian (Ehlers 1994). Hypotheses of their genesis are e.g. discussed by Huuse and Lykke-Andersen (2000), Smed (1998), Piotrowski (1997) or Praeg (1996).

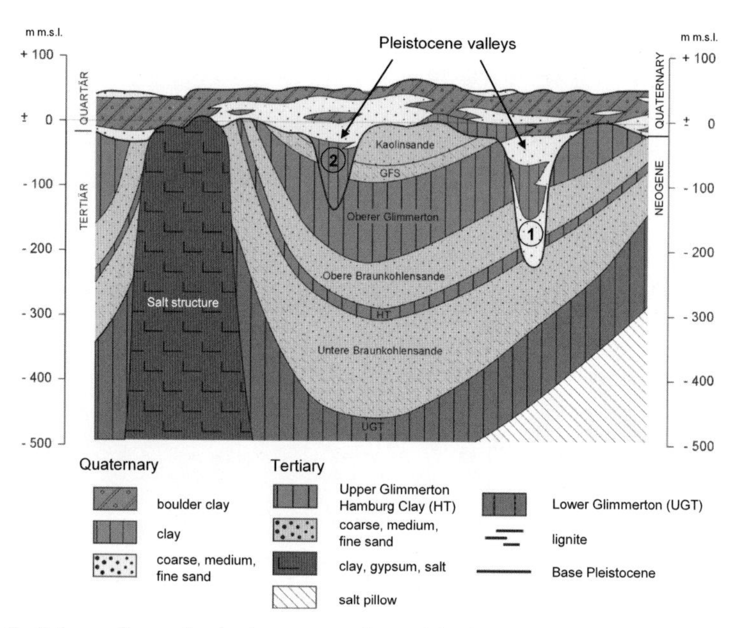

Fig. 12.8. Schematic geological cross section with characteristic structures as salt-domes and buried valleys from Northern Germany (after Gabriel et al. 2003, with permission from Elsevier)

A schematic geologic cross-section (Fig. 12.8) shows the characteristic geological structures for Northern Germany. For the water supply the Neogene and Quaternary sand layers yield the important groundwater reservoirs. Critical points for groundwater management and protection are areas where the normal layering is disturbed e.g. by glacial valleys, by faults (not sketched in Fig. 12.8) or where the groundwater bearing layers are tilted by salt dome uplift.

The two valleys in Fig. 12.8 exemplarily show different hydraulic characteristic: valley 1 incises the upper Miocene mica clay (Oberer Glimmerton) as well as the upper and lower Miocene lignite sands (Braunkohlensande) separated by a confining clay layer (Hamburger Ton HT). The valley itself has a sandy fill and in spite of the clayey covering layer (Lauenburger Ton) the valley enables a hydraulic connection between the upper and lower aquifers of the Braunkohlensande. Due to this hydraulic connection to the Braunkohlensande the valley sands may be important groundwater reservoirs. In contrast, valley 2 is filled completely with clay or till and acts as hydraulic barrier in the sandy aquifer.

12.2.3 Methods

To be mapped with geophysical methods the physical parameters (electrical conductivity, density and elastic constants) of the valley and the surrounding material must differ. In the case of Northern German Quaternary and Tertiary sediments we have to deal with alternating strata of high porous and water saturated sands of different grain sizes, till and clay. So the contrasts will be smaller than in areas where buried valleys are in bedrock environment. But the sediments should be resolvable by resistivity and electromagnetic methods due to their varying resistivities (Fig. 12.1c,f). There are also good conditions for reflection seismic methods due to the velocity and impedance contrasts (Fig. 12.1b). The refraction seismic method will be applicable only for the very near-surface section as the requirement of increasing velocity will not be given for greater depth. The density contrast between the valley fill and the surrounding sediments may affect gravity. From the magnetic susceptibility of the sediments no measurable contrast was expected; e.g. results of the aeromagnetic survey in the Cuxhaven area (Siemon et al. 2001) show no hint to the Cuxhaven Valley. But Fichler et al. (2005) recently show an impressing example of Quaternary channels mapped by 3D seismic as well as aeromagnetic data from the North Sea offshore south Norway. Integrated geophysical studies of buried valley aquifers are the best approach as shown by several authors (e.g. Wolfe and Richard 1996, Poulsen and Christensen 1999, Holzschuh 2002, Gabriel et al. 2003, Jørgensen et al. 2003b, Sandersen and Jørgensen 2003). Anyhow, to demonstrate the geophysical effects the examples in the following are ordered by the method.

Seismic and VSP

Sedimentary layering in general is good visualized by seismic methods. The first hints to buried valleys in Northern Germany – aside from boreholes and geophysical logging for hydrologic surveys – came from seismic exploration for hydrocarbons. The resolution in the near surface area was poor and was not the aim of these surveys; the results in general are not published (Frisch and Kockel 1993). But with the survey parameters (e.g. shotpoint and receiver spacing) adapted to the near surface area the buried valleys may be imaged well (e.g. Meekes et al. 1990, Wiederhold et al. 1998). The quality of high resolution shallow reflection survey depends on the field parameters and selection of an appropriate seismic source. With the technical developments of the last years a good choice of seismic sources and seismographs is available.

Examples from two surveys in Northern Germany are shown corresponding to the valley types sketched in Fig. 12.8 (Wiederhold et al. 2002). The acquisition parameters of both surveys are: a Vibrator source with a linear sweep of 10 sec length, frequency range 40 to 160 Hz, and 4 vertical stacks (Buness et al. 1997, Buness and Wiederhold 1999). Source and receiver spacing was 10 m, 47 channels were recorded with a Geometrics StrataView Seismograph. The CMP spacing is 5 m and the mean fold is about 23. An almost standard seismic processing sequence was applied to the data with the final processing step of depth conversion after time migration (Yilmaz 2001). The seismic section across the Ellerbek Valley – situated north of Hamburg – clearly shows the bottom of the valley by a high amplitude reflection signal as well as both walls with about 30° slope (Fig. 12.9). The valley depth is about -350 m m.s.l. The Tertiary layers outside the valley are undisturbed and the seismic amplitudes correlate well. Inside the valley the upper part is characterized by undisturbed reflection signals but the part below -150 m m.s.l. shows more or less chaotic or short reflection segments.

The interpretation of these signals is enabled or improved by logging and vertical seismic profiling (VSP). A groundwater observation well – 318 m deep - about 750 m south of the line in Fig. 12.9 is used for this. The seismic impulse source system Sissy (Wiederhold et al. 1998) is used as source, the receiver spacing is 4 m. The VSP raw data (Fig. 12.10a) show P wave signals of good quality. The shear wave signal is distorted by a tube wave with high amplitudes and a velocity of about 650 m/s. After wavefield separation the reflected P waves are revealed and the depth of their origin correlates well with lithological boundaries (black dots in Fig. 12.10b). A reflection of the valley rim, that should have another slope due to the dip, is not observed. The seismic depth section in Fig. 12.10c connects two boreholes and is parallel to the section in Fig. 12.9; the corresponding CMP range in Fig. 12.9 is 200-330 with the eastern rim of the valley.

The corridor stack helps linking the borehole data to the surface seismic data. The lithologic column shows the Quaternary fill of sand and clayey

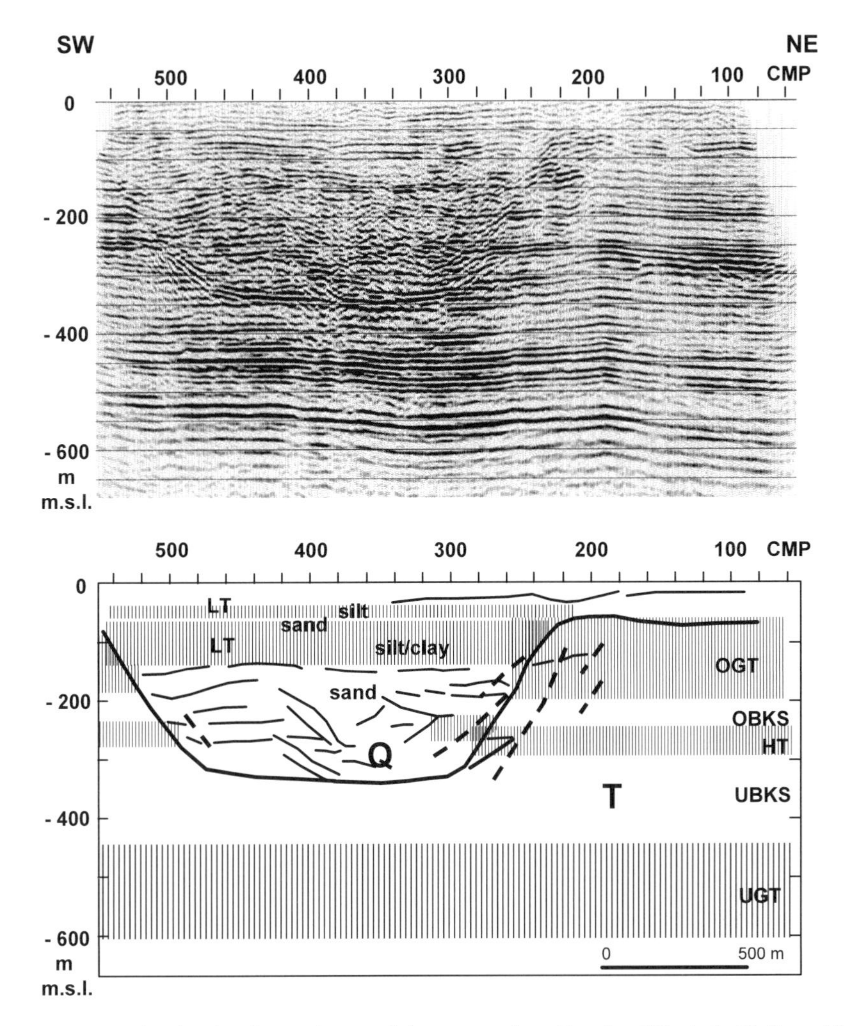

Fig. 12.9. Seismic depth section and interpretation for the Ellerbek Valley. The seismic datum is m.s.l. (the ground level elevation is about 15 m m.s.l). Clayey layers are hatched. Q = Quaternary, T = Tertiary, LT = Lauenburger Ton, OGT = Upper Glimmerton, OBKS = Upper Braunkohlensande, HT = Hamburger Ton, UBKS = Lower Braunkohlensande, UGT = Lower Glimmerton

or silty layers, that are well distinguished by the electrical conductivity where the clayey material shows high values. There is also a correlation between seismic velocities and lithology. The silt layer at about 95-150 m depth belonging to the Lauenburger Ton (LT) gives a velocity of 1600 m/s compared to 1700 m/s in the upper and deeper sand layers. With the till at

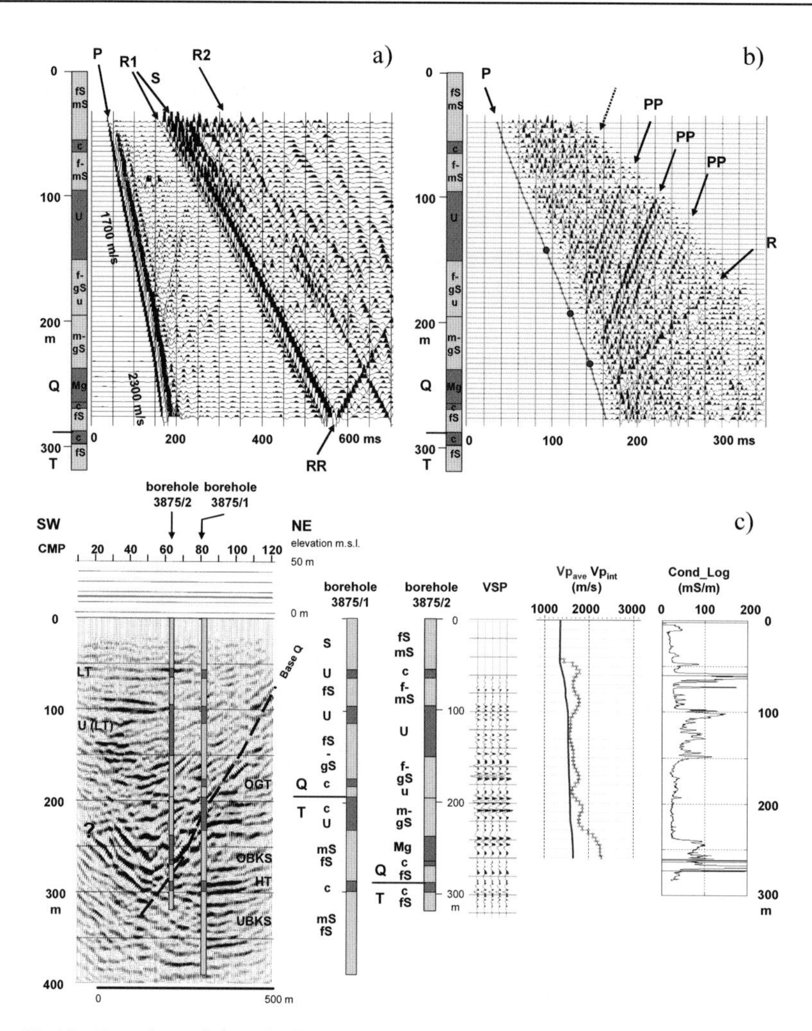

Fig. 12.10. Results of borehole logging (VSP and induction log in borehole 3875/2) and reflection seismic section (Ellerbek Valley). **a**) VSP raw data (P = compressional wave, S = shear wave, R1, R2 = tube wave, RR = reflected tube wave), **b**) upgoing wavefield (P = first breaks of P wave, PP = reflected P waves, R = reflected tube wave), **c**) from left to right: seismic depth section (datum is ground level), lithological logs (light grey: sandy material: fS = fine sand, mS = medium sand, gS = coarse sand; dark grey: c = clay, U = silt, Mg = till), corridor stack from VSP, seismic velocity (average and interval), electrical conductivity log

about 240 m the velocity rises to 2300 m/s. Especially the clayey or silty material of the LT can be identified by high electric conductivity and its top by seismic reflections.

The seismic section across the Trave Valley, near the city of Lübeck, is unimpressive compared to the preceding one (Fig. 12.11). The depth is only -100 m m.s.l. and the slope of the rims is lesser. But also here the valley is imaged well. Tertiary layers outside the valley are undisturbed; inside the valley the reflectivity is poor. A second line connects the first with a borehole, where logging and VSP measurements were done (Fig. 12.12). The glacial till in the valley is divided into two parts by the seismic velocities as well as in the electrical conductivities: in the upper part velocities above 2000 m/s and conductivity values greater 60 mS/m were measured but below 90 m depth these values decrease. These different parts hint to lithological changes in the till. No reliable reflections are observed in the VSP.

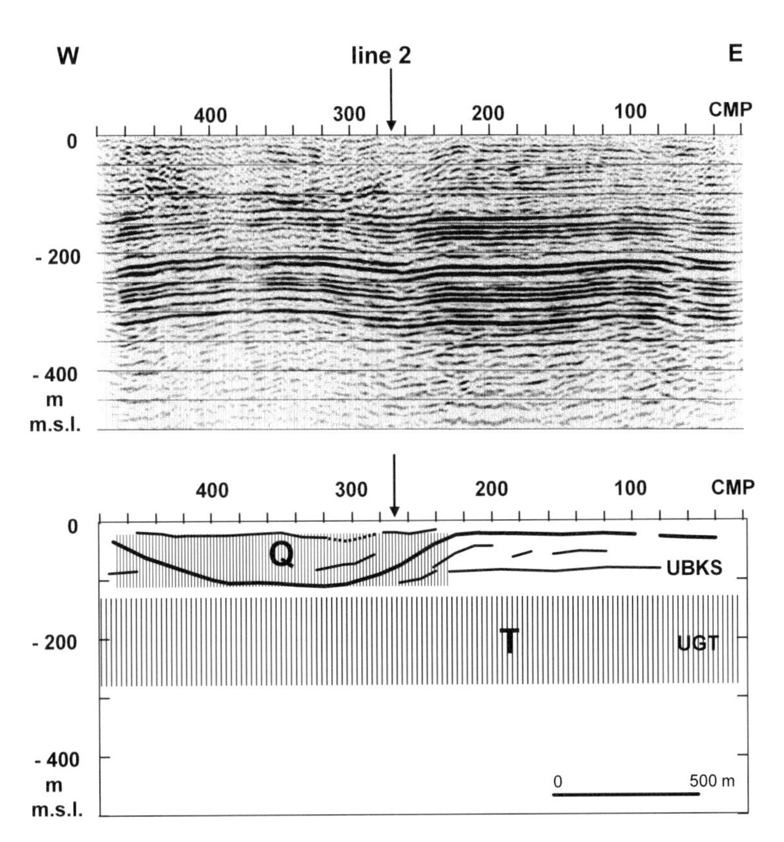

Fig. 12.11. Seismic section and interpretation for the Trave Valley. The seismic datum is m.s.l. (the ground level elevation is about 15 m m.s.l). Clay or till layers are hatched. Other abbreviations see Fig. 12.9

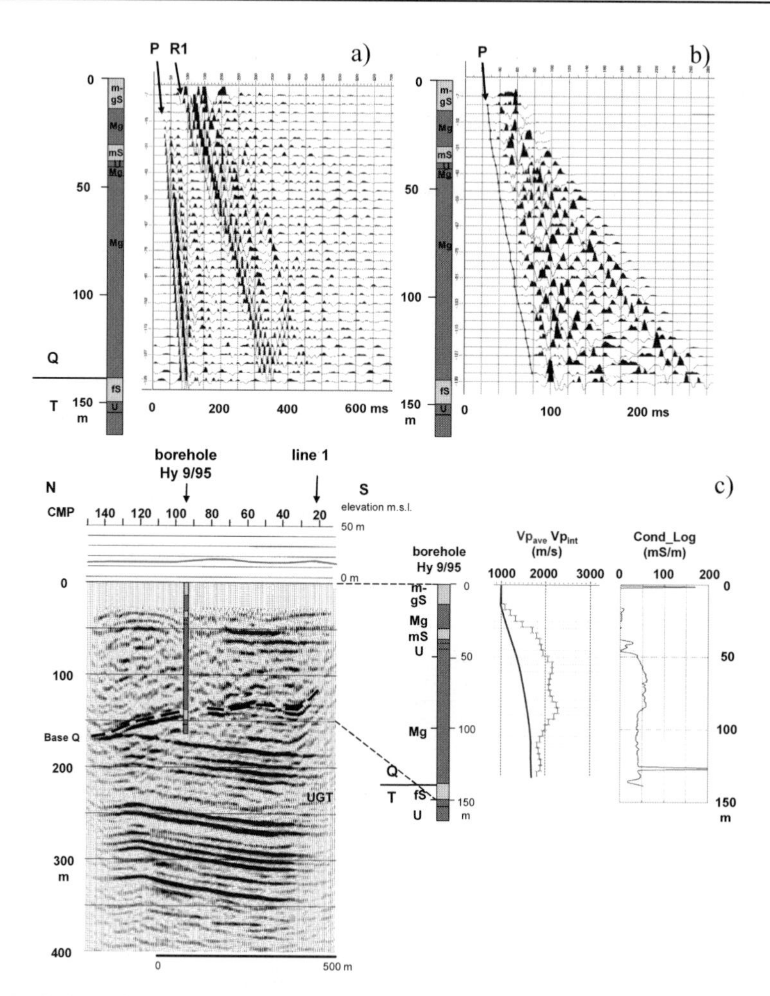

Fig. 12.12. Results of borehole logging (VSP and induction log in borehole Hy 9/95) and reflection seismic section (Trave Valley). **a**) VSP raw data (P = compressional wave, R1 = tube wave), **b**) upgoing wavefield (P = first breaks of P wave), **c**) from left to right: seismic depth section (datum is ground level), lithological logs (light grey: sandy material: fS = fine sand, mS = medium sand, gS = coarse sand; dark grey: U = silt, Mg = till), corridor stack from VSP, seismic velocity (average and interval), electrical conductivity log

The different seismic velocities of the valley fill at Ellerbek and Trave also have an effect on the seismic sections. So in Fig. 12.9 a slight pull down of the reflections below the valley is observable and in Fig. 12.11 a slight pull up. These effects that were not cared for by migration and depth conversion that was done only with a simple velocity model now give a hint on the valley fill. Jørgensen et al. (2003b) also show a clear distribution of velocities as function of sediment type and discuss the velocity effects by high content of till material in buried valleys.

Resistivity and electromagnetic methods

From the variety of resistivity methods the electric profiling or electromagnetic methods are well-tried for hydrogeological problems. E.g. Poulsen and Christensen (1999) give an example of mapping a buried valley in Denmark by the transient electromagnetic method TEM (Fig. 12.13). They use a PROTEM 47 recorder and transmitter from Geonics and the soundings were made in the central loop configuration with a 40 m time 40 m square transmitter loop. A total of 126 transient soundings were made, the inversion was performed using a 1D earth model. The valley is incised in Tertiary clay and from the depth estimation of the top of this very good conductive clay obtained by TEM measurements the course of the valley is observable. The maximum depth is about 120 m. There is a distinct difference in the resistivity values of the Quaternary sands in the valley (30-200 Ωm) and the Tertiary mica clay (<20 Ωm) beneath.

A pulled array transient electromagnetic method (PATEM) is developed by the HydroGeophysics Group of the University of Aarhus (Sørensen et al. 2005) that provides rapid and dense lateral coverage and is used in hydrogeological surveys and especially in buried valley mapping in Denmark (Danielsen et al. 2003) with the same geologic characteristic as the above reported survey after Poulsen and Christensen (1999). To cover even larger areas the newest development of the HydroGeophysics Group is a TEM system operated from a helicopter (SkyTEM, Sørensen and Auken 2004).

Over the last 10 years large parts of Denmark have been surveyed with the TEM method and in the western part of Denmark about 700 km of buried valley structures have been imaged (Jørgensen et al. 2003a, Sandersen and Jørgensen 2003).

The depth penetration of the frequency domain electromagnetic method is generally lesser than of TEM but in Germany very good results in mapping buried valleys were achieved by an aeroelectromagnetic survey (AEM or HEM) in the Cuxhaven region (Siemon et al. 2001, Siemon 2005). The valley is incised more than 300 m into Tertiary sediments. Its bottom is not detectable by AEM but a good conductive clay layer (Lauenburger Ton) on its top clearly marks the valley in the resistivity maps (see Chap. 5 Fig. 5.6).

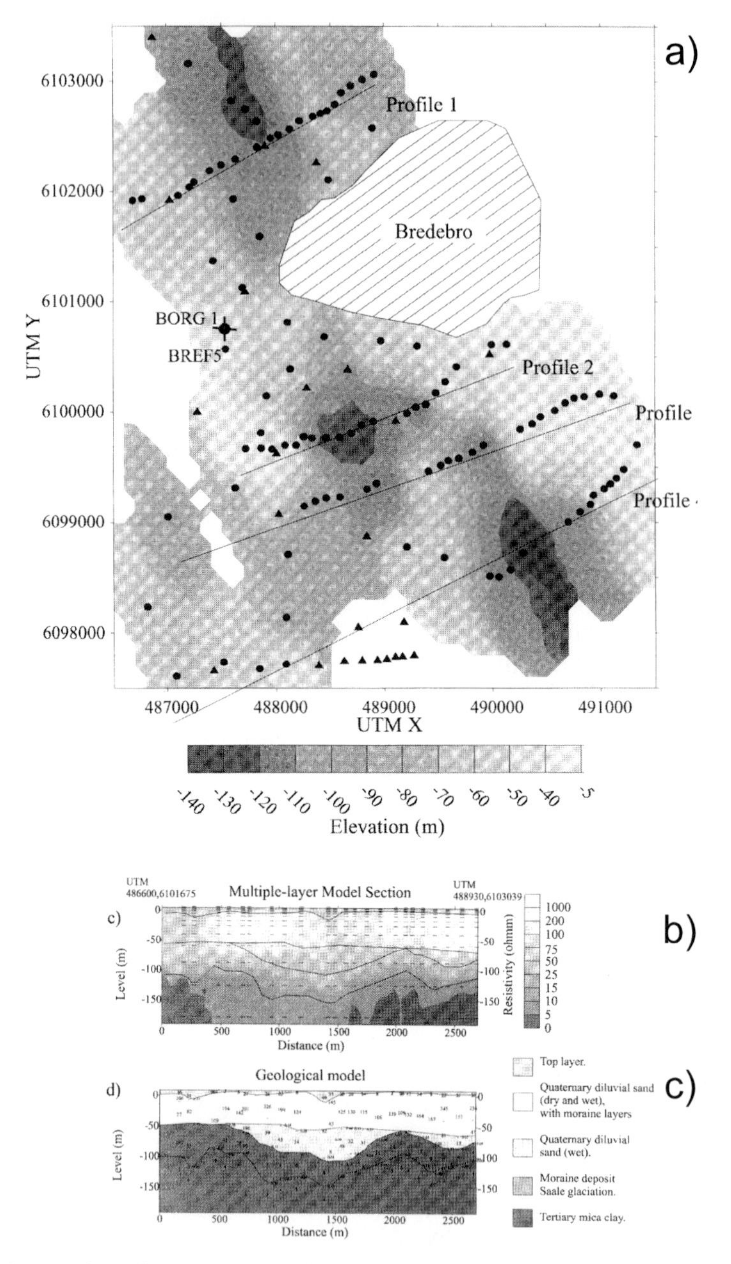

Fig. 12.13. Results of TEM survey in the Bredebro region from Poulsen and Christensen (1999, with permission from EEGS ES): **a**) Map of elevation of the topmost good conductor. The darker shades of grey correspond to the deeper part of the valley. Locations of TEM soundings are indicated by dots; **b**) resistivity section of profile 1; **c**) interpreted geological model

Gravity

Poulsen and Christensen (1999) and Friborg and Thomsen (1999) show that buried valleys may be also mapped by microgravimetric surveys. Fig. 12.14 shows the residual Bouguer anomaly for the same area as the TEM survey in Fig. 12.13. The investigation area in the Bredebro region in Denmark was approximately 25 km^2 and it was covered with a dense net of gravity measurements with approximately 100 m between stations. The contour interval is 0.025 mGal. The buried valley becomes apparent by the general positive anomaly trending north-northwest/south-southeast. The Quaternary coarse sands inside the valley show a higher density than the Tertiary clays in the surroundings.

Also in the Ellerbek area the valley can be mapped by gravity measurements along the same traverse as the seismic section in Fig. 12.9 (Gabriel et al. 2003, Gabriel 2005). The residual field shows here a distinct negative anomaly of 0.5 mGal at most (Fig. 12.15). Without structural information from seismic interpretation the gravity anomaly would be modelled by two geological units and assumed valley sediments of 400 m depth with a density contrast of 50 kg/m^3. A second profile 700 m south of the first one reveals similar anomalies indicating that the valley might be mapped by gravity measurements.

Residual field and model of the Trave Valley (Fig. 12.11) are shown in Fig. 10.5 (Chap. 10). The till filled valley incised in sand shows a clear positive anomaly.

In the Cuxhaven Valley area (Fig. 5.6) Gabriel et al. (2003) and Gabriel (2005) found very complex residual anomalies. They are partly superimposed by high-frequency negative anomalies caused by shallow structures. These negative anomalies correlate well with a near-surface clay layer that is also seen in the resistivity map of the AEM survey. These anomalies seem to reduce the probable positive anomaly of the Cuxhaven Valley.

12.2.4 Discussion and Conclusion

In the preceding only a short insight is given in the buried valley theme. None of the 4 mentioned studies meets the same geologic environment as the others. Some characteristic features of the four sites are summarized in Table 12.1. All the valleys show clear geophysical indications with each method. Nevertheless the interpretation is supported or often only reached by complementing with at least one other method. Because every method

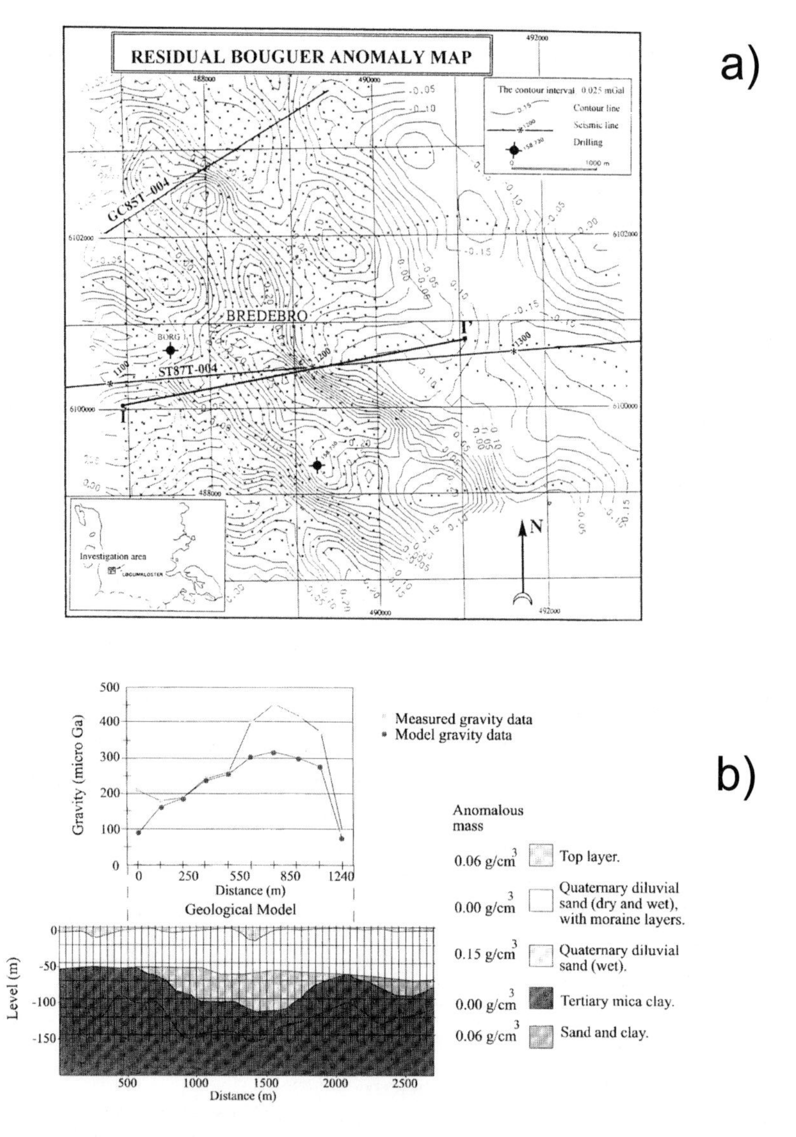

Fig. 12.14. Results of gravity survey in the Bredebro region from Poulsen and Christensen (1999, with permission from EEGS ES): **a**) contoured residual Bouguer anomaly map; **b**) very simple gravity model based on the TEM profile 1 in Fig. 12.13

has its advantages but also drawbacks an integrated approach is strongly recommended.

Electromagnetic methods turn out to be convincing for mapping buried valleys if a good conductive layer is embedded in less conductive material

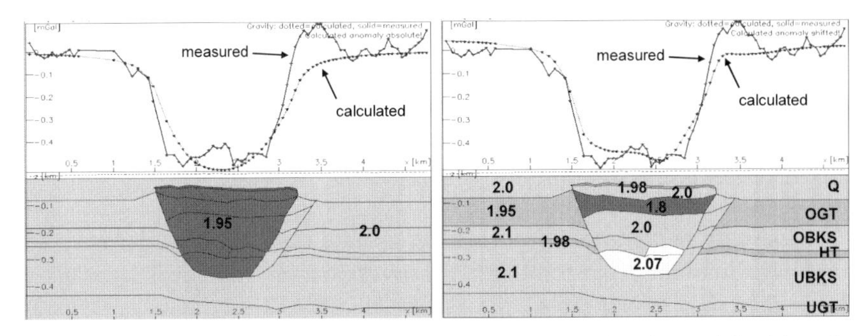

Fig. 12.15. Gravity cross section Ellerbek Valley: left: simple model with a density contrast of 50 kg/cm3; right: more sophisticated model based on structural information from seismic data in Fig. 12.9 (after Gabriel et al. 2003); density values in g/cm3, other abbreviations see Fig. 12.9

or vice versa. In the majority of cases the top of the good conductor will be detectable. The penetration depth depends on the frequency (AEM) or the magnetic moment of the transmitters (TEM). With TEM a depth of 250 m may be reached (Jørgensen et al. 2003a). Thus the bottom of the deep valleys is out of reach. In densely populated areas electric noise will tamper the results and make the interpretation difficult or impossible. The resistivity-depth sections bear the uncertainty of equivalent electrical response as the combination of layer resistivities and thicknesses produce undistinguishable electric sounding responses. An advantage is that the electrical conductivity is related to hydraulic conductivity (see Chap. 15) and thus may give an indication to vulnerability.

With the gravity effect depth and dimension of the valley should be resolvable. So gravity may support EM mapping in problematic areas. A problem with gravity is the separation of the regional and residual anomaly. The quantitative interpretation of the residual anomaly results from forward modelling where constraints on structure and density values from borehole, seismic or other methods are necessary.

The seismic section reveals the most detailed structural image and is able to resolve also the deep valleys. A problem of high resolution seismic may occur by near surface inhomogeneities (like peat) that absorb seismic energy and reduce the energy transmission into the ground. Buried sources may solve the problem but will raise the costs of the survey.

For all methods the connection to borehole information and geophysical well logs is essential for the interpretation. At present the investigations focus on structural investigation. The information content of seismic amplitudes, phases and velocities is not fully exploited and comprises much more hydrogeophysical important information. But true amplitude, amplitude variation with offset or prestack depth migration processes are time consuming (in fixing the input parameters). Also the potential of shear waves is not really used at the present time. However the combination of the shear wave velocities Vs with compressional wave velocities Vp is the key for derivation of elastic constants and Vs/Vp ratio that is sensitive to lithology and interstitial fluid.

For the very near surface information ground penetrating radar may add useful information. Also tomographic inversion of first break arrivals of seismic data may be valuable.

Further systematically geophysical investigations in six defined buried valley test areas are conducted until the end of 2006 in the EU project BurVal. Structural, hydraulical and physical properties will be integrated into 3D models. Results and findings will be given at www.burval.org.

Table 12.1. Summary of buried valley sites

Valley	Ellerbek[1)]	Trave[2)]	Cuxhaven[3)]	Bredebro[4)]
Depth [m]	<400	<150	<400	<150
Width [km]	1.6	1.2	1.5	1.25
Sedimentary fill	Top layer silt and clay; other: sand	till	Top layer: clay; other: sand	sand
Surrounding	Tertiary clay and sand	Tertiary sand	Tertiary sand and clay	Tertiary clay
Mapping by TEM	---	---	survey in process Feb. 2005	good
Mapping by AEM	survey in process 2005	---	good	---
Residual Bouguer anomaly	-0.5 mGal	+0.7 mGal	+0.15 mGal	+0.3 mGal
Velocity effect	pull down	pull up	pull down	not reported
Methods employed	2D seismic, VSP, gravity, AEM (only single line)	2D seismic, VSP, gravity	2D seismic, 2D gravity, 2D AEM, 2D magnetic , electrical soundings	2D gravity, 2D TEM, 2D seismic

Case studies are reported by [1)] Wiederhold et al. 2002, Gabriel et al. 2003, [2)] Wiederhold et al. 2002, [3)] Gabriel et al. 2003, Siemon et al. 2001, Wiederhold et al. 2005, [4)] Poulsen and Christensen 1999. For locations see Fig. 12.7

12.3 A Large-scale TEM survey of Mors, Denmark

F. Jørgensen, P. Sandersen, E. Auken, H. Lykke-Andersen, K. Sørensen

This chapter is a short version of a case study presented in Jørgensen et al. (2005).

Large areas of Denmark have been covered by TEM (Transient Electro-Magnetic) surveys since the early 1990s in order to outline aquifers and the vulnerability of water resources. This work has resulted in an intense development in the use of the TEM method for hydrogeological investigations. Optimized handling of instruments, deletion of noise-infected and coupled soundings and the identification of defective instruments have led to significant improvements of survey results. Experience shows that surveys of large areas with dense data coverage provide a solid base for the construction of geological models.

This case demonstrates that various types of geological structures can be defined and outlined, and that the understanding and knowledge of near-surface geology can be significantly improved by a detailed study of TEM data.

12.3.1 Study area – the island of Mors

The study area is confined to the island of Mors, which is situated in the northwestern part of Jutland, Denmark (Fig. 12.16). The island covers an area of about 360 km^2. It is 10 - 15 km wide and about 35 km long with a SSW-NNE trend.

The sub-Quaternary strata comprise Upper Cretaceous white chalk (Maastrichtian) and limestone (Danian) covered by clays and diatomite (Paleocene and Eocene). This is followed by micaceous clay of Oligocene age and, to the south, also by Miocene clay, silt and sand (Gravesen 1990, 1993). The Palaeogene sediments are in general 50 - 250 m thick, but locally they are thinner or even absent. They have in many places been subject to deformation during the Quaternary glaciations, but the most pronounced impact on the Palaeogene topography is by a series of incised Quaternary valleys. These valleys have mainly been filled with thick sequences of glacial deposits. Where no valleys exist, only relatively thin layers of glacial origin cover the pre-Quaternary formations.

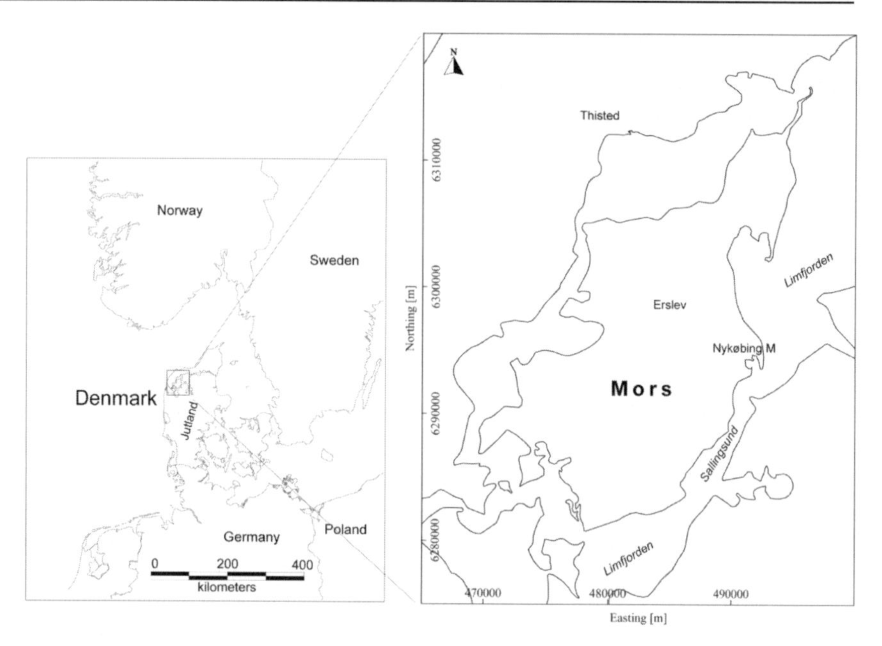

Fig. 12.16. Location map

The overall structure of the Tertiary and Quaternary formations on Mors is mainly controlled by 1) the Mors salt diapir, 2) glaciotectonic deformation during the Quaternary glaciations and 3) extensive systems of incised buried valleys.

The Mors salt diapir is located in the central part of the island (Elsam and Elkraft 1981, Larsen and Baumann 1982). The diapir, which is elliptical in horizontal views, covers an area of about 75 km^2 with its long axis trending E-W. Upper Cretaceous white chalk and Danian limestone, together with Palaeogene clay and diatomite, are uplifted over the salt diapir and along its flanks as a result of halokinetic movements. The top of the Zechstein salt is at a depth of more than 400 - 500 m (Madirazza 1977, Elsam and Elkraft 1981). Erosion has removed some of the pre-Quaternary deposits over the central parts of the diapir so that Quaternary sediments cover the Upper Cretaceous white chalk here.

Systems of buried valleys formed by subglacial meltwater erosion have been documented at several locations on Mors by a combination of borehole logs and TEM data (Jørgensen and Sandersen 2004). Most of the valleys are incised into pre-Quaternary clay-dominated sediments and over the salt diapir the valleys are eroded into the chalk and limestone. The valleys are filled with glaciofluvial deposits, glaciolacustrine deposits, interglacial deposits and tills. Extensive glacial erosion is also evident above the central part of the salt diapir where Tertiary sediments are absent (Elkraft and Elsam 1981; Gravesen 1990, 1993).

The thickness of the Quaternary succession varies significantly as a result of glacial erosion and deformation. The deposits in the buried valleys are commonly more than 150 m thick but where no valley erosion has taken place they can be less than 20 m thick. The cover is very thin or absent above the salt diapir. The majority of the Quaternary deposits are tills, but large amounts of glaciolacustrine clay and glaciofluvial sand and gravel are also present (Gravesen 1990, 1993).

12.3.2 Hydrogeological mapping by the use of TEM

Compared with other mapping methods, the TEM method is relatively cheap, it has a large penetration depth and at the same time it gives both structural and lithological information. It cannot, however, provide the same amount of structural detail as the more costly seismic methods (Jørgensen et al. 2003b).

Because the TEM method mainly resolves and quantifies low-resistive layers, the values for layers with resistivities exceeding 80 - 120 Ωm are not precisely determined; they just have a high resistivity. Furthermore, for certain combinations of layer thicknesses and resistivities, equivalence problems make it impossible to determine either the exact thickness or exact resistivity of the layers (Fitterman et al. 1988).

Like all electromagnetic diffusion methods, TEM has a decreasing resolution capability with depth. As a rule of thumb, layers with a thickness of 15 - 20 m can be resolved in the shallow part, while layers must be more than 20 - 50 m thick to be resolved at 100 m depth. Individual geological layers will commonly be averaged into one model layer because geological layers are normally too thin to be resolved. In general, 3-5 layers can be satisfactorily resolved in a TEM sounding. Furthermore, under normal circumstances the upper 10 - 20 m will be averaged into one layer, due to principal difficulties in recording data at very early decay times. The above-mentioned numbers are highly dependent on the resistivity of the layers.

The lateral resolution capability of 3D structures also decreases with depth. At depths of about 100 m the area from which data is obtained is more than 300 - 400 m in diameter. For depths of 25 m the area is only 75 - 100 m in diameter. This implies that 3D structures are less resolvable and become more diffuse with depth. Modelling of 2- and 3D structures has been thoroughly discussed by e.g. Newman et al. (1986), Goldman et al. (1994), Danielsen et al. (2003) and Auken et al. (2004).

In order to facilitate geological interpretations the inverted models are typically presented in thematic maps and cross-sections. The resistivities, as estimated by the inverted models, do not provide unambiguous information about lithology in the subsurface layers, and they have to be interpreted by the use of different types of thematic maps. The most commonly used thematic maps are interval resistivity maps and maps of depths to low-resistive layers. Interval resistivity maps are normally produced in 10 or 20 m elevation intervals, where average resistivities are calculated within each interval. Such short intervals are required for an optimized interpretation because some layer boundaries, especially those defining the surface of the low-resistive layer, are normally very well determined. A complete succession of intervals covering the subsurface down to the maximum depth of penetration has proved to be a useful way to visualize the TEM survey results as a basis for geological interpretations (GeoFysikSamarbejdet 2003, Jørgensen et al. 2003a). The interval resistivity maps are typically supplemented with depth control maps of a low-resistive deep-seated layer, if such a layer exists. This depth control map shows the elevation of a selected layer, where the selection is the deepest layer in all inverted models with resistivities below a given level.

A cross-section presentation is often used when TEM data are compared with other types of data or if detailed studies of the resistivity models are needed in selected areas. Just like borehole logs, TEM-models are most commonly shown on the sections by narrow vertical panels (e.g. GeoFysikSamarbejdet 2003; Jørgensen et al. 2003b; Sandersen and Jørgensen 2003). Another useful way to display the TEM survey results on cross-sections is to transfer a dissected succession of interval resistivity maps onto the sections.

Experience from geological interpretation of TEM surveys is primarily gained when their results are compared with other types of data sets, especially for boreholes. Large numbers of TEM surveys carried out in Denmark have contributed significantly to the understanding of how the resistivity images of the subsurface can be interpreted in geological terms (Auken et al. 2003; GeoFysikSamarbejdet 2003; Jørgensen et al. 2003b, 2003a; Sandersen and Jørgensen 2003; Jørgensen and Sandersen 2004).

The hurdle for the geological interpretation of TEM-data is the conversion from the modelled layer resistivities to the lithology of the layers and from layer geometry (as revealed by correlations between soundings) to structural reality. This conversion requires several aspects to be considered: 1) identification and exclusion of coupled and otherwise noise-infected soundings, 2) the vertical and horizontal resolution capability, 3) resistivity values of lithologies in the survey area, 4) ion content of the pore water. The principal means for achieving geological interpretation are: 1) correlation with independent information from other data sets, 2) evaluation of the various thematic maps (as mentioned in the previous paragraph) in terms of independently derived and well documented geological models.

The lithological interpretation of the TEM survey on Mors is based on empirical comparisons between TEM data and borehole logs. A list of typical resistivities for some of the most common sediments occurring on Mors and in adjacent areas is shown in Table 12.2. The resistivities are estimated for freshwater saturated sediments.

Table 12.2. Estimated resistivity values for freshwater saturated sediments in the Mors region. Modified from GeoFysikSamarbejdet (2003) and Jørgensen et al. (2003a)

Sediments	Resistivity
Meltwater sand and gravel	> 60 Ω
Clay till	25 - 50 Ωm
Glaciolacustrine and marine clay	10 - 40 Ωm
Palaeogene clay	1 - 10 Ωm
Palaeogene diatomite	10 - 40 Ωm
Maastrichtian white chalk	30 - 100 Ωm
Danian limestone	> 80 Ωm

12.3.3 Data collection and processing

A total of 2904 TEM soundings were collected on Mors during the period 1998 – 2002. The field survey, data processing and modelling were carried out by the consulting company Dansk Geofysik for Viborg County (Dansk Geofysik 2002). The survey was part of an ongoing hydrogeological survey campaign designed to investigate aspects of groundwater resources. The average density of TEM soundings is around 11 per km^2. In four high-priority areas the density is 12 - 16 soundings per km^2. These areas are situated in the far northern part, in the northern middle part, in the southern middle part and in the far southern part of the island (Fig. 12.17).

The TEM instrumentation used for the survey is the Geonics PROTEM 47 system and the conventional 40x40 TEM configuration was applied

(Dansk Geofysik 2002). All collected TEM data have been reported to the national database of geophysical data GERDA (http://gerda.geus.dk), and the TEM data used here is an extract from GERDA provided by Viborg County.

The extracted TEM data from GERDA has been carefully examined in order to remove noise-infected TEM soundings. A total of 249 TEM soundings have been classified as noise-infected and useless, primarily due to local transmitter coupling to man-made conductors or high levels of background noise (Sørensen et al. 2001). Mainly capacitive coupled soundings with oscillatory data have been removed (see examples in Danielsen et al. 2003, Figs. 12.19 and 12.20). Galvanic coupled data, however, can be impossible to identify, and a limited number of coupled soundings might therefore still be present in the dataset. After removal of the identified coupled soundings the remaining 2655 soundings constituted the basis for further interpretation.

All soundings have been inverted by the surveyors using 1D inversion modelling code EM1DINV (HydroGeophysics Group 2004a; 2004b). The selected models contain the fewest possible layers within accepted limits of fit between model data and observed data. All models contain 2 - 5 layers. No a priori information from borehole logs or other geological information was used during the inversion process or in the selection of models. The models therefore represent objective datasets, without biased interpretation at this stage of the data handling process.

The geological modelling presented here is mainly based on interval resistivity maps representing intervals of 10 m. Elevation maps of low-resistive layers are also used. The maps of low-resistive layers are generated on the following basis: 1) The layer resistivity of the selected layer must be less than 10 Ωm, 2) the layer below must not exceed 10 Ωm, and 3) the layer search routine is conducted from the top downwards. The 10 Ωm-limit is chosen because the Palaeogene clay layers in the area normally exhibit resistivities below 10 Ωm.

Kriging interpolation with cell spacing of 125 m and a search radius of 600 m is used for the gridding of the thematic maps. Colour scales with reddish colours for high-resistive layers and bluish colours for low-resistive layers are used, whereas for the elevation maps reddish colours show high levels and bluish colours show low levels.

To support the geological interpretation of the TEM data all relevant and available data and information are taken into account. Examples are borehole logs from the national well database "Jupiter" (GEUS 2003), results from field investigations in clay pits and coastal cliffs, as well as other geological investigations and relevant literature.

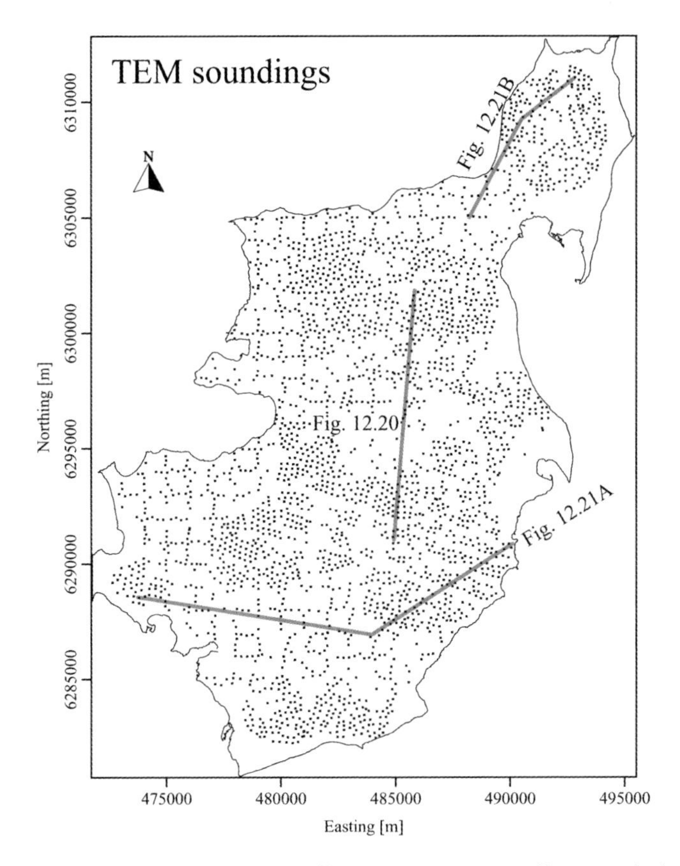

Fig. 12.17. Map showing all TEM soundings on Mors. Grey lines mark the cross-sections of Figs. 12.20 and 12.21. Scale: 5 km between axis ticks. Coordinate system: UTM zone 32

12.3.4 Results and discussions

The results of the TEM survey are outlined in Figs. 12.18 and 12.19, where thematic maps of the entire island are shown. An elevation map of the deepest low-resistive layer (<10 Ωm) is shown in Fig. 12.18A and four selected maps of interval resistivity are shown in Fig. 12.19A. Sketches of the geological structures seen in the individual thematic maps are shown separately in Figs. 12.18B and 12.19B. These sketches illustrate layer boundaries and geological structures in the selected intervals. They do not show the overall interpretations, but outline only what can be observed directly on the individual maps. They are meant as a guide for the descriptions, as are the letters and numbers annotating the various features.

The elevation map in Fig. 12.18 shows the deepest low-resistive layer (resistivity lower than 10 Ωm). From borehole logs it appears that, in most of the island, this layer can be correlated with Palaeogene clay, but occurrences of saline groundwater in sand/gravel or chalk/limestone may locally influence the image of the low-resistive layer. The map therefore shows the topography of the Palaeogene surface with local influence by saline groundwater. Such occurrences are mainly seen in the layers over the salt diapir. The elevation of the low-resistive layer ranges from more than 170 m b.s.l. to approximately 50 m a.s.l.

The four interval resistivity maps (Fig. 12.19A) show a vertical distribution from generally low resistivities in the deeper layers to high resistivities in the shallow layers. This illustrates the general succession of Tertiary and Quaternary sediments in the area, gradually changing from clay dominated layers at depth to more sandy layers in the upper parts.

The salt diapir as seen in the TEM data

Two curved depressions appear on the elevation map of the deepest low-resistive layer in the central part of Mors (Fig. 12.18) forming an elliptical shape open to the east and the west. The curved depressions (a) are more than 150 m deep and between 5 and 10 km in length. To the east they merge into a long linear depression (structure 1), and towards the west they terminate abruptly. The curved depressions are flanked by prominent ridges (Fig. 12.18, structure d).

Fig. 12.20 shows a cross-section through the area described above (for location, see Fig. 12.18). The two curved depressions (structure a) appear as dipping layers with very high resistivities (>120 Ωm), between which two more or less flat-topped bodies of low resistivity can be seen (structures b and c). The ridges of structure d occur as dipping layers resting on structure a, and the elongate depressions outside the ridges (structure 2) are, at least in the southern part, seen as a V-shaped structure with varying resistivity. The impression given by the cross-section is that the high-resistive structures of dipping layers converge towards a common point at around 100 m a.s.l.

The structures described above are also seen in the interval resistivity maps in Fig. 12.19. In the highest level (0 - 10 m a.s.l.) the structures of a and d are for instance seen as curved bodies with high and low resistivities, respectively. In the interval 10 - 20 m b.s.l., the structures of a and d become more distinct. Between 40 and 50 m b.s.l. the low-resistive layer begins to dominate the succession. Structure a is even more distinct, and structure d gradually merges into the low-resistive layer in general. In the lowermost slice of 90 - 100 m b.s.l. the low-resistive layer dominates. Due to the dipping layers the distance between the arcs of structure a gradually increases downwards through the intervals. In the uppermost slice, the maximum distance is around 3.5 km while at the deepest level it expands to about 5.5 km.

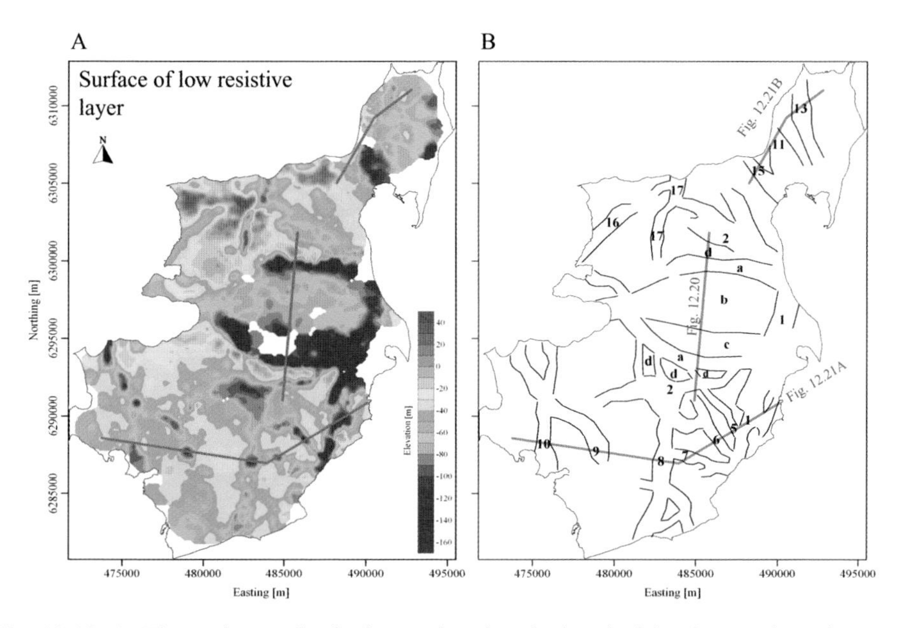

Fig. 12.18. A. Thematic geophysical map showing the level of the deepest layer in the TEM models with resistivities lower than 10 Ωm. Cross-sections in Figs. 12.20 and 12.21 are shown as thick grey lines. **B**. Outlines of structures and geological elements referred to in the text are shown on the sketch with black lines, numbers and letters. Scale: 5 km between axis ticks. Coordinate system: UTM zone 32

Interpretation of the salt diapir

The Mors salt diapir was previously described on the basis of mainly low resolution seismic surveys (e.g. Elkraft and Elsam 1981; Larsen and Baumann 1982), deep exploratory drillings (Elkraft and Elsam 1981) and the national archive of borehole logs (Gravesen 1990, 1993). Several boreholes penetrate the upper parts of the succession in the area, and from this it can be concluded that structure a corresponds to Danian limestone, and structures b, c and e correspond to Maastrichtian white chalk (compare

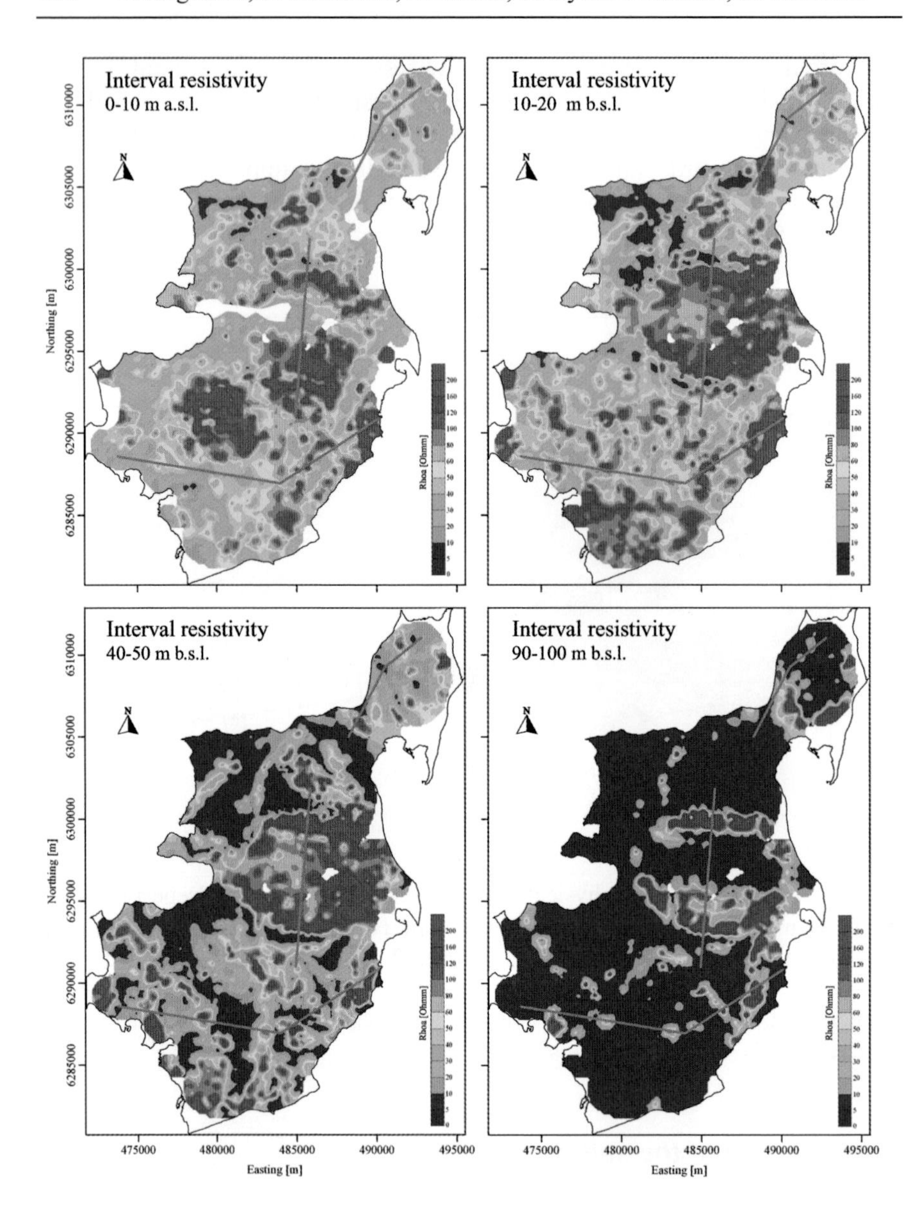

Fig. 12.19. A. Average resistivities in selected intervals. The TEM soundings do not penetrate more than about 30-50 m into the Palaeogene layers (blue), and the data coverage is therefore sparse in the low resistive areas of the deepest interval of 90-100 m b.s.l. Cross-sections of Figs. 12.20 and 12.21 are shown as thick grey lines

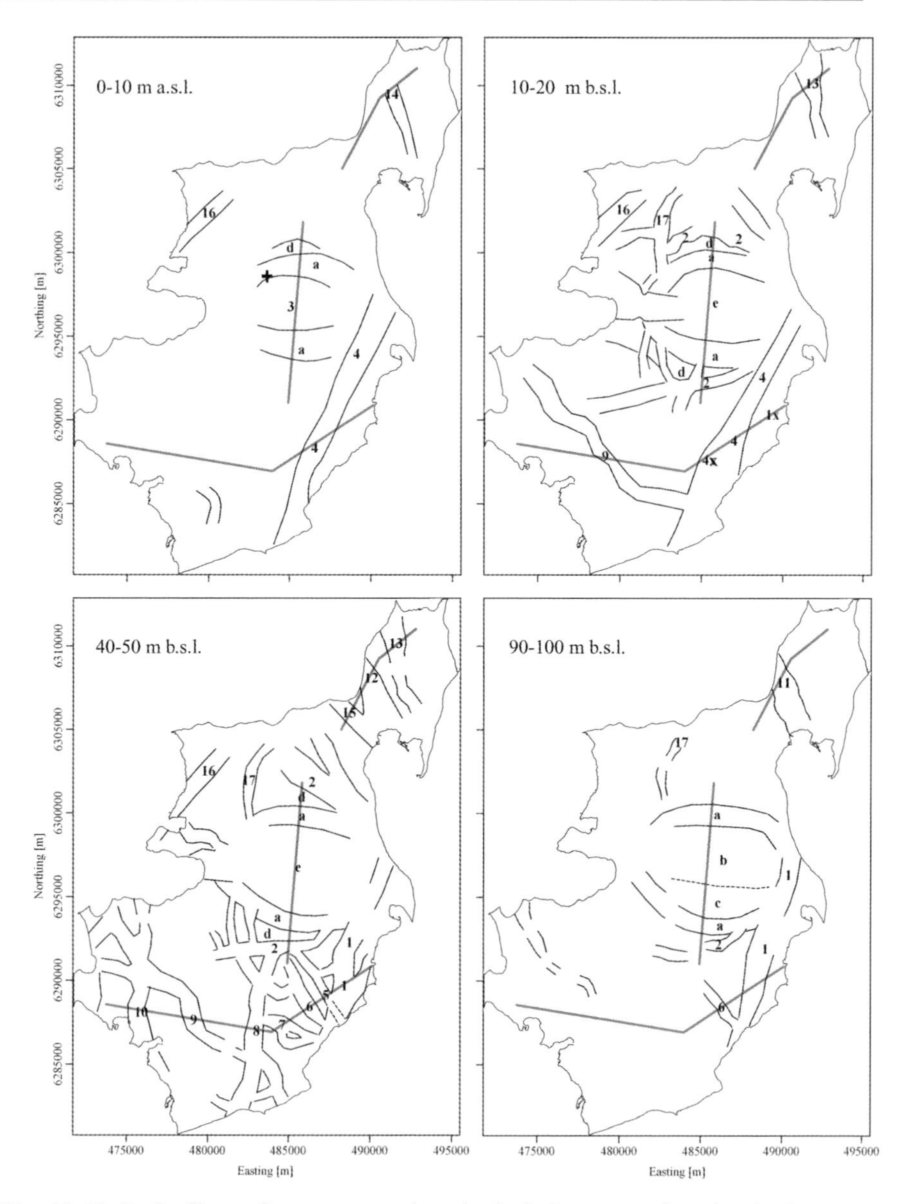

Fig. 12.19. B. Outlines of structures and geological elements referred to in the text are shown on the sketches with black lines, numbers and letters. The + on the interval of 0-10 m a.s.l. shows the location of a chalk pit at Erslev. Scale: 5 km between axis ticks. Coordinate system: UTM zone 32

with superimposed boreholes on the lower panel, Fig. 12.20). The boundary between the Danian and the Maastrichtian is exposed at about 10 m a.s.l. in a chalk pit at Erslev (Andersen and Sjørring 1992), and this level is consistent with the uppermost interval in Fig. 12.19. The location of the pit is marked with a cross on this interval map, and it is seen to correlate exactly with the inner/lower limit of structure a. Structure d corresponds to Palaeogene layers, locally glaciotectonically deformed into large fold structures (Pedersen 2000).

The structures of 1 and 2 correspond to various glacigenic sediments, but structure 4 appears more homogeneous and consists of meltwater clay, although described as tertiary clay in some borehole logs. Apart from relatively thin, scattered occurrences of Postglacial sediments, structure 3 also corresponds to glacigenic sediments, but this structure is mainly composed of clay till. The glacigenic sediments extend downwards into structure e, but this cannot be seen in the TEM images at deeper levels because here the sediments gradually become more sandy, resulting in an insignificant electrical contrast to the underlying chalk of medium to high resistivities. In Fig. 12.20 the boundary between glacial sediments and chalk is estimated from borehole data and marked by a dashed line. The structures 1, 2, 3 and 4 can be characterized as buried valleys, as described by Jørgensen and Sandersen (2004). The low resistivities of structures b and c indicate that the chalk contains highly saline pore water which has not been leached by circulating groundwater. The salt-freshwater boundary is located between 50 and 80 m b.s.l. over the northern part of the salt diapir (structure b), but to the south it seems to be located as deep as 120 m b.s.l. (structure c). The Danian limestone (structure a) is in general expected to be more coarse-grained and fractured than the white chalk. The hydraulic conductivity is therefore significantly higher in the limestone and the leaching of saline pore water may be effective to greater depths here. The deepest level of fresh groundwater in the limestone cannot be detected by the TEM soundings because only high resistivities occur within the assumed penetration depth of 120 - 140 m b.s.l. From the section (Fig. 12.20) it can be directly deduced that the stratigraphic thickness of the Danian limestone is about 80 m.

Buried valleys as seen in the TEM data

In the interval resistivity maps (Fig. 12.19A), valley structures appear as elongated structures defined by resistivity contrasts to their surroundings (compare with Fig. 12.19B). These structures mainly show moderate to high resistivities, but fairly low resistivities also occur, especially at the higher levels. The deepest incised valley structures are clearly expressed in the surface of the low-resistive layer as blue/green elongate features (Fig. 12.18). The buried valleys vary with respect to both dimension and orientation. The depths are found to be from a few tens of meters to more than about 140 m. The widths are normally between 500 and 1500 m, but some are up to 3000 m wide. The trends of the valleys are mostly restricted to the orientations N-S, SE-NW and E-W. In valley intersections it can occasionally be observed that younger valleys erode the infill of older valleys.

Beside the thematic maps two cross-sections presented in Fig. 12.21 (see Fig. 12.19 for location) demonstrate the occurrence and nature of the valleys. The cross-sections show an array of V- and U-shaped buried valleys, with infill of mainly high-resistivity sediments. The buried valleys on the cross-sections appear both as relatively shallow structures (structures 1x, 4, 4x, 5, 12, 13, 14, 15) or as deep structures (structures 1, 6, 7, 8, 9, 10, 11). The deep buried valleys are incised more than 80 meters into the deep low resistivity layer. The deep valleys seem to terminate upwards at a level corresponding to the present-day sea level. The exact depths of valley 1 and 11 are uncertain because no TEM soundings have been able to detect the valley floor.

Only a fraction of the buried valleys located in the TEM survey have been described above. Most of the valleys are outlined in the thematic maps in Fig. 12.18 and Fig. 12.19, but not all identified valleys are shown due to limitations in the number of thematic maps presented here. Some valleys, for example, only appear in limited intervals at levels between the selected intervals. Furthermore, only valleys with a significant resistivity contrast between the infill and the surroundings can be detected by the TEM method (Jørgensen et al. 2003a). It is also necessary that the valley fill consists of relatively homogeneous sediments without sections with varying resistivities, leading to inconsistent resistivity contrasts.

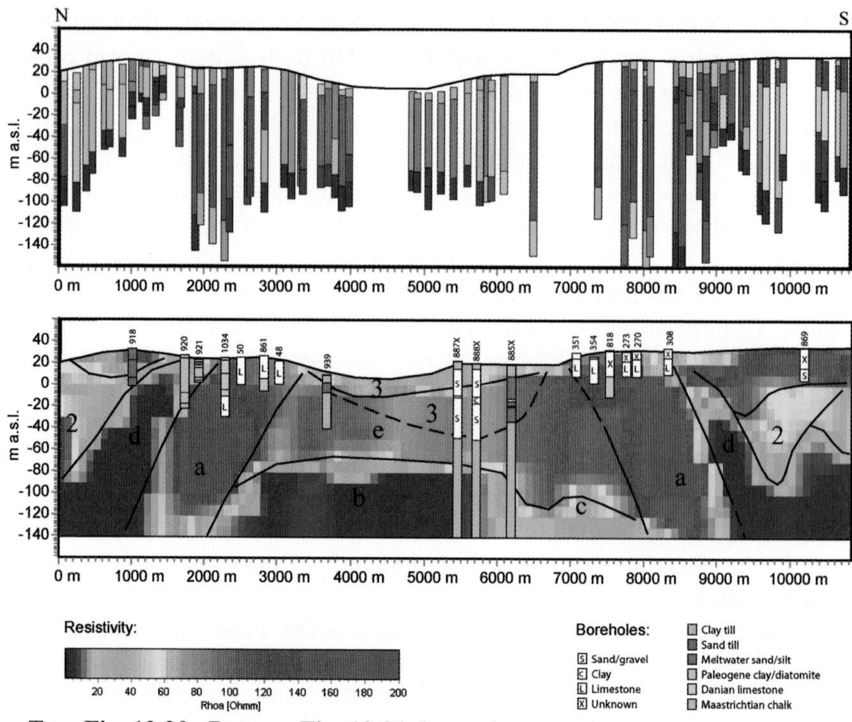

Top: **Fig. 12.20.**, Bottom: **Fig. 12.21.** Legends see next page

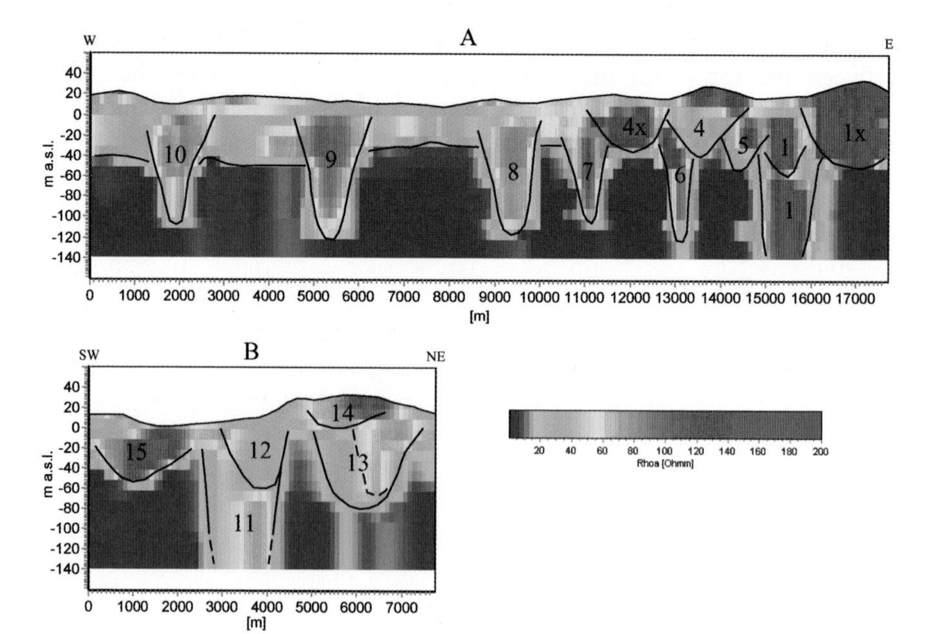

Fig. 12.20. Cross-section across the Mors salt diapir. Upper panel: single resistivity models of each TEM sounding situated within less than 300 m from the section line shown in Figs. 12.17, 12.18 and 12.19. Lower panel: the succession of interval resistivity grids as they are cut vertically by the cross-section. The TEM soundings do not penetrate more that about 30-50 m into the low resistivity layers (blue). Geological features referred to in the text are marked on the lower panel with black lines, numbers and letters. Lithological data from borehole logs are projected onto the section from distances of up to about 500 m. The annotated borehole numbers with prefix "37" refer to the national archive of borehole logs (GEUS 2003). Vertical exaggeration: 15 x

Fig. 12.21. Two cross-sections (A and B) showing buried valleys in the southern and the northern part of Mors respectively. The panels show the successions of interval resistivity grids as they are cut vertically by the cross-sections. Interpretations and geological elements referred to in the text are marked by black lines, numbers and letters. Note the considerable vertical exaggeration (22 x)

Interpretation of the buried valleys

As described above many structures appearing in the thematic maps can be identified as buried valleys, solely on the basis of their shapes. However, it is necessary to combine the TEM data with lithological information from boreholes in order to make a proper interpretation. Information from the national well database (Gravesen 1990, 1993; GEUS 2003) is therefore taken into account and compared with the TEM data. Most boreholes are poorly described lithologically and only few of them reach depths of more than 50 m. Nevertheless, many borehole logs provide useful information about the valley infill as well as the incised substratum.

The deep-seated low-resistive layer (Fig. 12.18) correlates generally well with Palaeogene clay in the borehole logs. However, in places where the diatomaceous Fur Formation (Pedersen and Surlyk 1983) dominates the upper part of the Palaeogene, a discrepancy between the Palaeogene surface and the low-resistive layer may occur. The reason is that the Fur Formation normally exhibits resistivities higher than 10 Ωm (see Table 12.2). In the coastal areas, saline pore water at deep levels may lead to erroneous geological interpretations. The low resistivities, e.g. below structure 1x, can therefore be a result of saline pore water, and the valley may be deeper than it appears on the TEM images. Valley fill of high resistivity (> 60 Ωm) is in general dominated by meltwater sand and gravel; medium resistivities are normally layers dominated by clay till or mixed deposits, whereas layers with low-to-medium resistivities are dominated by lacustrine clays.

Viborg County has recently performed three deep exploration drillings in buried valleys on Mors (Krohn et al. 2004). The three drillings are situated within the valley structures 1, 2 and 17 (see Fig. 12.19B). Except for some few meters of glaciomarine sediments around present-day sea level in one of the valleys (structure 1), all valley infill material is described as being glaciofluvial and glaciolacustrine sediments and till.

The widespread occurrence of lacustrine and marine clays in the valleys indicates that at least parts of the valleys were exposed during late glacial or interglacial periods.

Valley generations and relative ages

As seen in the cross-sections in Fig. 12.21, different valleys appear to incise each other (e.g. structures 1 and 1x; 4 and 6; 13 and 14; structures 11 and 12) thus indicating the presence of more than one generation of buried valleys on Mors. Signs of valley that incise other valleys can also be seen on the thematic maps, for example in Fig. 12.19 where the narrow valley of structure 5 crosses structure 1. Here, deposits of low-resistivity extend across the valley of structure 1 indicating that it is crosscut by structure 5 (see dashed lines in Fig. 12.19B, 40 - 50 m b.s.l). The presence of different valley generations also appears where young buried valleys confined to the shallow part of the subsurface overlie older, deep-seated valleys, as seen in the cross-sections in Fig. 12.21. In such instances, infill deposits of young valleys can be detected in higher levels of the TEM-survey, independently of the deeper lying valley deposits. The most convincing example demonstrated in the thematic maps is where the valley of structure 4, which contains low-to-medium resistive layers, lies above the buried valleys of structures 1, 5, 6 and 7 (see Fig. 12.19).

The orientations of the buried valleys on Mors appear to cluster within three groups: N-S, SE-NW and E-W (Fig. 12.22). It is proposed that these preferred orientations represent individual valley generations formed by multiple glaciations. The N-S orientation can furthermore be divided into at least two generations located at different vertical levels and stratigraphic positions. All indications of age relationships are examined and compared for the different generations, and the found age relationship is shown in Fig. 12.22. The E-W generation is most likely older than the two N-S generations, and the SE-NW generation is found to be younger than the first N-S generation, but older than the second one.

12.3.5 Conclusions

Detailed examination of the TEM survey on Mors shows that large-scale TEM surveys are able to add important new and detailed information to the construction of geological models, thus giving a more comprehensive understanding of the formation of the subsurface layers and structures. These capabilities of the TEM method for geological mapping have been developed over the last few years and are in particular the results of enhancements in TEM instrumentation, data processing, data handling and data presentation. As a consequence the geological understanding and interpretation of TEM data has been improved.

Apart from offering excellent images of the Palaeogene clay surface, the TEM survey are also able to resolve structures and layers situated in the overlying succession. Geological structures are detected by the resistivity distributions as observed in surface maps and interval resistivity maps, while the type of lithology is indicated by the resistivity values. In this way structural and lithological information is attained from the same data set, thus adding important information for interpretation of the geological environment.

The morphology of the layers over the Mors salt diapir is clearly demonstrated as an ellipsoid-shaped structure by the TEM survey. Palaeogene clay rests upon Danian limestone which partly covers the Maastrichtian chalk. Saline pore water is present in the chalk 50 - 80 m b.s.l. Numerous buried valleys dissect the entire island, including the layers over the salt diapir. The valleys are widespread and can be divided into different generations, each having their own preferred orientation. In places where they overlie or cross each other relative ages of valley generations can be estimated. Four generations have been identified, but more may exist. Most of the valleys are presumed to be old, most likely formed during the Elsterian glaciation or earlier. The buried valleys are mainly filled with glacial sediments, but interglacial sediments also occur. Many valleys contain lacustrine and marine clays in their upper parts up to levels between sea level and 30 m a.s.l.

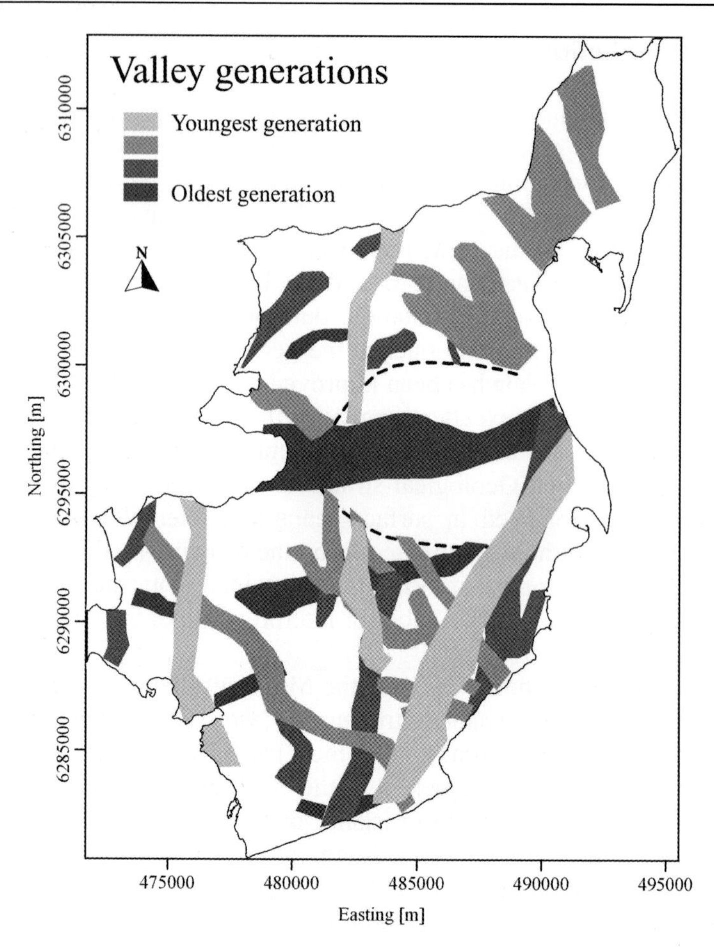

Fig. 12.22. Map of all the identified buried valleys divided into four proposed valley generations of different ages. Age relationships are indicated, and the location of the salt diapir is marked with a hatched line. Scale: 5 km between axis ticks. Coordinate system: UTM zone 32

12.4 Groundwater prospection in Central Sinai, Egypt

M.A. Mabrouk, N.M.H. Abu Ashour, T.A. Abdallatif, A.A. Abdel Rahman

12.4.1 Introduction

The growing activities in the central part of Sinai Peninsula have caused an ever increasing need for water. According to the data of the drilled boreholes in that area, groundwater exists within the fractures of the carbonate rocks at different depths. The complex structural pattern in that area made it difficult to follow up the water bearing layers along their extensions.

Resistivity sounding is a suitable technique that can be applied in exploring for groundwater in anisotropic aquifers such as carbonate rocks. The application of this technique depends on following up the changes in the resistivity of the layers. The resistivity variation within a single layer can be considered as an expression of the changes in porosity (primary and/or secondary). Moreover, delineation of the major and minor structural elements as well as the associated fractured zones is of great help in understanding the conditions of groundwater occurrence.

In the present work, the area lying to the north of Nekhl town (Fig. 12.23) was selected for groundwater exploration by applying resistivity sounding. Emphasis has been given to the low lying areas at Abu Hamth and El Bruk localities. The investigated area lies between latitudes 29° 55' and 30° 15' N and longitudes 33° 40' and 33° 50' E.

The present geophysical work has been carried out with the objectives of delineating the subsurface geologic setting i.e. the horizontal and vertical variations in lithology and the structural elements that affect the succession, delineating the water bearing layers in the study area and studying the impacts of the structural elements on the groundwater occurrences. As an example, results from the Abu Hamth area are presented in this chapter.

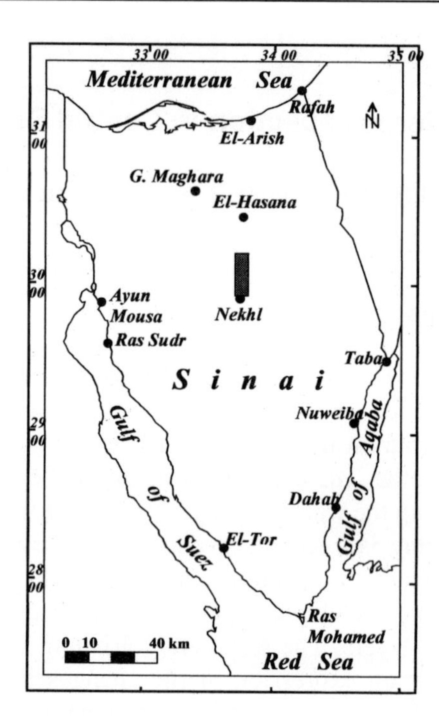

Fig. B12.23. Location map of the study area

12.4.2 Geological and hydrogeological aspects

Central Sinai lies within the arid belt of Egypt. The area is characterized by hot summer, mild winter, relatively little amount of rainfall, mainly in winter and spring, and high evaporation rates.

Central Sinai slopes regionally northward and is characterized by several types of low and high relief landforms, which were developed as a result of combined influences of endogenetic and exogenetic processes. The prevailing geomorphic patterns are represented by elevated plateaus, elongated dome-like hills, cuestas, hog-backed massifs, fault scarps, morphotectonic depressions and hydrographic basins.

According to the previous geological studies and information of some existing wells, the generalized lithostratigraphic succession in the study area can be described from base to top as follows:

Jurassic rocks: The Jurassic sequence constitutes the cores of the domal structural areas at central and northern Sinai. The thickness of marine sediments of the Jurassic sequence increases gradually northwards whereas the thickness of clastic sediments increases southwards.

Cretaceous rocks: The Lower Cretaceous rocks are represented by Malha Formation. This formation shows a general increase in thickness and variation in facies from continental sandy facies in the south into marine facies northwards.

The Upper Cretaceous succession is represented, in the investigated area, by the Cenomanian, Turonian and Senonian rocks. The Cenomanian rocks are composed of limestone, marl and shale. These rocks constitute also the core of El Bruk anticline. The Cenomanian rocks were encountered in several drilled wells with variable thickness at different depths. The Turonian rocks are similar in lithology to the underlying Cenomanian facies and are represented by Wata Formation that consist of marly, yellowish brown dolomitic and soft chalky limestone. The Senonian rocks are subdivided into two lithostratigraphic rock units, from base to top, Matulla Formation and Sudr Formation. Matulla Formation is composed of white limestone with shale and marl interbeds. Sudr Formation is composed of massive snow-white chalk.

Tertiary: The Paleocene section shows a uniform lithology of greenish-grey shale (Esna Shale Formation) overlying the Senonian chalk. The Eocene rocks are represented by the Thebes Formation (Lower Eocene), which is exposed at the environs of Nekhl and is composed of massive flinty limestone. In the subsurface, the Eocene deposits are often absent in many boreholes. The Oligocene occurrences are represented by basaltic dykes that cut all the above-mentioned succession.

Quaternary: The Quaternary deposits unconformably overly different rock strata of the preceding discussed stratigraphic units and are distinguished into alluvium and aeolian deposits. The alluvial deposits are composed of calcareous loamy sand, while the aeolian deposits are represented by drift sand accumulations in the form of sand sheets or moving sand dunes.

Structurally, central and northern Sinai is subdivided into four structural units (Fig. 12.24) as follows: (1) the stable foreland, (2) the gently folded zone, (3) the fractured or Shear zone, and (4) the strongly folded province.

Hydrogeologically, precipitation represents the main source of groundwater recharge. In the study area different rock units, belonging to different ages, are developed into water bearing formations. El Ghazawi (1989) and Hassanin (1997) classified the water bearing formations, according to the stratigraphic sequence, from younger to older, as follows:

1 the Quaternary alluvium aquifer,
2 the Eocene limestone aquifer,
3 the Upper Cretaceous fractured limestone aquifers

(Senonian aquifer, Turonian aquifer, Cenomanian aquifer)
4 the Lower Cretaceous aquifer (Nubia type sandstone).

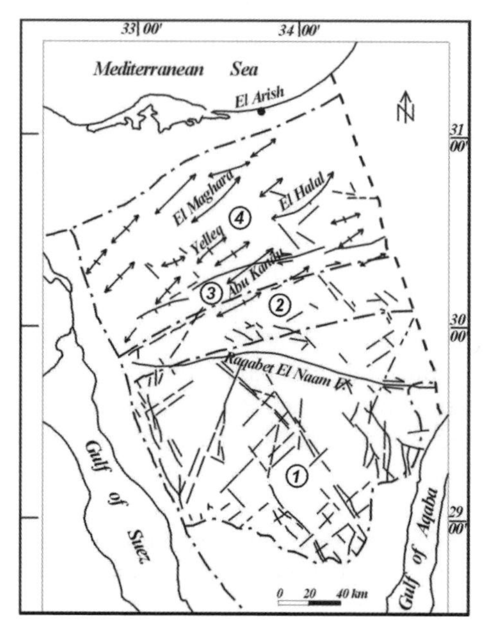

Fig. 12.24. Structural map of Central Sinai (after Shata 1956); 1: the stable foreland, 2: the gently folded zone, 3: the fractured zone, 4: the strongly folded zone

12.4.3 Field work and interpretation

A reasonable coverage for the study area was reached by a total of 46 vertical electrical soundings (VES) with 21 soundings in the Abu Hamth area (Fig. 12.25). The Schlumberger 4-electrode configuration was applied in the geoelectrical measurements with current electrode separation (AB) of 4000m. This electrode separation proved to be sufficient to reach the required depth that fulfils the aim of the study. Use was made of direct current resistivity meters (Terrameter SAS 300C and Terrameter SAS 4000) to carry out the electrical measurements. Both instruments directly measure the resistance (R) at each electrode separation with high accuracy.

Some sounding curves of the Abu Hamth area are shown in Fig. 12.26. For the quantitative interpretation of the resistivity sounding data, use was made of the computer program RESIX-PLUS, ver.2.39 (Interpex, 1996). The software is an interactive, graphically oriented, forward and inverse modeling program for interpreting the resistivity curves in terms of a layered earth model.

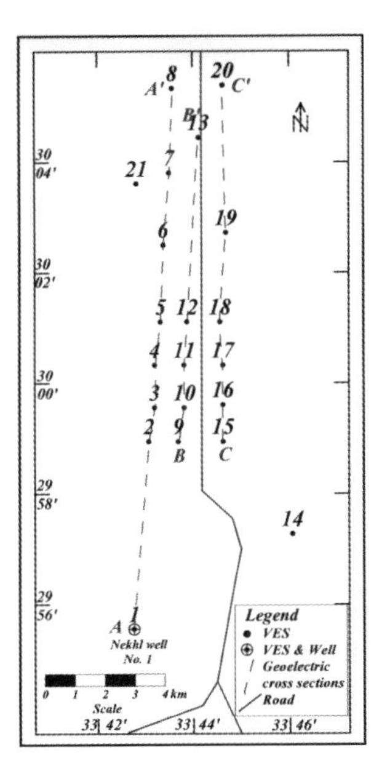

Fig. 12.25. Location map of the sounding stations and electrical cross sections in the investigated sites

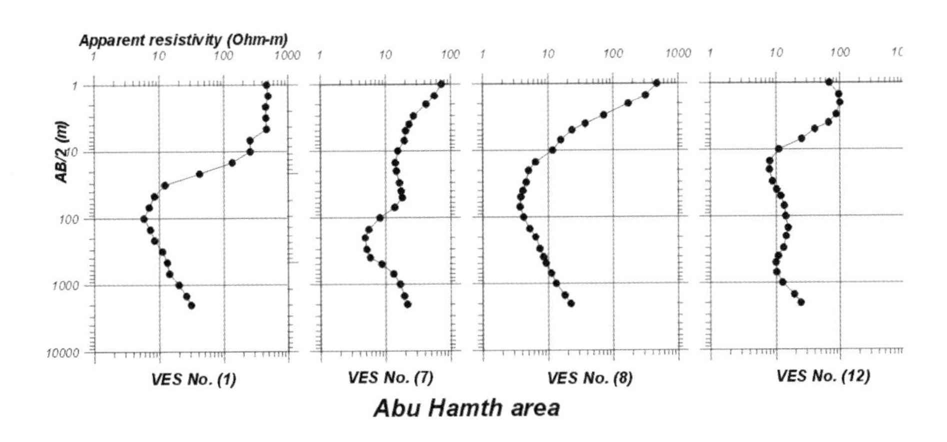

Fig. 12.26. Samples of the field resistivity sounding curves at the investigated sites

To reach optimum correlation between the resistivity layers and the predominant geologic units, some successive thin resistivity layers (mostly the uppermost ones) have been grouped together in one layer. The resistivity of such a layer is plausibly expressed in terms of the average transverse resistivity as calculated from the resistivities and thicknesses of the group of thin layers.

The general geologic setting and relevant structural elements in the investigated site Abu Hamth are visualized and described in view of a number of generated electrical profiles crossing the concerned sites in different directions.

The interpretation of the resistivity soundings led to the detection of nine layers. Some of these layers have not been detected at all sounding stations. The distribution of the resistivities and thicknesses of the nine encountered layers, from the surface downwards, is given as follows:

(1) The resistivity of the compiled group of near surface thin layers is expressed in terms of the average transverse resistivity. The layer shows diversity of the resistivity values (15- 500 Ωm) that characterize and are typically indicative of alluvial deposits (gravel, sand, marl and clay). This layer reaches its maximum thickness at VES 5 (16.7 m) and does not exceed 1m at VES 8.

(2) The second layer is characterized by relatively low resistivity values (2-14 Ωm). This resistivity values represent fine-grained materials such as clays, shales or marls. This layer has not been detected at soundings 1, 8 and 21. The thickness of this layer differs greatly from one locality to another where its maximum thickness (29.5 m) has been recorded at VES 2 and its minimum thickness (0.8 m) has been found at VES 13.

(3) The third layer extends all-over the study area (except for the locations of VES 1 and 11). The resistivity range of this layer (14.4- 98 Ωm) is interpreted as marly limestone. The variation of the resistivity within this range indicates a lateral lithologic change from marl to limestone. The extremely high values may occur due to the effect of dry fractures in the limestone. The exceptional resistivity value (165 Ωm), recorded at VES 12, clearly reflects this phenomenon. The thickness of this layer is, more or less, uniform where it varies from 9.5 m at VES 16 to 12.25 m at VES 9. It tends to decrease towards the northern soundings to reach 3.8 m at VES 8.

(4) The fourth layer attains low resistivity values that range from 0.6 Ωm to 15.7 Ωm. This resistivity range represents the Esna Shale. This layer extends allover the investigated site with relatively great thickness ranging from 43 to 59 m. Exceptionally, this thickness may decrease as at VES 11 (9.6 m) or increase as at VES 8 (75.2 m).

(5) The fifth layer exhibits resistivities ranging from 30.5 Ωm to 103 Ωm. An exceptional low resistivity value of 15 Ωm has been recorded at VES 11. This resistivity range represents, according to geologic evidence, the upper part of chalky limestone layer related to the Senonian (Upper Cretaceous). The thickness of the layer is nearly uniform throughout the area where it ranges from 46 m to 53.6 m. At VES 14 the layer decreases in thickness to reach 23.8 m.

(6) The sixth layer is characterized by low resistivity that changes within narrow range (0.3- 1.6 Ωm). At VES 14, the layer showed higher resistivity value (17 Ωm). According to the geologic information, this resistivity represents a layer that is mainly composed of argillaceous limestone. The anomalous resistivity value at VES 14 is due, mainly, to the decrease of the argillaceous material within this layer at the location of this sounding. The layer exhibits a uniform thickness ranging from 12 m to 16.5 m.

(7) The seventh layer attains resistivity values that vary from 32 Ωm to 175 Ωm. Nevertheless, at most of the soundings the resistivity of this layer lies within the range from 40 to 80 Ωm. This range is greatly similar to that of the fifth layer. According to the data of the wells, this layer corresponds to the lower part of the chalky limestone layer, which has the same composition of the fifth layer. The base of this layer has not been reached at many of the sounding stations. However, the detected thickness at the other soundings has been found to be uniform (23- 24 m).

(8) The eighth layer has been recorded at six soundings with relatively low resistivity (5.5 - 21.5 Ωm). Disregarding the extreme resistivity values, the resistivity of this layer in most cases lies within the range from 12 to 16 Ωm. According to the geological data, this layer corresponds to soft chalky limestone related to the Lower Senonian (Upper Cretaceous). The thickness of the layer, where it is detected, ranges from 37.4 to 45.4 m.

(9) The ninth layer attains resistivity variations from 32.6 to 95 Ωm. However, at most of the soundings the layer is characterized by a resistivity within the range from 50 to 60 Ωm. As recorded in the nearby wells, this layer corresponds to the upper part of the Turonian dolomitic limestone. This zone is not recorded as water bearing in Nekhl well or Hamth well, which has been specially drilled and designed to utilize the Nubian sandstone aquifer. However, the relatively low resistivities indicate possibility of groundwater occurrence. The base of this layer has not been reached.

The vertical and horizontal extensions of the detected layers along with the structural elements that affected the succession are illustrated through three electrical cross sections (Fig. 12.27).

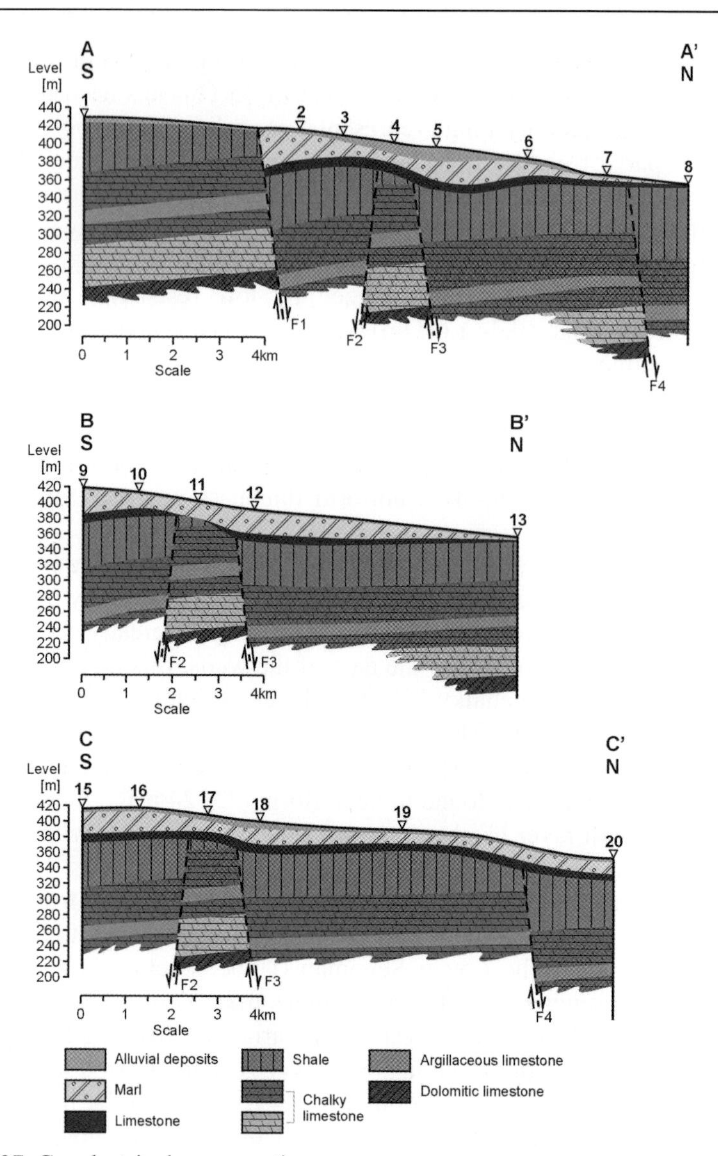

Fig. 12.27. Geoelectrical cross sections

As shown on the location map (Fig. 12.25), the cross section AA' extends from sounding 1 (at Nekhl well) across Wadi El Hamth to sounding 8 at the northern part of the investigated site. The cross section BB' extends parallel to AA' from sounding 9 to sounding 13. The third one CC' runs also in south-north direction from sounding 15 to sounding 20. The cross sections indicate that the first, second and third layers show general decrease in their thicknesses towards north. The layers show regular regional dip towards the south. Most of the layers extend along the cross sections with nearly uniform thickness except for the fourth layer due to structural and erosional processes. Four normal faults have been found to affect the succession. The faults F1, F3 and F4 throw down towards north, whereas F2 throws down towards the south and forms with F3 a horst structure. The fault F1 affected the whole succession, except for the surface alluvial deposits, whereas the faults F2, F3 and F4 affected the succession, which is older than the upper three layers. This means that the fault F1 is younger than the other faults and the upper three layers were successively deposited on the erosion surface of the underlying faulted layer.

In order to provide better insights into the structural configuration in the investigated site, the lower surface of the fourth layer (corresponding to the Esna shale) is selected to draw a structural contour map (Fig. 12.28 left). From this map, it is obvious that the fault F1 strikes in ENE-WSW direction, whereas the faults F2 and F3 strike nearly in E-W direction and the fault F4 strikes in WNW-ESE direction. According to the relative displacements of the identified faults different structural highs and lows had

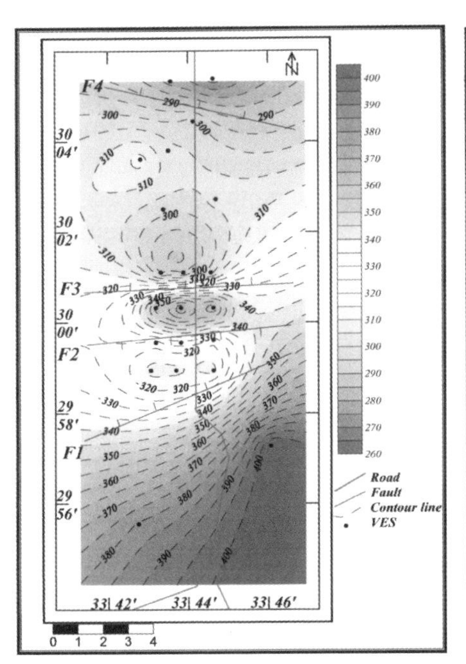

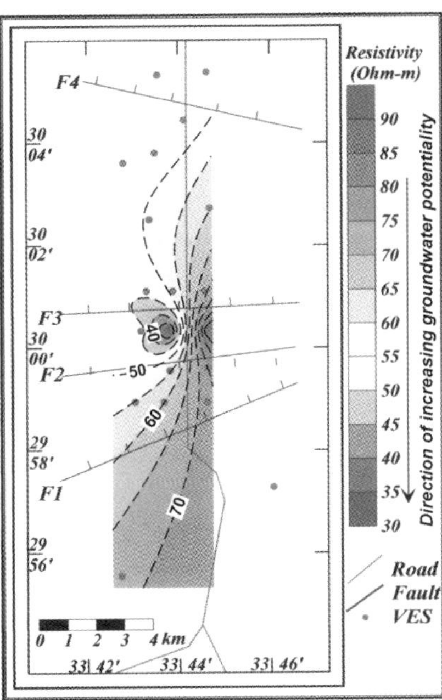

Fig. 12.28. Left: Structural contour map for the lower surface of the fourth layer, right: groundwater potentiality in layer 9

been developed. The faults F1, F3 and F4 throw down nearly towards north, whereas the fault F2 throws down to the south. A graben is developed between F1 and F2 followed by a horst between F2 and F3.

12.4.4 Groundwater occurrence

The geophysical results have been used to map the spatial extension of the layers which were known as water bearing from borehole data. The groundwater potentiality of these layers throughout the investigated sites has been evaluated in view of the distribution of the resistivity exhibited by the concerned layers. This has been achieved through isoresistivity maps for the water bearing layers in each of the investigated sites. Furthermore, the identified structural elements have been illustrated together with the isoresistivity maps in order to indicate their impact on the groundwater occurrences.

As a rule of thumb, values around the recorded resistivity of the water bearing layer at the location of a well are here considered to represent high groundwater potentiality. On the other side, the groundwater potentiality decreases at zones of low resistivity values in clay rich layers, where the clay content is the resistivity controlling factor or at zones of high resistivity values in clay free layers where the fracture density and degree of water saturation are the resistivity controlling factors.

The geophysical results indicate that groundwater possibly occurs in the upper part of the Turonian dolomitic limestone (the ninth layer). This layer is characterized, at most of the soundings, by resistivity values ranging between 50 Ωm and 60 Ωm with anomalous high and low resistivity values. The groundwater potentiality within this layer has been determined by making use of the isoresistivity contour map (Fig. 12.28, right).

The groundwater potentiality within the concerned layer increases by decreasing the resistivity values due to increasing the density of the water saturated fractures. Facies change to materials of low resistivity such as clay is excluded according to the lithologic description of the concerned layer (dolomitic limestone) in the wells. Based on the same concept, high resistivity values are considered to represent zones of less fracture density. From Fig. 12.28 (right) it is obvious that the groundwater occurrence is restricted to a small area at the horst structure, which is bounded by the faults F2 and F3. Elsewhere, groundwater almost does not exist. Generally, the layer shows poor groundwater potentiality. It is also evident that the highly fractured zone initiated under the effect of the two adjacent faults contributes to the groundwater occurrence.

12.5 References

Andersen S, Sjørring S (1992) Geologisk Set. Det nordlige Jylland. En beskrivelse af områder af national geologisk interesse. Geografforlaget, 208 pp

Auken E, Jørgensen F, Sørensen K (2003) Large-scale TEM investigation for groundwater. Exploration Geophysics, 34:188-194

Auken E, Christiansen AV, Jacobsen L, Sørensen KI (2004) Laterally Constrained 1D-Inversion of 3D TEM Data. 10th meeting EEGS-NS, Utrecht, The Netherlands, EEGS-NS. Extended Abstracts Book

Bahloul F, Blindow N, Lange M (1999) Kartierung von gespannten Aquiferen mit EMR: Anwendungsmöglichkeiten und Grenzen. Proc 5th DGG-Seminar Umweltgeophysik, Mitteilungen der Deutschen Geophysikalischen Gesellschaft, Sonderband II/99:124-134

Buness H, Bram K, Druivenga G, Grüneberg S (1997) A vibrator system for shallow high-resolution reflection seismics. 59th Mtg Eur Assn Geosci Eng, Session:P154

Buness H, Wiederhold H (1999) Experiences with a vibrator system for shallow high-resolution seismics. 61st Mtg Eur Assn Geosci Eng, Session:4042

Danielsen JE, Auken E, Jørgensen F, Søndergaard V, Sørensen KI (2003) The application of the transient electromagnetic method in hydrogeophysical surveys. Journal of Applied Geophysics 53:181-198

Dansk Geofysik (2002) Geofysisk kortlægning på Mors. Transiente elektromagnetiske (TEM) sonderinger. July 2002, 14 pp

Ehlers J (1994) Allgemeine und historische Quartärgeologie. Ferdinand Enke Verlag, Stuttgart

El Ghazawi MM (1989) Hydrogeological studies in northeast Sinai, Egypt. PhD thesis, Fac of Sci, El Mansoura Univ

Elkraft, Elsam (1981) Disposal of high-level waste from nuclear power plants in Denmark. Salt dome investigations. Report prepared by Elsam and Elkraft, June 1981. Volume II, geology, 411 pp

Fichler C, Henriksen S, Rueslaatten H, Hovland M (2005) North Sea Quaternary morphology from seismic and magnetic data – indications for gas hydrates during glaciation? Petroleum Geoscience (accepted)

Fitterman DV, Meekers JA, Ritsema IL (1988) Equivalence behavior of three electrical sounding methods as applied to hydrogeophysical problems. Proc of the 50th Ann Mtg European Assn Expl Geophys

Friborg R, Thomsen S (1999) Kortlægning af Ribe Formation. Et fællejysk grundvandssamarbejde. Popular rapport. Ribe Amt, Ringkjøbing Amt, Viborg Amt, Århus Amt, Vejle Amt, Sønderjyllands Amt. Tønder, ISBN 87-7486-360-6

Frisch U, Kockel F (1993) Geotektonischer Atlas von Nordwest-Deutschland – Das Tertiär. Unpublished report, BGR, Hannover

Gabriel G (2005) Gravity investigation of buried Pleistocene subglacial valleys. Near Surface Geophysics (accepted)

Gabriel G, Kirsch R, Siemon B, Wiederhold H (2003) Geophysical investigation of buried Pleistocene subglacial valleys in Northern Germany. Journal of Applied Geophysics 53:159-180

GeoFysikSamarbejdet (2003) Anvendelse af TEM-metoden ved geologisk kortlægning GeoFysikSamarbejdet. Århus Universitet, 72 p. Available at http://www.gfs.au.dk

GEUS (2003) The national well data archive. Viborg Amt. (www.geus.dk/jupiter)

Goldman M, Tabarovsky L, Rabinovich M (1994) On the influence of 3-D structures in the interpretation of transient electromagnetic sounding data. Geophysics 59:889-901

Gravesen P (1990) Geological map of Denmark 1:50.000. Kortbladet 1116 I Thisted. Geologisk basisdatakort. Geological Survey of Denmark. Map series no. 13

Gravesen P (1993) Geological map of Denmark 1:50.000. Kortbladet 1116 II Nykøbing Mors. Geological basic data map. Geological Survey of Denmark. Map series no. 21

Hassanin AM (1997 Geological and geomorphological impacts on the water resources in central Sinai, Egypt. PhD thesis, Fac of Sci, Ain Shams Univ

Holzschuh J 2002 Low-cost geophysical investigations of a paleochannel aquifer in the Eastern Goldfields Western Australia. Geophysics 67:690–700

Huuse M, Lykke-Andersen H (2000) Overdeepened Quaternary valleys in the eastern Danish North Sea: morphology and origin. Quat Science Reviews 19:1233–1253

HydroGeophysics Group, University of Aarhus (HGG) (2004a) SiTEM/Semdi processing and inversion package. Web document, accessed September 2004. Available at http://www.hgg.au.dk

HydroGeophysics Group, University of Aarhus (HGG) (2004b) Manual for the inversion program em1dinv. Web document, accessed September 2004. Available at http://www.hgg.au.dk

Jørgensen F, Sandersen PBE, Auken E (2003a) Imaging buried Quaternary valleys using the transient electromagnetic method. Journal of Applied Geophysics 53:199-213

Jørgensen F, Lykke-Andersen H, Sandersen PBE, Auken E, Nørmark E (2003b) Geophysical investigations of buried Quaternary valleys in Denmark: an integrated application of transient electromagnetic soundings, reflection seismic surveas and exploratory drillings. Journal of Applied Geophysics 53:215-228

Jørgensen F, Sandersen P (2004) Kortlægning af begravede dale i Jylland og på Fyn. Opdatering 2003-2004. De jysk-fynske amters grundvandssamarbejde. Vejle Amt, Watertech, July 2004, 178 p (Download at www.buried-valleys.dk)

Jørgensen F, Sandersen PBE, Auken E, Lykke-Andersen H, Sørensen K (2005) Contributions to the geological mapping of Mors, Denmark – a study based on a large-scale TEM survey. Accepted for publication in Bull Geol Soc Denmark. In press

Krohn C, Kronborg C, Nielsen OB, Knudsen KL, Sørensen J, Kragelund A (2004) Boring DGU. Nr. 37.1241, 37.1242 og 37.1248. Report No. 04VB-01. University of Aarhus, Denmark. Internal report. 64 pp

Larsen G, Baumann J (1982) Træk af Mors salthorstens udvikling. DGF Årsskrift for 1981:151-155

Larsen G, Kronborg C (1994) Det mellemste Jylland. En beskrivelse af områder af national geologisk interesse. Geografforlaget, 272 pp

Mabrouk MA (1991) Contribution to the hydrogeologic setting of Dara area, Gulf of Suez, Egypt: an electrical resistivity sounding approach. Egypt J Geol 34:339-352

Madirazza I (1977) Zechstein bassinet og saltstrukturer i Nordjylland med særligt henblik på Nøvling og Paarup. Dansk Geol Foren, Årsskrift for 1976:57-68

Marion D, Nur A, Yin H, Han D (1992) Compressional velocity and porosity in sand clay mixtures. Geophysics 57:554-563

Meekes JAC, Scheffers BC, Ridder J (1990) Optimization of high-resolution seismic reflection parameters for hydrogeological investigations in the Netherlands. First Break 8:263-270

Morgan NA (1969) Physical properties of marine sediments as related to seismic velocities. Geophysics 34:529-545

Newman GA, Hohmann GW, Anderson WL (1986) Transient electromagnetic response of a three-dimensional body in a layered earth. Geophysics 51:1608-1627

Palmer D (1981) An introduction to the generalized reciprocal method of seismic refraction interpretation. Geophysics 46:508-1518

Pedersen SAS (2000) Superimposed deformation in glaciotectonics. Bulletin of the Geological Society of Denmark 46:25-144

Pedersen GK, Surlyk F (1983) The Fur Formation, a late Paleocene ash-bearing diatomite from northern Denmark. Bulletin of the Geological Society of Denmark 32:43-65

Piotrowski JA (1997) Subglacial hydrology in north-western Germany during the last glaciation: groundwater flow, tunnel valleys, and hydrological cycles. Quat Science Reviews 16:169-185

Poulsen LH, Christensen NB (1999) Hydrogeophysical mapping with the transient electromagnetic sounding method. European Journal of Environmental and Engineering Geophysics 3:201-220

Praeg D (1996) Morphology, stratigraphy and genesis of buried Mid-pleistocene tunnel-valleys in the southern North Sea basin. PhD Thesis, University of Edinburgh

Sandersen PBE, Jørgensen F (2003) Buried Quaternary valleys in western Denmark – occurence and inferred implications for groundwater resources and vulnerability. Journal of Applied Geophysics 53:229-248

Schaumann G (1997) Anwendung der Transientenelektromagnetik zur Grundwassersuche in Namibia. Z angew Geol 43:70-74

Schulz-Rincke V, Kirsch R, Scheer W, Utecht T (1997) Geophysikalische Meßverfahren bei der Kartierung tertiärer Grundwasserleiter am Rande von Salzstöcken. Z angew Geol 43:154-158

Sen PN, Goode PA, Sibbit A (1988) Electrical conduction in clay bearing sandstones at low and high salinities. J Appl Phys 63:4832-4840
Shata A (1956) Structural development of the Sinai Peninsula, Egypt Bull Inst Desert 6:117-157
Siemon B (2005) Ergebnisse der Aeroelektromagnetik zur Grundwassererkundung im Raum Cuxhaven-Bremerhaven. Z Angew Geol 1/2005 in press
Siemon B, Voß W, Röttger B, Rehli HJ, Pielawa J (2001) Forschungsvorhaben "Detaillierte aerogeophysikalische Landesaufnahme" (DAGLA). Messgebiet Cuxhaven, Mai 2000. Technischer Bericht BGR, Archiv-Nr. 121 236
Smed P (1998) Die Entstehung der dänischen und norddeutschen Rinnentäler (Tunneltäler) – Glaziologische Gesichtspunkte. Eiszeitalter und Gegenwart 48:1-18
Stackebrandt W, Ludwig AO, Ostaficzuk S (2001) Base of Quaternary deposits of the Baltic Sea depression and adjacent areas. Brandenburgische Geowiss Beitr 1/2001:13-19
Sørensen KI, Auken E (2004) SkyTEM – a new high-resolution helicopter transient electromagnetic system. Exploration Geophysics 35:191-199
Sørensen KI, Auken E, Christensen NB, Pellerin L (2005) An integrated approach for hydrogeophysical investigations: new technologies and a case history. Society of Exploration Geophysicists, Near Surface Geophysics, 2 (in press)
Sørensen KI, Thomsen P, Auken E, Pellerin L (2001) The effect of Coupling in Electromagnetic Data. EEGS-Proceedings, Birmingham, England, 108-109
THOR Geophysical GmbH (1989) Water Supply of Port Sudan. Unpublished report (Kirsch R, Lütjen H), Kiel
TNO (1976) Geophysical well logging for geohydrological purposes in unconsolidated formations. Groundwater Survey TNO, The Netherlands Organisation for Applied Scientific Research, Delft
Wiederhold H, Buness H, Bram K (1998) Glacial structures in northern Germany revealed by a high-resolution reflection seismic survey. Geophysics 63:1265-1272
Wiederhold H, Agster G, Gabriel G, Kirsch R, Schenck PF, Scheer W, Voss W (2002) Geophysikalische Erkundung eiszeitlicher Rinnen im südlichen Schleswig-Holstein. Z Angew Geol 1/2002:13-26
Wiederhold H, Gabriel G, Grinat M (2005) Geophysikalische Erkundung der Bremerhaven-Cuxhavener Rinne im Umfeld der Forschungsbohrung Cuxhaven. Z Angew Geol 1/2005 in press
Wolfe PJ, Richard BJ (1996) Integrated geophysical studies of buried valley aquifers. Journal of Environmental and Engineering Geophysics 1:75-84
Yilmaz Ö (2001) Seismic data analysis: processing, inversion and interpretation of seismic data. In: Doherty SM (ed) Investigations in Geophysics no 10. Society of Exploration Geophysicists, Tulsa

13 Aquifer structures: fracture zones and caves

Kord Ernstson, Reinhard Kirsch

13.1 Hydraulic importance of fracture zones and caves

Fracture zones and caves (karst caves, volcanic caves) can play an important role for groundwater supply and generally in hydrogeological and environmental geological practice. In most cases fracture zones are considered hydraulic conductors, but they may sometimes also act as hydraulic barriers preventing flow across them (Committee on Fracture Characterization and Fluid Flow et al. 1996). The porosity of the fractures is called secondary porosity. Rock material can contain smaller fissures, e.g. by contraction while cooling, or larger fractures by tectonic movements along fault zones (Fig. 13.1). Fissured rocks have similar petrophysical properties as primary-porous material, so in principle the same geophysical techniques as for the exploration of water reservoirs in primary-porosity material can be applied (Chap. 12).

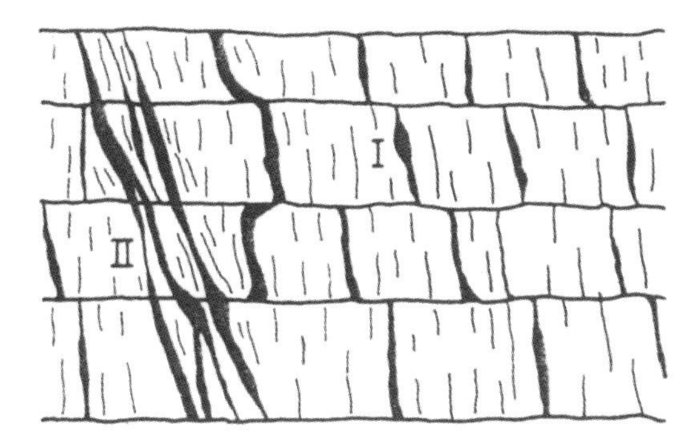

Fig. 13.1. Fissures (I) and fractures (II) in rocks (Schneider 1988)

In addition, fracture zones are a special target for geophysical and hydrogeological exploration, because in general, hydraulic and petrophysical properties of fracture zone and host material are strongly different. Although extending over large distances, the width of fracture zones is mostly narrow. Moreover, the dip angle of fracture zones must be taken into account for the siting of wells (Fig. 13.2).

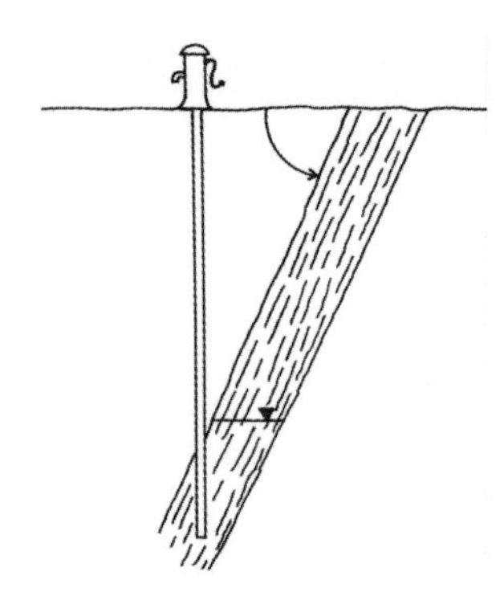

Fig. 13.2. Well location considering the dip angle of the fracture zone

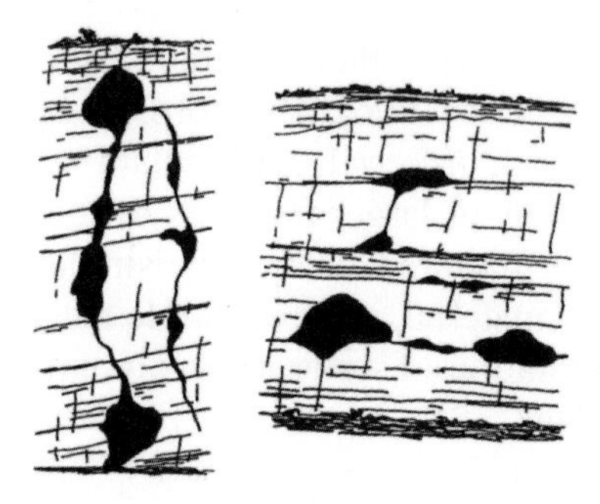

Fig.13.3. Karst systems oriented along fractures (left) or layer boundaries (right) (after Matthess and Ubell 1983, with permission from Borntraeger Verlag)

Frequently, the strike direction of fracture zones is known from tectonic considerations. Fracture zones can often be detected as lineation structures in satellite imagery or on air photos. However, for a successful groundwater exploration this remote mapping must be backed by airborne or ground geophysical surveys. Even a localisation error of the fracture zone as small as 10 m can result in a dry borehole (van Lissa et al. 1992).

Karst and volcanic caves are a further important class of aquifers. Formed by dissolution of carbonate rocks (limestone, dolomite), karst caves and channels are found mainly along tectonic fractures or horizontal layer boundaries (Fig. 13.3). Enhanced dissolution of limestone occurs in coastal areas in the transition zone between seawater and freshwater leading to widespread karst caves, e.g., in the Yucatan peninsula (Mexico) and on the Bahamas (Mylroie and Carew 1990). Apart from lava tubes, large voids in volcanic rocks are related with pillow basalt formation, with highly vesicular, broken zones at the top and bottom of lava flows, and with volcanic caldera collapse.

Fracture zones, karstic and volcanic caves can be aquifers of great yield. As a result of the high hydraulic conductivity of these aquifers a fast contaminant spread is possible. Also widespread saltwater intrusions may occur.

13.2 Geophysical exploration of fracture zones: seismic methods

Fracture zones, in general, are characterized by reduced seismic velocities due to open fissures compared to the host material. If the fracture zone coincides with tectonic rupture, a depth discontinuity of layer boundaries can occur. The possible influence of a fracture zone on the seismic wave field is sketched in Fig. 13.4: refracted waves can be reflected, diving waves can be reflected, refraction and reflection horizons can show a depth discontinuity. Moreover, due to enhanced absorption of seismic energy within the fracture zone, this zone can be a low-pass filter for seismic waves.

Fig. 13.5 shows a seismogram from a seismic refraction survey, where the geophone spread crosses a narrow fracture zone. Low seismic velocities inside the fracture zone lead to a travel time step of the refraction arrivals (marked by an arrow).

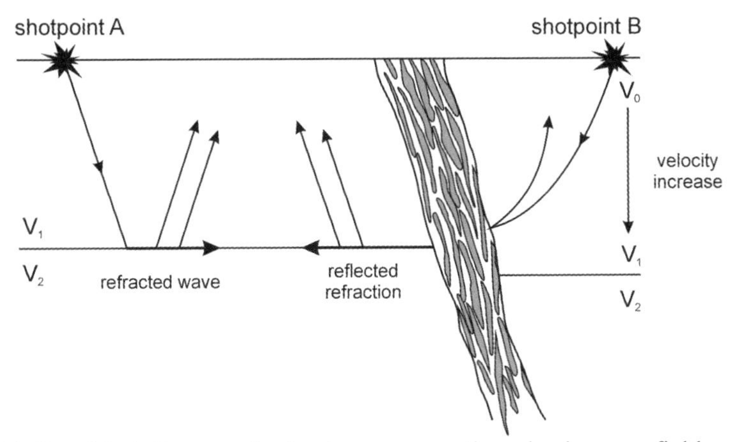

Fig. 13.4. Possible influence of a fracture zone on the seismic wave field: reflection of refracted waves (left, shotpoint A), reflection of diving waves (right, shotpoint B), depth discontinuity of refraction and reflection horizons due to tectonic movement at the fracture zone

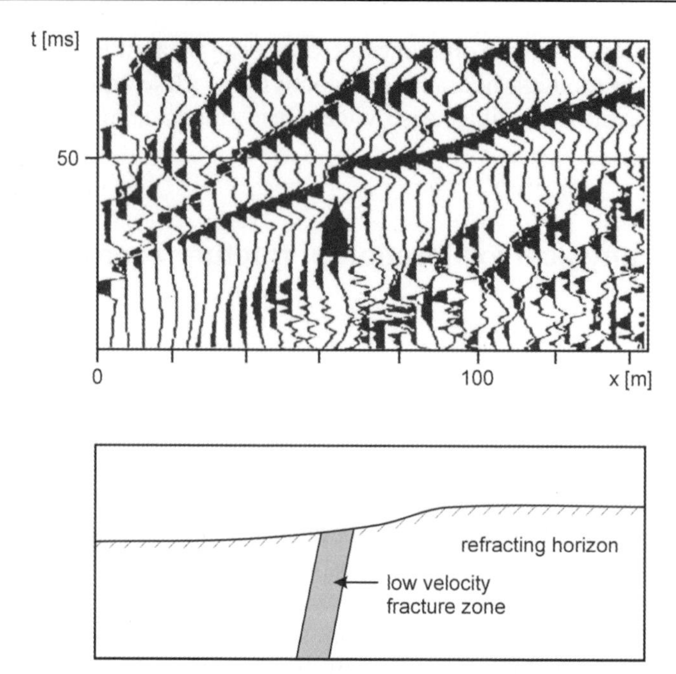

Fig. 13.5. Top: seismogram from a seismic refraction survey crossing a narrow fracture zone (after Wolfert 1992), arrow: travel time step of refraction arrivals due to low-velocity fracture zone, below: sketch of refracting horizon with low velocity fracture zone

In reflection seismic data, fracture zones associated with tectonic movements are indicated by interrupted or depth-shifted reflection horizons. An example of a high resolution shallow seismic survey is shown in Fig 13.6. Shotpoint and receiver distance was 2 m, a Buffalo gun was used as seismic source. The fracture zone is indicated by terminated reflection horizons, and it is associated with a fault dividing Quaternary sediments and greywacke (Wise et al. 2003). Fracture zone identification by depth shifted or terminated reflection horizons is shown by Shtivelman et al. (1998). Here a sledge hammer was used as seismic source with 2.5 m shotpoint and receiver spacing (Fig. 13.7). Especially in crystalline basement rocks, sub-horizontal fracture zones, their size and orientation can be studied by vertically travelling p waves (see, e.g., Green and Mair 1983).

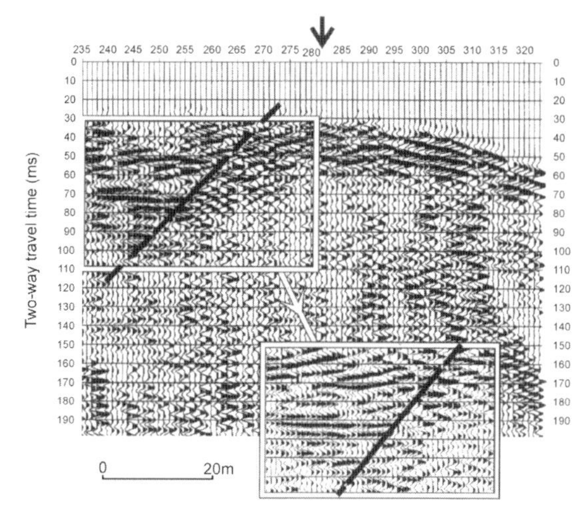

Fig 13.6. Stacked seismic section crossing a fault (trace spacing is 1 m). The fracture zone is indicated by terminated reflection horizons. Inset shows part of section after f-k migration leading to a clearer image (after Wise et al. 2003, with permission from Elsevier)

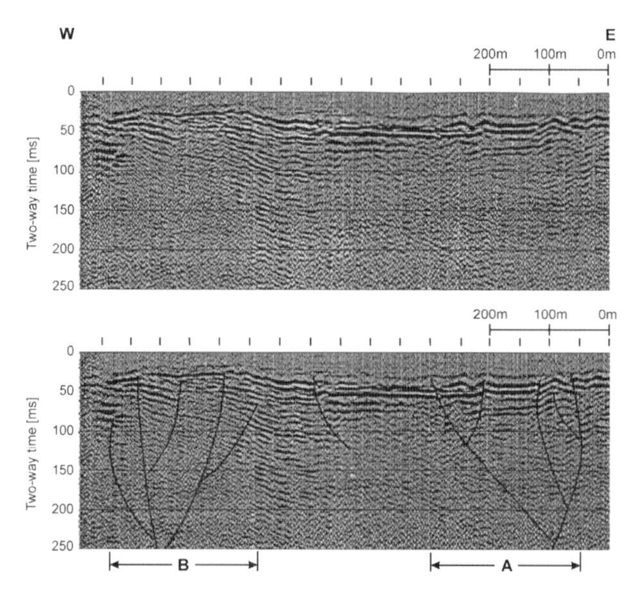

Fig. 13.7. Shallow seismic reflection survey for fracture zone detection, top: stacked seismic section (1.25 m trace spacing), below: fracture interpretation by depth-shifted or terminated reflection horizons. A and B indicate extensions of known fracture zones (Shtivelman et al. 1998, with permission from SEG)

Reduced seismic velocities for p- and s-waves, e.g. detected by inversion of first arrival travel times, can be used as indicators for fracture zones. If p- and s-wave velocities are known, dynamic elastic parameter as bulk modulus k, modulus of elasticity E, shear modulus μ, and Poisson´s ratio ν can be calculated after:

$$K = \rho \cdot (v_p^2 - \frac{4}{3} v_s^2) \tag{13.1}$$

$$E = \rho \cdot v_s^2 \cdot \frac{3v_p^2 - 4v_s^2}{v_p^2 - v_s^2} \tag{13.2}$$

$$\mu = \rho \cdot v_s^2 \tag{13.3}$$

$$\nu = \frac{v_p^2 - 2 \cdot v_s^2}{2 \cdot (v_p^2 - v_s^2)} \tag{13.4}$$

ρ = density

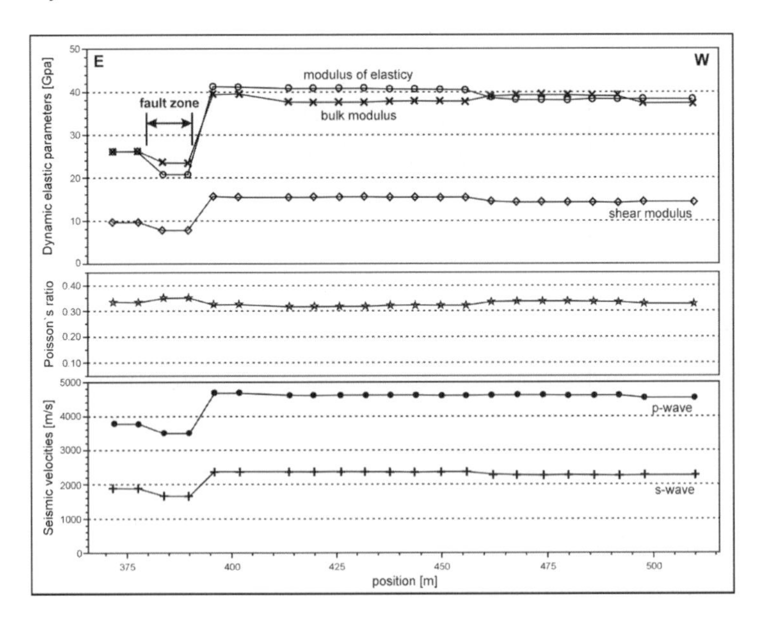

Fig. 13.8. Measured seismic velocities and calculated dynamic elastic parameter along a profile crossing a fault zone (Seren et al. 2002)

An example of a velocity analysis in terms of dynamic elastic parameter is shown in Fig. 13.8 (Seren et al. 2002). A fault zone is situated between 380 m and 390 m. Its left boundary is indicated by a drop of p- and s-wave velocities down to about 92% (the even larger jump to higher velocities on the right side seems to be caused by material change). The calculated drop of modulus of elasticity is about 80%; thus this parameter seems to be more sensitive for fracture zone detection than seismic velocities alone. Poisson's ratio shows an increase at the fault location.

This observation is backed by synthetic data. Based on the velocity-porosity relation obtained by Han et al. (1986), dynamic elastic parameters are calculated in relation to porosity (Fig. 13.9). It comes out that the modulus of elasticity is more sensitive to porosity changes than p- and s-wave velocities. This leads to the assumption that the decrease in velocities and dynamic elastic parameter at the location of the fracture zone is caused by increased porosity.

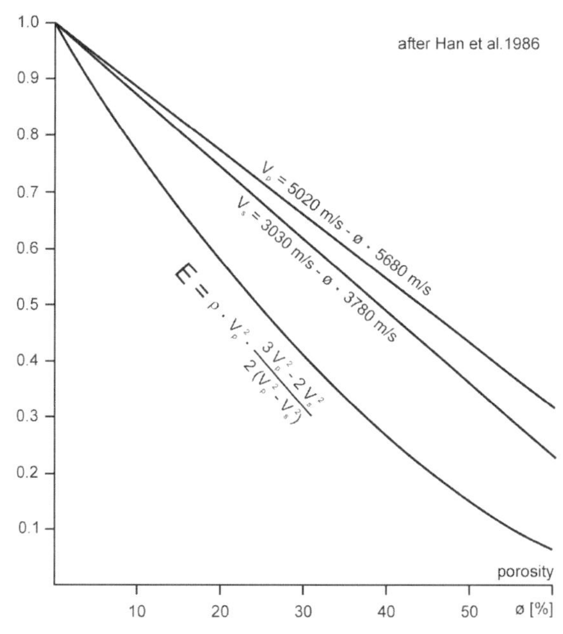

Fig. 13.9 Relative changes of seismic velocities calculated after Han et al. (1986) and modulus of elasticity related to porosity (all quantities normalized to zero-porosity material)

13.3 Geophysical exploration of faults and fracture zones: geoelectrical methods

Fracture zones are frequently associated with tectonic faults, and that is why the geophysical location of fracture zone aquifers is frequently confined to the location of faults. Moreover and different from seismic measurements, faults can in many cases be more easily found by low-expense electrical methods. Hence, geoelectrical techniques for the evaluation of fracture zone aquifers may be applied as shown in Fig. 13.10.

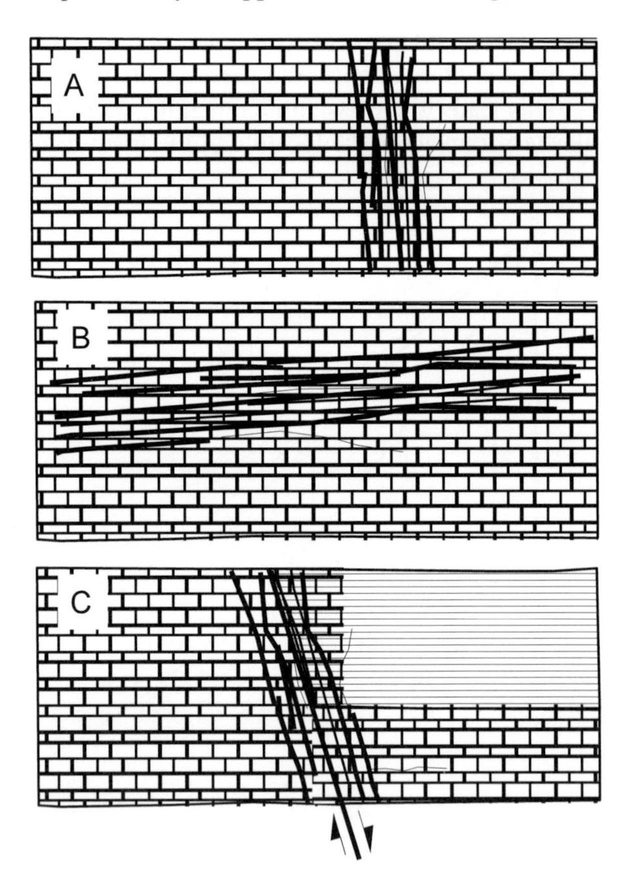

Fig. 13.10. Models of fracture zone aquifers for geoelectrical surveys: steeply dipping fracture zones for horizontal-profiling methods (A), sub-horizontal fracture zones for depth sounding methods (B), and fracture zones associated with faults for both depth sounding and horizontal-profiling methods (C)

Somewhat confusing, both resistivity and the reciprocal electrical conductivity are used as measuring quantities especially with geoelectrical profiling methods. As a consequence, resistivity (preferentially galvanic techniques) and conductivity (preferentially induction techniques) may bring about "negative" and "positive" anomalies over the same fracture zone.

Due to open water-filled fissures, the resistivity within a fracture zone is in general lower than the resistivity of the host rock. Depending on geological conditions, this is partly valid even for dry fracture zones due to enhanced weathering at open fissures. Clay and loam in fissures from weathering in crystalline and carbonate rocks and from infiltration lead to low resistivities, but on the other hand may strongly reduce hydraulic conductivities. Large scale fracture zones therefore are indicated by low resistivities or high electrical conductivities, as it is shown in Fig. 13.11.

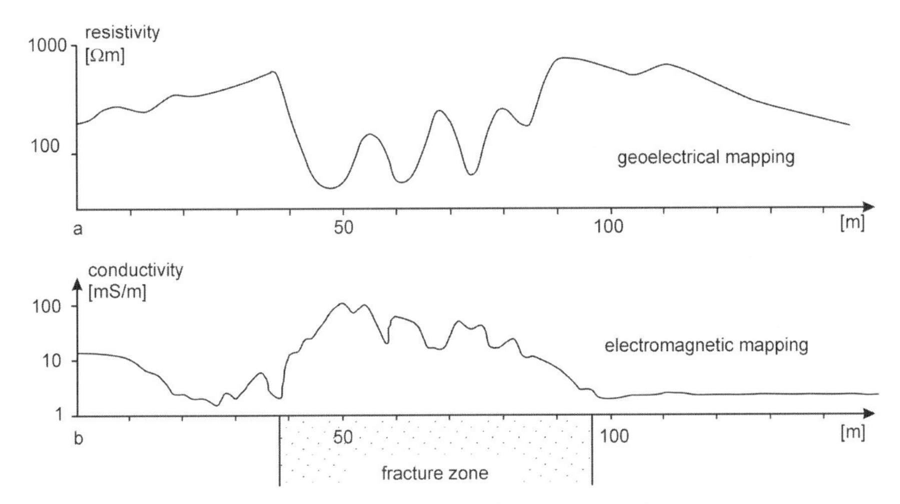

Fig.13.11. Electromagnetic (EM 31) and electrical (gradient array) survey for mapping of an extended fracture zone (Kirsch 1998, with permission from GRL GmbH, Leipzig)

EM measurement for fracture zone detection should be done on parallel profiles, only anomalies that correlate over the profiles should be taken as fracture zone indicators. Results of an EM 34 survey with two parallel profiles 20 m apart are shown in Fig 13.16, anomalies at 10 m, 60 m and 160 m correlate well. At these locations also anomalies at a square array survey were found (Fig. 13.13).

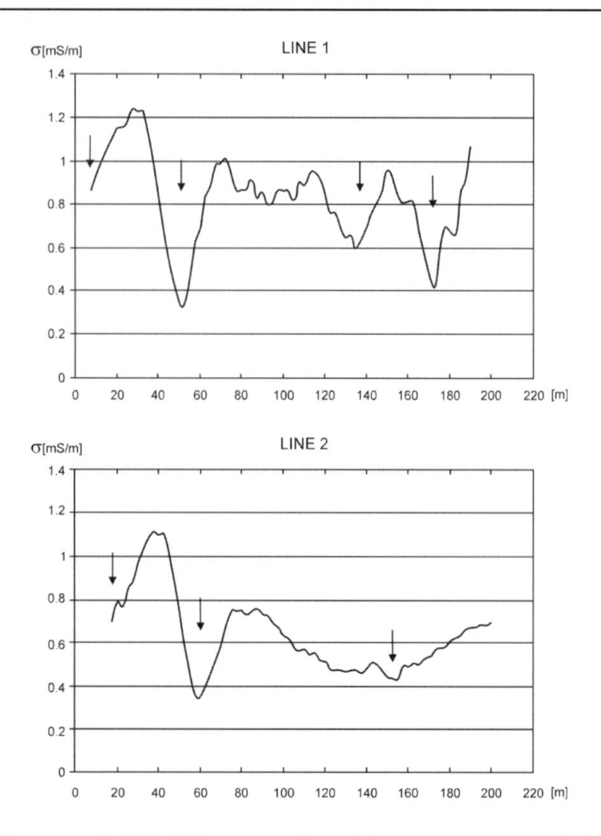

Fig. 13.12. Two parallel EM 34 survey lines 20 m apart, assumed fracture zones are indicated by arrows (Powers et al. 1999, with permission from USGS)

The fractures of a fracture zone often are aligned along the tectonic stress direction, which results in an anisotropy of resistivity. Here the anisotropy paradoxon is essential when a 4-point array with lined-up electrodes (e.g., Schlumberger, Wenner) is used. Due to channelling of electrical currents in the aligned fractures the measured *apparent* resistivity parallel to the fracture direction is larger than the *apparent* resistivity measured perpendicular to the fractures (contrary to the *true* resistivities).

An electrical configuration being sensitive for anisotropy is the square array configuration, where current and potential electrodes form a square (for description see Chap. 3). For each point of the profile, two resistivity measurements of the normal and two resistivity measurements of the crossed array are done. From this four apparent resistivity values, a mean resistivity and a secondary porosity is calculated (Chap. 3). Along the profile, fracture zones are indicated by low mean resistivities and high secondary porosities (Fig. 13.13).

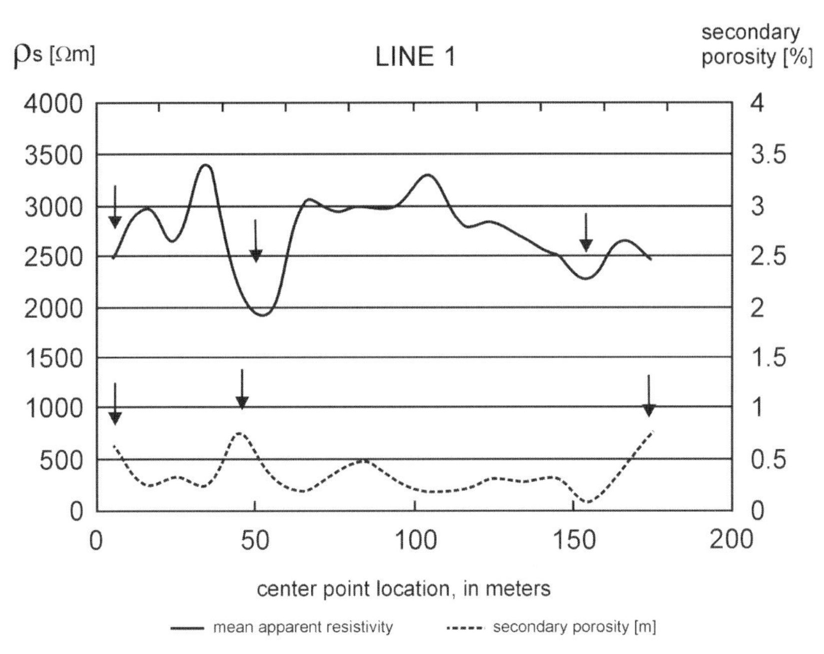

Fig. 13.13. Results of a square array survey, fracture zones (arrows) are indicated by low mean apparent resistivities and high secondary porosities (Powers et al. 1999, with permission from USGS)

A more detailed picture of the resistivity structure of fracture zones can be obtained by 2-D electrical mapping. As an example, Fig. 13.14 shows an apparent-resistivity pseudosection from electrical imaging along a profile over deep-reaching karst structures in limestones and dolomites. Frequently, pseudosections are modelled by computer-based 2D inversion techniques to bring out true resistivity sections. As long as field measurements, however, do not prove the two-dimensional nature of the resistivity underground structure, those modelling procedures are of only very limited value. Also the so-called smooth-inversion technique that relates the continuously varying apparent resistivities of the pseudosection with continuously varying underground true resistivities is in many cases far from being geologically realistic. For blocks defined by sharp boundaries and more or less homogeneous resistivities (lenses, faulted blocks and others), the smooth inversion method reasonably is not optimal (see, e.g., discussion by Loke et al. 2003) and in the case of modelling, e.g., caves simply inadequate.

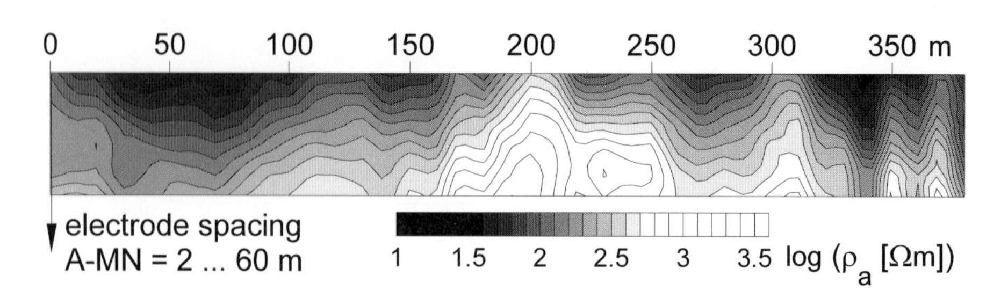

Fig. 13.14. Profile of electrical imaging over karstified Jurassic limestones and dolomites exhibiting deep-reaching fracture zones of low apparent resistivities. Pseudosection of apparent resistivities, pole-dipole (Halfschlumberger) configuration

Frequently, faults and steeply dipping conductive fracture zones are studied more rapidly and more economically by electromagnetic induction measurements than by galvanic resistivity surveys. The most common methods are the slingram (or dipole induction) method and the VLF method. Without connecting the ground, electromagnetic coupling enables even continuously moving digital data acquisition, and single-person VLF equipments have been developed which allow an additional synchronous earth magnetic field measurement.

Typical slingram anomalies over thin dipping sheet conductors as models for fracture zones or other steeply dipping conductivity zones are shown in Fig. 13.15. The anomaly curves of the field example in Fig. 13.16 widely correspond with the theoretical models, and asymmetries and the shift from the zero line obviously are an overburden effect.

Decreasing the dip in Fig. 13.15 to become zero, the horizontal-sheet conductor and its EM response (Fig. 13.17) may serve as a model for a fault between rocks of strongly differing resistivities (Fig. 13.18).

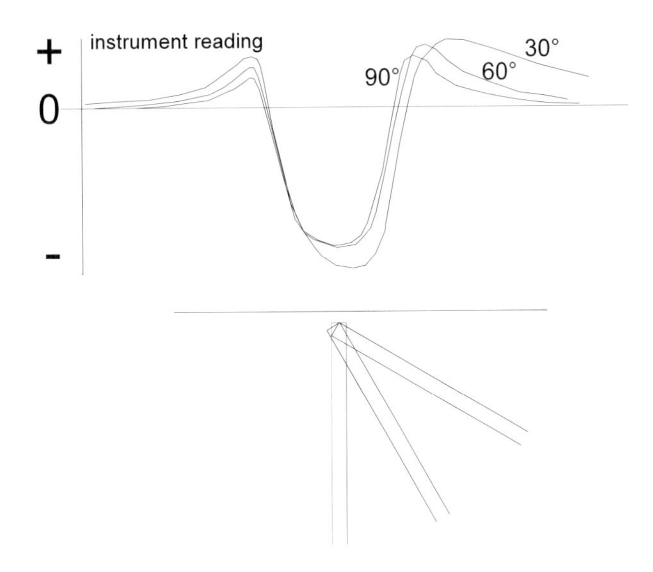

Fig. 13.15. The thin vertical and dipping sheet conductor as an EM model for fracture zones, and related frequency domain slingram anomalies (top). Modified and simplified from Keller & Frischknecht (1970)

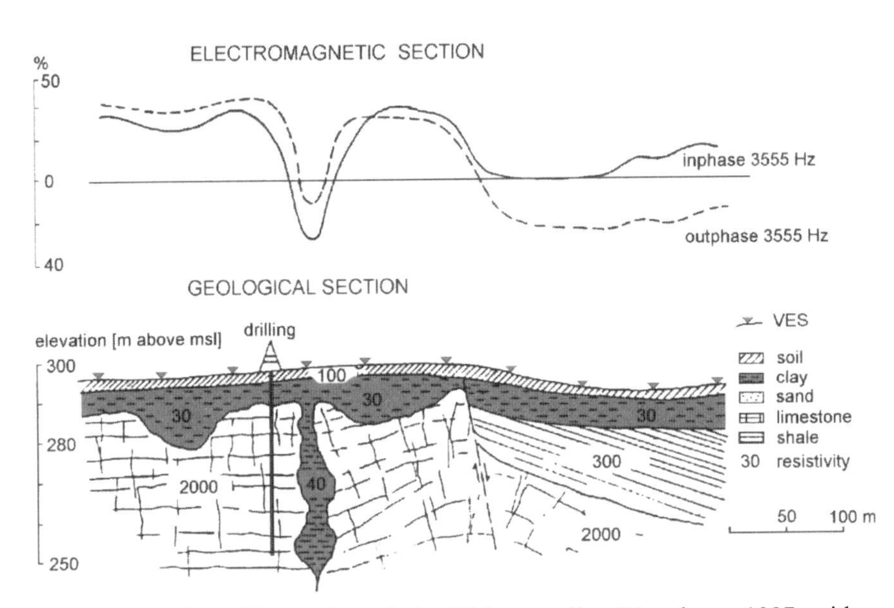

Fig. 13.16. Mapping of karst channels by EM-anomalies (Vogelsang 1987, with permission from Blackwell Publishing)

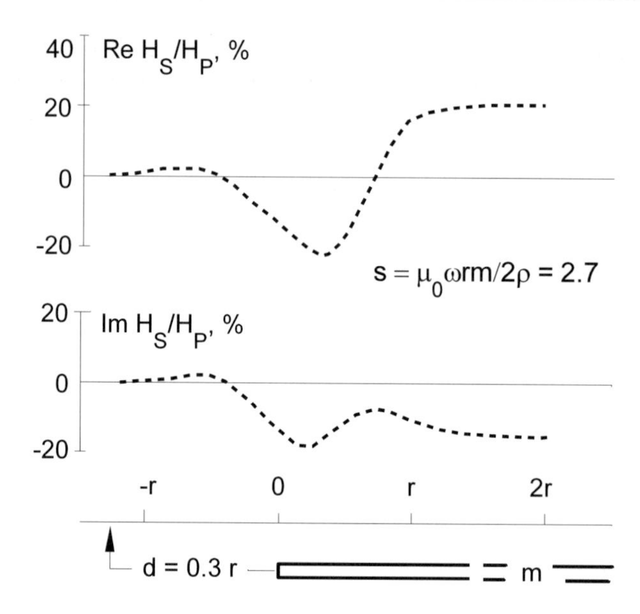

Fig. 13.17. Horizontal-loop EM response over horizontal-sheet conductor. s = induction parameter; $\omega = 2\pi\nu$, ν frequency, r loop separation, ρ sheet resistivity. Modified from Keller & Frischknecht (1970)

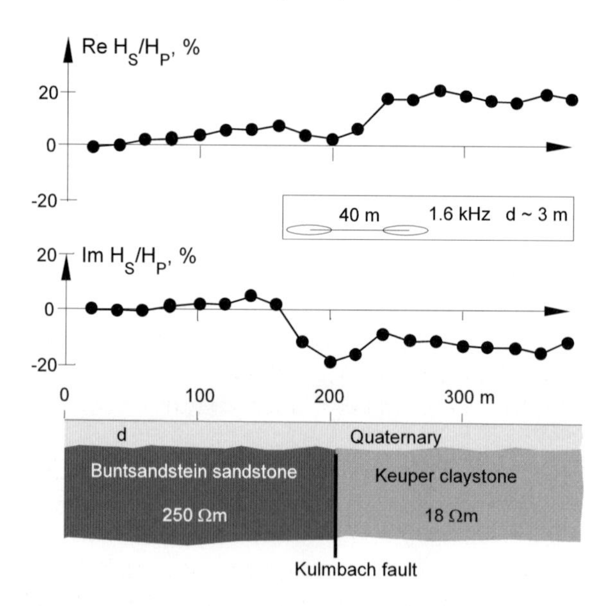

Fig. 13.18. EM34 horizontal-loop profiles over a fault. The inphase and out-of-phase anomalies resemble the response of the horizontal sheet conductor in Fig. 13.17. For comparison of the VLF anomaly see Fig. 13.21

As shown in Fig. 13.15, a dipping fracture zone leads to an asymmetric signal, and the shape can be used for an assessment of the dip angle of the fracture zone. However, it should be kept in mind that lateral heterogeneities of the near surface rock resistivity can also lead to asymmetric anomalies over a fracture zone (compare Fig. 13.16), especially when the fracture zone is associated with a fault separating different rock materials (compare Fig. 13.18). The influence of near-surface heterogeneities is demonstrated also by a slingram EM 34 profile measured with 3 coil separations and transmitter frequencies (Fig. 13.19). Only the anomaly at about 220 m is related with a known fracture zone, whereas the other anomalies of comparable amplitudes must be attributed to overburden resistivity variations. For 0.4 kHz, the fracture zone is rarely detectable, which may possibly be related with both reduced induction and increased depth penetration at the low frequency, and a closure of the fractures at depth.

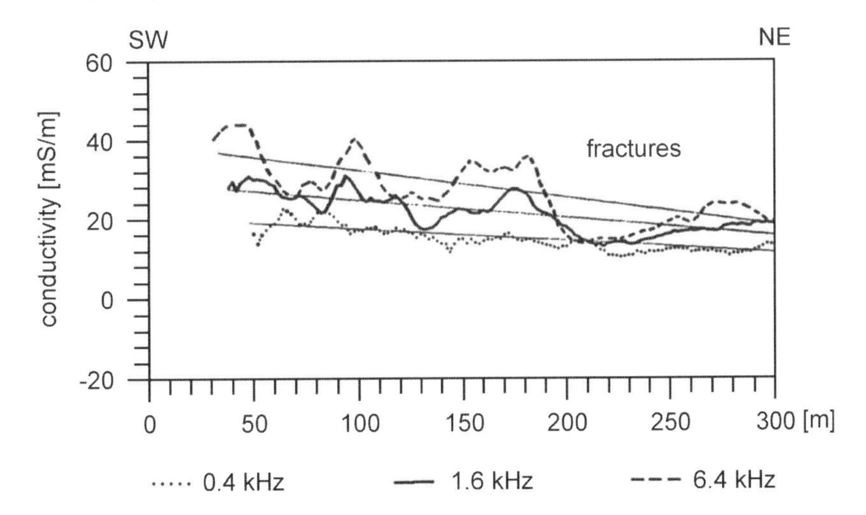

Fig. 13.19. Complex slingram anomalies from an EM34 survey for groundwater prospecting purposes; horizontal dipoles with three coil separations and transmitter frequencies (Wolfert 1992). For explanation see text

In Fig. 13.19, the electromagnetic response is given in conductivity units mS/m which is the instrument reading of the EM 34 equipment. Apparent resistivities and conductivities may be calculated from the inphase and outphase EM signals, but one should be aware that the true resistivity distribution in and around a fracture zone may be basically different from the apparent-parameter readings. This is very similar with galvanic resistivity soundings over narrow fracture zones.

In the search for fracture zone aquifers in basement rocks VLF surveys are widely used, but also in hard-rock sedimentary areas they can serve to rapidly locate faults and fracture zones. The typical VLF inphase anomalies for both situations (maximum or minimum dip angle over a fault, a crossover from positive to negative dip angles over a fracture zone) are shown in Fig. 13.20. In practice, the simple theoretical model curves are seldom realized (Fig. 13.21, Fig. 13.22). This may be related with the influence of overburden layering, a more complex structure of the fault and the fracture zone, or a traverse not perpendicular to the strike. Filtering of the data (Fig. 13.22) may sometimes help to clarify the layering, but in general VLF should be considered a rough and rapid reconnaissance tool only. And if VLF is used too schematically following the "equation" dip angle crossover = fracture zone = borehole, a dry well may be the consequence (Fig. 13.23).

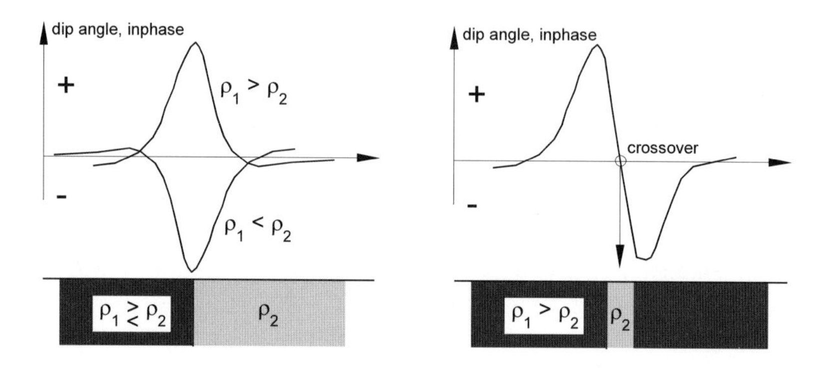

Fig. 13.20. VLF dip angle curves over vertical resistivity contact and sheet-like conductor: models of a fault and a fracture zone

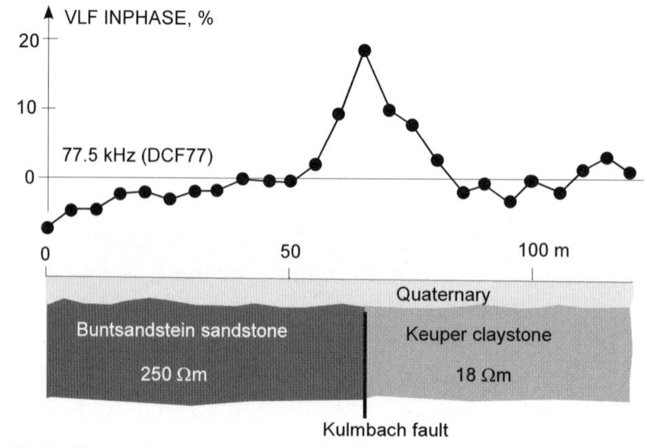

Fig. 13.21. VLF dip angle anomaly over a fault. The resistivities of the sandstone and the claystone have been modelled from Schlumberger VES

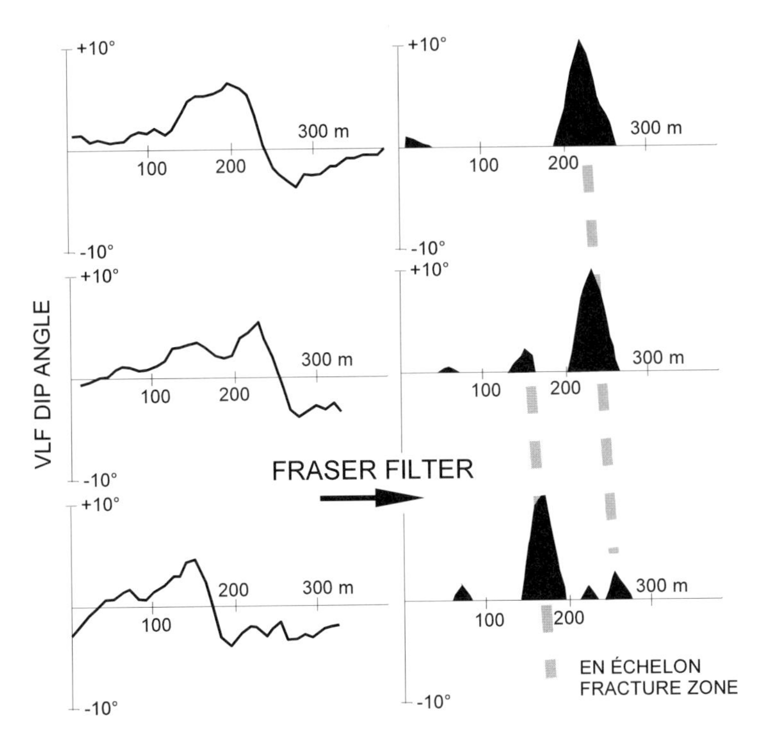

Fig. 13.22. Parallel VLF dip-angle profiles across a fracture zone in granite. By Fraser filtering, en-échelon fracture structure becomes more evident

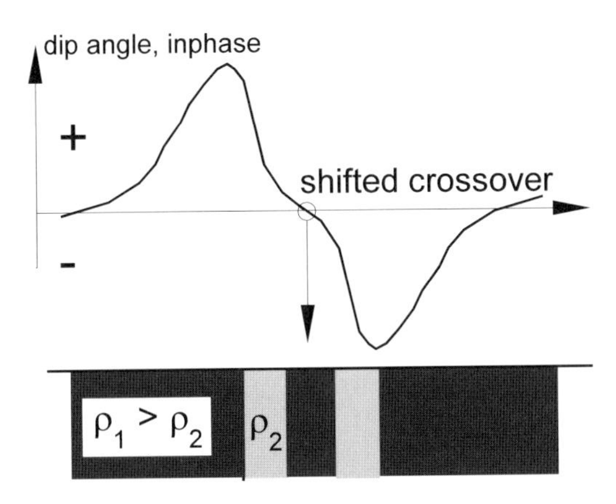

Fig. 13.23. VLF dip angle anomaly over two vertical-sheet conductors exhibiting a shifted crossover. A borehole for groundwater exploitation that is placed with regard to the crossover only, will go wrong

13.4 Geophysical exploration of fracture zones: GPR

Similar to the detection of fracture zones with seismic methods, in GPR surveys the offset of horizons can be used to locate a fracture zone. The high depth resolution of GPR techniques due to the wavelength of radar signals in the centimeter range enable the detection of smaller offsets (Fig. 13.24). Additionally, diffraction signals produced by heterogeneities within the fracture zone can be used for a localisation (Liner and Liner 1997).

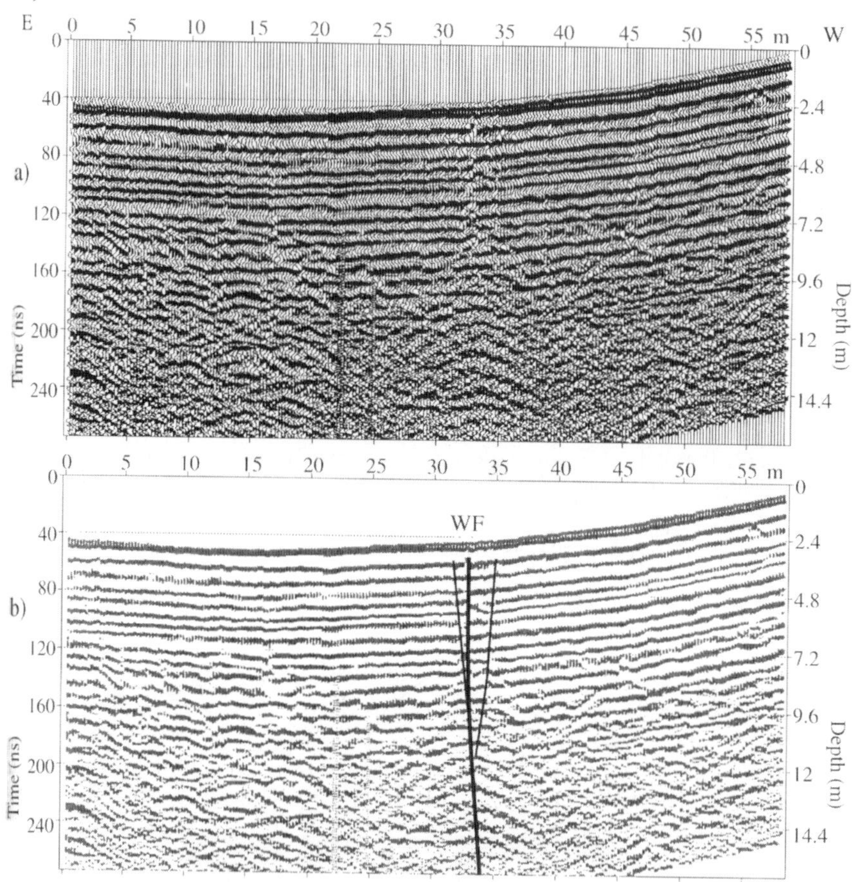

Fig. 13.24. Fault zone detection by GPR measurements using a 200 MHz Pulse EKKO instrumentation (Bano et al. 1998), depth conversion by a constant velocity of 12 cm/ns

13.5 Exploration of faults and fracture zones: Geophysical passive methods (self-potential, gravity, magnetic, geothermal and radioactivity methods)

Although seismic and geoelectric surveys play a dominant role in the exploration of faults and fracture zone aquifers, the importance of other geophysical techniques for special purposes and in special situations should not be underestimated.

Self-potential measurements enable a direct access to underground hydraulic conditions, if streaming potentials are considered. Fracture zones that are frequently associated with prominent groundwater flow may be a rewarding subject for self-potential surveys, and many drinking-water wells have successfully been drilled within streaming potential anomalies (Fig. 13.25). Usually, geophysical borehole measurements carried out in these drill holes confirmed the groundwater flow direction as predicted by the positive or negative sign of the self-potential anomaly.

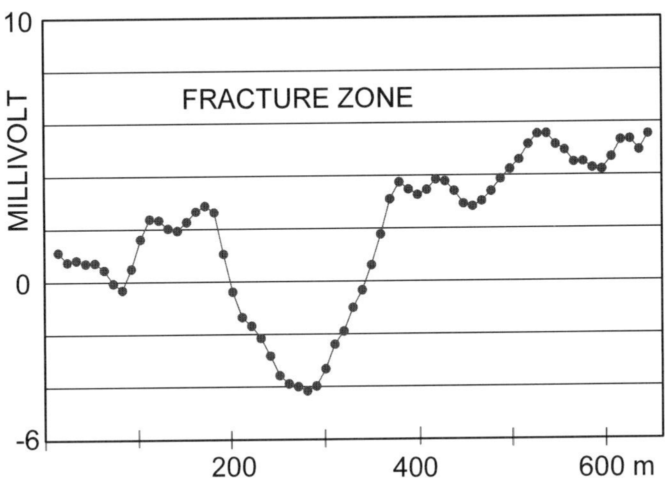

Fig. 13.25. Self-potential anomaly over broad fracture zone in Jurassicsandstone. The negative sign indicates groundwater infiltration. A drinking-water well was successfully drilled in the center of the anomaly

Among the geophysical exploration methods gravimetry is normally put in second place which is not justified at all. Originally applied especially to large-scale studies, gravimetry has meanwhile become an interesting tool in geotechnical and environmental geophysics, hydrogeological aspects included. Modern ultrahigh-sensitive gravimeters and GPS surveys enable rapid data acquisition, and the ability to work practically uninfluenced by traffic, power lines, supply grids and densely built-up areas, makes gravimetry indispensable in many cases. Frequently, gravimetry can successfully be used as an economic reconnaissance technique prior to a far more expensive seismic survey. Results of a gravity survey integrated in a comprehensive geophysical campaign for geothermal groundwater exploration are shown in Fig. 13.26. Deep boreholes were planned to be placed near a supposed and partly known fault (Fig. 13.26A). From the gravity measurements and additional resistivity soundings (VES and electrical imaging) the fault proved to be rather a tectonic graben which is clearly seen in the Bouguer anomaly map of Fig. 13.26A. Moreover, modelling of the gravity anomaly (Fig. 13.26B) showed that the graben must be considered a broad fracture zone with reduced density in order to account for the amplitude of the negative anomaly.

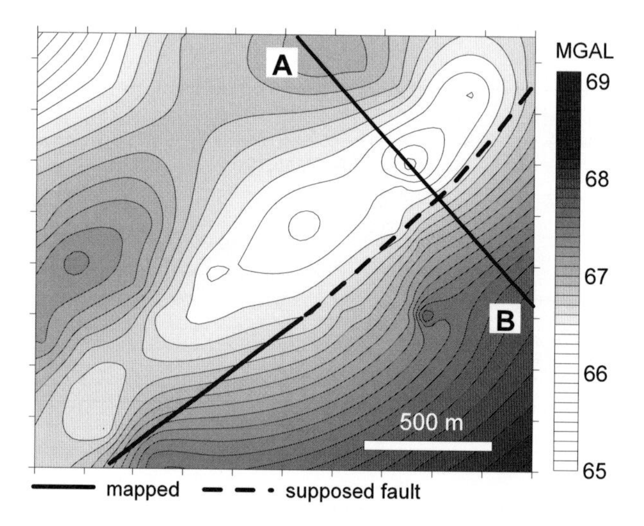

Fig. 13.26A. Bouguer gravity anomaly map from a geophysical campaign for hydrothermal groundwater exploration

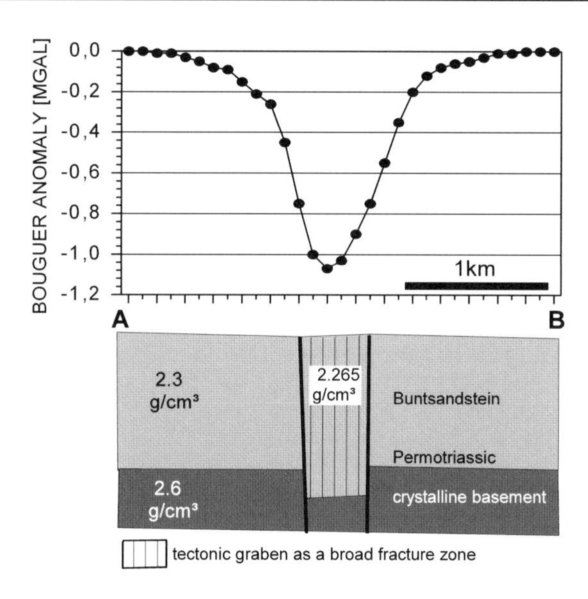

Fig. 13.26B. Bouguer gravity profile taken from the map in Fig. 13.26A after removal of a regional gradient (upper), and simplified model of geological interpretation (lower)

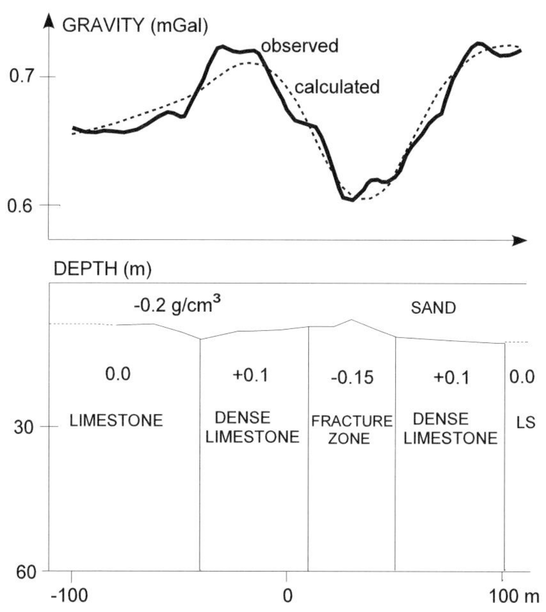

Fig. 13.27. Bouguer anomalies and gravity model for a vertical fracture zone in limestone. Modified from Stewart and Wood (1990)

A fracture zone on a smaller scale and its gravimetric signature are shown in Fig. 13.27.

In volcanic areas and in regions where basement rocks are included in hydrogeological research and groundwater exploration, earth magnetic field measurements may play an important role for clarifying the general geological layering and for delineating tectonic structures. Examples are well-known from the wide-spread literature. Here we add a somewhat unusual example from hydrogeological practice. For a long time, the high yield of a spring located midst of thick Jurassic claystones was considered an enigma, until a magnetic measurement provided a solution (Fig. 13.27).

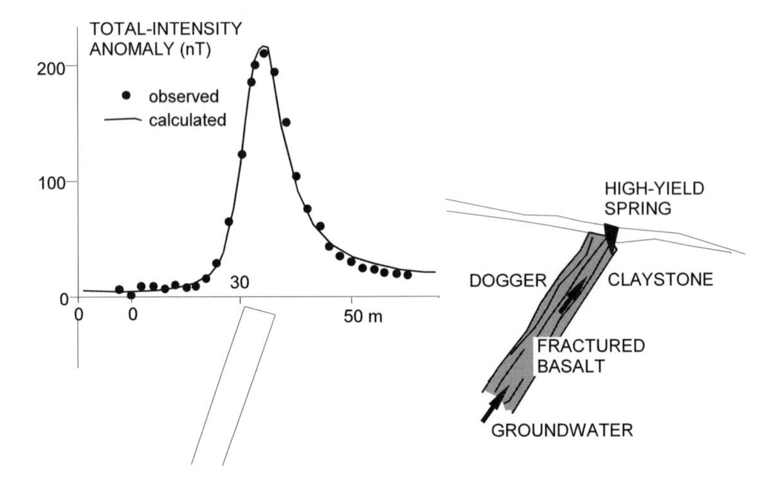

Fig. 13.28. Unusually high-yield spring in an area of thick Jurassic claystones is explained by a buried dike of fractured basalt serving as a hydraulic path. The existence of the dike and its shape are verified by a distinct magnetic total-field anomaly and model calculations

Because of their versatility and taking into account the modern high-sensitive equipments for rapid and economic data acquisition, magnetic measurements can successfully be applied also in common sedimentary environments for locating faults and fracture zones. In Fig. 13.28, an example of the investigation of deep-reaching fracture zones by magnetic measurements in a karst area is shown. The magnetic anomalies preferentially following the known tectonic directions result from enhanced magnetization of Tertiary material infiltrating the fracture zones in Jurassic limestones.

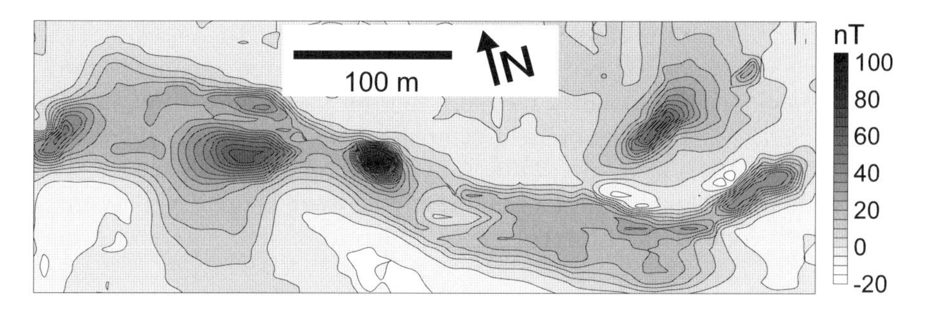

Fig. 13.29. Anomalies of the vertical component of the earth's magnetic field in an area of karstified Jurassic limestones. The magnetic anomalies delineate deep-reaching fracture zones filled by magnetite-bearing sandy and loamy material

Near-surface temperature measurements may be helpful to locate deep-reaching fracture zone aquifers enabling heat convection. The geothermal method is successfully applied not only to thermal water exploration but also to the search for drinking water. The significant temperature anomalies in Fig. 13.30 are related with karst groundwater ascending along a fracture zone from several 100 m depth. The water is used for the provision of drinking water.

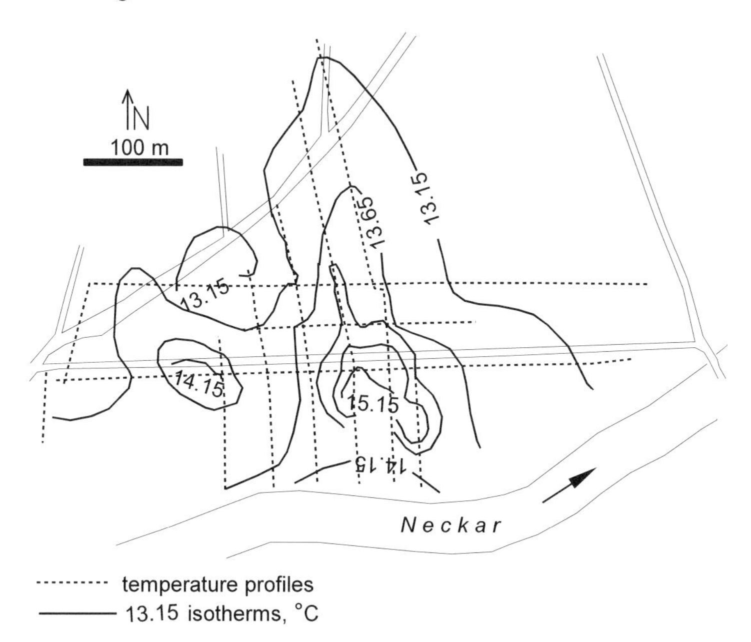

Fig. 13.30. Temperature anomalies (at 1.5 m depth) related with groundwater ascent from deep-seated karst aquifer. Simplified from Kappelmeyer (1985)

Although technically laborious, radioactivity measurements in the form of radon soil gas analyses may be interesting for the location of faults and fracture zones, and results of field campaigns are increasingly published (Levin 2000, Nielson et al. 1990). A typical example is shown in Fig. 13.31, and the authors suggest a relationship between the radon concentration and the width of the fractures.

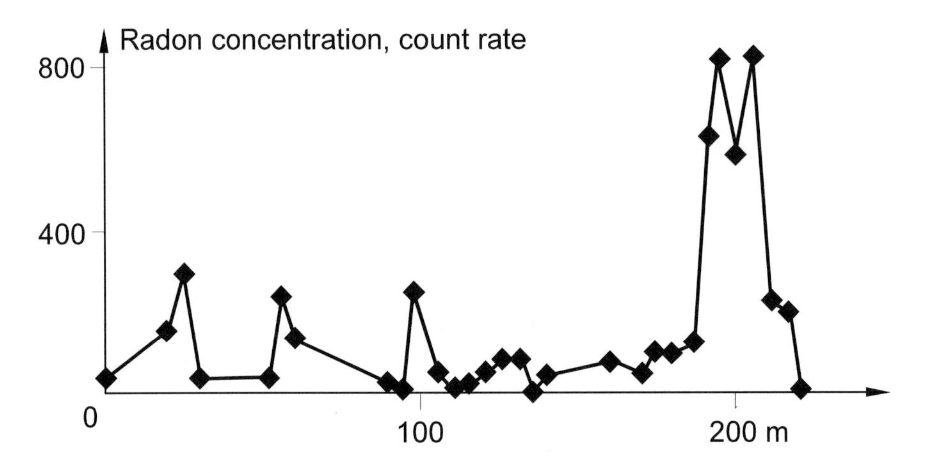

Fig. 13.31. Radon soil gas anomalies over fracture zones in sandstone and mudstone bedrock, Guanyinyan study area, China. Modified from Wu et al. (2003)

13.6 Geophysical exploration of caves

The exploration of caves for groundwater exploitation purposes is part of the general problem, namely the detection of underground voids (caverns, galleries, shafts, burial chambers etc) by geophysical methods. Much research has been done on this subject with the outcome that frequently a geophysical solution has *a priori* to be excluded. From the physical point of view, the assessment of the problem is most easily understandable for caves exposed to geophysical potential fields. By relatively simple calculations the response can be obtained for geoelectrical, gravity, magnetic and geothermal measurements. Correspondingly and irrespective of the contrast in petrophysical parameters (resistivity, density, magnetization and thermal conductivity of air and water in the cave and of the host rock), a cave cannot be detected if it is located at depths much larger than the vertical size of the cave. A non-conductive sphere ($\rho = \infty$) as a model for an air-filled cave may be detected in a resistivity gradient array only if $z/R \leq 1.2$, z = depth of upper boundary and r = radius of the sphere (Mundry 1985). A horizontal cylindrical cave full of water (midpoint at 12 m depth, 6 m diameter, 2.4 g/cm^3 host rock density) exhibits a maximum gravity anomaly of no more than 40 microgals (μG), which may possibly be detected only by a microgravity survey (Arzi 1975).

Somewhat different conditions are given with travel-time methods like seismic or ground penetrating radar (GPR). Similar to the potential methods, a purely physical limit exists, which is here the signal wavelength in relation to the size of the cave. In addition, the amount of energy reflected from the cave plays a significant role. This depends on the incident energy related with the seismic or radar source and the attenuation in the host rock, and on the reflectivity at the interface (contrast in density and seismic velocity, and dielectric constant). Hence, a successful detection of a cave by seismic or GPR methods can hardly be predicted in many cases.

A relatively new possibility of the detection of water-filled underground voids is given by the magnetic resonance sounding method (SNMR) and a successful application to saturated karst caves explored and mapped by speleologists has been reported (Vouillamoz et al. 2003). As can be expected from the measuring principle (see Chap. 8), the amount of water in relation to the depth is a threshold of the detectabilty.

Widely independent of physical limitations by potential field configurations and wavelength-related resolution, caves may geophysically be detected indirectly. Karst caves originating from the dissolution in carbonate rocks are preferentially associated with tectonic structures and increased jointing and rock fracturing. Hence, the geophysical methods suited for the detection and delineation of faults and fracture zones may serve also for the exploration of caves if their existence or the extension of known voids are suspected. A very typical example of a geophysical indirect location of caves is shown in Fig. 13.32. The underground voids verified by boreholes are related with reduced Radon concentrations in the soil over them, which is explained by water down-seeping in fractures thus preventing radon upward propagation.

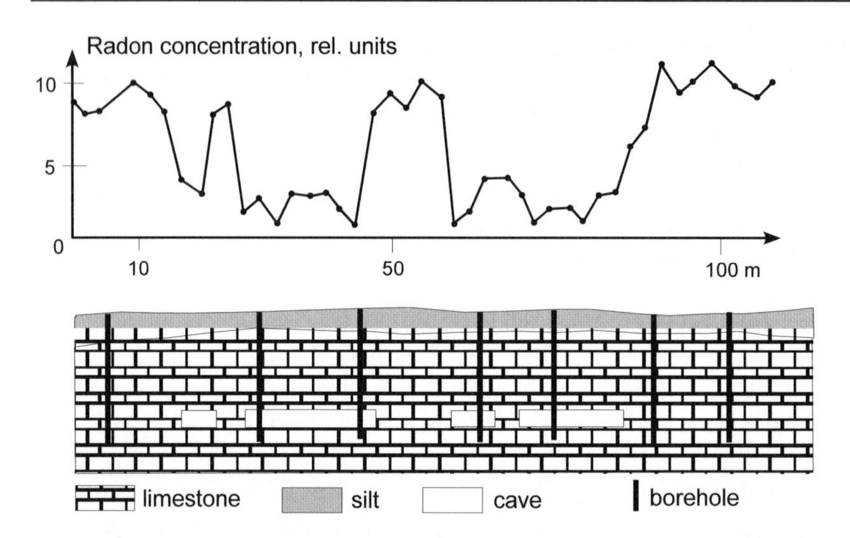

Fig. 13.32. Radon soil gas lows associated with underground voids. Modified from Bondarenko et al. (1983)

Similarly, suspected caves may be explored by geoelectrical resistivity or induction methods, or by self-potential surveys. Although theoretically caves may be measured by the disturbance of the terrestrial heat flow field (Chap. 9, Fig. 9.14A), the amplitudes of observed temperature anomalies over caves generally favour the convection model (Fig. 9.14B) and thus an indirect detection (Militzer et al. 1985).

13.7 References

Arzi AA (1975) Microgravity for engineering applications. Geophys Prosp 23:408-425

Bano M, Audru JC, Nivière B, Begg J, Berryman K, Henrys S, Maurin JC (1998) Application of Ground Penetrating Radar to Investigation of the Wellington Fault (NZ). Proceedings of the IV meeting of the Environmental and Engineering Geophysical Society, Barcelona, pp 655-658

Bondarenko VM, Victorov GG, Demin NV, Kulkov BN, Lumpov EE, Christich VA (1983) New methods in engineering geophysics. Press Nedra, Moscow

Committee on Fracture Characterization and Fluid Flow, U.S. National Committee for Rock mechanics (1996) Rock Fractures and Fluid Flow. Contemporary Understanding and Applications. National Academy Press, Washington, DC

Green AG, Mair JA (1983) Subhorizontal fractures in a granitic pluton: their detection and implications for radioactive waste disposal. Geophysics 48:1428-1449

Han DH, Nur A, Morgan D (1986) Effects of porosity and clay content on wave velocities in sandstones. Geophysics 51:2093-2107

Kappelmeyer O (1985) Geothermik. In: Bender F (ed) Angewandte Geowissenschaften, Band II, Methoden der Angewandten Geophysik und mathematische Verfahren in den Geowissenschaften, Enke, Stuttgart, pp 435-490

Keller GV, Frischknecht FC (1970) Electrical Methods in Geophysical Prospecting. Pergamon Press, Oxford

Kirsch R (1998) Geophysikalische Oberflächenverfahren. In: Ma ttheß G (ed) Die Grundwassererschließung. Bornträger, Stuttgart, pp 314-334

Levin M (2000) The radon emanation technique as a tool in ground water exploration. Borehole Water J 46:22-26

Liner CL, Liner JL (1997) Application of GPR to a site investigation involving shallow faults. The Leading Edge 16:1649-1951

Loke MH, Acworth I, Dahlin T (2003) A comparison of smooth and blocky inversion methods in 2D electrical imaging surveys. Exploration Geophys 34:182-187

McNeill JD (1980) EM34-3 survey interpretation techniques. Technical Note TN-7, Geonics Ltd, Mississauga

McNeill JD (1990) Electromagnetic Methods. In: Ward SH (ed) Geotechnical and Environmental Geophysics, vol 1. Soc Expl Geophys, Tulsa, pp 191-218

Mattheß G, Ubell K (1983) Lehrbuch der Hydrogeologie, Bd. 1: Allgemeine Hydrogeologie, Grundwasserhaushalt. Bornträger, Stuttgart

Militzer H, Oelsner C, Weber F (1985) Geothermik. In: Militzer H, Weber F. (eds) Angewandte Geophysik, Band 2, Geoelektrik - Geothermik - Radiometrie - Aerogeophysik. Springer, Wien New York - Akademie, Berlin, pp 215-278

Mundry E (1985) Gleichstromverfahren. In: Bender F (ed) Angewandte Geowissenschaften, Band II, Methoden der Angewandten Geophysik und mathematische Verfahren in den Geowissenschaften, Enke, Stuttgart, pp 299-338

Mylroie JE, Carew JL (1990) The flank margin model for dissolution cave development in carbonate platforms. Earth Surface Processes and Landforms 15:413-424

Nielson DL, Cui Linpei, Wards SH (1990) Gamma-Ray Spectrometry and Radon Emanometry in Environmental Geophysics. In: Ward SH (ed) Geotechnical and Environmental Geophysisc, Vol. 1: Review and Tutorial, SEG, Tulsa, pp 219-250

Powers CJ, Singha K, Haeni FP (1999) Integration of Surface Geophysical Methods for Fracture Detection in Bedrock at Mirror Lake, New Hampshire. In: Morganwalp DW, Buxton HT (eds) Proceedings of Technical Meeting of U.S. Geological Toxic Substances Hydrology Program, Charleston, South Carolina. USGS Water Resources Investigations Report 99-4018C 3:757-768

Schneider H (1988) Die Wassererschließung. Vulkan Verlag, Essen

Seren S, Kleberger J, Simsek O (2002) Engineering Geological and Geophysical Investigation at the Dam Site Cine/Turkey. Proc. 8th meeting EEGE-ES:13-16

Shtivelman V, Frieslander U, Zilberman E, Amit R (1998) Mapping shallow faults at the Evrona playa site using high-resolution reflection method. Geophysics 63:1257-1264

Stewart M, Wood J (1990) Geologic and geophysical character of fracture zones in a tertiary carbonate aquifer, Florida. In: Ward SH (ed) Geotechnical and

Environmental Geophysisc, Vol. II: Environmental and Groundwater, SEG, Tulsa, pp 235-243

van Lissa RV, van Maanen HRJ, Odera FW (1992) The use of remote sensing and geophysics for groundwater exploration in Nyanza province - Kenya. In: GEONICS Ltd, Groundwater Exploration Applications, Mississauga

Vogelsang D (1987) Examples of electromagnetic prospecting for karst and fault systems. Geophysical Prospecting 34:604-618

Vouillamoz JM, Legchenko A, Albouy Y, Bakalowicz M, Baltassat JM, Al-Fares W (2003) Localization of saturated karst aquifer with magnetic resonance sounding and resistivity imagery. Ground Water 41: 578-586

Wise DJ, Cassidy J, Locke CA (2003) Geophysical imaging of the Quaternary Wairoa North Fault, New Zealand: a case study. Journal of Applied Geophysics 53:1-16

Wolfert RL (1992) Vergleichender Einsatz geophysikalischer Methoden zur Kartierung von Klüften im Raum Linares/Nuevo Leon (Mexiko) unter Berücksichtigung der Wasserprospektion. Diploma thesis, Institut für Geophysik, Christian-Albrechts-Universität Kiel

Wu Y, Wang W, Xu Y, Liu H, Zhou X, Wang L, Titus R (2003) Radon concentration: A tool for assessing the fracture network at Guanyinyan study area, China. Water SA 29:49-53

14 Groundwater quality - saltwater intrusions

Reinhard Kirsch

14.1 Definition

Saline groundwater can lead to serious problems for water supply, especially under arid climatic conditions. To define saline groundwater, Grube et al. (2000) proposed a concentration of chloride exceeding 250 mg/l. Since the electrical conductivity of water is strongly related to the salt concentration, the EU-directive on water quality (1998) sets an upper limit of 2500 μS/cm for the electrical conductivity of drinking water.

14.2 Origin of saltwater intrusions

In addition to anthropogenic effects (e.g. fertiliser, waste deposits) the main origins of saline groundwater are

* seawater intrusions in coastal areas
* salt domes
* enhanced mineral concentration in groundwater under arid conditions.

When aquifers in coastal areas have hydraulic contact with seawater, **seawater intrusions** occur due to higher density of saltwater. In the aquifer, freshwater is underlain by saltwater leading to a more or less well defined salt/freshwater interface. Freshwater reservoirs on islands are lens shaped (Fig 14.1). If hydrostatic equilibrium and homogeneous aquifers can be assumed, an estimate of the depth of this interface is given by the classical formulation of Ghijben (Drabbe and Ghijben 1888) and Herzberg (1901):

$$\rho_{salt} \cdot g \cdot z = \rho_{fresh} \cdot g \cdot (z + h) \tag{14.1}$$

ρ_{salt} density of saltwater

ρ_{fresh} density of freshwater

When the density of seawater and the depth to the water table are known, the depth of salt/freshwater interface can be calculated. This is only valid for well balanced groundwater conditions.

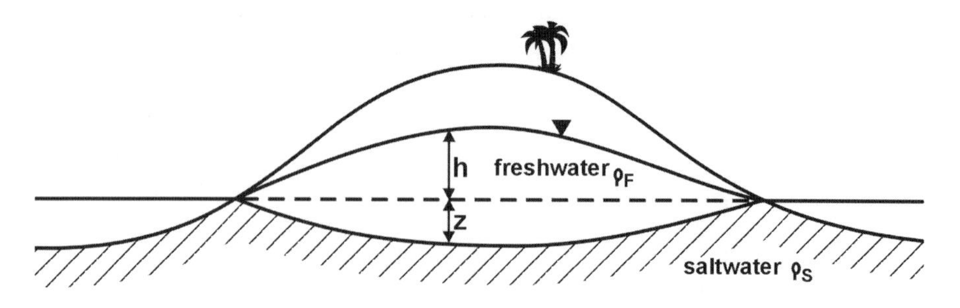

Fig. 14.1. Freshwater lens on a small island

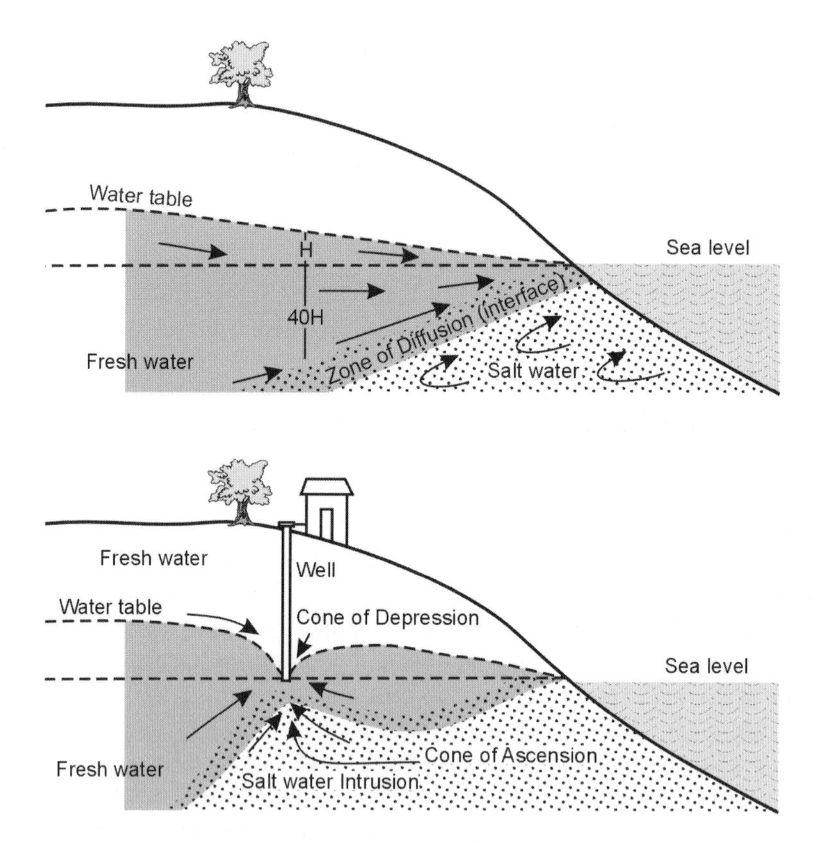

Fig.14.2. Balanced (top) and disturbed (below) salt/freshwater interface in a coastal area (after Keller 1988)

If the equilibrium between freshwater and saltwater is disturbed by high pumping rates for water supply, then the depth of the salt/freshwater interface is lowered and an enforced saltwater intrusion can occur (Fig. 14.2).

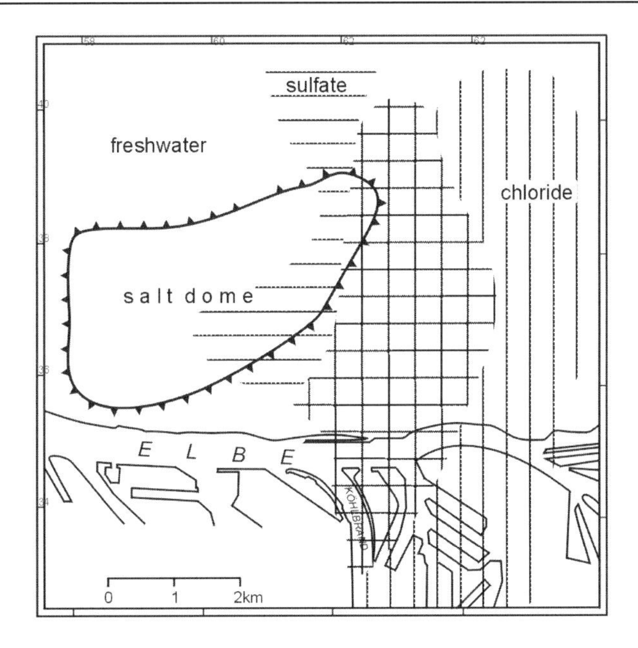

Fig. 14.3. Saline groundwater in the vicinity of a salt dome in the Hamburg downtown area (Löhnert 1967)

In the Cuxhaven-Bremerhaven area of North-Western Germany, the test field CAT (coastal aquifer test field) is being operated by the GGA-Institut (Hannover) which enables research to be carried out on saltwater intrusions (Siemon and Binot 2003).

Salt domes are wide spread underground structures. Groundwater contact to the salt body can lead to saline groundwater (Fig. 14.3). As a consequence, saline groundwater generally occurs in deeper aquifers in salt dome regions. In Schleswig-Holstein salt/freshwater interface is found at depths ranging from a few meters to several hundred meters. Salt water sources at the surface apart from salt domes can be explained with large scale thermohaline convection systems (Jahnke et al. 2004).

Under **arid climate conditions** enhanced salt concentration in the near surface subsoil can occur. Salt is transported by dust and rainwater, or it is dissolved from rocks due to high groundwater temperatures up to 30°C (Matthess 1994). Depending on soil and ion type, high salt concentration in certain depth ranges can occur (Blume et al. 1995). As a consequence, salt concentration of pore water in the vadose zone can be higher than in the saturated zone leading to low electrical resistivities. For example, Demirel (1997) found pore water resistivities of less than 1 Ωm in unsaturated soil samples from Milet (Western Turkey). During the rain season the salt deposits of the vadose zone can enter the groundwater reservoir. Recently, naturally enhanced Nitrate concentrations up to 600 mg/l were recorded in the groundwater in Southern Africa (Stadler et al. 2004).

14.3 Electrical conductivity of saline water

Depending on the content of Cl^- ions, water can be classified as freshwater (< 150 mg/l or ppm), brackish water (150 - 10000 mg/l), or saltwater (>10000 mg/l). Strongly related to ion content is the electrical conductivity or the resistivity of water (Fig. 14.4). A rough estimate of the electrical according to McNeill (1980) can be obtained through:

$$\sigma[\mathrm{mS/m}] = 96500 \cdot \Sigma C_i \cdot M_i \tag{14.2}$$

C_i = concentration of ions (gram equivalent/1000 l)

M_i = mobility of ions $\frac{m/s}{V/m}$

Typical ion mobility at 25°C is (McNeill 1980):

H^+:	$36.2 * 10^{-8}$	Na^+:	$5.2 * 10^{-8}$
K^+:	$7.6 * 10^{-8}$	OH^-:	$20.5 * 10^{-8}$
SO_4^-:	$8.3 * 10^{-8}$	Cl^-:	$0.9 * 10^{-8}$
NO_3^-:	$7.4 * 10^{-8}$	HCO_3^-:	$4.6 * 10^{-8}$

For technical reasons the content of ions which precipitate under normal conditions (e.g. Ca^{2+} or Mg^{2+}) is important as a criterion for the water quality (Fig. 14.5). This criterion is called hardness and is defined as follows:

$$H_{tot} = 0.14 Ca^{2+}\left[\frac{mg}{l}\right] + 0.23 Mg^{2+}\left[\frac{mg}{l}\right] \tag{14.3}$$

The mobility of ions depends strongly on the temperature resulting in a temperature dependence of electrical conductivity (TNO 1976, referring to electrical conductivity at 18°C):

$$\sigma_t = \sigma_{18}[1 + a(t-18) + b(t-18)^2] \tag{14.4}$$

Values for coefficients a and b are (TNO 1976):

NaCl:	$a = 226 * 10^{-4}$	$b = 84 * 10^{-6}$
$MgSO_4$:	$a = 238 * 10^{-4}$	$b = 92 * 10^{-6}$
$NaNO_3$:	$a = 220 * 10^{-4}$	$b = 75 * 10^{-6}$

As coefficients a are quite similar and coefficients b are two magnitudes smaller, Eq. 14.4 can be simplified to:

$$\sigma_t = \sigma_{18}[1 + 0.023 \cdot (t - 18)] \tag{14.5}$$

This temperature effect must be kept in mind when water conductivities are interpreted in terms of water quality. A chloride concentration of 200 mg/l at 18°C results in an electrical conductivity of 600 μS/cm or a resistivity of 17 Ωm. The mean temperature (with minor seasonal changes) for Middle European near surface groundwater is 9°C leading to a resistivity of 21 Ωm, whereas groundwater of identical ion concentration under tropical conditions at 28°C has a resistivity of 12.5 Ωm only.

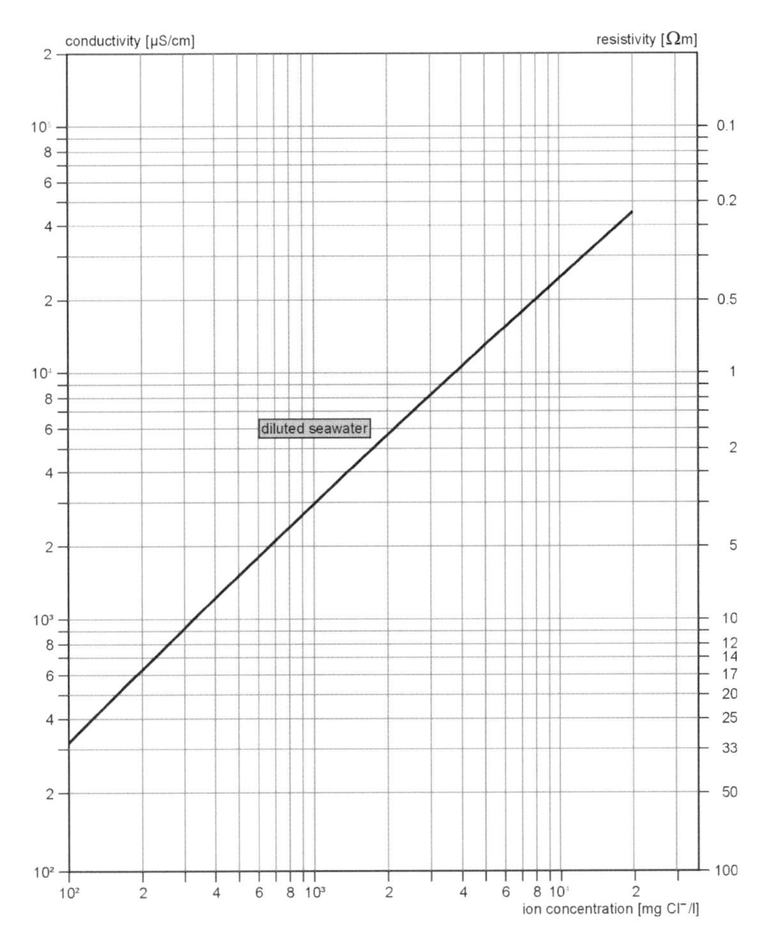

Fig. 14.4. The conductivity of saltwater (at 18°C) in relation to the Cl^- content (TNO 1976, with permission from TNO)

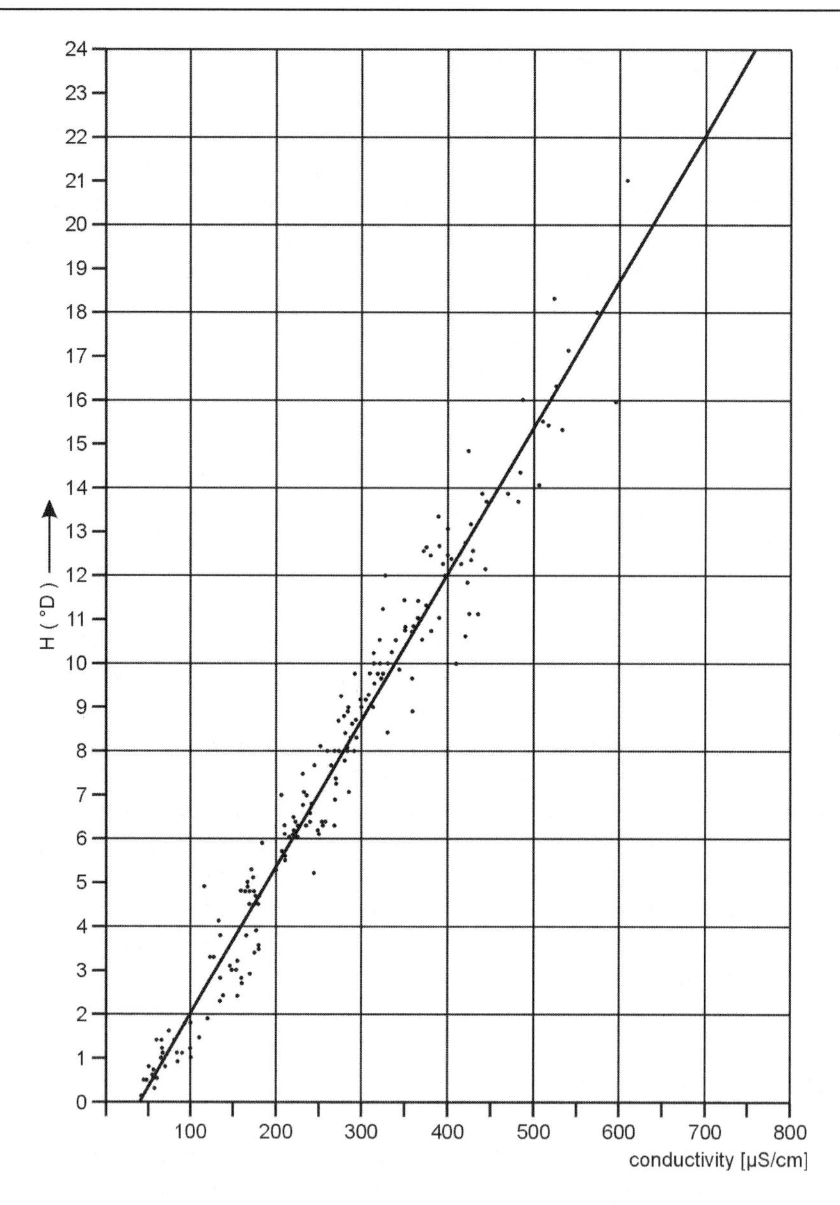

Fig. 14.5. The conductivity of saltwater (at 18°C) in relation to the hardness (TNO 1976, with permission from TNO)

14.4 Exploration techniques

Due to the high electrical conductivities, saltwater intrusions are "classical" targets for electrical or electromagnetic measurements (Barker 1990, Goldstein et al. 1990). Since the 70es, VES measurements were carried out by the Bundesanstalt für Geowissenschaften und Rohstoffe (BGR, Hannover) to map freshwater lenses on small North Sea islands (Repsold 1990). Meanwhile, the BGR operates a Helicopter EM system in order to cover large areas. In the CAT field region, not only saltwater intrusions were mapped with this system, but also freshwater occurrences in the tidal flats were found by increased electrical resistivities (Siemon et al. 2001).

Interpretation problems for VES can occur due to the high resistivity contrasts from the high resistive topsoil to the low resistive saline aquifer. Embedded layers with freshwater at medium resistivities are often not detected in the sounding curve or can be misinterpreted. As mentioned above, the salt concentration in the upper subsoil can also lead to low resistivities. Further interpretation problems arise from the overlapping resistivities of saltwater saturated sand and clay. Lühr (1984) measured the following resistivities at the North Sea coastal area of Schleswig-Holstein (Northwestern Germany): clay, silt (freshwater saturated): 7-16 Ωm, clay, silt (saltwater saturated): 1-3 Ωm, fine sand (freshwater saturated): 27-70 Ωm, medium - fine sand (saltwater saturated): 3-12 Ωm.

Resistivity methods are also used to monitor the development of salt/freshwater interfaces after flooding or while pumping (examples shown in Fig. 14.9). Slater and Sandberg (2000) made 2-D DC-measurements at the North Atlantic shoreline to monitor tidal fluctuations of salt/freshwater interface and found a phase lag between high tide and maximum of saltwater influence.

14.5 Field examples

Next, field examples of mapping or monitoring of saltwater intrusions in coastal regions and under arid climate conditions shall be shown. The measurements in the North Sea coastal region (North Frisian Islands) were done by Riewert Ketelsen of the former Geological Survey of Schleswig-Holstein, now State Agency for Nature and Environmental conservation (LANU), whereas the measurements in the Red Sea Province of Sudan were carried out by the author for THOR Geophysical GmbH (Kiel).

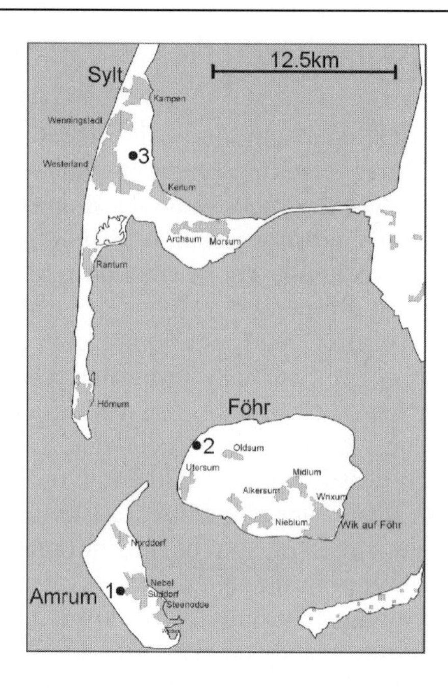

Fig. 14.6. North Frisian Islands, the numbers refer to the project areas (see text)

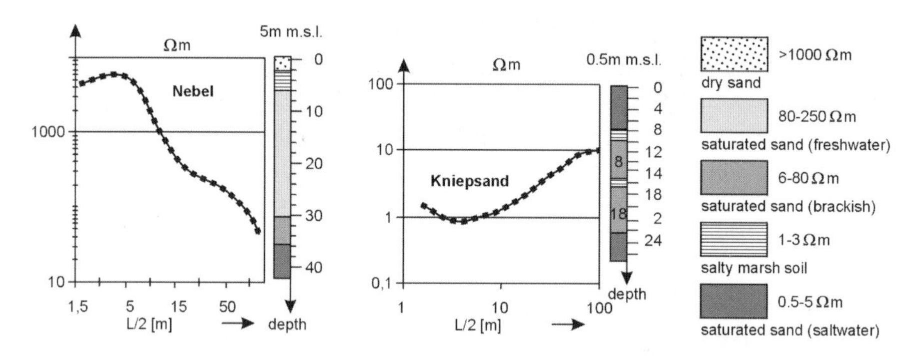

Fig. 14.7. Sounding curves from the North Frisian Island Amrum (Ketelsen and Kirsch 2004)

14.5.1 Saltwater intrusions in the North Sea region

The water supply of the North Frisian Islands Sylt, Amrum, and Föhr (Fig. 14.6) is mainly based on Quaternary aquifers. Due to the increasing water demand, e.g. due to tourism, the observation of salt/freshwater interface is essential to enable a safe water supply. Sounding curve examples from the island Amrum (location 1 in Fig. 14.6) are shown in Fig. 14.7. The VES location Nebel lies 700 m behind the shoreline and shows freshwater bearing sands covered by salty clay, whereas at the VES location Kniepsand (a sandbank 600 m offshore) two layers of brackish water were found. Although this sandbank is flooded several times per year, freshwater influence from the island leads to brackish water instead of saltwater.

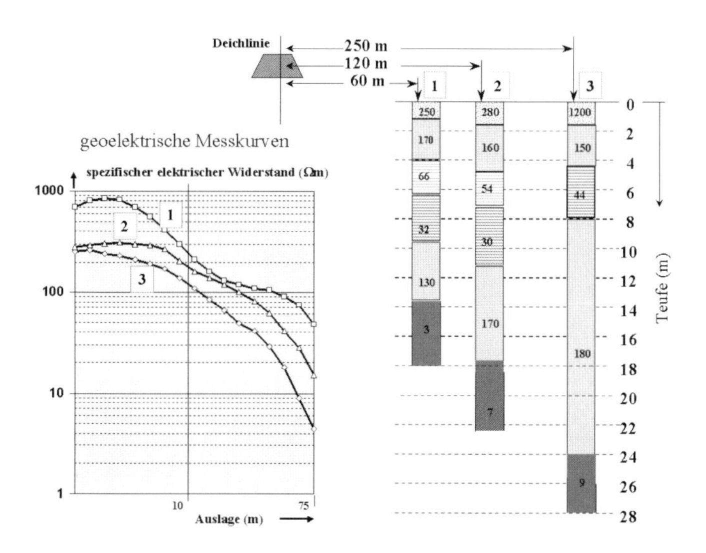

Fig. 14.8. Sounding curves from the North Frisian Island Föhr (Ketelsen and Kirsch 2004)

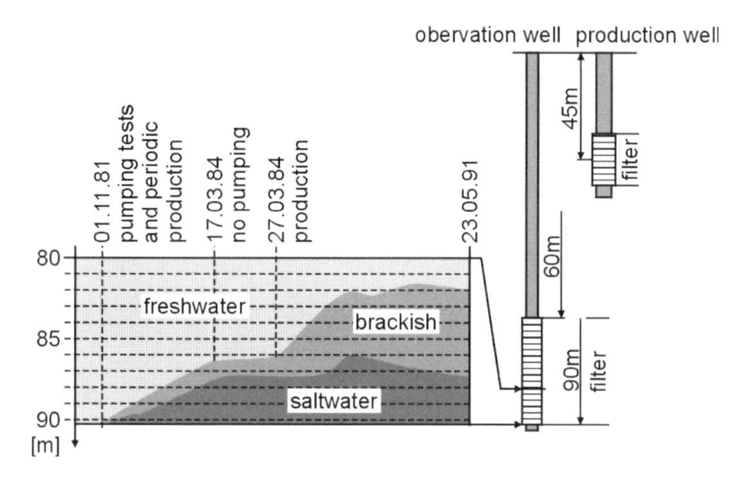

Fig. 14.9. Depth of brackish and saltwater in an observation well influenced by pumping (Ketelsen and Kirsch 2004)

For an assessment of freshwater reservoirs on the island of Föhr (location 2 in Fig. 14.6) VES were made along a profile to determine the dip of the salt/freshwater interface (Fig. 14.8). Terrain heights at all locations were 2 m above msl, depths to the groundwater table were 1 m below surface. A freshwater aquifer was detected by resistivities ranging from 130 – 180 Ωm. The salt/freshwater interface was dipping in inland direction from 14 – 28 m below surface. This is shallower than expected after Eq. 14.1.

To monitor the salt/freshwater interface in the surrounding of newly established water production wells near to the town of Westerland (Sylt, location 3 in Fig. 14.6), electrical conductivity of groundwater was measured in an observation well in the vicinity of the production wells (Fig. 14.9). During 16 months of pumping tests and periodical pumping, the saltwater table was lowered by 3 m and a layer of brackish water of one meter thickness was formed. This configuration remained stable for a period of about one year without pumping. A new start of production lead to a thickening of the brackish zone, whereas the saltwater table remained more or less stable. Permanent pumping brought a new equilibrium with ca. 5 m of the brackish zone.

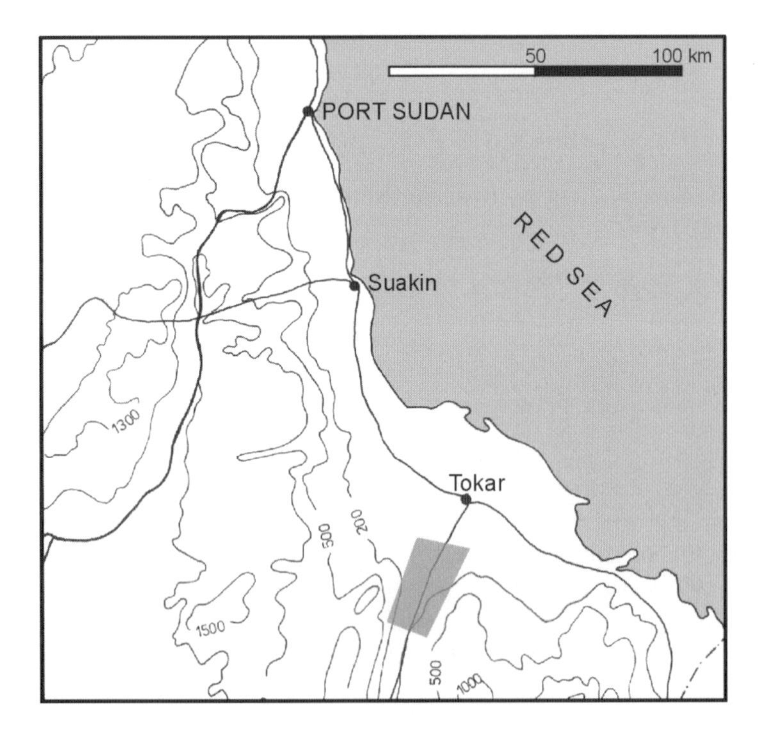

Fig. 14.10. Project area Khor Baraka, Red Sea Province, Sudan

14.5.2 Saline groundwater in the Red Sea Province, Sudan

To improve the water supply for the growing town of Port Sudan, a refraction seismic and VES survey was done in 1989 in the Khor Baraka area south east of Port Sudan (Fig. 14.10). The general manager of the project was the Rhein-Ruhr Ingenieurgesellschaft (Dortmund), contractor for the geophysical survey was THOR Geophysical GmbH (Kiel). The geophysical works were based on VES measurements done by BGR (Fielitz 1979) and on seismic measurements carried out by the Institut für Geophysik, University of Kiel (Stümpel 1980).

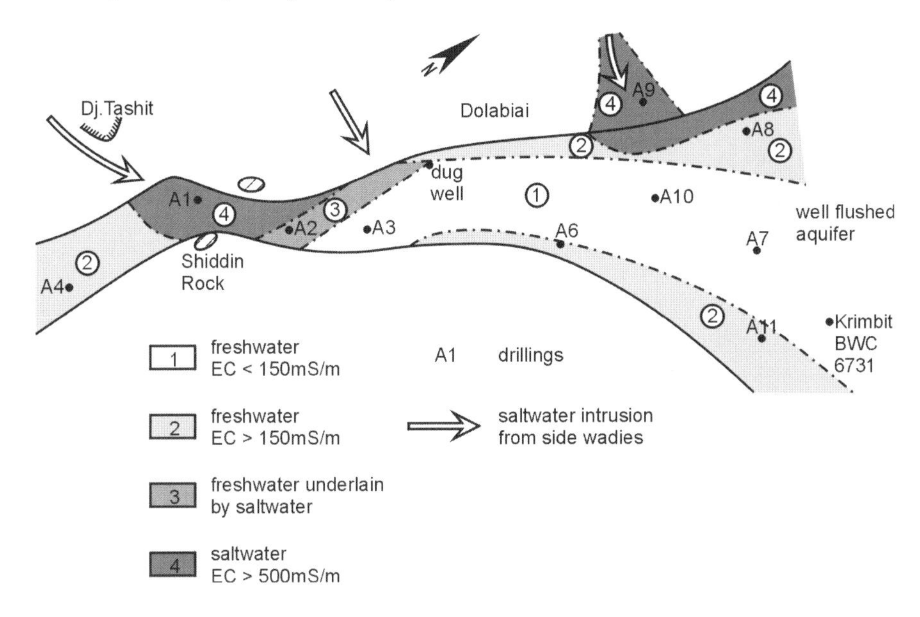

Fig. 14.11. Electrical conductivity of groundwater in the Khor Baraka valley (Langsdorf 1981, with permission from DGG)

The geological and hydrogeological conditions in the project area were described by Langsdorf (1981). The wadi Khor Baraka begins in the mountain range of Eritrea and reaches into the coastal plain of the Red Sea Province of Sudan. Basement rock is mainly basalt, whereas the sedimentary cover consists of silt, sand, and gravel. Due to heavy rainstorms in the Eritrea mountain range, Khor Baraka is flooded periodically. The high evaporation of more than 4000 mm/a leads to saline groundwater. Fig. 14.11 shows the electrical conductivity of the groundwater based on measurements in wells (Langsdorf 1981). Freshwater can only be expected in the central part of the valley where high groundwater recharge occurs because of flooding.

The depth to the basement and thus to the lower boundary of the aquifer in the project area was obtained by 30 km of refraction seismic measurements (THOR 1989). The interpolation of basement depth leads to a 3-D picture of the basement structure as shown in Fig. 14.12.
A total of 38 VES with 100 m electrode spacing (AB/2) were measured in the project area to map saline groundwater. A sounding curve example is shown in Fig. 14.13. Probably due to the salt concentration in the topsoil, low resistivities were found in the topsoil. The fourth layer with 40 Ωm resistivity was identified as freshwater aquifer in comparison to the drilling results. High groundwater temperatures of about 25°C result in relatively low resistivities of the freshwater aquifer.

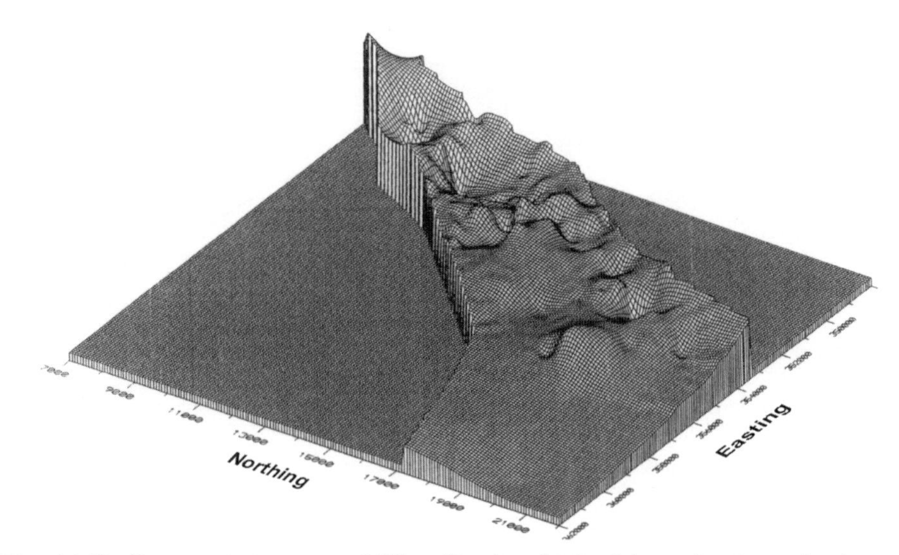

Fig. 14.12. Basement structure of Khor Baraka obtained by refraction seismic measurements (THOR 1989)

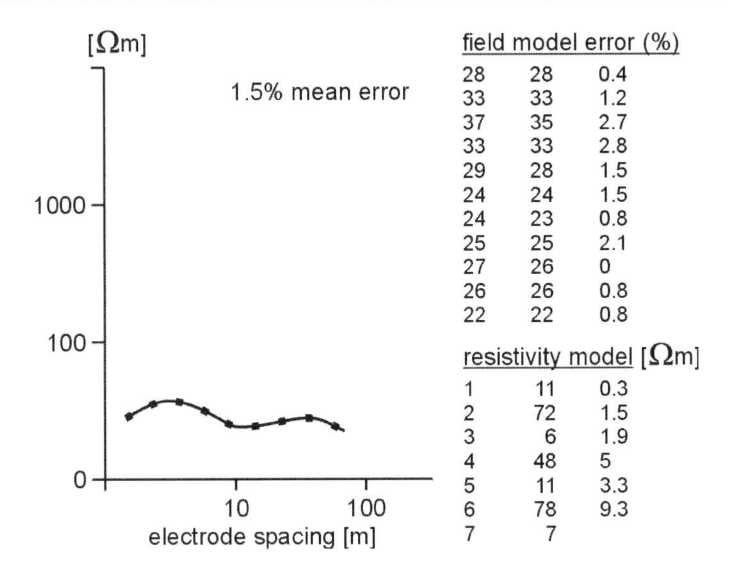

Fig. 14.13 Example of an electrical sounding curve (THOR 1989)

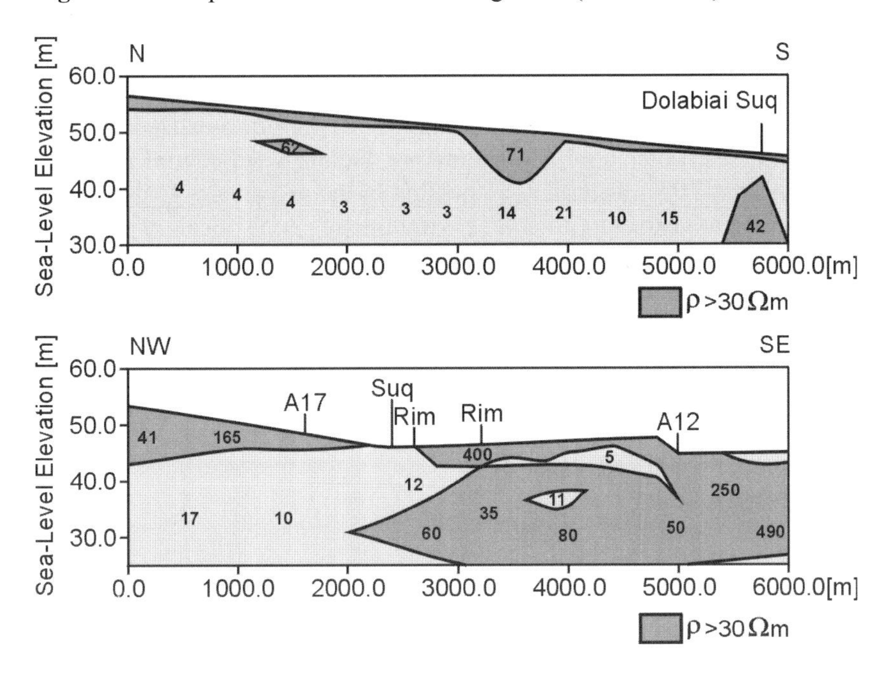

Fig. 14.14. resistivity profiles parallel (A) and perpendicular (B) to Khor Baraka (THOR 1989)

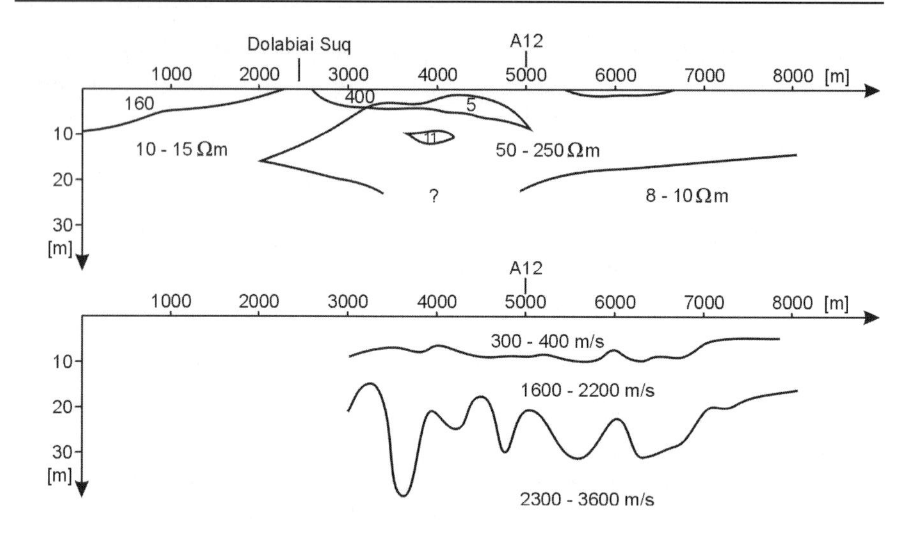

Fig. 14.15. Comparison of refraction seismic and resistivity results

Examples for the resistivity profiles are shown in Fig. 14.14. The profiles are crossing at the market place (suq) of the small village of Dolabiai. Profile A runs parallel to the Khor Baraka, whereas profile B is crossing the valley. Low resistivities are dominating. Resistivities exceeding 30 Ωm indicating freshwater aquifers are mainly found in the central part of the valley. A small freshwater lens is situated in the suq area of Dolabiai. Here a well was dug years ago, of course without any geophysical mapping.

A comparison of the resistivity structure from Fig. 14.14 (B) and the results of refraction seismic measurements is shown in Fig 14.15. A low resistive layer (8 - 10 Ωm) is clearly identified as basement rock with p-wave velocities ranging from 2300 – 3600 m/s. Fractures or pores of the basement rocks are also filled with highly saline water.

14.6 References

Barker R (1990) Investigation of groundwater salinity by geophysical methods. In: Ward SH (ed) Geotechnical and Environmental Geophysics, vol 2, Soc Expl Geophys, Tulsa, pp 201-211

Blume H-P, Yair A, Yaolon DH (1995) An initial study of pedogenic features along a transect across longitudinal dunes and interdune areas, Nizzana Region, Negev, Israel. In: Blume HP, Berkowicz M (eds) Arid Ecosystems, Advances in Geoecology 28. Catena Verlag, Cremlingen pp 51-64

Demirel F (1997) Kombinierter Einsatz verschiedener geo- und bodenphysikalischer Methoden und Modellrechnungen an einem antiken Wasserleitungssystem in Milet (Türkei). PhD-Thesis, Institut für Geophysik, Christian-Albrechts-Universität Kiel

Drabbe J, Ghijben WB (1888) Nota in Verband met de voorgenomen putboring nabij Amsterdam. Kon Ins Ing Tijdschr:8-22

EU (1998) Richtlinie 98/83/EG des Rates der Europäischen Gemeinschaften vom 3. November 1998 über die Qualität von Wasser für den menschlichen Gebrauch. Amtsblatt der Europäischen Gemeinschaften, 5.12.98:32 – 55

Fielitz K (1979) Reconnaissance of groundwater and surface water resources in the coastal area of Sudan (Eastern Red Sea Province), Chapter 3.3.2: Geoelectrical Investigations. Unpublished report, Bundesanstalt für Geowissenschaften und Rohstoffe, Hannover

Goldstein NE, Benson SM, Alumbaugh D (1990) Saline groundwater plume mapping with electromagnetics. In: Ward SH (ed) Geotechnical and Environmental Geophysics, vol 2. Soc Expl Geophys, Tulsa, pp 17-27

Grube A, Wichmann K, Hahn J, Nachtigall KH (2000) Geogene Grundwasserversalzung in den Porengrundwasserleitern Norddeutschlands und ihre Bedeutung für die Wasserwirtschaft. Schriftenreihe des DVGW-Technologiezentrums Wasser (TZW), Karlsruhe

Herzberg B (1901) Die Wasserversorgung einiger Nordseeheilbäder. J Gasbeleucht u Wasserversorg 44:815-819

Jahnke C, Magri F, Tesmer M, Bayer U, Möller P, Pekdeger A, Voigt H-J (2004) Großskalige tiefreichende thermohaline Strömungsprozesse im Norddeutschen Becken. Proc Fachsektion Hydrogeologie der DGG, Schriftenreihe der Deutschen Geologischen Gesellschaft 32:148

Keller EA (1988) Environmental Geology. Merrill Publishing Company, Toronto, London, Melbourne

Ketelsen R, Kirsch R (2004) Zur geophysikalischen Erkundung von Versalzungszonen im Grundwasser. Meyniana 56:21-45

Langsdorf W (1981) Mineralisierung des Grundwassers eines Wadis unter wüstenhaftem Klima (Khor Baraka/Sudan). Z dt geol Ges 132:637-646

Löhnert E (1967) Grundwasserversalzungen im Bereich des Salzstockes von Altona-Langenfelde. Abhdl . Verhdl naturw Verein Hamburg 11:29-46

Lühr B (1984) Bestimmung der Salz-Süßwassergrenze in einem Gebiet der Schleswig-Holsteinischen Westküste mit geophysikalischen Untersuchungsverfahren. Diploma Thesis, Institut für Geophysik, Christian-Albrechts-Universität Kiel

Matthess G (1994) Lehrbuch der Hydrogeologie, Bd.2: Die Beschaffenheit des Grundwassers. 3rd Edition. Bornträger, Stuttgart

McNeill JD (1980) Electrical conductivity of soils and rocks. Technical Note TN-5. Geonics Ltd, Mississauga

Repsold H (1990) Geoelektrische Untersuchungen zur Bestimmung der Süßwasser/Salzwasser-Grenze im Gebiet zwischen Cuxhaven und Stade. Geologisches Jahrbuch C56:3-37

Siemon B, Röttger B, Eberle D (2001) Airborne geophysical investigation of salt-water intrusions and coastal aquifers in Northwest Germany. Proceedings 7th Meeting Environmental and Engineering Geophysics:228-229

Siemon B, Binot F (2003) Aerogeophysikalische Erkundung von Salzwasserintrusionen und Küstenaquiferen im Gebiet Bremerhaven-Cuxhaven - Verifizierung der AEM-Ergebnisse. In: Hördt A, Stoll J (eds) Protokoll über das 19. Kolloquium "Elektromagnetische Tiefenforschung", 01.-05.10.2001; Burg Ludwigstein:319-328

Slater LD, Sandberg SK (2000) Resistivity and induced polarization monitoring of salt transport under natural hydraulic gradients. Geophysics 65:408-420

Stadler S, Hoyer M, Hötzl H, Himmelsbach T (2004) Ungeklärte Nitratanreicherung im Grundwasser der Kalahari Botswanas. Proc Fachsektion Hydrogeologie der DGG, Schriftenreihe der Deutschen Geologischen Gesellschaft 32:100

Stümpel, H (1980) Water supply for Port Sudan. Institut für Geophysik, Christian-Albrechts-Universität Kiel (unpublished report)

THOR Geophysical GmbH (1989): Water Supply of Port Sudan. – unpublished report (Kirsch R, Lütjen H), Kiel

TNO (1976) Geophysical well logging for geohydrological purposes in unconsolidated formations. Groundwater Survey TNO, The Netherlands Organisation for Applied Scientific Research, Delft

15 Geophysical characterisation of aquifers

Reinhard Kirsch, Ugur Yaramanci

15.1 Definition of hydraulic conductivity and permeability

Hydraulic conductivity (or alternatively permeability) characterizes the dynamic behaviour of an aquifer to allow for fluid flow and so is a key parameter for hydrogeology, strongly influencing, e.g., the yield of wells, the velocity of contaminant spread, or the consolidation behaviour of soil under an applied load.

Hydraulic conductivity is defined by the velocity of fluid flow through the underground material under the influence of an applied pressure gradient. In the classical experiment of Darcy (1803-1858), pressure gradient is given by the slope h/l of a rock sample (Fig. 15.1), the flow velocity v is then proportional to the slope (Mattheß and Ubell 1981):

$$v = K \cdot \frac{h}{l} \cdot \qquad (15.1)$$

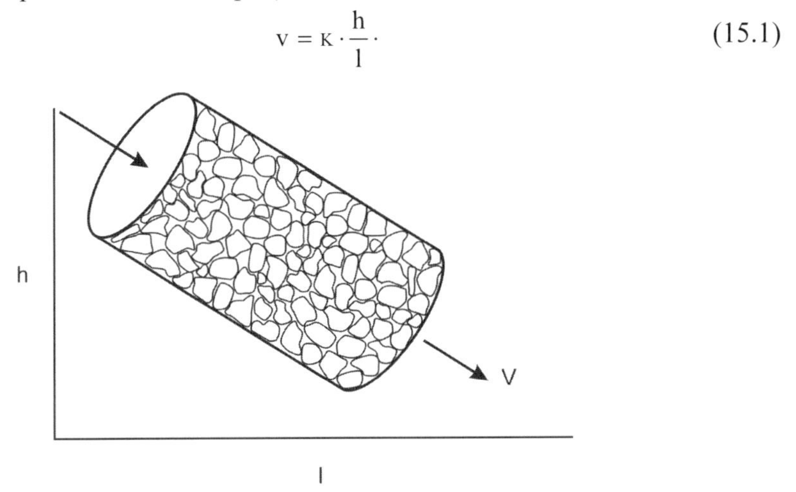

Fig. 15.1. Darcy´s law, fluid flow through a soil sample with slope h/l

The hydraulic conductivity K (german expression: k_f-value) has the unit of velocity (m/s or cm/s). Normalising K by the viscosity η of the fluid and replacing the slope by the applied pressure gradient, then permeability κ is obtained:

$$v = \frac{\kappa}{\eta} \cdot \text{gradP} \qquad (15.2)$$

The unit of permeability is area (m^2), however, traditional unit is Darcy (D) related to SI units by 1 D = 0.98710^{-12} m^2. For water filled pore space at 20°C, permeability and hydraulic conductivity are related by: 1 m/s = 1.03 * 10^5 D.

Typical hydraulic conductivities and corresponding permeabilities for unconsolidated sediments are (after Matthess and Ubell 2003):

	K [m/s]	κ [Darcy]
gravel	10^{-2}-1	10^3-10^5
clean sand	10^{-5}-10^{-2}	1-10^3
silty sand, fine sand	10^{-8}-10^{-5}	10^{-3}-1

The hydraulic conductivity of an aquifer (sand, gravel) may range over four decades. Therefore, strong lateral changes of hydraulic conductivity can be assumed for aquifers with lateral changes of its grain size distribution. So, an assessment of hydraulic conductivity based on geophysical data is a useful tool for hydrogeological modelling or for defining of well locations.

Aquifers are not only characterised by hydraulic conductivity, but also by transmissivity (product of hydraulic conductivity and aquifer thickness) and diffusivity (ratio of transmissivitiy and storage coefficient). The storage coefficient is defined in Chap. 12 (Eqs. 12.3 – 12.4).

15.2 Hydraulic conductivity related to other petrophysical parameter

There is no simple relation between hydraulic conductivity and other petrophysical parameter. Beside porosity, the hydraulic conductivity is influenced by effective porosity, grain size distribution, pore size distribution, shape of pore channels, tortuosity, constrictivity, and internal surface area. However, to make it more complicated, all these parameter are interrelated too.

With some generalisations, it is possible to relate hydraulic conductivity to one or two parameter of this list. "Classical" methods of hydraulic conductivity estimation are based on grain size distribution as, e.g., Seelheim (1880, after Matthess and Ubell 2003)

$$k = 0.00357 \cdot d_{50}^2 \qquad (15.3)$$

with d_{50} grain diameter (in cm) of the 50% fraction of the grain size distribution.

Pekdeger and Schulz (1975) analysed samples from sand pits in Schleswig-Holstein and found most reliable results for hydraulic conductivity estimation using the method of Hazen (1895, after Matthess and Ubell 2003) in the modified version after Beyer (1964, after Hölting 1996):

$$k = C \cdot d_{10}^2 \tag{15.4}$$

while factor C depends on unconformity U (=d_{60}/d_{10})

U	C
1.0 - 1.9	$110 * 10^{-4}$
2.0 - 2.9	$100 * 10^{-4}$
3.0 - 4.9	$90 * 10^{-4}$
5.0 - 9.9	$80 * 10^{-4}$
10.0 - 19.9	$70 * 10^{-4}$
> 20	$60 * 10^{-4}$

(Hölting 1996).

Grain size distribution β_0 and porosity ϕ are combined in the formulation of Terzaghi (1925) (in cm^2):

$$k = \beta_0 \cdot d_{10}^2 \cdot [(\phi - 0.13)/(1-\phi)^{1/3}]^2 \tag{15.5}$$

β_0 reflects the grain form: $\beta_0 = 800 \rightarrow$ rounded grains
$\beta_0 = 460 \rightarrow$ angular grains.

Grain size distribution and porosity are also combined in the formulation of Berg (1970, after Nelson 1994) (in mD):

$$k = 80.8 \cdot \phi^{5.1} \cdot d^2 \cdot e^{-1.385p} \tag{15.6}$$

with d = mean grain diameter [mm]
p = degree of sorting = $d_{90} - d_{10}$.

Fig. 15.2 shows the influence of porosity, mean grain diameter, and sorting on hydraulic conductivity from laboratory measurements. Poor sorting and small grain diameter reduce hydraulic conductivity.

Kozeny (1928) and Carman (1937) developed the following fundamental porosity- hydraulic conductivity relation

$$k = \frac{\phi \cdot r_{hy}^2}{f \cdot T} \tag{15.7}$$

with: ϕ = porosity
T = tortuosity
r_{hy} = effective hydraulic radius, inverse of specific inner surface
f = form factor (dimensionless, 1.7 - 3.0).

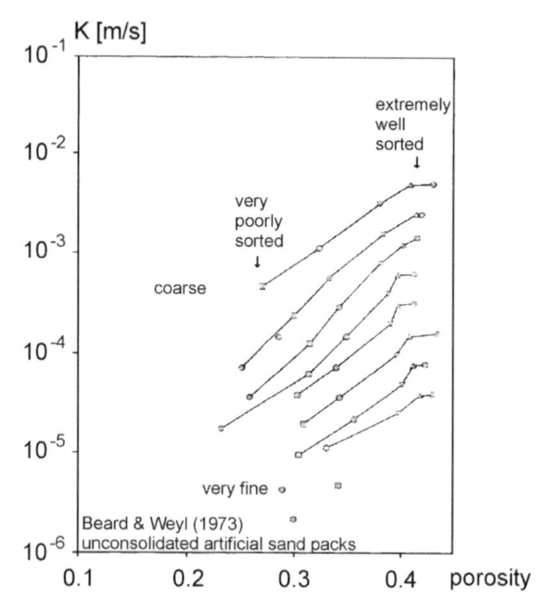

Fig. 15.2. Hydraulic conductivity of sands in relation to porosity, grain size, and sorting (after Beard and Weyl 1973, with permission from AAPG)

Based on the Kozeny-Carman relation, Georgi and Menger (1994) developed the formulation

$$k = \frac{r_{hy}^2}{f \cdot T^2} \cdot \frac{\phi^3}{(1-\phi)^2} \tag{15.8}$$

The Kozeny-Carman relation was further modified by Pape et al. (1998) to the following form:

$$k = \frac{\phi \cdot r_{eff}^2}{8 \cdot T} \tag{15.9}$$

r_{eff} = effective radius of pore channel.

An outline of porosity– hydraulic conductivity relations based on fractal pore models for sandstone is given by Pape (2003).

Marotz (1968) relates effective porosity (drainable pore space, see Chap. 12) to hydraulic conductivity and found the following relation at sandstone samples (Fig. 15.3):

$$\phi_{eff} = 25.5 + 4.5 \ln k \tag{15.10}$$

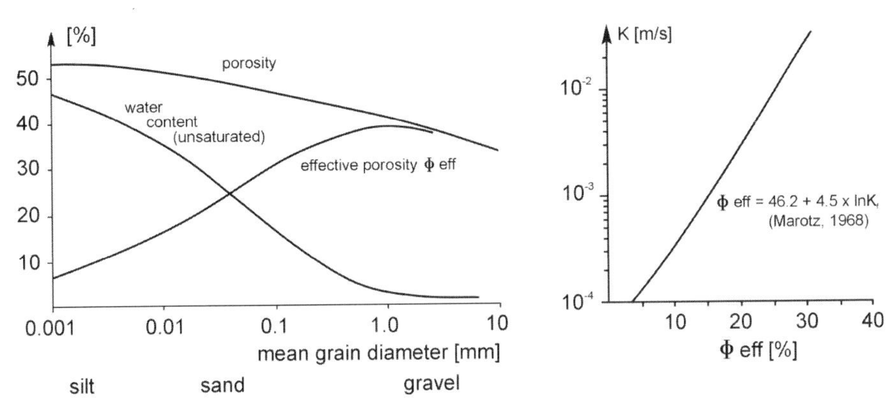

Fig. 15.3. Porosity and effective porosity of unconsolidated sediments (after Matthess and Ubell 2003) and hydraulic conductivity related to effective porosity (after Marotz 1968)

Porosity and effective porosity are linked by the content of undrainable pore water S_{wirr} (irreversible water saturation, $S_{wirr} = \phi - \phi_{eff}$). Timur (1968) found a relation between hydraulic conductivity, porosity and S_{wirr} (in mD)

$$S_{wirr} = 3.5 \cdot \frac{\phi^{1.26}}{K^{0.35}} - 1 \tag{15.11}$$

15.3 Geophysical assessment of hydraulic conductivity

As shown before, hydraulic conductivity is not easily linked to porosity as geophysical parameters are. Therefore, no straight hydraulic conductivity - resistivity or hydraulic conductivity - seismic velocity relations can be expected. However, an attempt to enable a geophysical way for interpolation of hydraulic conductivities valid at least for a limited project area should be made.

15.3.1 Resistivity

The relation between complex resistivity and hydraulic conductivity is discussed in details in Chap. 4. In the following, only the real part of resistivity which can be determined by electrical soundings is taken into account. The close relation of electrical formation factor F to porosity (Archies law) and tortuosity (see Eq. 1.8, Chap. 1) makes an attempt to find relations between hydraulic conductivity and resistivity or hydraulic conductivity and formation factor reasonable. Field and laboratory results are reported by many authors with puzzling results. So, e.g., one group of authors like Urish (1981), Frohlich and Kelly (1985), Huntley (1986), and Leibundgut et al. (1992) found positive correlation between hydraulic conductivity and formation factor, while other authors like Worthington (1975), Heigold et al. (1979), and Biella et al. (1983) reported negative correlation (Fig. 15.4).

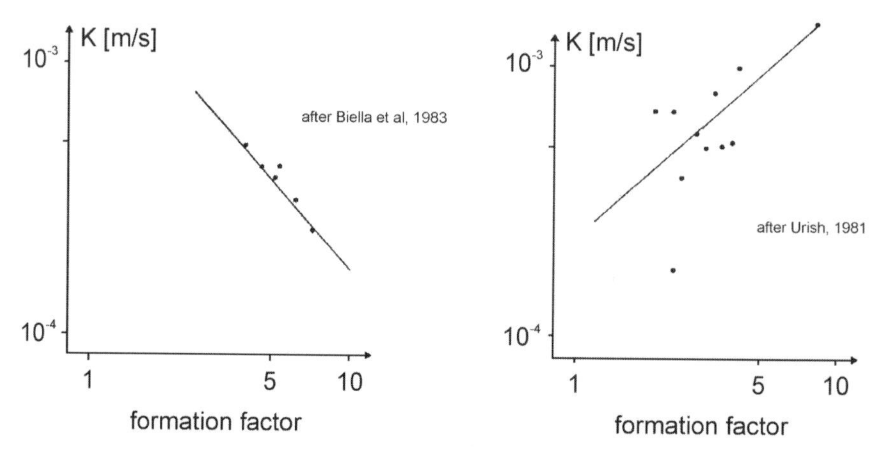

Fig. 15.4. Negative and positive correlation between electrical formation factor and hydraulic conductivity after Biella et al. (1983) and Urish (1981)

A compilation of resistivity – hydraulic conductivity relations is given by Mazác et al. (1985, 1990), Fig. 15.5. Within one sediment group (gravel, coarse sand, etc) resistivity and hydraulic conductivity are inversely correlated. As porosity and resistivity (or formation factor) are inversely correlated too, a positive correlation exists between porosity and hydraulic conductivity, as it is indicated by, e.g., the Kozeny-Carman relation (Eq. 15.6). However, if the sediment groups are compared, then positive correlation between resistivity and hydraulic conductivity is observed leading to negative correlation between porosity and hydraulic conductivity. This is in accordance with Fig. 15.3 which shows that well sorted coarse sediments like gravel have smaller porosities than well sorted fine sediments, although effective porosity and hydraulic conductivity of coarser sediments is higher.

This is backed by laboratory experiments of Biella et al. (1983). They used artificial sediments of increasing uniform grain sizes from 0.2 to 8 mm which were used to produce 2-component sediment mixtures, e.g., consisting of material with grainsize 1 mm and 8 mm. Different percentage of fine and coarser material lead to different porosities. For all mixtures of grain compositions electrical formation factor was linear related to porosity (Fig. 15.6). However, different correlations of hydraulic conductivity and porosity as well as of hydraulic conductivity and formation factor were obtained for the different mixtures (Fig. 15.7). Samples taken arbitrarily from the different mixtures would show no correlation.

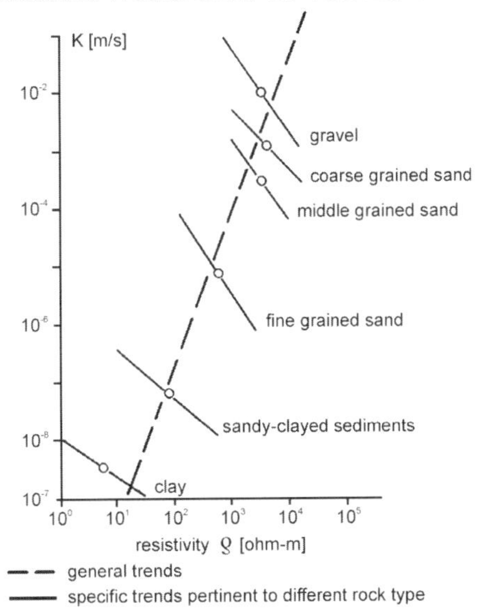

Fig. 15.5. Correlation of hydraulic conductivity and resistivity for sediment groups (after Mazác et al. 1985, 1990, with permission from SEG)

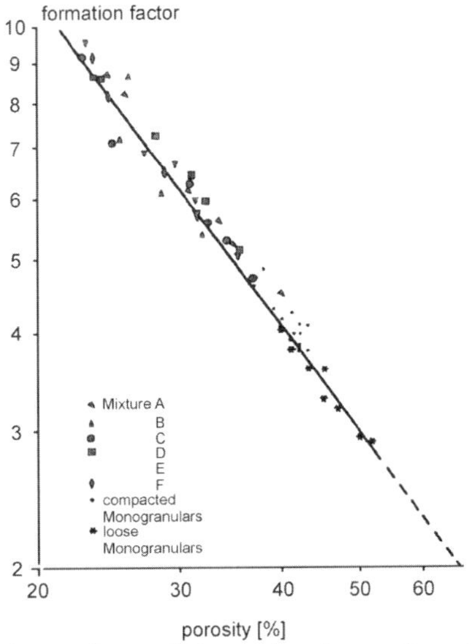

Fig. 15.6. Correlation of porosity and formation factor for artificial sediment samples (after Biella et al. 1983), best fit of data was by $F = 1.15 \cdot \phi^{-1.42}$ or $F = \phi^{-1.54}$

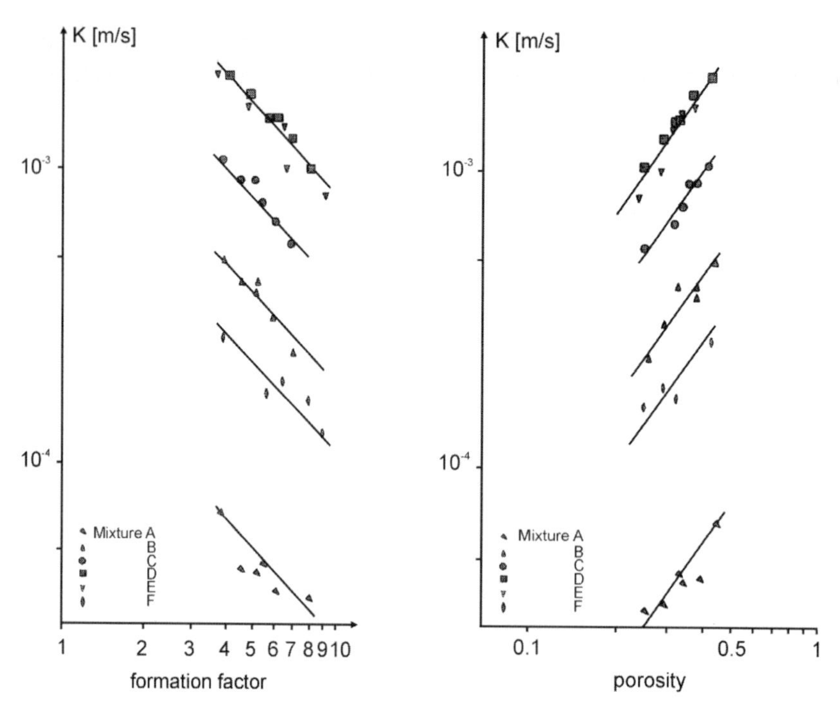

Fig. 15.7. Correlation of porosity and hydraulic conductivity (left) and formation factor and hydraulic conductivity (right), although no general correlation is obvious, clear correlation is obtained within the groups (after Biella et al. 1983)

15.3.2 Seismic velocities

Seismic velocities, as shown in chapter B1, are strongly related to porosity. After Gassmann (1950), porosity is linked to seismic velocities by the porosity dependence of bulk modulus

$$\frac{K_{sat}}{K_m - K_{sat}} = \frac{K_{us}}{K_m - K_{us}} + \frac{K_{fl}}{\phi \cdot (K_m - K_{fl})} \tag{15.12}$$

with K_{sat} = bulk modulus of saturated material
K_{us} = bulk modulus of unsaturated material
K_m = bulk modulus of rock matrix
K_{fl} = bulk modulus of pore fluid

Bulk modulus of saturated and unsaturated material can be obtained from p- and s-velocities and density ρ by

$$K_{sat,usat} = \rho \cdot (v_{psat,usat}^2 - \frac{4}{3} v_s^2) \tag{15.13}$$

However, no significant influence of hydraulic conductivity on seismic velocities is reported, as shown, e.g., by the relation found at sandstone samples (Klimentos 1991):

$$v_p = 5.27 - 5.4 \cdot \phi - 2.54 \cdot C + 0.001 \cdot k \tag{15.14}$$

So, for a hydraulic conductivity assessment, the same problems with unknown porosity – hydraulic conductivity relations exist as shown above. As a consequence, like for resistivity, positive as well as negative hydraulic conductivity – seismic velocity relations are found. Fechner (1998) made seismic tomography in a hydrogeological test fields with known hydraulic conductivity distributions and found both types of correlation (Fig. 15.8). In the test field "Horkheimer Insel" (Baden-Württemberg) with fluviatile gravels and boulder of river Main hydraulic conductivity is negatively correlated to velocity, this is comparable to the sediment groups in Fig. 15.5 (Mazác et al. 1985, 1990). In the testfield "Belauer See" (Schleswig-Holstein) a positive correlation was found. This field consists of sander sediments of Saale and Weichsel glaciation, this sediments are not as uniform as those of the test field "Horkheimer Insel". Here a regression as log k = 0.004332 v_p – 12.825 was found.

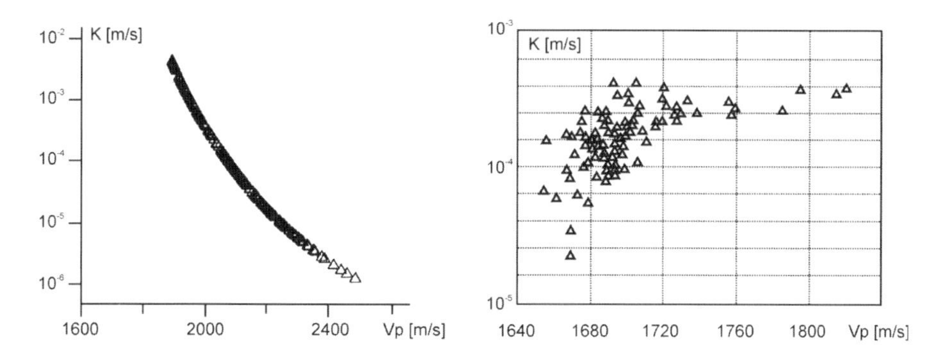

Fig. 15.8. Correlation of p-wave velocity and hydraulic conductivity; left: "Horkheimer Insel" (Baden-Württemberg), right: "Belauer See" (Schleswig-Holstein) (Fechner 1998)

15.3.3 Nuclear resonance decay times

Recently the well proven laboratory investigation technique for rock properties and in particular for porosities and hydraulic conductivities i.e. nuclear magnetic resonance (NMR) is available in field scale called as Surface NMR or Magnetic Resonance Sounding (MRS) (Yaramanci et al.

1999, Legtchenko et al. 2002, Lubcynski and Roy 2005). It is possible to map the hydrogen proton relaxation behaviour of individual layers and areas where the initial amplitude of the relaxation is related to water content i.e. porosity in case of full saturation and the decay time is related to the hydraulic conductivity. Due to technical limitations the relaxation can not be recorded for very early times and in conclusion the determined water content relates to the mobile (extractable) part of the water and therefore, the determined porosity relates to effective porosity.

In MRS (see Chap. 8) usually the free induction decay time T_2^* is recorded. For frequencies used in MRS to match the Larmor frequency (in the range of 1 – 3 kHz worldwide depending on the strength of local Earth magnetic field) free induction decay time T_2^* equals to the transversal decay time T_2. Occasionally it is possible to record the longitudinal relaxation decay time T_1 using multiple NMR excitations which is more accurate but needs larger measurement time.

Very early in MRS applications it was realized that the decay time T_2^* correlates well with the material grain size and thus with hydraulic conductivities. The correlation is based on a large number of field data with MRS and corresponding grain size analyses on relevant material (Schirov et al. 1991) and is given as (decay time in ms): Sandy clays <30, clayey sands and very fine sands 30-60, fine sands 60-120, medium sands 120-180, coarse and gravelly sands 180-300, gravel deposits 300-600, surface water bodies 600-1500.

The correlation indicates a simple relation of hydraulic conductivity and decay times (Yaramanci et al. 1999) based on the usual grain size vs. hydraulic conductivity relations in hydrogeology

$$k \approx T^4 \qquad (15.15)$$

Here k is in m/s and T standing for free induction decay time T_2^* is in s. This relation, though purely empirical, is well proven by many measurements in particular in the usual porosity ranges encountered in different types of aquifers. In fact it does not contain porosity explicitly as usual in relationships otherwise for hydraulic conductivity. For that it is not valid for very low porosity material i.e. for small mobile water content. Decay times which can be detected by MRS are ranging approximately from 0.03 s to 1 s correspond to hydraulic conductivities from $6\ 10^{-6}$ m/s to 1 m/s.

In laboratory NMR as well as in borehole NMR very often the special experimental relation of permeability to decay times is found as (Kenyon 1997):

$$\kappa = c \cdot \phi^4 \cdot T^2 \qquad (15.16)$$

Here the permeability κ is in mD and related to hydraulic conductivity k_f by: 1 m/s = $1.03 \cdot 10^5$ D. The decay time T stands for both decay times of longitudinal relaxation, T_1, and transversal relaxation, T_2, and is in ms. The constant c depends on the surface relaxivity of the mineral grains forming the matrix and therefore, strongly site specific. Usual values for c are 4,5 m^2/s^2 for sandstones mainly quartzite in matrix and 0,1 m^2/s^2 in limestones. T stands for both decay times of longitudinal relaxation T_1 and transversal relaxation T_2.

As it is clear from Eq. 15.16 the decay time T with power of 4 has more influence on hydraulic conductivity than porosity with power of 2. The reason is that the decay time is directly related to the pore size by

$$\frac{1}{T} = \rho \cdot \frac{S}{V} = \rho \frac{1}{r_{hy}} \tag{15.17}$$

with S the inner surface, V pore volume, r_{hy} effective hydraulic radius and ρ surface relaxivity. Pore size mainly controls the hydraulic flow.

There are some attempts to adapt Eq. 15.16 in a more generalized form for MRS as:

$$\kappa = c \cdot \phi^a \cdot T^b \tag{15.18}$$

Hereby the porosity needs to refer to the NMR-porosity which is not the total porosity as usual in laboratory NMR, but the porosity corresponding to the mobile water content as found by MRS. In some applications of MRS like in sands, clayey sands, limestones a = 1 and b = 2 has been found very suitable. Thus there are not many systematic studies yet to establish a reliable relationship i.e. classifying structures and lithologies according to their a and b which basically reflects the pore structure as well as c reflecting the mineral composition of the material. Although the use of Eq. 15.16 is well established in laboratory and borehole NMR it is difficult to adapt it directly for MRS measurements.

Another way of estimating hydraulic conductivities from NMR properties is to use the mobile water content (= effective porosity) by using more general relationships relating hydraulic conductivity to porosity and other structure parameter like hydraulic radius or equivalently internal surface as given in Eqs. 15.5 – 15.10. The size of the needed structural parameters may be available from analyses on core material or even by well-logging at representative location at the site.

15.4 Case history: Hydraulic conductivity estimation from SIP data

Andreas Hördt

Spectral induced polarisation is discussed in details in Chap. 4. Here the application of SIP measurements for the estimation of hydraulic conductivity shall be demonstrated. A SIP survey was carried out at the Krauthausen hydrogeological test site. The site is operated by the Forschungszentrum Jülich and the has been described by Vereecken et al. (2000). Fig. 15.9 gives an overview of the average lithology. It is characterised by a 1m surface soil layer, followed by an aquifer of about 10 m thickness. The aquifer is heterogeneous and consists of fluvial gravel and sands. A clay layer at 11 m depth forms the base of the aquifer.

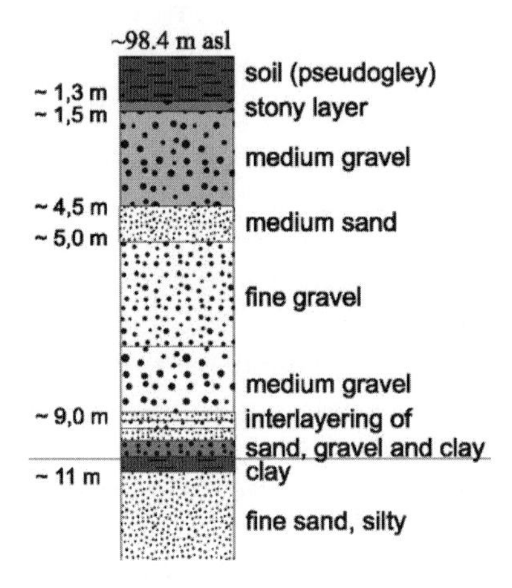

Fig. 15.9. Average lithology at the Krauthausen test site (after Döring 1997)

One particular profile will be discussed here that crosses a borehole location where detailed grain size information is available. The data were recorded using the SIP 256C equipment (Radic 2004). Each electrode is connected to a remote unit which controls the current injection or voltage measurement, digitizes the data and transfers them to a PC through a fiber optics cable. We used 32 electrodes at 2 m spacing, switching them in a dipole-dipole configuration. For each transmitter-receiver pair, the data consist of apparent resistivity and phase vs. frequency. Fig. 15.10 shows an example data set.

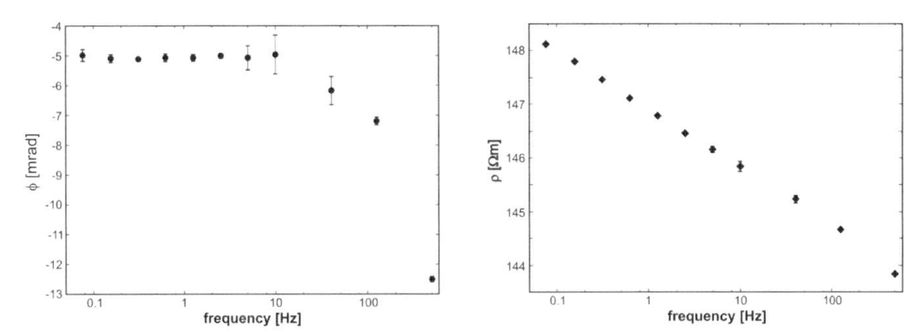

Fig. 15.10. A complex apparent resistivity data set from the Krauthausen test site. The transmitter and receiver dipoles are 2m in length, the spacing is 24m. Left panel: Phase shift vs. frequency. Right panel: Magnitude vs. frequency (from Hördt et al., 2005)

The apparent resistivity (left panel) decreases with frequency. This is the expected behaviour, because the pore space is assumed to be able to store electrical charges, similar to a capacitance in an electrical circuit.

The effect is called membrane polarisation and is described by Börner (this volume, Chap. 4). The phase, given in milliradians, is negative, because the voltage lags behind the injected current. It is fairly constant over a wide frequency range at low frequencies. This behavior has been observed in laboratory measurements and seems to be typical of many unconsolidated sediments (Börner et al. 1996). The strong decrease at frequencies above 10 Hz is due to electromagnetic effects, i.e. induction and capacitive coupling (Radic 2004). In general, it is assumed that frequencies are sufficiently low such that these effects can be ignored. In that case, any variation with frequency is only due to variation of the intrinsic subsurface conductivity. If the high frequencies need to be used, electromagnetic effects have to be corrected for. Here, because the phase is constant over a wide range, the result does not strongly depend on frequency and we proceed with single frequency data at 0.3125 Hz.

From the 2-D inversion of the data along the profile we obtain a model for the intrinsic magnitude and phase of the complex conductivity (Fig. 15.11). The inversion algorithm is based on the code described by Kemna (2000) and Kemna et al. (2004) with an important modification. The smoothness constraint was replaced by a regularisation suggested by Portniaguine and Zhdanov (1999), combined with an idea presented by Yi et al. (2003). The regularisation allows sharp contrasts and supports images that display well separated zones or layers, with smooth parameter variations within the zones. The model in Fig. 15.10 looks reasonable. Layer boundaries may clearly be identified and nicely correspond with the boundaries in the average lithology. Both conductivity and phase increase below 11m, where clay and fine sand form the base of the aquifer.

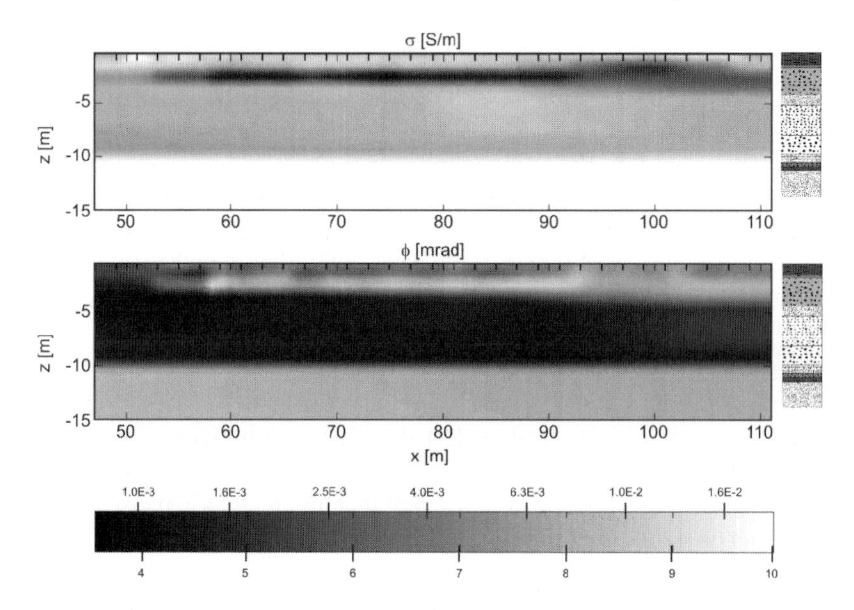

Fig. 15.11. 2-D inversion model (conductivity magnitude and phase at 0.3125 Hz) along profile b32. The upper grey shaded scale indicates conductivity magnitude in S/m, the lower scale is phase in mrad. The average lithology was inserted for comparison

To estimate hydraulic conductivity, we used the approach suggested by Börner et al. (1996). It is based on empirical relationships derived from laboratory measurements on unconsolidated sediments. Formation factor F is calculated mainly from the real part and imaginary part of conductivity:

$$F = \frac{\sigma_W}{\left[\mathrm{Re}(\sigma) - \mathrm{Im}(\sigma)\right]/1} \tag{15.19}$$

where σ_w is the pore fluid conductivity (0.1 S/m). The factor l is the ratio between imaginary and real part of the surface conductivity. It is used as an adjustable parameter to correct for the contribution of surface conductivity to the measured real conductivity. The second parameter is the specific inner surface area S_{por} given by:

$$S_{por} = 8.6 \cdot 10^4 \cdot \mathrm{Im}(\sigma) \tag{15.20}$$

where σ is in S/m and S_{por} in 1/μm.

The formation factor and specific inner surface area then enter into a Kozeny-Karman type of equation to calculate hydraulic conductivity:

$$k_f = \frac{1}{F S_{por}{}^{c}} \tag{15.21}$$

where k_f is in m/s and S_{por} in 1/μm. Exponent c is an adjustable parameter found to be in the range between $2.8 < c < 4.6$.

The equations were applied to the 2-D inversion result to obtain an estimate of the hydraulic conductivities (Fig. 15.12). The average lithology is nicely reflected in the k_f-section. The sharp decrease at 11 m corresponds with the known base of the aquifer. In the aquifer between 4 and 11 m, k_f varies between 10^{-4} and $5 \cdot 10^{-3}$ m/s. This agrees roughly with the values determined from tracer tests, where the average values of different layers vary between $7 \cdot 10^{-4}$ m/s and $2 \cdot 10^{-3}$ m/s (Vereecken et al., 2000). Above 4 m, the k_f values seem to increase significantly. However, the sediments in that depth range have a heterogeneous grain size distribution. The empirical equations were not derived for this type of lithology, and thus the results are likely to be invalid. Above 2 m, we have unsaturated conditions, and the equations do not apply either.

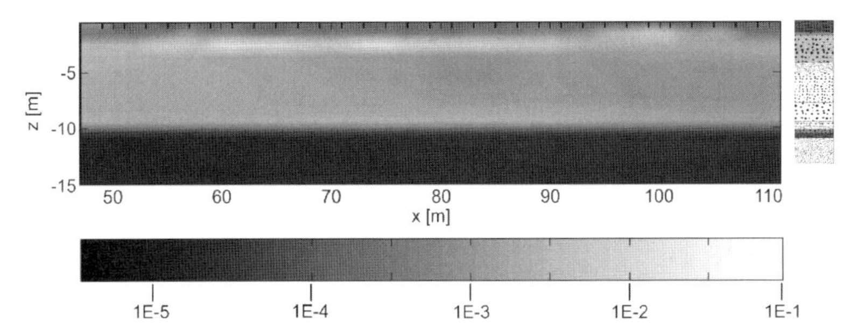

Fig. 15.12. Hydraulic conductivity (k_f) section calculated from the inversion result shown in Fig. 15.11 (after Hördt et al., 2005). The grey shaded bar indicates k_f in m/s. Borehole b32 is at 76 m on the profile. The average lithology is inserted on the right

In order to estimate the k_f-values, the adjustable parameters in the empirical equations had to be determined. We used the grain size data from borehole 32, which is at 76 m on the profile. The result is shown in Fig. 15.13. There are several equations to calculate hydraulic conductivity from grain size data. From the uniformity of the grain size distribution we decided to use the Seiler-10 equation (Seiler, 1973):

$$k_f = C_{10}(u) d_{10}^2 \tag{15.22}$$

where C_{10} are tabulated coefficients depending on the uniformity, and d_{10} is 10th percentile of the cumulative grain size distribution. The values were averaged along depth to obtain data at the points where SIP results are available. Parameters c and l were then calculated to minimize the difference between SIP-estimates and grain size results. Both curves show a similar behaviour with depth. There is a maximum in k_f between 2 and 4 m, a local minimum at 6 m, an increase towards 10 m, and a decrease towards the base. The structure is more pronounced in the grain size results, but the overall agreement can be considered satisfactory.

The case history shows that the method has some potential to provide useful information on hydraulic conductivity from spectral IP data on the field scale. The estimated k_f-section corresponds with the overall lithology, and the values are in the expected range. The section is not dominated by either the amplitude or the phase data, which means that measurement of both parameters seems to be essential. Clearly, more case histories and additional work will be necessary before a routine application may be considered. This includes the consideration of information from the full frequency spectrum, a 3-D inversion, and further laboratory work to obtain a better understanding of the spectra under different conditions.

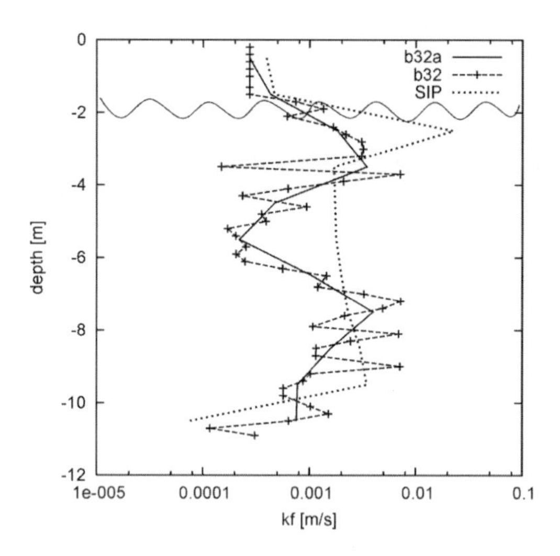

Fig. 15.13. Hydraulic conductivities calculated from SIP data compared with those from grain size analyses in borehole 32 (after Hördt et al., 2005). B32: Grain size analysis results using a Seiler-d10 equation (Seiler 1979). B32a: B32 averaged over 1m, SIP: k_f calculated from SIP results after Börner et al. (1996) using c=3.07 and l=0.015. The wiggly line denotes the position of the water table, which usually varies between 1-2 m

15.5 References

Beard DC, Weyl PK (1973) Influence of texture on porosity and permeability of unconsolidated sand. American Association of Petroleum Geologists Bulletin 34:943-961

Berg R (1970) Method for determining permeability from reservoir rock properties. Transactions, Gulf Coast Association of Geological Societies 20:303-317

Biella G, Lozej A, Tabacco I (1983) Experimental study of some hydrogeophysical properties of unconsolidated porous media. Ground Water 21:741-751

Börner FD, Schopper JR, Weller A (1996) Evaluation of transport and storage properties in the soil and groundwater zone from induced polarization measurements. Geophys. Prosp. 44: 583-601

Carman PC (1937) Fluid flow through granular beds. Trans Inst Chem Eng 15:150-166

Döring U (1997) Transport der reaktiven Stoffe Eosin, Uranin und Lithium in einem heterogenen Grundwasserleiter. Berichte des Forschungszentrums Jülich 3354. Forschungszentrum Jülich

Fechner T (1998) Seismische Tomographie zur Beschreibung heterogener Grundwasserleiter. Tübinger Geowissenschaftliche Arbeiten, C40

Frohlich RK, Kelly WE (1985) The relation between hydraulic transmissivity and transverse resistance in a complicated aquifer of glacial outwash deposits. Journal of Hydrology 79:215-229

Gassmann F (1950) Über die Elastizität poröser Medien. Vierteljahreszeitschrift der Schweizer naturforschenden Gesellschaft, 1

Georgi DT, Menger SK (1994) Reservoir quality, porosity and permeability relationships. Proc. 14th Mintrop Seminar, Münster

Heigold PC, Gilkeson RH, Cartwrigt K, Reed PC (1979) Aquifer transmissivity from surfial electrical methods. Ground Water 17:338-345

Hölting B (1996) Hydrogeologie. Enke, Stuttgart

Hördt A, Blaschek R, Kemna A, Suckut J, Zisser N (2005) Hydraulic conductivity from spectral induced polarisation measurements – a case history, 18th Annual Meeting Symposium on the Application of Geophysics in Engineering and Environmental Problems (SAGEEP)

Huntley D (1986) Relations between permeability and electrical resistivity in granular aquifers. Ground Water 24:466-474

Kemna A (2000) Tomographic Inversion of Complex Resistivity. PhD thesis, Universität Bochum

Kemna A, Binley A, Slater L. (2004) Crosshole IP imaging for engineering and environmental applications. Geophysics 69:97-107

Kenyon WE (1997) Petrophysical principles of application of NMR logging. The Log Analyst March-April: 21-43

Klimentos T (1991) The effects of porosity-permeability-clay content on the velocity of compressional waves. Geophysics 56:1930-1939

Kozeny JA (1928) Die Durchlässigkeit des Bodens. Der Kulturtechniker 35:478-486

Legtchenko AV, Baltassat JM, Beauce A, Bernard J (2002) Nuclear Magnetic Resonance as a geophysical tool for hydrogeologists. Journal of Applied Geophysics 50:21-46

Leibundgut C, De Carvalho Dill A, Maloszewki P, Müller I, Schneider J. (1992) Investigation of solute transport in the porous aquifer of the test site Wilerwald (Switzerland). Steir Beitr z Hydrogeologie 43:229-250

Lubcynski M, Roy J. (2005) MRS contribution to hydrogeological system parameterisation. Near Suface Geophysics 4: (in print)

Marotz G (1968) Technische Grundlagen einer Wasserspeicherung im natürlichen Untergrund. Verlag Wasser und Boden, Hamburg

Mattheß K, Ubell G (1981) Allgemeine Hydrogeologie – Grundwasserhaushalt. Borntraeger, Berlin, Stuttgart

Mazác O, Císlerová M, Kelly WE, Landa I, Venhodová D (1990) Determination of hydraulic conductivities by surface geoelectrical methods. In: S.H. Ward, (ed.): Geotechnical and Environmental Geophysics, Vol. II:125 - 131, Soc. Expl.Geophys., Tulsa, Ok, USA

Mazác O, Kelly WE, Landa I (1985) A hydrogeophysical model for relations between electrical and hydraulic properties of aquifers. Journal of Hydrology 79:1-19

Nelson PH (1994) Permeability-porosity relationships in sedimentary rocks. The log analyst:38-62

Pape H (2003) Fractal relation between porosity and permeability: theory and verification. In: Clauser C (ed) Numerical simulation of reactive flow in hot aquifers. Springer, Heidelberg

Pape H, Clauser C, Iffland J (1998) Permeability prediction based on fractal pore-space geometry. Geophysics 64:1447-1460

Pekdeger A, Schulz HD (1975) Ein Methodenvergleich zur Laborbestimmung des kf-Wertes von Sanden. Mayniana, 27:35-40

Portniaguine O, Zhdanov MS (1999) Focusing geophysical inversion images. Geophysics 64:874-887

Radic T (2004) Elimination of Cable effects while Multichannel SIP measurements. 10th European meeting of Env. Eng. Geophys., Utrecht

Schirov M, Legtchenko A, Creer G (1991) A new direct non-invasive groundwater detection technology for Australia. Exploration Geophysics 22:333-338

Seiler K-P (1979) Durchlässigkeit und Porosität von Lockergesteinen in Oberbayern. Mitteilung zur Ing.- u. Hydrogeologie 9:105-126

Terzaghi K (1925) Erdbaumechanik auf bodenphysikalischer Grundlage. Leipzig, Wien

Timur A (1968) An investigation of permeability, porosity, and residual water saturation relationships for sandstone reservoirs. The Log Analyst:8-17

Urish DW (1981) Electrical resistivity-hydraulic conductivity relationships in glacial outwash aquifers. Water Resources Research 17:1401-1408

Vereecken H, Döring U, Hardelauf H, Jaekel U, Hashagen U, Neuendorf O, Schwarze H, Seidemann R (2000) Analysis of solute transport in a heterogeneous aquifer: the Krauthausen field experiment. Journal of Contaminant Hydrology 45:329-358

Worthington PF (1975) Quantitative geophysical investigations of granular aquifers. Geophys. Surv.:313-366

Yaramanci U, Lange G, Hertrich M (2002) Aquifer characterisation using Surface NMR jointly with other geophysical techniques at the Nauen/Berlin test site. Journal of Applied Geophysics 50:47-65

Yaramanci U, Lange G, Knödel K (1999) Surface NMR within a geophysical study of an aquifer at Haldensleben (Germany). Geophysical Prospecting 47:923-943

Yi M-J, Kim J-H, Chung S-H (2003) Enhancing the resolving power of least-squares inversion with active constraint balancing. Geophysics 68:931-941

16 Groundwater protection: vulnerability of aquifers

Reinhard Kirsch

16.1 General

The protection of groundwater reservoirs is given by the covering layers, also called protective layers. Surface water percolates through the protective layers leading to groundwater recharge. During this percolation process contaminant degradation can occur by mechanical, physicochemical, and microbiological processes. An effective groundwater protection is given by protective layers with sufficient thickness and low hydraulic conductivity leading to high residence time of percolating water (Fig. 16.1). In the following, protective layers are regarded as homogenous bodies which can be characterized by bulk properties like hydraulic conductivity. Inhomogeneities as sandy intrusions or fissures in clay which can lead to preferred pathways for percolation are not taken into account, although they are quite common (Douma et al. 1990).

16.2 Vulnerability maps

Vulnerability of an aquifer is defined as the sensitivity of groundwater quality to an imposed contaminant load, which is determined by the intrinsic characteristics of the aquifer (Lobo-Ferreira 1999). This is different to the expression pollution risk which depends on vulnerability as well as on the existence of pollutants entering the subsurface.

Groundwater protection requires information on groundwater vulnerability. Maps showing the lateral distribution of well protected and poorly protected aquifers are therefore essential for spatial development. These maps are called vulnerability maps.

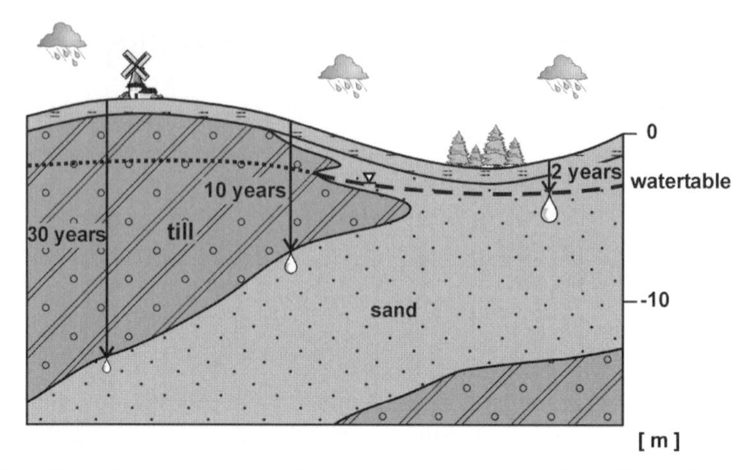

Fig. 16.1. Groundwater protective layers, percolation time (2 – 30 years) depends on material and thickness of protective layers

Vulnerability maps can display geological parameter. An early set of maps was produced for The Netherlands, showing on a 1:400000 scale the lateral distribution of depth to groundwater table, clay content, cation exchange capacity, and percolation time of surface water (RIVM 1987). Cation exchange capacity CEC is calculated from clay content C and content of organic substance OS by CEC [mMol/Ha] = 1.5 * OS + 0.6 * C.

The residence time T for percolating water was calculated for the unsaturated vadose zone after $T = D * \Theta / I$, and for the saturated zone down to the aquifer actually in use after $T = (D^2 * \phi) / (k * \Delta h)$ with:

D = thickness of vadose or saturated zone

Θ = water content of vadose zone

I = mean infiltration velocity, calculated after:
yearly precipitation – 0.8 * yearly evaporation

ϕ = porosity

k = vertical hydraulic conductivity

Δh = pressure difference saturated/vadose zone.

Vulnerability maps had to be produced in the scope of the EC water framework directive for all European countries. In Schleswig-Holstein for example, the northernmost state of the Federal Republic of Germany, maps were produced on a 1:100000 scale showing the depth to groundwater table and thickness of clayey layers (LANU 2003).

Another group of vulnerability maps are based on one single parameter to quantify vulnerability. Methods in use are, e.g., the DRASTIC method by the US Environmental Protection Agency, the AVI method by the Canadian Prairie Provinces Water Board, and the SGD method by the board of directors of the German state geological surveys. These methods are characterized as follows in short form.

DRASTIC is based on a set of hydrogeologic factors related to contaminant spread as: **D**epth to water, net **R**echarge, **A**quifer media, **S**oil media, **T**opography (hill slope), **I**mpact of the vadose zone media, and hydraulic **C**onductivity of the aquifer (Osborn et al. 1998). A rating from 1 – 10 based on the range of values is assigned to each of the hydrogeologic factors. To obtain the so called DRASTIC index for a given location, the ratings of the hydrogeologic factors are multiplied by a relative weight ranging from 1 – 5 according to the importance of the hydrogeologic factor. DRASTIC index is then given by

$$D_{INDEX} = 5 \cdot D + 4 \cdot R + 3 \cdot A + 2 \cdot S + T + 5 \cdot I + 3 \cdot C \qquad (16.1)$$

DRASTIC indices are ranging from 23 to 230, a high index means high aquifer vulnerability. Vulnerability categories in terms of DRASTIC indices are (Wei 1998):

< 100	low vulnerability
100 – 180	moderate vulnerability
> 180	high vulnerability.

As the ratings of the hydrogeologic factors depend on the local or regional conditions, DRASTIC delivers only relative vulnerability.

AVI Aquifer Vulnerability Index (Van Stempvoort et al. 1992): This method quantifies vulnerability by hydraulic resistance to vertical flow of water through the protective layers. Hydraulic resistance c is defined by:

$$c = \sum_{i} d_i / K_i \qquad (16.2)$$

where d_i, K_i are thickness and hydraulic conductivity of each protective layer.

Typical values for K, based on Freeze and Cherry (1979) as used by Van Stempvoort et al. (1992), are sand: 10 m/d, silt: 10^{-1}m/d, and massive till (mixed sand-silt-clay): 10^{-5}m/d. As the K-values for sand are several magnitudes higher than those for clayey layers, hydraulic resistance as defined above is dominated by clayey layers.

As K has the unit length/time (e.g. m/d or m/s), the dimension of c is time. Following Van Stempvoort et al. (1992) this can be used as a rough estimate of vertical traveltime of water through the unsaturated layers, however, important parameter controlling the traveltime as hydraulic gradient, diffusion, and sorption are not considered here.

Expressing c in years, the log c can be used for a classification like:

log c < 1	extremely high vulnerability
log c = 1-2	high vulnerability

log c = 2-3 moderate vulnerability
log c = 3-4 low vulnerability
log c > 4 extremely low vulnerability.

SGD system: The board of directors of the state geological surveys (SGD) of the Federal Republic of Germany launched a vulnerability quantification system based on the cation exchange capacity CEC of the protective layers (Hölting et al. 1995). The cation exchange capacity is related to clay content and hydraulic conductivity. High cation exchange capacity refers to high clay content and low hydraulic conductivity.

Vulnerability is quantified by a protection function S_G (Schutzfunktion) calculated by:

$$S_G = \sum_i d_i \cdot G_{Li} \qquad (16.3)$$

where d_i, G_{Li} are thickness and cation exchange capacity code (Punktzahl) of each covering layer. This cation exchange capacity code is ranging from 5 (coarse gravel) to 500 (fat clay).

Vulnerability is quantified by the protection function:

S_G = 500 – 1000 poor aquifer protection, high vulnerability
S_G = 1000 – 2000 moderate aquifer protection, moderate vulnerability
S_G = 2000 – 4000 high aquifer protection, low vulnerability
S_G > 4000 extremely high aquifer protection, extremely low vulnerability.

Protection function is also related to percolation time:

S_G = 500 – 1000 percolation time several months – 3 years
S_G = 1000 – 2000 percolation time 3 – 10 years
S_G = 2000 – 4000 percolation time 10 – 25 years
S_G > 4000 percolation time > 25 years.

Comparisons of different techniques for vulnerability assessment are presented by Gogu et al. (2003) for limestone aquifers with karst features and by Magiera (2002) for Quaternary aquifers. All vulnerability mapping systems have in common that they are based on 1D drilling results. GIS techniques can be applied to produce maps, but a sufficient density of drillings is required. In the following, the use of geophysical techniques to enable a better interpolation between drillings is demonstrated.

16.3 Electrical conductivity related to hydraulic resistance, residence time, and vulnerability

The key parameters hydraulic conductivity (for AVI-method) and cation exchange capacity (for SGD-method) are closely related and are linked by the clay content of the material. Clay minerals are characterized by high cation exchange capacity (an example for this relationship is shown in Chap. 1, Fig. 1.14). As clay grains are several magnitudes smaller than sand grains, in a clay sand mixture clay particles can block the narrow channels connecting pores spaces and so reduce the hydraulic conductivity effectively. An example for the correlation of clay content and hydraulic conductivities of soils is shown in Fig. 16.2.

The dominating influence of clay content on electrical resistivity or conductivity was demonstrated in Chap. 1. Increasing clay content leads to decreasing electrical resistivity or to increasing electrical conductivity (Fig. 16.3). Although this can be seen as a general trend only (as other parameter like porosity, conductivity of pore water, and relation between cation exchange capacity and clay content can change vertically and horizontally), electrical and hydraulic conductivity correlate inversely.

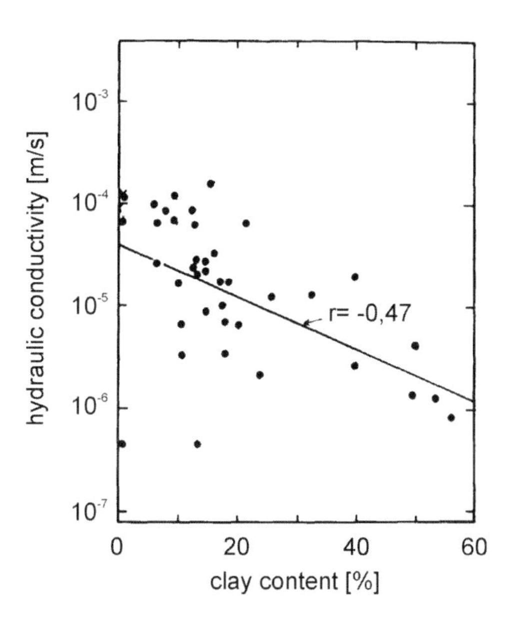

Fig. 16.2. Hydraulic conductivity vs. clay content for soil samples of depth range 0.8-1.2 m (after Scheffer and Schachtschabel 1984, permission from Elsevier GmbH, Spektrum Akademischer Verlag)

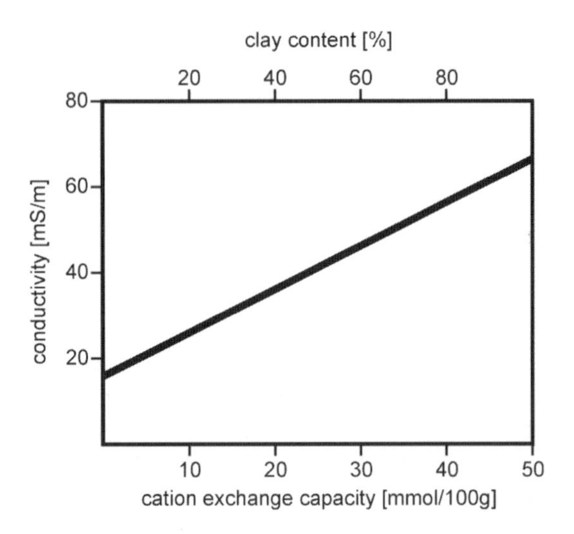

16.3. Electrical conductivity in relation to clay content and cation exchange capacity after Sen et al. (1988) and Günzel (1994), a porosity of 30% and an electrical conductivity of pore water of 100 mS/m were assumed

Of special interest for vulnerability assessment are the groundwater covering layers above the water table, i.e. the unsaturated zone. An example of the electrical conductivity of unsaturated soils is shown in Fig. 16.4. Sandy and clayey soil samples were saturated with water and dried while electrical conductivity was measured (Durlesser 1999). For clayey soils constant or even increasing electrical conductivities were observed for decreasing saturation degrees, until a saturation of about 70% was reached. The increase of conductivity with decreasing water content can be explained by the increasing ion concentration of the pore water while drying. Due to the high specific surface area of clayey soils, the water content of these layers in the unsaturated zone is high, at least under humid climatic conditions. So, for clayey layers, similar electrical conductivities can be assumed in the saturated and the unsaturated zone.

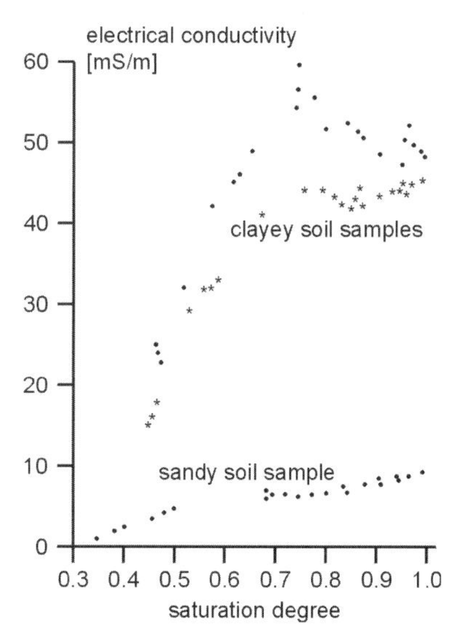

Fig. 16.4. Electrical conductivity of clayey soil samples (clay content: 30%) and of a sandy soil sample (clay content: 14%) related to saturation degree (after Durlesser 1999)

AVI and SGD vulnerability assessments are based on residence time of percolating water. In their time-of-travel concept Kalinsky et al. (1993) relate percolation time through unsaturated protective layers to electrical resistivity. Percolation time through a layer of thickness d is related to volumetric water content Θ and groundwater recharge rate q by:

$$t = \frac{d \cdot \Theta}{q} \tag{16.4}$$

If the protecting layer consists of different layers with thicknesses d_i and water contents Θ_i, then the percolation time is given by:

$$t = \frac{1}{q} \sum_i d_i \cdot \Theta_i \tag{16.5}$$

Water content Θ, porosity Φ, and electrical conductivity σ are related by the formalism of Mualem and Friedman (1991) as shown in Chap. 1:

$$\sigma_0 = \sigma_{WATER} \cdot \frac{\Theta^{2.5}}{\phi} + \sigma_{SURFACE} \tag{16.6}$$

with $\sigma_{SURFACE} = 2.3 * C - 0.021$ (in mS/cm) as obtained by Rhoades et al. (1990).

For a well field in Lincoln, Nebraska, water content of samples from protective layers were measured and compared with vertical electrical sounding VES results. Also from the water content data, time-of-travel was calculated for several locations of the well field and related to measured electrical conductance $S = d/\rho$ (Dar Zarrouk parameter). Within the error range, a clear correlation was obtained (Fig. 16.5). Using S instead of resistivity ρ for the determination of percolation times has the advantage that Dar Zarrouk parameters can be determined from electrical sounding curves without equivalence problems (see Chap. 3).

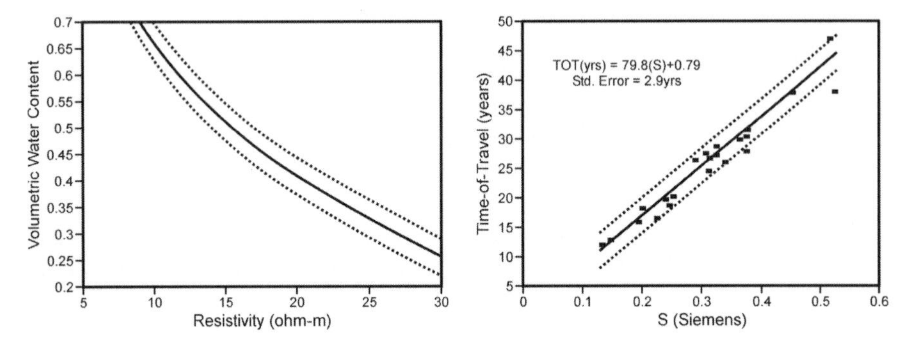

Fig. 16.5. Water content vs electrical resistivity (left) and time-of-travel vs electrical conductance (right) as obtained from VES data (after Kalinski et al. 1993, permission from Elsevier)

16.4 Vulnerability maps based on electrical conductivity

The relation between soil parameters like clay content, cation exchange capacity, and hydraulic conductivity to electrical resistivity enables a vulnerability assessment based on geoelectrical or electromagnetic measurements. Results of these measurements can be used as an estimate of clay content and hydraulic conductivity of the soil, or to interpolate vulnerability quantified by rating systems like AVI between drillings. For this, electrical or electromagnetic mapping systems are required to cover larger areas.

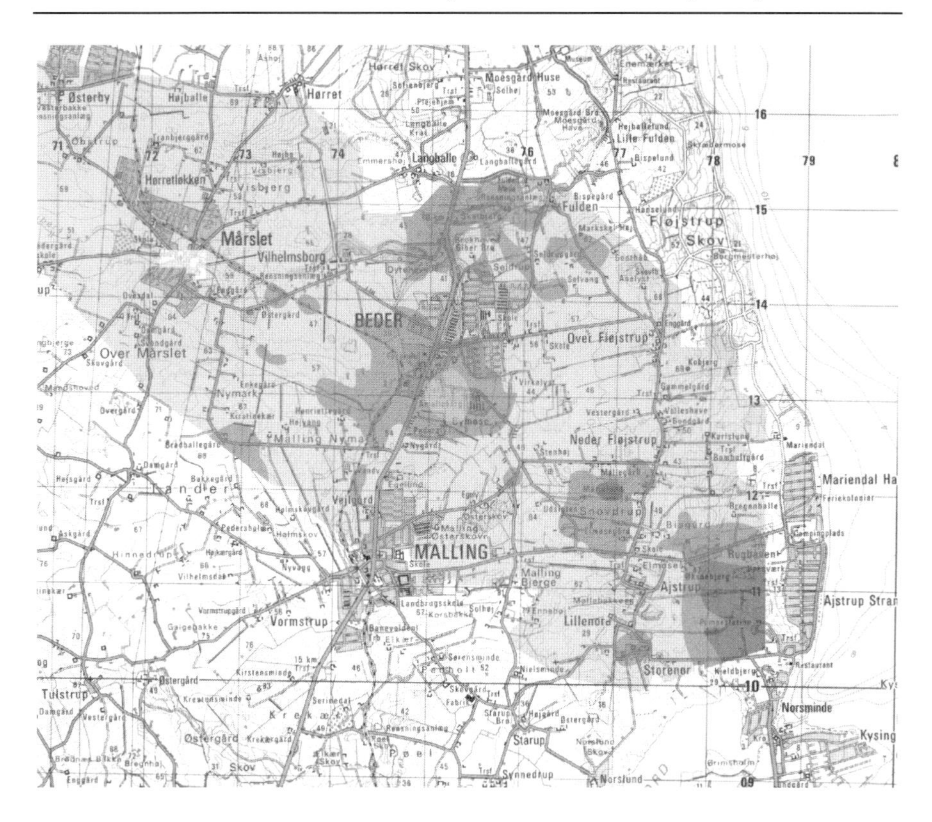

Fig. 16.6. Vulnerability map based on apparent resistivity (Wenner array, a=30m) obtained by PACEP measurements. High resistivities (dark grey) are caused by sandy inclusions in the clayey protective layers (light grey) resulting in decreased vulnerability (Thorling and Sørensen 1995)

An example for fast land based mapping systems is the Pulled Array Continuous Electrical Profiling (PACEP) which has been developed by the Hydrogeophysics Group, Aarhus University. An electrode array is towed across the field behind a small vehicle and measurements with three sets of electrodes with different separations are performed continuously and simultaneously while actively towing the electrode array. A system with 3 electrode separations (10 m, 20 m, and 30 m Wenner) has been used extensively since 1988 and has now measured more than 10000 km of profile. Measurements are typically made at 1m intervals along the survey lines, with the distance between lines being 50-300 m. With the PACEP method, two people can complete 10 to 15 km of profile in one day. At present, an electrode array with eight electrode spacing (PACES) has been developed, which enables continuous geoelectrical soundings in Wenner and dipole-dipole configurations with electrode distances between 2 and 30 m.

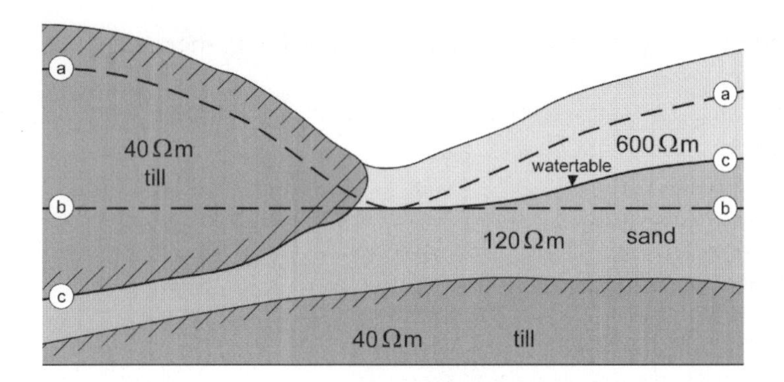

Fig. 16.7. Options of datum levels for vulnerability maps: a) constant depth below ground (as groundwater table roughly follows topographic relief); b) constant depth referring to mean sea level, e.g. main depth of groundwater table; c) top of groundwater table as indicated by electrical resistivity (60–200 Ωm)

The resistivity data are presented as contoured apparent resistivity maps, one for each electrode configuration. An early example is shown in Fig. 16.6 for a Wenner electrode spacing of 30 m. High resistivities (dark grey) are caused by sandy material with increased vulnerabilities. These "windows" for contaminants correlate with an increase nitrate content of shallow ground water (Auken et al. 1994, Thorling and Sørensen 1995).

Airborne systems are able to cover large areas. In the scope of a Danish-German groundwater mapping project in the border area (Christensen et al. 2002) an airborne EM survey was flown by the Bundesanstalt für Geowissenschaften und Rohstoffe. The main purpose of this survey was to map near surface geology and aquifers. Additionally, the data set was used for vulnerability assessment.

The concept of vulnerability assessment used in this project is similar to concepts of AVI or SGD: as the electrical conductivity is related linear to the cation exchange capacity, the cation exchange capacity code G_{Li} in Eq. 16.3 is replaced by the electrical conductivity σ_i to form an integrated electrical conductivity IEC (Kirsch et al. 2003):

$$IEC = \sum_i \sigma_i \cdot d_i \tag{16.7}$$

The same expression is used by Casas et al. (2005) to quantify vulnerability. Electrical conductivity σ and thickness d of each layer are results of the inversion of airborne EM data. For the definition of IEC a datum level must be defined to which IEC is calculated. Options for this datum level are, e.g. (Fig. 16.7):

a) constant depth below ground (as groundwater table roughly follows topographic relief)

b) constant depth referring to mean sea level, e.g. mean depth of groundwater table
c) top of groundwater table as detected by electrical resistivity (60–200 Ωm).

As an example, a map showing IEC based on option a (30 m below ground) is shown in Fig. 16.8. Regions with high IEC (clay rich moraine material) referring to high groundwater protection are clearly divided from sandy regions with low IEC.

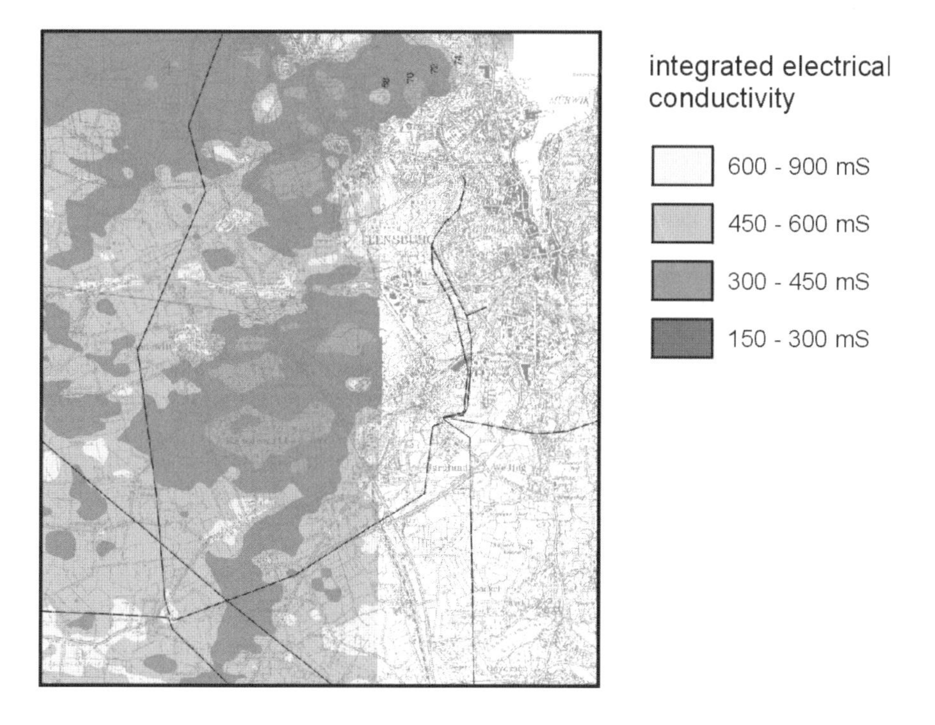

Fig. 16.8. Vulnerability map based on integrated electrical conductivity for the Flensburg area (Danish-German border region)

Vulnerability maps based on geoelectrical or electromagnetic measurements can give an overview on the groundwater protection of the area. Small scale and more detailed measurements can be done on demand, e.g. within the surrounding of well fields. Geophysical results can also be used to interpolate AVI or SGD data, when VES or EM measurements are done at drilling locations where the drilling results were used for a SGD quantification. Integrated conductivity can be adjusted to the SGD scheme at those locations. VES, TEM, or PACEP in the area between drillings can then be used to interpolate protection functions.

16.5 References

Auken E, Christensen NB, Sørensen K, Efferso F (1994) Large scale hydrogeological investigations in the Beder Area - a case study. Proc Symposium on the Applications of Geophysics to Engineering and Environmental problems:615-627

Casas A, Himi M, Tapias JC (2005) Sensibility analysis of electrical imaging method for mapping aquifer vulnerability to pollutants. Near Surface Geophysics Conference, Palermo

Christensen P-F, Christensen S, Friborg R, Kirsch R, Rabbel W, Röttger B, Scheer W, Thomsen S, Voss W (2002) A geological model of the Danish-German Border Region. Meyniana 54:73-88

Douma J, Helbig K, Schocking F, Tempels J (1990) Shear-wave splitting in shallow clays observed in a multi-offset and walk-around VSP. Geologie en Mijnbouw 69:417-428

Durlesser H (1999) Bestimmung der Variation bodenphysikalischer Parameter in Raum und Zeit mit elektromagnetischen Induktionsverfahren. PhD-Thesis, Technische Universität München

Freeze RA, Cherry JA (1979) Groundwater. Prentice Hall, Englewood Cliffs, New Jersey

Gogu RC, Hallet V, Dassargues A (2003) Comparison of aquifer vulnerability assessment techniques. Application to the Néblon river basin (Belgium). Environmental Geology 44:881-892

Günzel F (1994) Geoelektrische Untersuchung von Grundwasserkontaminationen unter Berücksichtigung von Ton- und Wassergehalt auf die elektrische Leitfähigkeit des Untergrundes. PhD thesis Ludwig-Maximilians-Universität München

Hölting B, Härtlé T, Hohberger K-H, Nachtigall KH, Villinger E, Weinzierl W, Wrobel J-P (1995) Konzept zur Ermittlung der Schutzfunktion der Grundwasserüberdeckung. Geolugisches Jahrbuch C63:5–24

Kalinski RJ, Kelly WE, Bogardi I, Pesti G (1993) Electrical resistivity measurements to estimate travel times through unsaturated ground water protective layers. Journal of Applied Geophysics 30:161-173

Kirsch R, Sengpiel K-P, Voss W (2003) The use of electrical conductivity mapping in the definition of an aquifer vulnerability index. Near Surface Geophysics 1:3-20

LANU (2003) Umsetzung der EG-Wasserrahmenrichtlinie. Unpublished report, Landesamt für Natur und Umwelt des Landes Schleswig-Holstein, Flintbek

Lobo-Ferreira JP (1999) The European Union experience on groundwater vulnerability assessment and mapping. COASTIN A Coastal Policy Research Newsletter 1:8-10

Magiera Ph (2002) GIS-gestützte Bewertung der Verschmutzungsempfindlichkeit des Grundwassers [GIS-based groundwater vulnerability assessment]. Geologisches Jahrbuch SC3

Mualem Y, Friedman SP (1991) Theoretical prediction of electrical conductivity in saturated and unsaturated soil. Water Resources Research 27:2771-2777

Osborn NI, Eckenstein E, Koon KQ (1998) Vulnerability Assessment of Twelve Major Aquifers in Oklahoma. Oklahoma Water Resources Board, Technical Report 98-5

RIVM (1987) Kwetsbaarheid van het Grondwater. Rijksinsituut voor Volksgezondheid en Milieu hygiene, Staatsuitgeverij, ΄s-Gravehage

Rhoades JD, Manteghi NA, Shouse PJ, Alves WJ (1989) Soil electrical conductivity and soil salinity: new formulations and calibrations. Soil Sci Soc Am J. 53:433-439

Scheffer F, Schachtschabel P (1984) Lehrbuch der Bodenkunde. Enke Verlag, Stuttgart

Sen PN, Goode PA, Sibbit A (1988) Electrical conduction in clay bearing sandstones at low and high salinities. J. Appl. Phys. 63:4832-4840

Thorling L, Sørensen K (1995) Effect of "windows" on groundwater quality mapped by geophysical methods. Proc. XX General Assembly European Geophysical Society, Hamburg

Van Stempvoort D, Ewert L, Wassenaar L (1992) Aquifer vulnerability index: a GIS-compatible method for groundwater vulnerability mapping. Canadian Water Resources Journal 18:25-37

Wei M (1998) Evaluating AVI and DRASTIC for assessing groundwater pollution potential in the Fraser Valley. CWRA 51st Annual Conference Proceedings, Victoria, BC

17 Groundwater protection: mapping of contaminations

Reinhard Kirsch

17.1 The brownfields problem

Large areas of all countries that are covered with abandoned industrial sites and waste dumps are a relict of earlier stages of industrialization and consumer society. In addition large areas are covered with military training areas, airfields, barracks, and depots which are a relict of the cold war. From the commercial point of view these areas are called brownfields, because their expansion, redevelopment, or use may be complicated through the presence or potential presence of a hazardous substance, pollutant, or other contaminants. From the viewpoint of groundwater protection these areas can be seen as potential sources of contamination which have to be mapped and classified. Often it is necessary to avoid drilling which could mobilize hazardous material, so geophysical mapping is the first choice for a reconnaissance of those areas.

Typical underground structures found at brownfields are:

- industrial relicts such as building foundations, storage tanks, pipelines, all more or less damaged
- industrial or municipal waste deposits (also called landfills)
- contaminated soil and groundwater.

The use of geophysical techniques at these sites might be complicated due to:

- strong lateral and horizontal heterogeneities
- high electrical conductivities at waste deposits (depending on the type of waste)
- concentrated or randomly distributed metal parts
- partly sealed ground.

The United States Environmental Protection Agency launched a reference guide on the use of geophysical techniques (EPA 1993). In Europe, several research projects on mapping or remediation of waste dumps and former industrial sites including geophysical methods were conducted, e.g.

- the waste dump research program of the German state of Baden-Württemberg (LFU 1993, Vogelsang 1993, 1994)

- the joint project to investigate underground conditions of waste dumps, organized by the Federal Institute for Geosciences and Natural Resources (BGR) (Knödel et al. 1997)
- the EU project NORISC (Network Oriented Risk-assessment by In-situ Screening of Contaminated sites, 2001-2003) (e.g. Perk et al. 2004).

Examples shall follow that explain and illustrate the application of geophysical techniques for

- the mapping of waste deposits
- the mapping of abandoned industrial sites
- the mapping of groundwater contamination.

In most cases it is necessary to characterize the geological and hydrogeological conditions of brownfield areas. Techniques shown in the chapters on pore and fracture zone aquifers (Chaps. 12 and 13) can be applied.

17.2 Mapping of waste deposits

In recent years, waste of all kinds was deposited without treatment in hollows like ponds or abandoned open pit mines. In many cases, no indications for these waste deposits were found on the surface. A localization of these waste deposits is necessary for an assessment of the contamination risk for soil and groundwater.

A certain contrast of physical properties of waste and surrounding material is necessary for the geophysical mapping of waste deposits. Sewage slurries and municipal waste of sufficient moisture content show high electrical conductivities exceeding 100 mS/m, whereas building rubble has a high magnetization due to the incorporated metallic components. As a consequence resistivity or magnetic methods are normally used to map waste deposits.

Simple examples for the use of electromagnetic measurements to localize waste deposits are shown in Figs. 17.1 and 17.2. In the example of Fig. 17.1, municipal waste was deposited in quaternary sands with a shallow covering layer and partly saturated with groundwater. An electromagnetic survey (Geonics EM 31) showed conductivities exceeding 120 mS/m indicating the waste body. The electromagnetic cross section (Fig. 17.2) through a waste deposit (industrial slurries and municipal waste) in an abandoned sandstone quarry was obtained by the inversion of three-frequency EM results (Geonics EM 34). High electrical conductivities indicate the waste body as well as a fracture zone filled with contaminated water.

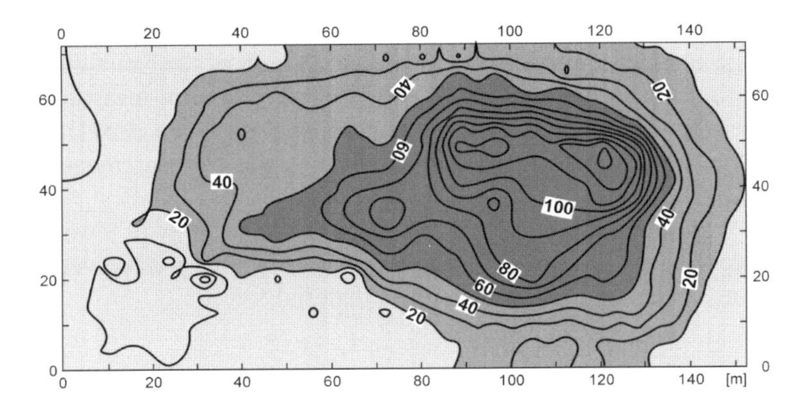

Fig. 17.1. EM 31 survey indicating a deposit of municipal waste in quaternary surrounding ,conductivities in mS/m (Stoll 1994)

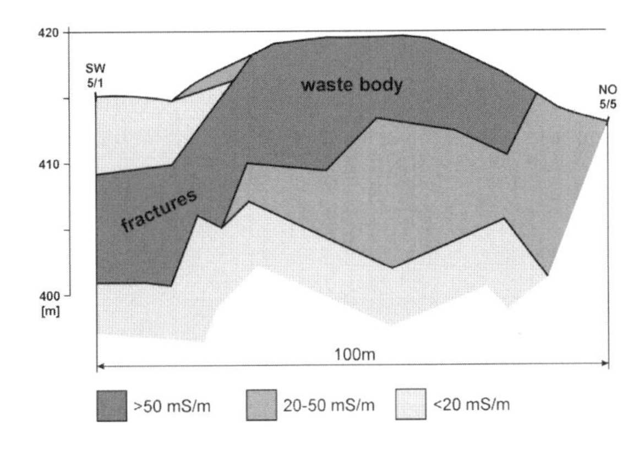

Fig. 17.2. Cross section through a waste body (industrial slurries, municipial waste) embedded in sandstone after inversion of EM34 results, high conductivities at the left side indicate a fracture zone being a potential pathway for contaminated water (THOR 1988)

Frequency domain EM measurements enable continuous measurements which allow the mapping of small scale underground heterogeneities like waste deposits. Instruments for continuous time domain electromagnetic (TEM) measurements are in the test phase (Hydrogeophysics Group, Aarhus University). Taking reduced lateral resolution into account, stationary TEM measurements can be applied for waste deposit mapping (Schaumann 2001). Fast mapping of waste deposits is also possible with modern radio magnetotelluric (RMT) techniques (Tezkan 1998).

Conventional VES measurements normally can not be applied to the waste deposit mapping due to the strong lateral discontinuities, the application of VES is mainly restricted to the determination of the resistivity conditions within the surrounding. However, 2-D geoelectrical measurements in Wenner or Dipole-Dipole configuration can enable a similar view through the waste deposit as multi-frequency EM measurements (e.g. Sweeney 1984).

A high resolution reflection and refraction seismic survey of a waste deposit is reported by De Iaco et al (2003). Low seismic p-wave velocities (200-400 m/s) were found for the landfill material.

17.3 Mapping of abandoned industrial sites

Abandoned industrial sites are mainly characterized by remnants of production facilities (pipelines, tanks, cable, building foundations) and contaminated soil. If geoelectrical or electromagnetic measurements are possible without perturbation by power lines or metallic objects, resistivity mapping can show an overview on the previous installations of that place. Fig. 17.3 shows, as an example, an apparent resistivity map of a former railway work, the remnants of buildings in the subsurface are indicated by high resistivities.

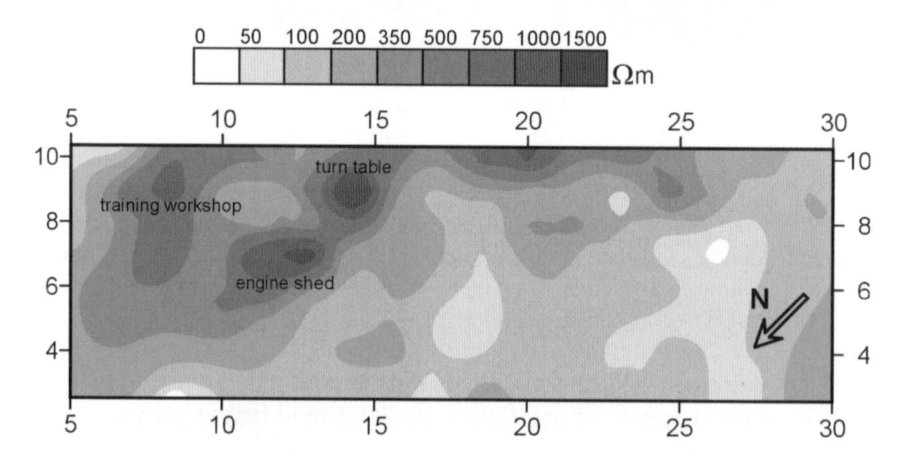

Fig. 17.3. Resistivity survey (AB = 3m) of a former railway work (Roosen and Tezkan 1998), remnants of buildings in the underground are indicated by high resistivities

Radar measurements enable a detailed view on the structure of shallow former underground installations. Fig. 17.4 shows GPR results of a former tannery, now covered with military barracks. Former pits for leather production were located in the GPR sections by interruptions of the natural layering of the ground. These locations are of particular interest because anthrax bacillus from animal skins might be found there. This bacillus can survive several decades in the soil. A typical disturbance for the GPR measurements at sealed ground is also shown in Fig. 17.4, steel reinforcements in concrete plates lead to a series of diffraction hyperbola in the radargram.

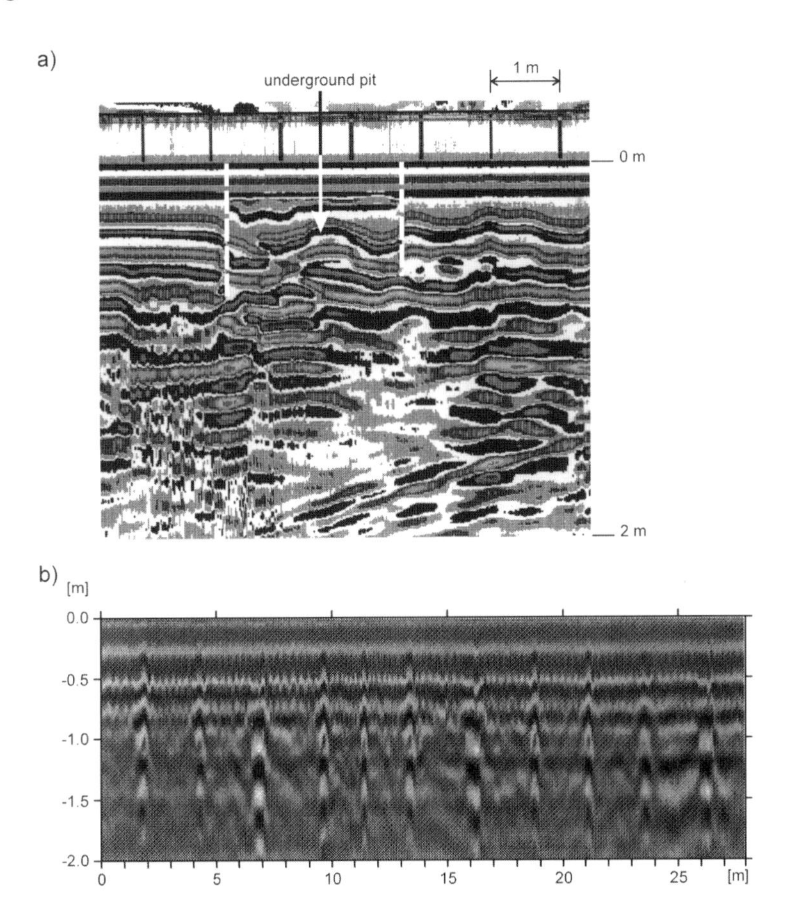

Fig. 17.4. Results of GPR measurements on a former leather production site, now military barracks. a) underground pits for tannery purposes (80 MHz antenna), b) diffraction hyperbola by steel reinforcements of a concrete plate (THOR 1996)

The lateral extent of underground structures can be visualized by GPR amplitude maps or time slices. This technique requires densely spaced (e.g. 0.5 m) GPR profiles. Amplitudes of the travel time range of relevant reflections are integrated, interpolated between the GPR profiles, and are displayed. The GPR amplitude map in Fig 17.5 shows, as an example, underground relicts of a former military airfield.

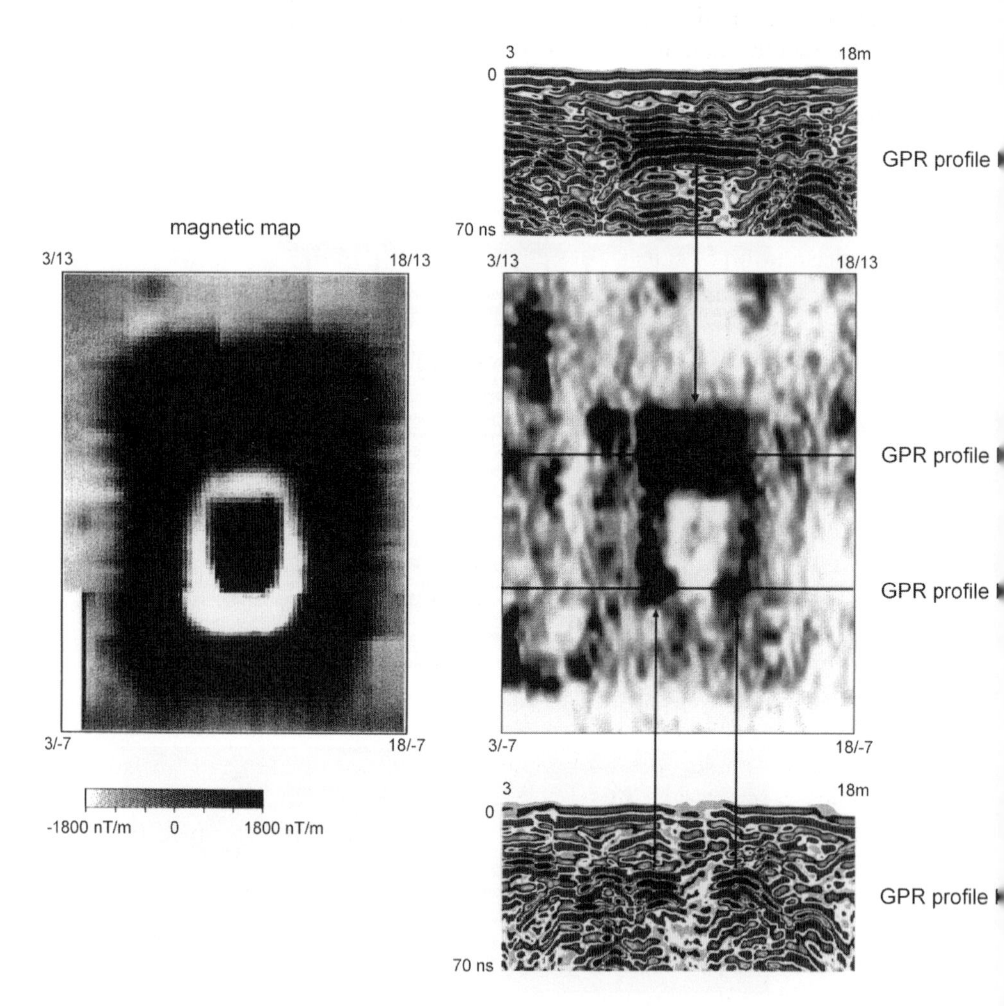

Fig. 17.5. GPR amplitude map (120 MHz antenna) of a fuel tank in a concrete shelter, not visible on the surface. For comparison: magnetic anomaly map of this site (Lorra et al. 1997)

Soil contamination, e.g. through mineral oil or slags, is common on previous industrial sites. Considered here are only the contaminations in the vadose zone above the groundwater table, whereas the groundwater contaminations will be treated in the next paragraph. Hydrocarbons such as diesel fuel normally have low conductivities. However, long term hydrocarbon pollution in the vadose zone can undergo biodegradation processes under aerobic conditions resulting in high conductivities or low resistivities due to an increase of cation exchange capacity (Godio and Naldi 2003).

Processes related to biodegradation are discussed by Sauck (2000). Main processes influencing resistivity are:

- the production of organic and carbonic acids by aerobic and anaerobic bacteria
- the dissolution of salts from native soils due to reduction of pH
- the LNAPL-water mixture in the pore space leading to water-filled capillaries which provide conductive paths.

These processes result in a reduction of resistivity instead of the expected increase.

Laboratory experiments on the influence of hydrocarbon concentration in unsaturated sand on resistivity (Gajdoš and Kráĭ 1995) show a resistivity decrease at low and an increase at higher hydrocarbon concentrations (Fig. 17.6). An explanation would be that at low hydrocarbon concentrations pore water at the grain surface is replaced by hydrocarbons and concentrated in the center of the pore leading to an increase of free ion density and a decrease of resistivity. Electrical current pathways are interrupted at higher hydrocarbon concentration by the nearly isolating hydrocarbons leading to a resistivity increase. No effect on resistivity was observed with hydrocarbon concentrations of about 10%.

This can explain the sometimes complicated pattern of resistivity anomalies connected to hydrocarbon pollutions. Grinat (2001), e.g., observed very high resistivities in the vadose zone above an up to 0.85 m thick kerosene layer on the groundwater table (Fig. 17.7).

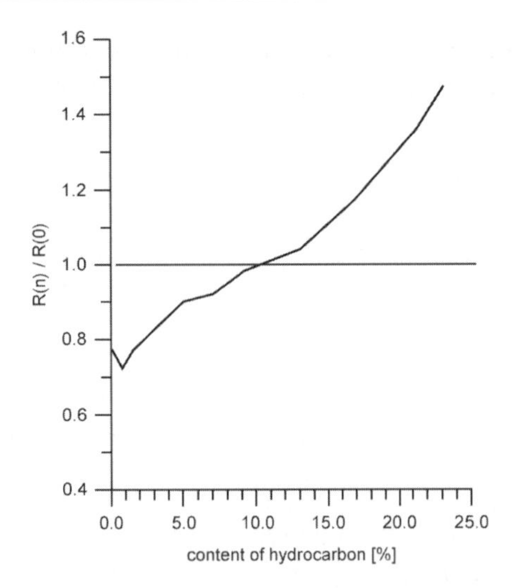

Fig. 17.6. Influence of hydrocarbon concentration in unsaturated sand on resistivity (Gajdoš and Kráĭ 1995), R(n) resistivity measured, R(0) resistivity at zero hydrocarbon concentration

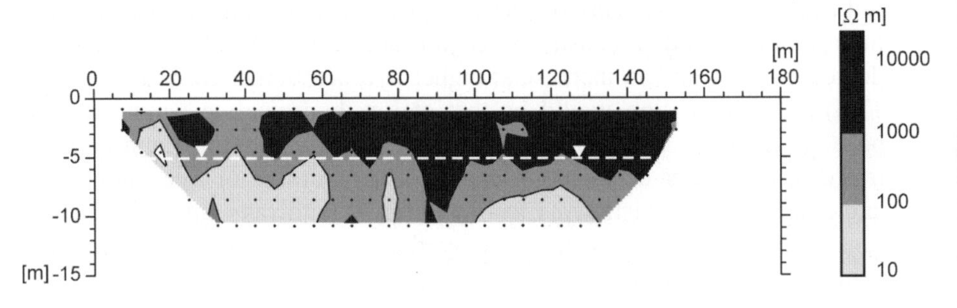

Fig. 17.7. Resistivity cross section (Wenner configuration, inversion results) of a kerosene contaminated area on an airport, high resistivities in the vadose zone above a kerosene layer floating on the groundwater table (Grinat 2001)

17.4 Mapping of groundwater contaminations

Organic and anorganic contaminants can be the origin of groundwater pollution within the surrounding area of waste deposits and abandoned industrial sites. Geophysical mapping strategies of contaminant plumes depend on the contaminant-groundwater interaction: anorganic contaminants dissolve, whereas organic contaminants are in general not dissolvable and so do not affect the physical properties of the groundwater.

17.4.1 Anorganic contaminants

Groundwater pollution through anorganic contaminants leads to an increase in the electrical conductivities as can be observed in waste deposits. Fig 17.8 shows the vertical and horizontal distribution of conductivity measured by direct push techniques within the surrounding of a waste deposit (Bock 1997). The mapping of contaminant plumes in groundwater is therefore similar to the mapping of saltwater intrusions where electrical or electromagnetic techniques can be applied. Similar to the mapping of saltwater intrusions, interpretation problems can occur due to clayey layers of high conductivity. Additional problems, as in general given in brownfields geophysics, are possible due to man made structures in the subsoil like metal parts or cables.

In Fig. 17.8, an increase in aquifer conductivity with depth was observed leading to the assumption that contaminated groundwater do not mix totally with uncontaminated groundwater. Due to the higher density contaminated groundwater can be concentrated in the lower parts of the aquifer. A consequence for the planning of monitoring wells is sketched in Fig. 17.9. A contaminant plume in the groundwater from a waste deposit sinks to the clayey bottom of the aquifer and flows around a buckle of the aquifer base. Therefore, a water sample from a monitoring well placed on this buckle contains clean groundwater. The structure of the aquifer base should be considered for the planning of observation well. The geophysical reconnaissance of the aquifer structure, depending on the geological conditions, could be done by 2-D resistivity or shallow seismic reflection techniques.

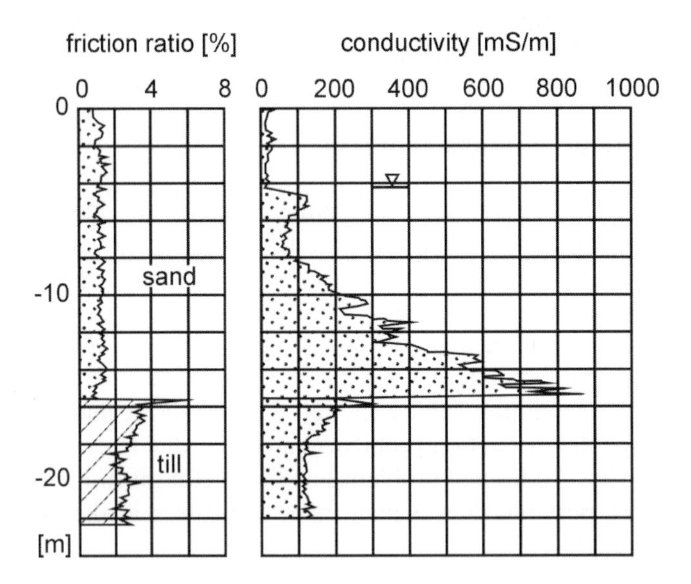

Fig. 17.8. Results of direct push measurements within the surrounding of a waste dump in Schleswig-Holstein showing conductivity and friction ratio (for the discrimination of sand and till, see Chap. 11) of the material, the conductivity of aquifer is increasing with depth due to contaminated water (Bock 1997, permission from BGR)

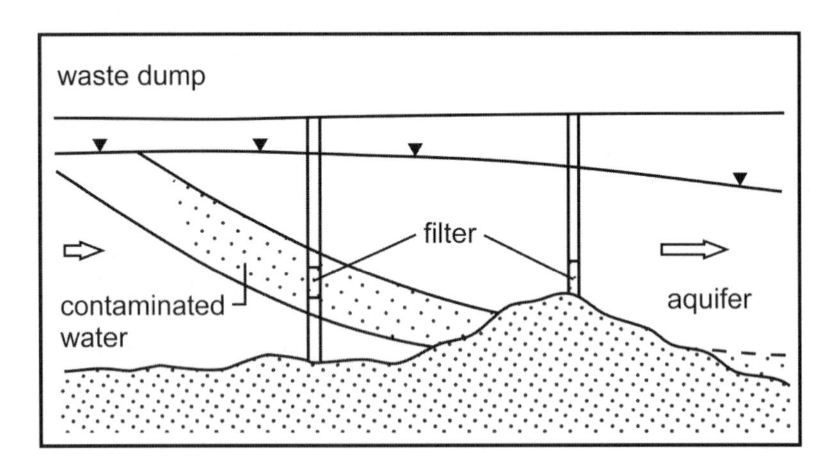

Fig. 17.9. Due to the higher density, contaminated water from a waste dump sinks to the bottom of the aquifer, so that well contaminated water was found only in the left monitoring (Bock 1985)

17.4.2 Organic contaminants

Organic contaminants in groundwater only dissolve to a minor (but for health reasons and potential pollution important) extent, so that no influence on the groundwater conductivity can be expected. Organic contaminants can be divided through their density into two categories: LNAPL (light nonaqueous phase liquid) and DNAPL (dense nonaqueous phase liquid). The general spread of LNAPL and DNAPL is sketched in Fig. 17.10. LNAPLs are floating on the groundwater and moving with the groundwater flow, whereas DNAPLs sink to the bottom of the aquifer, their motion is mainly controlled by gravity. Typical examples for LNAPL and DNAPL are gasoline and PCB.

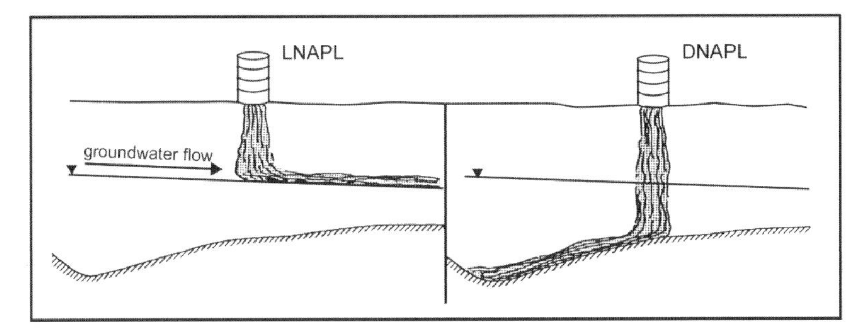

Fig. 17.10. Pathways for LNAPL (left) and DNAPL (right) in the underground

Since hydrocarbons have in general a lower electrical conductivity and a lower dielectric permittivity than water, a blanket of LNAPL with sufficient thickness floating on the groundwater table should be detectable through resistivity and GPR methods. An example is shown by Benson et al. (1997), where a gasoline spill was mapped with 2-D resistivity as a resistive zone on the groundwater table and in the vadose zone (Fig. 17.11). However, the moderate resistivity increase was only detectable in a low resistivity environment. Reference profiles outside the contaminated area are essential for the interpretation of the measurements.

The potential of SIP measurements for organic contaminant detection is discussed in details in Chap. 4.

GPR measurements of controlled and real LNAPL spills show contradictory results. Campell et al. (1996) observed, in a controlled experiment, a sharper GPR reflection of the water table in the contaminated area than outside and concluded, that the LNAPL plume on the watertable decreases the capillary fringe which normally affects the GPR reflection coefficient (Pipatpan and Blindow 2002). However, Sauck et al. (1998), observed a diffuse shadow zone in GPR signals within the LNAPL spilled zone in real field measurements of a fuel spill on a former military air field (Fig. 17.12).

This zone was also associated with electrical resistivities of less than 100 Ωm being in contrast to the high resistivities of hydrocarbons. The authors conclude that, as already mentioned in the paragraph on soil contamination, biodegradation of the fuel leads to free ions and so to reduced resistivities. The shadow zone would be caused by the high absorption of the GPR signal in the low resistive area of the LNAPL plume.

Biodegradation can change the sign of resistivity anomalies of a LNAPL spill from high resistivities of a fresh spill to low resistivities after biodegradation (see above). The contamination history of the project area should be taken into consideration for the development of the geophysical mapping concept.

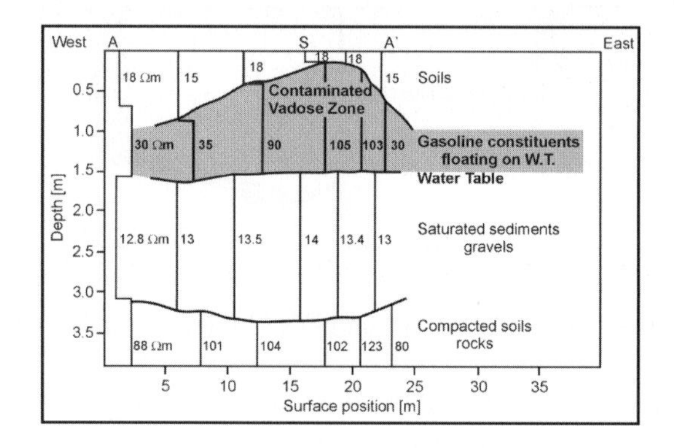

reference profile, no contamination

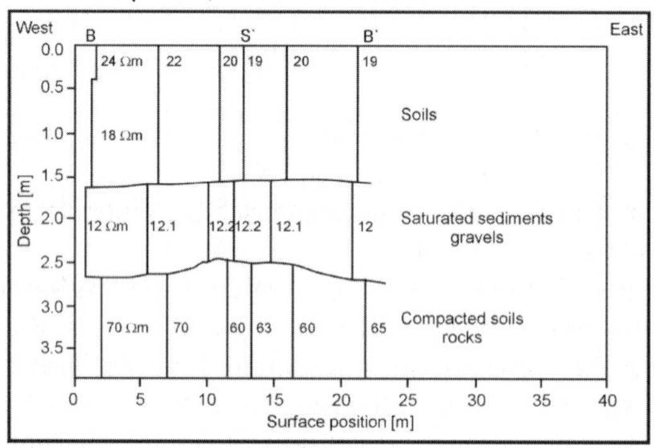

Fig. 17.11. Resistivity profile covering a gasoline spill (top) and undisturbed reference profile (bottom), after Benson et al. (1997, permission from SEG)

The migration paths of DNAPL plumes in the groundwater are complicated. They are influenced, as sketched in Fig. 17.10, by gravity and groundwater flow, but also by the permeability structure of the underground material and by the interaction with clay and quartz minerals. Enhanced DNAPL concentrations can be found in structural traps like paleochannels, fault zones or depressions of the aquifer bottom. These trap structures are the main target for geophysical mapping of DNAPL contaminated sites.

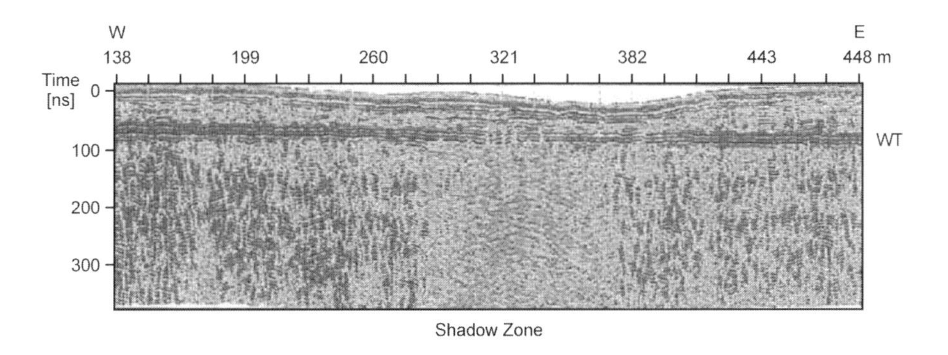

Fig. 17.12. Shadow zone for GPR signals due to low resistivities of a LNAPL plume after biodegradation (Sauck et al. 1998, permission from EEGS)

17.5 References

Benson AK, Payne KL, Stubben MA (1997) Mapping groundwater contamination using DC resistivity and VLF geophysical methods - a case study. Geophysics 62:80-86

Bock W (1985) Ausbreitung deponiebelasteten Wassers in quartären Sedimenten. Proc 5. Tagung Ingenieurgeologie:29-32, Kiel

Bock W (1997) Grundwasser und Ionenaustausch im Unterstrom einer Abfalldeponie. Z angew Geol 43:159-165

Campbell DL, Lucius JE, Ellefsen KJ, Deszcz-Pan M (1996) Monitoring of a controlled LNAPL spill using ground penetrating radar. in: Proceedings of the Symposium on the Application of Geophysics to Engineering and Environmental Problems pp 511-517, Keystone, Colorado

De Iaco R, Green AG, Maurer H-R, Horstmeyer H (2003) A combined seismic reflection and refraction study of a landfill and its host sediments. Journal of Applied Geophysics 52:139-156

EPA (1993) Use of Airborne, Surface, and Borehole Geophysical Techniques at Contaminated Sites. United States Environmental Protection Agency EPA/625/R-92/007

Gajdoš V, Kráľ V (1995) Influence of hydrocarbon pollution to soil conductivity. Proceedings of the Symposium on the Application of Geophysics to Engineering and Environmental Problems, Orlando, Fl

Godio A, Naldi M (2003) Two-dimensional electrical imaging for detection of hydrocarbon contaminants. Near Surface Geophysics 1:131-137

Grinat M (2001) Geoelektrische Messungen über einem Kerosinschaden auf dem Flughafen Schwerin-Parchim. Proc 7. DGG-Seminar "Ingenieur- und Umweltgeophysik", Mitteilungen der Deutschen Geophysikalischen Gesellschaft, Sonderband II/02:43-47

Knödel K, Krummel H, Lange G (1997) Handbuch zur Erkundung des Untergrundes von Deponien und Altlasten, Bd. 3 Geophysik. Springer, Berlin

LFU (1993): Das Modellstandortprogramm des Landes Baden-Württemberg. Erfahrungen für die Praxis der Altlastenbearbeitung. - Landesanstalt für Umweltschutz Baden-Württemberg / Materialien zur Altlastenbearbeitung, Band 12. Karlsruhe

Lorra S, Bruhn C, Kirsch R, Thomsen D (1997) Geophysikalische Untersuchungen in Lübeck-Blankensee. Unpublished report, Institut für Geophysik, Christian-Albrechts-Universität Kiel

Perk M, Tezkan B, Hördt A (2004) Interdisciplinary Waste Site Investigation in Balassagyarmat / Hungary (NORISC-Project). Near Surface 2004, proceedings of annual meeting of European Association of Geoscientists and Engineers, Utrecht

Pipatpan S, Blindow N (2002) The effect of the capillary fringe on GPR water table reflections. Proc 8. DGG-Seminar "Ingenieur- und Umweltgeophysik", Mitteilungen der Deutschen Geophysikalischen Gesellschaft, Sonderband III/03:26-35

Roosen H, Tezkan B (1998) Mehrdimensionale Interpretation von Schlumberger Sondierungen zur Erkundung von Restfundamenten des ehemaligen Bahnbetriebswerkes Düren. Proc 5. DGG-Seminar "Ingenieur- und Umweltgeophysik", Mitteilungen der Deutschen Geophysikalischen Gesellschaft, Sonderband II/99:48-62

Sauck WA, Atekwana EA, Nash MS (1998) High conductivities associated with an LNAPL plume imaged by integrated geophysical techniques. Journal of Environmental and Engineering Geophysics 2:203-212

Sauck WA (2000) A model for the resistivity structure of LNAPL plumes and their environs in sandy sediments. Journal of Applied Geophysics 44:151-165

Schaumann G (2001) Transientenelektromagnetische Messungen auf Mülldeponien – Untersuchung des Einflusses von 3D-Leitfähigkeitsvariationen und 1D-frequenzabhängiger Polarisierbarkeit. PhD Thesis, Technische Universität Braunschweig

Stoll H (1994) Geophysikalische Untersuchungen zur Abschätzung des inneren Aufbaus und der Tiefenerstreckung von Altablagerungen in quartären Lockersedimenten. Diplomarbeit, Institut für Geophysik, Christian-Albrechts-Universität Kiel

Sweeney, JJ (1984) Comparison of Electrical Resistivity Methods for Investigation of Ground Water Conditions at a Landfill Site. Ground Water Monitoring Review 4:52-59

Tezkan B (1998) Radiomagnetotellurische Erkundung einer Industriealtlast. Proc 5. DGG-Seminar "Ingenieur- und Umweltgeophysik", Mitteilungen der Deutschen Geophysikalischen Gesellschaft, Sonderband II/99:79-88

THOR Geophysikalische Prospektion GmbH (1988) Geophysikalische Erkundungen an Modellstandorten. Unpublished report (Kirsch R, Schade J, Utecht T), Kiel

THOR Geophysikalische Prospektion GmbH (1996) Geophysikalische Testmessungen auf dem Gelände der Hindenburg-Kaserne, Neumünster. Unpublished report, Kiel

Vogelsang D (1993) Geophysik an Altlasten. Springer, Berlin

Vogelsang D (1994) Environmental Geophysics: A Practical Guide. Springer, Berlin

Index

Printing: Krips bv, Meppel
Binding: Stürtz, Würzburg